MACROMOLECULAR CHEMISTRY AND PHYSICS

Macromolecular Chemistry and Physics

Editor:

Maria Emilova Velinova, Ph.D.

ArclerPress

www.arclerpress.com

Macromolecular Chemistry and Physics

Editor: Maria Emilova Velinova, Ph.D.

708 3rd Avenue, 6th Floor
New York
NY 10017
United States of America
www.arclerpress.com

ISBN: 978-1-68094-419-8

Library of Congress Control Number: 2016951138

About the Editor

Maria Emilova Velinova, Ph.D.

Maria Velinova is Ph.D. holder in Quantum chemistry at the University of Sofia since April 2012. Her major research experience is in the field of Computational Chemistry, especially in statistical mechanics methods applied to different sorts of biomolecules. Member of the Laboratory of Quantum and Computational Chemistry at the University of Sofia.

Contents

List of Contributors

Niels ten Brummelhuis
Department of Chemistry, Humboldt-Universität zu Berlin, Brook-Taylor-Str. 2, 12489 Berlin, Germany

Sarah Tempelaar
Department of Chemistry, University of Warwick, Coventry, CV4 7AL, UK

Laetitia Mespouille
Center of Innovation and Research in Materials and Polymers (CIRMAP), Laboratory of Polymeric and Composite Materials, University of Mons, Place du Parc 23, 7000 Mons, Belgium

Olivier Coulembier
Center of Innovation and Research in Materials and Polymers (CIRMAP), Laboratory of Polymeric and Composite Materials, University of Mons, Place du Parc 23, 7000 Mons, Belgium

Philippe Dubois
Center of Innovation and Research in Materials and Polymers (CIRMAP), Laboratory of Polymeric and Composite Materials, University of Mons, Place du Parc 23, 7000 Mons, Belgium

Andrew P. Dove
Department of Chemistry, University of Warwick, Coventry, CV4 7AL, UK

Jun-Jie Yan
CAS Key Laboratory of Soft Matter Chemistry, Department of Polymer Science and Engineering, University of Science and Technology of China, Hefei, 230026, Anhui, P. R. China

Jiao-Tong Sun
CAS Key Laboratory of Soft Matter Chemistry, Department of Polymer Science and Engineering, University of Science and Technology of China, Hefei, 230026, Anhui, P. R. China

Ye-Zi You
CAS Key Laboratory of Soft Matter Chemistry, Department of Polymer Science and Engineering, University of Science and Technology of China, Hefei, 230026, Anhui, P. R. China

De-Cheng Wu
Beijing National Laboratory for Molecular Sciences, State Key Laboratory of Polymer Physics & Chemistry, Institute of Chemistry, Chinese Academy of Sciences, Beijing 100190, China

Chun-Yan Hong
CAS Key Laboratory of Soft Matter Chemistry, Department of Polymer Science and Engineering, University of Science and Technology of China, Hefei, 230026, Anhui, P. R. China

Thomas H. Epps, III
Department of Chemical and Biomolecular Engineering and Department of Materials Science and Engineering, University of Delaware, Newark, Delaware 19716, USA

Rachel K. O'Reilly
Department of Chemistry, University of Warwick, Gibbet Hill, Coventry, CV4 7AL, UK

Fiona L. Hatton
Department of Chemistry, University of Liverpool, Crown Street, L69 7ZD, UK

Pierre Chambon
Department of Chemistry, University of Liverpool, Crown Street, L69 7ZD, UK

Alison C. Savage
Department of Chemistry, University of Liverpool, Crown Street, L69 7ZD, UK

Steve P. Rannard
Department of Chemistry, University of Liverpool, Crown Street, L69 7ZD, UK

Li-Fen Wang
Department of Applied Chemistry and Material Science, Fooyin University, 151 Chin-Hsueh Road Ta-Liao Hsiang, Kaohsiung, 83102, Taiwan

Jyun-Wun Chen
Department of Applied Chemistry and Material Science, Fooyin University, 151 Chin-Hsueh Road Ta-Liao Hsiang, Kaohsiung, 83102, Taiwan

Jyun-Wei Chen
Department of Applied Chemistry and Material Science, Fooyin University, 151 Chin-Hsueh Road Ta-Liao Hsiang, Kaohsiung, 83102, Taiwan

S. Deckers
Laboratory for Molecular Electronics and Photonics, University of Leuven, Celestijnenlaan 200D, 3001 Leuven, Belgium

S. Vandendriessche
Laboratory for Molecular Electronics and Photonics, University of Leuven, Celestijnenlaan 200D, 3001 Leuven, Belgium

D. Cornelis
Laboratory for Polymer Synthesis, University of Leuven, Celestijnen 200F, 3001 Leuven, Belgium

F. Monnaie
Laboratory for Polymer Synthesis, University of Leuven, Celestijnen 200F, 3001 Leuven, Belgium

G. Koeckelberghs
Laboratory for Polymer Synthesis, University of Leuven, Celestijnen 200F, 3001 Leuven, Belgium

I. Asselberghs
Imec, Kapeldreef 75, 3001 Leuven, Belgium

T. Verbiest
Laboratory for Molecular Electronics and Photonics, University of Leuven, Celestijnenlaan 200D, 3001 Leuven, Belgium

M. A. van der Veen
Catalysis Engineering, Chemical Engineering Department, Delft University of Technology, Julianalaan 136, BL 2826 Delft, The Netherlands

Vikram G. Kamble
Research Fellow, Proof and Experimental Establishment (DRDO), Balasore, Odisha-756025, India
St Joseph Engineering College, Mangalore, Karnataka-575028, India

Naveen Kumar L.
MTech Scholar, Sri Jayachamarajendra College of Engineering, Mysore, Karnataka -570006, India

Chakaphan Ngaowthong
Department of Materials and Production Engineering (MPE), The Sirindhorn International Thai – German Graduate School of Engineering (TGGS), King Mongkut's University of Technology North Bangkok, Bangkok, Thailand
Department of Agricultural Engineering for Industry, King Mongkut's University of Technology North Bangkok, Prachinburi Campus, Prachinburi, Thailand

Vilai Rungsardthong
Department of Agro-Industrial Technology, Faculty of Applied Science, King Mongkut's University of Technology North Bangkok, Bangkok, Thailand

Suchart Siengchin
Department of Materials and Production Engineering (MPE), The Sirindhorn International Thai – German Graduate School of Engineering (TGGS), King Mongkut's University of Technology North Bangkok, Bangkok, Thailand
Center of Innovation in Design and Engineering for Manufacturing (CoI-DEM), King Mongkut's University of Technology North Bangkok, 1518 Pracharaj 1, Wongsawang Road, Bangsue, Bangkok 10800, Thailand

Christian Heinzmann,
Adolphe Merkle Institute, University of Fribourg, Chemin des Verdiers 4, 1700 Fribourg, Switzerland

Christoph Weder
Adolphe Merkle Institute, University of Fribourg, Chemin des Verdiers 4, 1700 Fribourg, Switzerland

Lu Montero de Espinosa
Adolphe Merkle Institute, University of Fribourg, Chemin des Verdiers 4, 1700 Fribourg, Switzerland

Elizabeth Sánchez-Bustos
Instituto Tecnológico de Tijuana, Centro de Graduados e Investigación en Química, Apartado Postal 1166, 22000 Tijuana, BC, Mexico
Facultad de Ciencias Químicas e Ingeniería, Universidad Autónoma de Baja California, Calzada Universidad 14418, 22390 Tijuana, BC, Mexico

José M. Cornejo-Bravo
Facultad de Ciencias Químicas e Ingeniería, Universidad Autónoma de Baja California, Calzada Universidad 14418, 22390 Tijuana, BC, Mexico

Angel Licea-Claverie
Instituto Tecnológico de Tijuana, Centro de Graduados e Investigación en Química, Apartado Postal 1166, 22000 Tijuana, BC, Mexico

Sivanand S Pennadam
School of Pharmacy and Biomedical Sciences, Institute of Biomedical and Biomolecular Sciences, University of Portsmouth, Portsmouth, PO1 2DT UK

Keith Firman
School of Biological Sciences, Institute of Biomedical and Biomolecular Sciences, University of Portsmouth, Portsmouth PO1 2DT, UK

Cameron Alexander
School of Pharmacy and Biomedical Sciences, Institute of Biomedical and Biomolecular Sciences, University of Portsmouth, Portsmouth, PO1 2DT UK

Dariusz C Górecki
School of Pharmacy and Biomedical Sciences, Institute of Biomedical and Biomolecular Sciences, University of Portsmouth, Portsmouth, PO1 2DT UK

Richard Balint
School of Materials, Faculty of Engineering and Physical Sciences, University of Manchester, Manchester M1 7HS, UK

Nigel J. Cassidy
School of Physical and Geographical Sciences, Keele University, Stoke-on-Trent ST5 5BG, UK

Sarah H. Cartmell[a]
School of Materials, Faculty of Engineering and Physical Sciences, University of Manchester, Manchester M1 7HS, UK

Joanna Rydz
Bulgarian Academy of Sciences, Institute of Polymers, Acad. Georgi Bonchev St., Bl. 103A, Sofia 1113, Bulgaria
Polish Academy of Sciences, Centre of Polymer and Carbon Materials, 34 M. Curie-Sklodowska St., Zabrze 41-800, Poland

Wanda Sikorska
Polish Academy of Sciences, Centre of Polymer and Carbon Materials, 34 M. Curie-Sklodowska St., Zabrze 41-800, Poland

Mariya Kyulavska
Bulgarian Academy of Sciences, Institute of Polymers, Acad. Georgi Bonchev St., Bl. 103A, Sofia 1113, Bulgaria

Darinka Christova
Bulgarian Academy of Sciences, Institute of Polymers, Acad. Georgi Bonchev St., Bl. 103A, Sofia 1113, Bulgaria

Preface

In the last years, the macromolecular chemistry and physics made remarkable progress. From particular subject for engineers interested in specific fields, it developed into one of the fundamental discipline in the science. For this reason, the essential study of the physical and chemical properties of the polymeric materials is required, while their applications have very fast progress, crossing the boundaries between disciplines. All of the above is the inspiration for the set of research articles presented in this book that report the last updates and developments of macromolecular science and polymer engineering.

The book has been divided into three parts: synthesis, properties and applications of the polymeric materials. Every part presents interdisciplinary studies on the state-of-arts of each topic under consideration through a combination of overviews and original results.

The book begins with introducing the methods in which the macromolecular materials are synthesized. Polymer synthesis is a complex procedure and can take place in a variety of ways. Methods as natural photosynthesis, template, high-throughput and ring-opening polymerization are discussed.

The second part of the book looks at new materials with unique properties as optical non-linearity, adhesion and strength. Area of particular interest is the tensile strength, which is important for a material that is going to be stretched as the fibers.

The last part of the book contains researches of the new polymeric materials and their application. In particular, some of the most promising biodegradable polymers such as polylactide and polyhydroxyalkonoates are discussed. Synthetic pathways, favorable properties and utilization of these attractive polymer families are outlined. Other interesting application of the macromolecular materials is the drug delivery. As example of drug delivery vehicle is shown the core cross-linked star polymers and their potential application in cancer treatment. Smart biomaterial for tissue engineering, protein-polymer nano-machines and different polymer optical fibers for applications in textile fabrics are also reported.

Editor
Maria Emilova Velinova, Ph.D.

CHAPTER 1

CONTROLLING MONOMER-SEQUENCE USING SUPRAMOLECULAR TEMPLATES

Niels ten Brummelhuis

Department of Chemistry, Humboldt-Universität zu Berlin, Brook-Taylor-Str. 2, 12489 Berlin, Germany

ABSTRACT

The transcription and translation of information contained in nucleic acids that has been perfected by nature serves as inspiration for chemists to devise strategies for the creation of polymers with well-defined monomer sequences. In this review the various approaches in which templates (either biopolymers or synthetic ones) are used to influence the monomer-sequence are discussed.

INTRODUCTION

The monomer sequences in biopolymers such as nucleic acids and proteins are strongly linked to their function, and even small errors can yield non-functional macromolecules, or worse. Nature therefore expends significant resources on the (nearly) flawless synthesis of these polymers.[1] For example, DNA is replicated using DNA polymerases, which achieve error rates as low as 1 in 10^6 nucleotide additions, a rate which is further reduced by several orders of magnitude by proof-reading and repair mechanisms.[1] Presently, DNA is the primary information carrier used in nature.[2] RNA, which can be prepared by the transcription of DNA by RNA polymerases, in most cases serves as a middleman, conveying information from DNA to amino acid sequences in proteins (*e.g.* as messenger or transfer RNA; mRNA and tRNA), though RNA can also perform catalytic functions itself, and it has been suggested that RNA played a much bigger part in emerging life before the advent of proteins.[3–5]

The wide variety of functions that biopolymers perform with a limited number of monomers (*e.g.* the 21 proteinogenic amino acids in proteins) has led polymer chemists to conclude that an increased control over the monomer sequence could similarly imbue complex properties upon synthetic polymers, allowing them to be used for a plethora of novel applications.[6–13]

Various methods towards this ultimate goal of sequence control have been suggested, ranging from solid-phase synthesis, the coupling of presynthesised oligomers, ring-opening polymerisation of multifunctional ring systems and kinetic control.[7–11] Perhaps the most extensively investigated method, *i.e.* template polymerisation, is closer to the procedures perfected by nature though. Replication, transcription and translation are all based on templated synthesis of biopolymers; the monomer sequence is strictly predetermined by the template molecule (DNA in replication and transcription, mRNA in translation). In this review the use of various types of template polymerisation for the control over monomer sequence in synthetic polymers are discussed.

Citation: Niels ten Brummelhuis, Controlling monomer-sequence using supramolecular templates, DOI: 10.1039/C4PY01522C.

TEMPLATE POLYMERISATION

The term template polymerisation is widely used in literature to describe any polymerisation where a template is used to influence properties of the daughter polymer. Molecular weight, dispersity and tacticity can for example be influenced strongly by their polymerisation in the presence of a template (macro)molecule.[14–18]

Supramolecular interactions (such as Coulombic interactions or hydrogen bonding) between the template and monomers and/or the growing polymer are required for a template to exert control over any property of the polymer. In literature, the term templating is occasionally also used for systems where covalent bonds are used to influence the properties of polymers (in an intramolecular polymerisation), but these will not be considered here.

The nature of the template varies depending on the polymer characteristic that one wishes to influence. For the control of the molecular weight and dispersity of polymers, template polymers with well-defined molecular weight and dispersity are used. Ideally, daughter polymers with identical dispersity and degree of polymerisation are obtained.[14,16] Both small molecules and macromolecules can be used as templates to influence the tacticity of polymers. Small molecules (or metal-ion complexes) are used to supramolecularly influence the relative positions of monomers,[17] placing them in positions that promote *e.g.* iso- or syndiotactic incorporation of the monomers. Alternatively, a small molecule can be used to effectively make a monomer bulkier, thereby prohibiting the isotactic incorporation of monomers and yielding preferentially syndiotactic polymers.[17]

The best example of a macromolecular template for the control of the tacticity of polymers is based on the formation of stereocomplexes,[17]*e.g.* between isotactic and syndiotactic poly(methyl methacrylate), where either one induces the formation of the complementary polymer. This review focusses on the use of templates for the control of monomer sequence in synthetic polymers. This is not the only reason why template polymerisations have been studied though. Template directed control of monomer sequences has also been studied extensively to understand the fundamental requirements for the transfer of information between (bio-)macromolecules, and how this relates to the origin of life.[19] Reproduction is arguably the most crucial feature of living systems, and reproduction in nature is purely a templated process. Even before self-replicating nucleic acid systems arose, templating by clay crystals[20,21] or lipid membranes[22] could very well have catalysed the formation of nucleic acid containing polymers.

The questions chemists have been trying to answer about the origin of self-replicating nucleic acid systems, and about later, more complex, biological systems, are becoming increasingly relevant to material chemists in their quest for routes towards sequence-controlled polymerisations, since a thorough understanding of the replication, transcription and translation processes might allow for the adoption of these principles for the preparation of synthetic polymers with similarly well-defined monomer sequences. Efforts to adopt aspects of template polymerisation to synthetic polymers range from polymerisations that are strongly connected to biological systems to fully synthetic approaches. In the following sections the various approaches have been divided into (1) systems where biopolymers (*i.e.* nucleic acids) are used as template molecules and (2) fully synthetic systems.

BIOPOLYMER TEMPLATES

The use of biopolymers (or close analogues thereof) as templates is the most commonly used method in the sequence control of synthetic polymers. A number of very different methods based on biopolymers have been proposed. For clarity, they have been categorised as methods akin to replication/transcription or translation in natural systems, or as approaches fully unrelated to biopolymer synthesis.

Template Polymerisation Akin to Replication and Transcription

Scientists have extensively used nature's machinery for the creation of fully or partially synthetic polymers. In these approaches enzymes (DNA/RNA polymerases) are used to catalyse the transcription of a template nucleic acid polymer. This strategy is limited in its scope since the natural systems are highly specialised and only effectively catalyse the synthesis of nucleic acids that are similar in structure to DNA/RNA.[23,24] Fortunately many sources of polymerases are available and the specificity of the various enzymes varies greatly. Some examples of nucleic acids that can be polymerised using polymerases are shown in Fig. 1. Both the (deoxy)ribose group (Fig. 1a) and the phosphodiester group can be exchanged (Fig. 1b).

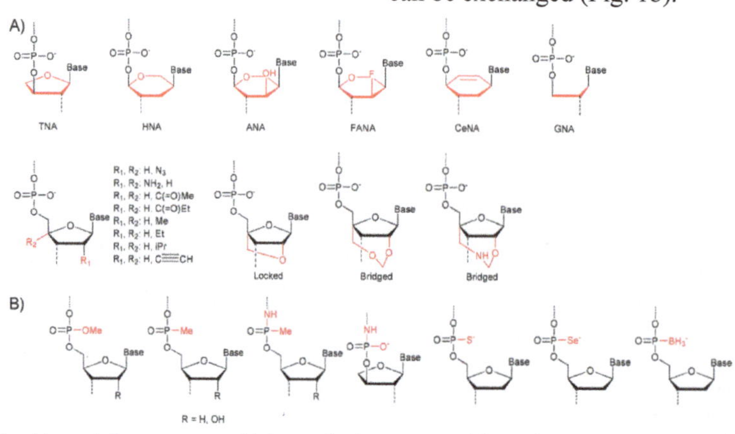

Figure 1: (A) Derivatives of the (deoxy)ribose group which can be incorporated by polymerase enzymes. TNA: threose NA, HNA: 1,5-anhydrohexitol NA, ANA: 1,5-anhydroatritol NA, FANA: 2′-deoxy-2′-fluoroarabino NA, CeNA: cyclohexenyl NA, GNA: glycerol NA. (B) Modification of the phosphodiester backbone. PNA: peptide NA.[23,24]

For TNA, HNA, ANA, FANA, CeNA and LNA heredity and evolution could even be shown, supporting the assumption that other nucleic acids besides RNA and DNA could have been involved in the rise of early life-forms.[25]

Several groups have attempted to extend the scope even further by modifying the polymerases to allow for a broader scope of monomers. In many cases though these efforts also result in less specific incorporation of the appropriate base pair.[26,27]

New functions can potentially be incorporated into nucleic acids by introducing functional groups. This is most reliably achieved by placing functional groups in the major and/or minor groove of the double-stranded helix. To avoid interference with the recognition between bases, any changes need to be performed on sites on the opposite side from the H-bonding face (Fig. 2a). A notable exception is presented by Szostak and co-workers.[28] They showed that the use of 2-thiothymidine (2-TT) instead of thymidine (T) lead to the suppression of mismatched pairs through stronger binding between 2-TT and adenine (A) and weaker unspecific binding to cytosine (C) and guanine (G), and therefore yields lower error rates in the enzyme-free replication of phosphoramidate DNA.

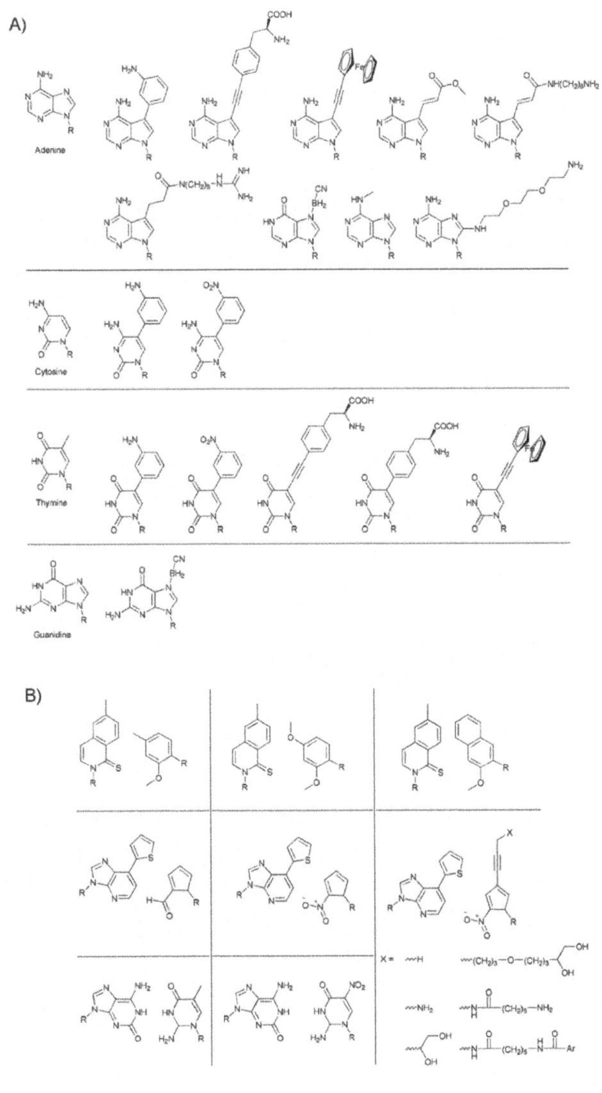

Figure 2: (A) Some examples of modified bases used to introduce functional groups into nucleic acids.[23] (B) Examples of artificial base pairs that are capable of being selectively incorporated into nucleic acids by PCR.[29]

Besides using modified versions of the four natural bases, in which recognition is limited to two base pairs, efforts have been made to devise completely new artificial recognition pairs which are orthogonal to the A/T and G/C recognition.[29] The base pairs shown in Fig. 2b could even be shown to be compatible with the polymerase chain reaction (PCR) procedure, in which they were specifically and efficiently incorporated into the nucleic acid chain.

The incorporation of functional groups and unnatural bases is still limited by the selectivity of the polymerases that are used. Liu and coworkers suggested using DNA ligases rather than polymerases for the transcription of a ssDNA template.[30] They were able to show that T4 DNA ligase was able to efficiently couple 5′-phosphorylated trinucleotides in the presence of a suitable single-stranded DNA (ssDNA) template. The presence of functional groups on the 5′-base did not effect the efficiency of the ligase significantly. Eight different functional groups could be incorporated in a sequence specific way using this strategy, and 10 to 50 three-nucleotide codons could be transcribed efficiently (full-length product yields >80%).

Enzyme-Free Replication and Transcription

Replication and transcription have also been extensively studied without the use of the highly optimised enzymes that nature has given rise to,[23,24] which has the advantage that also backbones very different from natural nucleic acids can be used, provided the template nucleic acid strand and the daughter polymer efficiently hybridise.

Enzyme-free replication and transcription of nucleic acids has been far more difficult than was originally assumed. Pioneering work by Orgel and co-workers showed that the template polymerisation of activated phosphate moiety containing nucleotides could be used to promote the formation of a small number of phosphoesterifications between mono-, di-, tri- or oligonucleotides, though mostly in rather poor yields.[31–39] Also, recognition between cytosine and guanine were shown to be far more effective as templates than adenine and thymine (or uracil), presumably because of the stronger interaction between C and G.

Using mononucleotides the incorporation of monomers was very error prone: error rates as high as 30% for G/C and >50% for A/T base pairing were found.[40] Enzyme-free template polymerisation with some accuracy and efficiency has become possible only using oligonucleotides. Similar observations were made when using non-natural nucleotides (*e.g.* PNA, ANA). Though a large body of work relating to the template catalysed ligation of two oligomers can be found in literature, using a wide variety of different chemical reactions and polymer backbones, these works do not provide a real control over monomer sequence and have been reviewed extensively elsewhere[41,42] and will therefore not be covered here.

A good example where oligonucleotides were coupled together was provided by Liu and co-workers.[43,44] They utilised PNA tetramers[42] to recognise four-base codons on a DNA template. Various 5′-amine (which serves as the initiator) DNA hairpin template molecules with predominantly AGTC codons and a single codon with a different sequence (ten codons in total) were used. The position of the differing codon was varied. In the presence of only NH_2-GACT-CHO PNA tetramer primarily the truncated product, with a length dependent on the position of the mismatched codon, were obtained. Even the presence of all possible PNA tetramers, except for the one complementary to the second codon, primarily yielded the truncated product. Only in the presence of NH_2-GACT-CHO and the PNA tetramer complementary to the second codon were full length products obtained in good yield.

In an extension of this work it was shown that 12 different PNA pentamers could be polymerised in a sequence-specific manner.[45] A suitable set of PNA tetramers which polymerize efficiently, selectively (*e.g.* with minimal out of frame hybridisation) and exhibit sufficient complexity could not be found. The authors used this system to show that functional PNA sequences can be selected from a library. This is done by displacing the PNA strand from the DNA template using a DNA polymerase to create the corresponding double stranded (ds) DNA, thereby allowing the PNA strand to interact with moieties in solution.

The group also investigated the influence of various side-chains on the α- or γ-carbon atoms of PNA pentamers.[43] 20 different groups could be placed on one or several of the PNA monomers in the NH_2-GGATT-CHO PNA pentamers that were used, and in most cases the full-length product was obtained in high yields (70–85%). Only the use of monomers (functionalised on the underlined thymidine residue: NH_2-GGA<u>T</u>T-CHO) functionalised with D-stereochemistry resulted in very low yields (~10%). The authors show this as a first proof that functional groups can be incorporated in PNA polymers. The work is summarised graphically in Fig. 3.

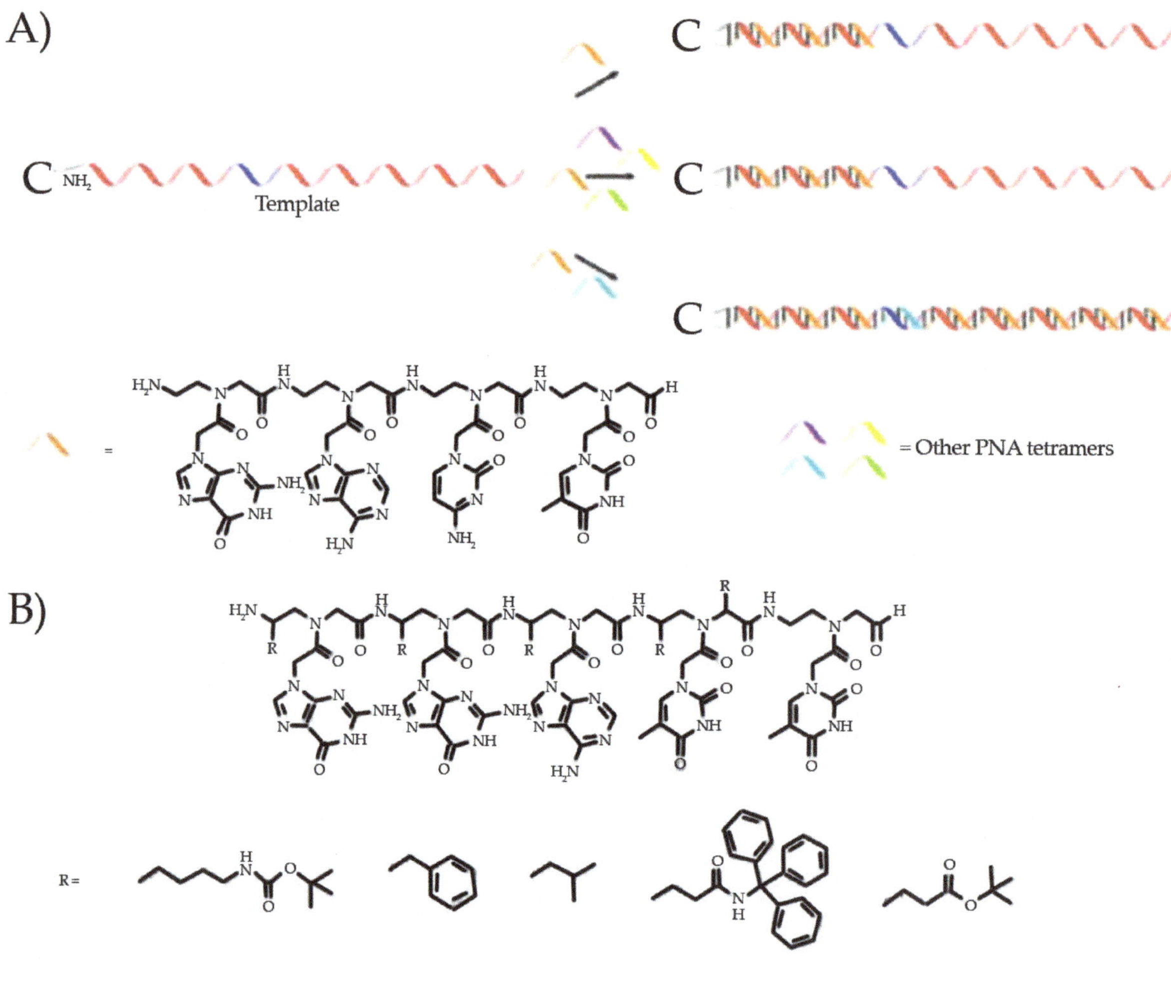

Figure 3: (A) Template polymerisation of PNA tetra- or pentamers using 5′-amine functionalised DNA as a template. Truncated products are mainly formed in the absence of the PNA aptamer complementary to all codons present in the template. (B) PNA aptamers in which functional groups were introduced in the positions marked with 'R'. In most cases a single functional group was introduced into an aptamer, but multiple functional groups could also be used.[43,44]

The overwhelming majority of investigations into enzyme-free template polymerisation using nucleic acids as templates have used the linkage of monomers/oligomers to control the monomer sequence of the daughter polymer. A novel approach was described by Ura *et al.*, who used DNA templates for the selective post-polymerisation functionalisation of peptides containing cysteine at every second position.[46] They used bases with a thioester linker that, through a transesterification were linked to the free thiol groups of the peptides. The presence of a DNA template molecule (*e.g.* C_{20} or T_{20}) was shown to promote the incorporation of the complementary base, with a selectivity of nearly 90%. Removal of the template and addition of another led to the enrichment of the newly complementary base (Fig. 4). Though no real control over the sequence of bases on the cysteine moieties could be shown yet, the methodology certainly allows for the incorporation of a functional group at every second amino acid, and the nature of the intermediary amino acids (AAs) (between the cysteine residues) in the precursor polymer could be varied.

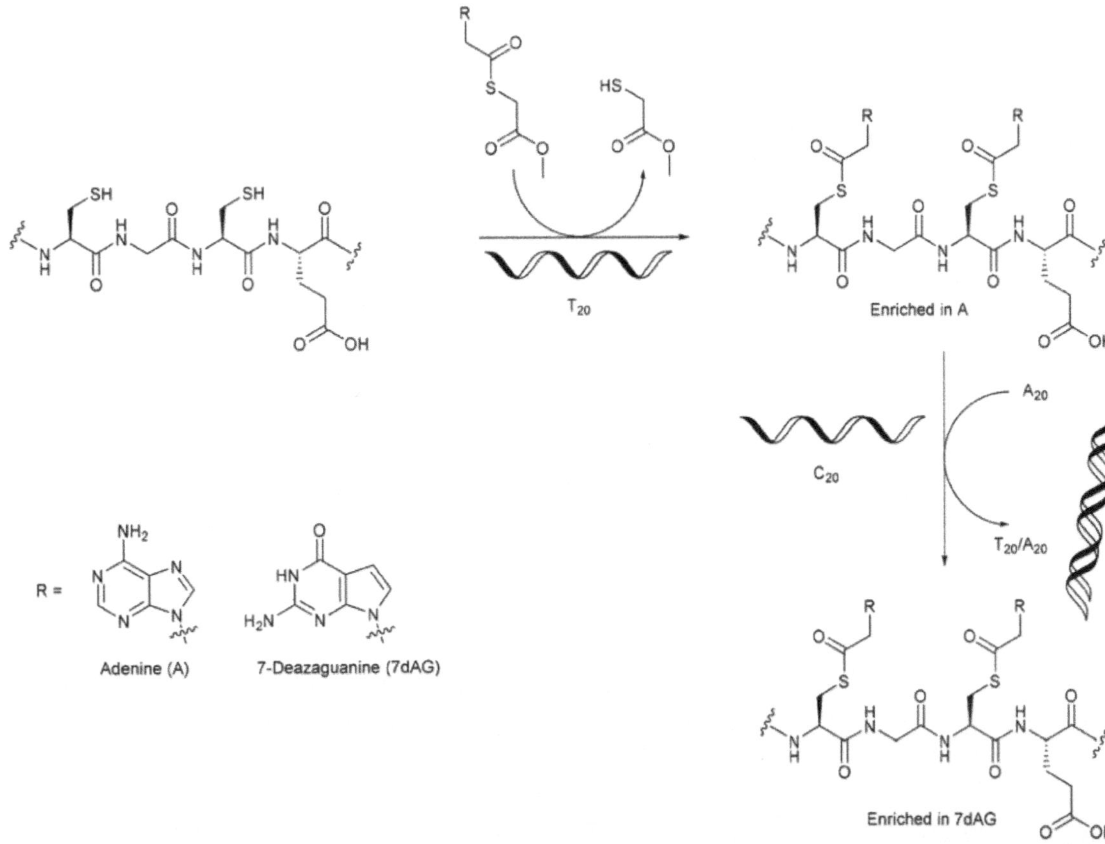

Figure 4: The DNA-template directed post-polymerisation functionalisation of cysteine-rich peptides.[46]

Template Polymerisation Akin to Translation

The translation of the genetic information carried by mRNA to amino acid sequences performed in ribosomes is a highly complex procedure, which is nicely illustrated by the large number of components that constitute a functional ribosome (57 proteins and three ribosomal RNA strands (rRNA) for bacterial ribosomes, and even more for eukaryotic ribosomes).[2] Nevertheless, ribosomes can also be used to produce synthetic polymers. The most common procedure is the incorporation of non-canonical amino acids into proteins.[47–51]

Early approaches towards the incorporation of non-canonical AAs used non-canonical aminoacyl tRNAs based on the desired AA and a recombinant tRNA to suppress a nonsense stop codon. In this way only a small number of non-canonical AAs could be incorporated *in vitro*. More advanced approaches, which allow for the incorporation of larger numbers of non-canonical monomers, uses a reconstituted translation system including purified ribosomes and a complex mix of recombinant proteins and cofactors instead of a cell lysate. The fact that various components, *e.g.* canonical aminoacyl-tRNAs, can be omitted, eliminates the risk of undesirable incorporation of the amino acid for which a given codon naturally codes, thereby allowing an alternative designation of codons to monomers.

Obviously, amino acids with altered side-chains can be incorporated, but in a few cases also the nature of the peptide backbone was changed. *N*-methylation, which in nature is often performed post-translationally, can also be realised using *N*-methylated amino acids during ribosomal translation.[52–56] Though the incorporation of longer alkyl chains was also attempted, the success was very limited. Besides the L-amino acids that are typically used in nature, also D-amino acids can be incorporated under the right conditions.[57,58] Scientist have even gone so far as to replace the amide bond in the peptide backbone with an ester bond.[59,60] Ohta *et al.*, following up on observations by Fahnestock and Rich,[61] were able to change the meaning of up to eight codons (seven α-hydroxy acids and formylated methionine assigned to the AUG start codon) using a flexizyme[62] system, and used these to prepare polyester sequences with lengths of up to 12 monomers (Fig. 5). Additionally, also lysine, aspartic acid and tyrosine could be incorporated in the polymer chain.

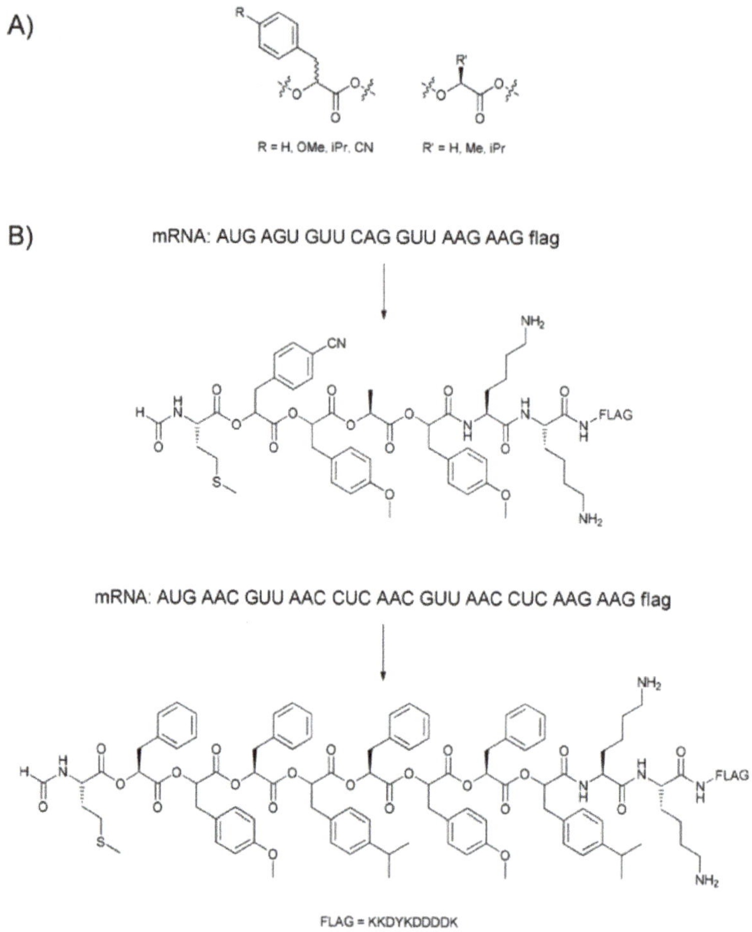

Figure 5: (A) The α-hydroxy acids used by Suga and co-workers. (B) Two examples of the *in vitro* translation of mRNA sequences into polyesters.[59,60]

The incorporation of non-canonical amino acids (or, more generally, monomers) is limited by the specificity that ribosomes exhibit. Some groups have attempted to modify the ribosome in such a way that more monomers can be incorporated in a well-defined manner, but the complexity of the ribosome makes this endeavour challenging.[55] Similarly, the incorporation of non-canonical AAs in peptides synthesised by nonribosomal peptide synthases can be facilitated by point mutations in these enzymes.[63]

Liu and co-workers recently presented a more synthetic procedure to translate nucleotide sequences in DNA to a sequence of polymer building blocks.[64] They used a PNA adapter covalently linked with polymer building blocks at the α- and ω-ends to recognise nucleotide codons. The formation of covalent bonds (copper-catalysed azide alkyne cycloaddition proved most efficient) between the polymer building blocks led to the extension of the hairpin DNA with up to 16 polymeric building blocks, though the yield of full-length product was reduced with longer templates. Homotelechelic monomers, *i.e.* monomers having two alkyne or two azide moieties at both ends, were used since heterotelechelic monomers yielded undesired side products due to cyclisation. The use of homotelechelic building blocks limits the number of possible products from a given number of PNA adapters though. Cleavage of the disulphide linkers between the PNA adapter and the polymer building blocks yielded the final polymer (Fig. 6).

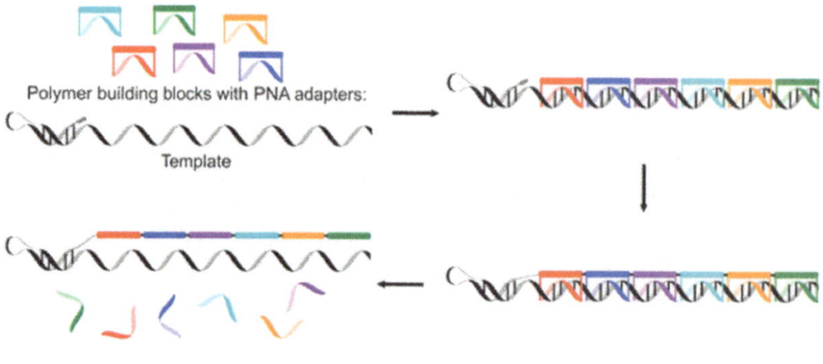

Figure 6: The use of a DNA template for the synthesis of sequence defined polymers by using a PNA adapter as an artificial tRNA.[64]

Novel Approaches

The hybridisation of two complementary single stranded DNA (ssDNA) strands can be used to place moieties (*e.g.* monomers) into close proximity to each other. Liu and co-workers, and more recently the combined groups of Turberfield and O›Reilly, used this principle extensively to create sequence-defined oligomers.

The first approach by Liu and co-workers used a DNA template with a 5′-amine group which consisted of three coding domains.[65] A DNA decamer, complementary to bases 21–30 of the template, bearing a 3′-amino acid on a base labile sulfone linker and a 5′-biotin moiety, was added. A peptide bond was formed, using EDC as the coupling reagent, between the terminal amine group of the template DNA and the carboxylic acid group on the DNA decamer. After breaking of the sulfone linker using basic conditions and removal of the DNA decamer using streptavidin-linked magnetic nanoparticles, another decamer, complementary to bases 11–20, or in the last step to bases 1–10, was added and the cycle was repeated to yield a tripeptide (Fig. 7a). The yield of the additions ranged from 52 to 82%, so that the overall yield of the tripeptide was rather poor.

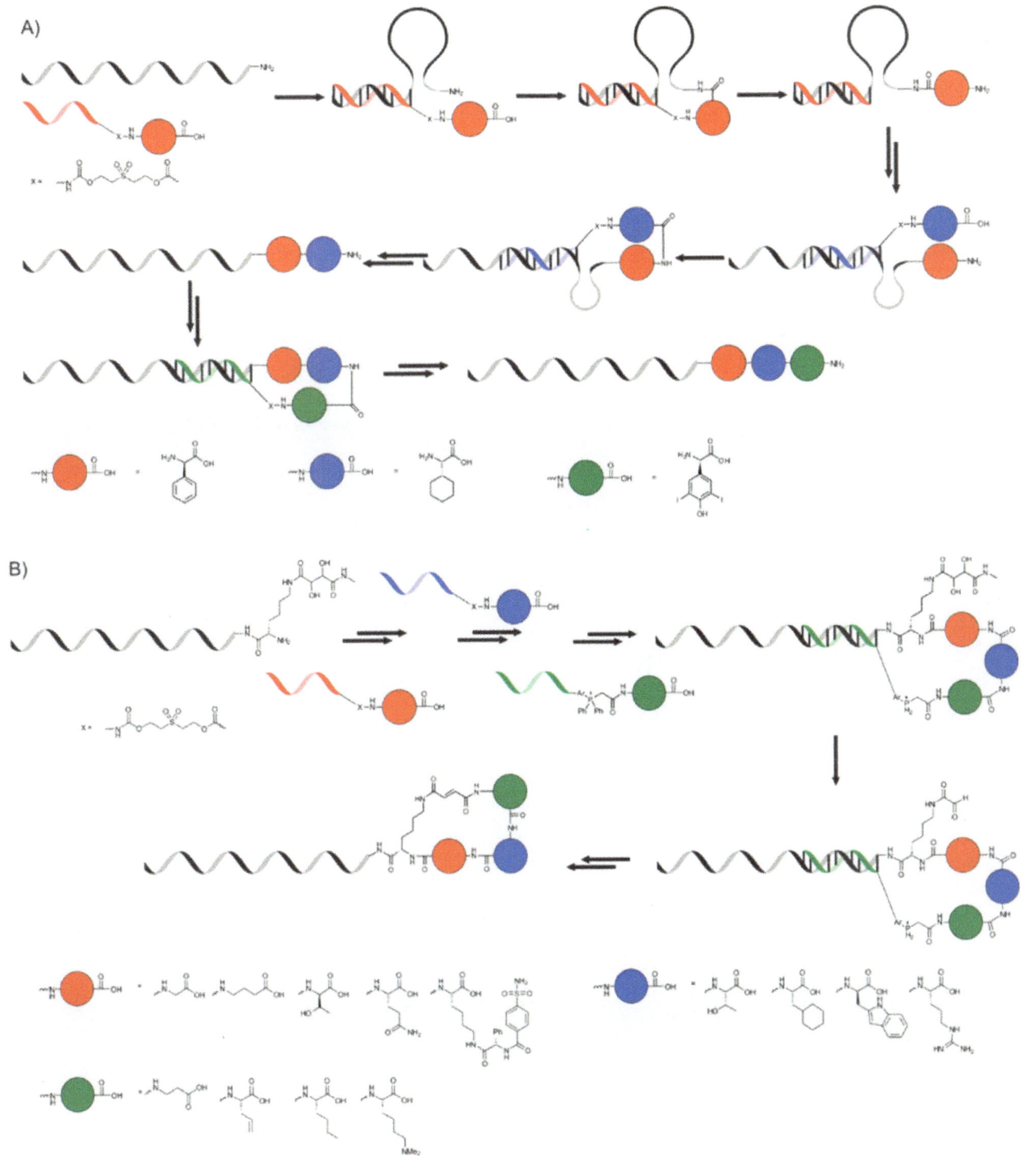

Figure 7: The synthesis of tripeptides (A) and peptide macrocycles (B) presented by Liu and co-workers.[65,66]

In a second publication they used this same strategy to produce macrocycles using an extra Wittig olefination reaction for ring closure.[66] An amino acid bearing a diol group on the 5′-terminus of the template DNA was used to provide the primary amine instead of a simple 5′-amine group. After three successive couplings, the last of which incorporated a phosphonium group, an aldehyde was generated by the oxidative cleavage of the diol group, making the Wittig olefination possible (Fig. 7b). The authors were able to produce a library of different macrocycles, and it even proved possible to prepare four different macrocycles in one pot using four different templates.

Liu and co-workers also showed that masking of the template using non-functional ssDNA of different lengths could be used to facilitate the one pot synthesis of tripeptides.[67] The template DNA bearing a 5′-amine was hybridised with two ssDNA masks of different lengths, thereby masking two of the three coding regions. Subsequently, three ssDNA fragments 3′-functionalised with non-natural amino acids or biotin were added to the mixture at 4 °C. Since only one of the coding regions (the shortest one) is left unmasked, only the first of the ssDNAs can bind and only one amino acid is transferred to the template DNA. The hybridisation of both the spent first ssDNA and the shorter mask with the template is selectively disrupted by increasing the temperature to 37 °C, using the low melting temperatures of this hybridisation. This allows for binding of the second ssDNA to the second coding region and the subsequent addition of the second amino acid. A further increase of the temperature to 62 °C leads to the release of the last mask and the second spent ssDNA fragment, so that hybridisation of the final and longest ssDNA is made possible and addition of the biotin moiety can take place, yielding the desired product in an overall yield of 38% (Fig. 8). The formation of oligomers in the absence of the masks yielded a mixture of all possible products. In addition to the formation of peptide bonds also Wittig olefination could be used as the coupling reaction in this approach.

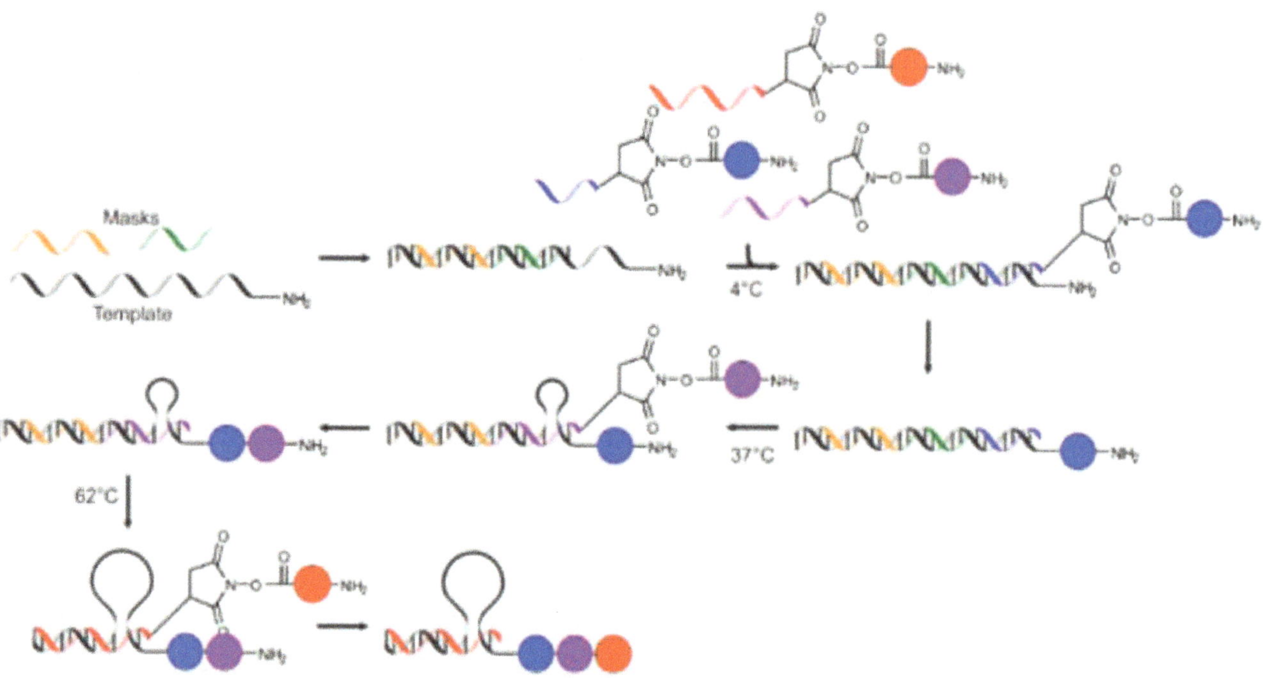

Figure 8: The stepwise addition of monomers is achieved through masking of part of the template ssDNA and stepwise removal of the masks using increasing temperature.[67]

Using ssDNAs of increasing lengths, all complementary to the same template molecule, also results in the replacement of the previous ssDNA through stronger binding (Fig. 9a). In this way hexamers could be prepared in an overall yield of 35%.[68]

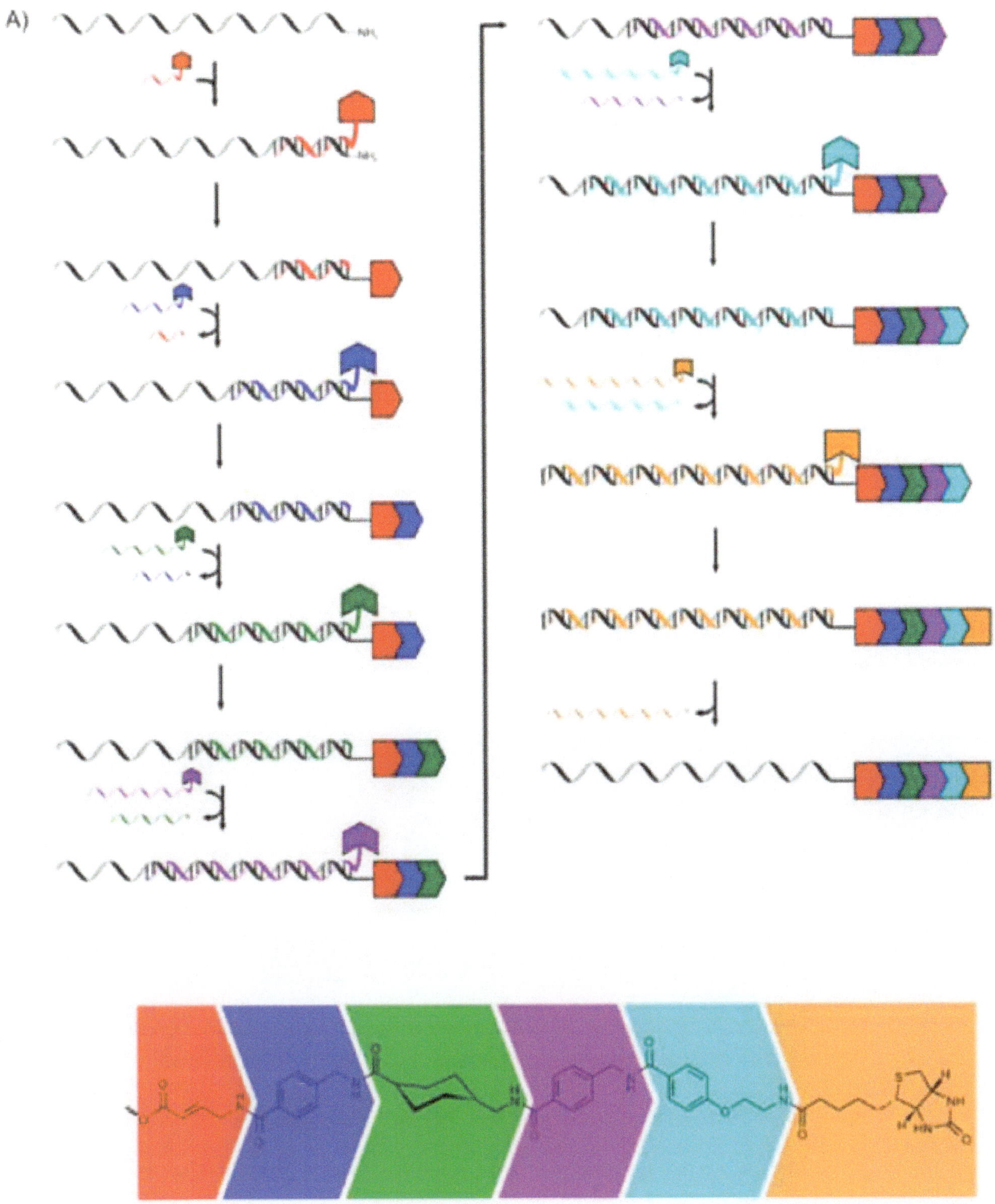

B)

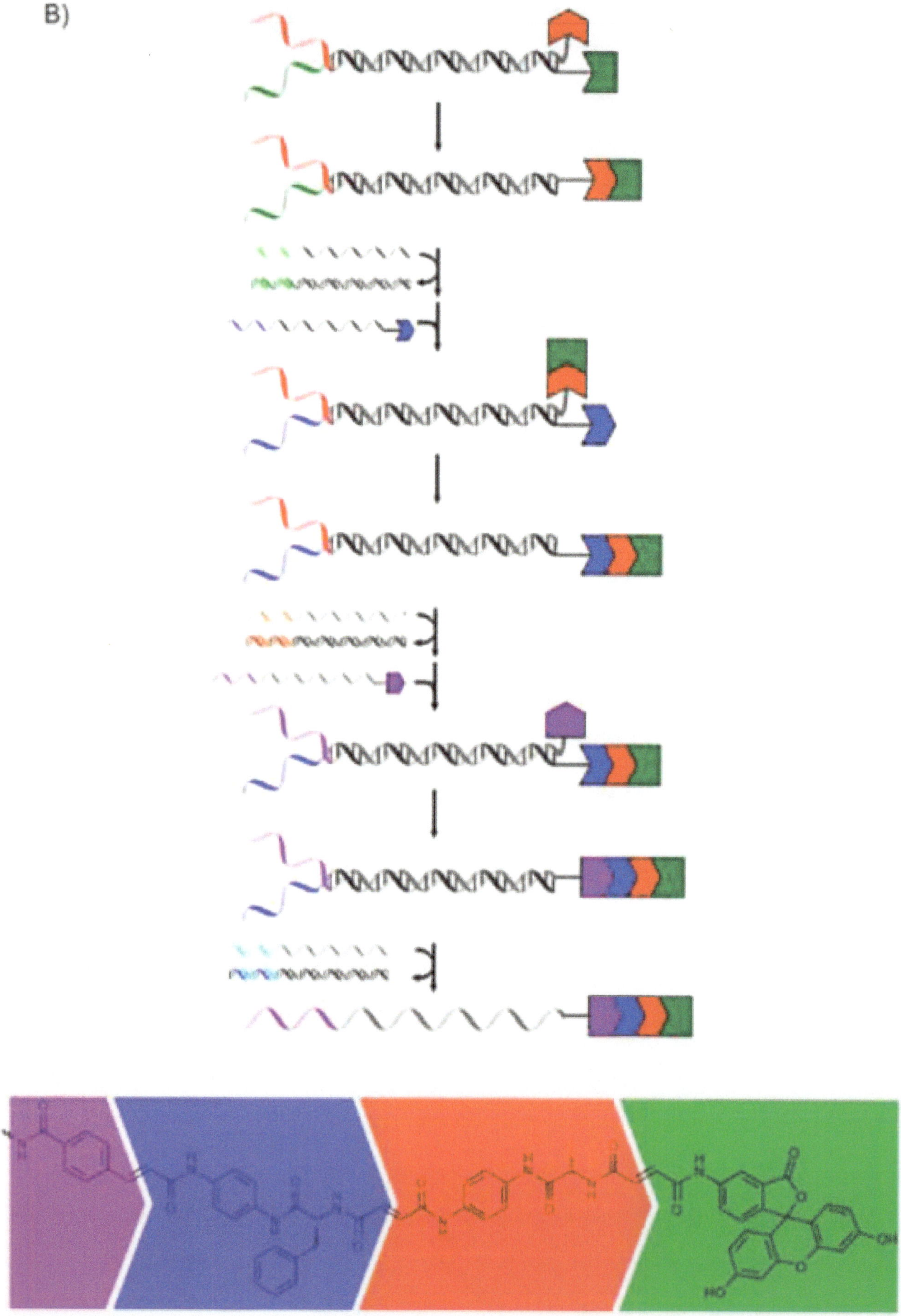

C)

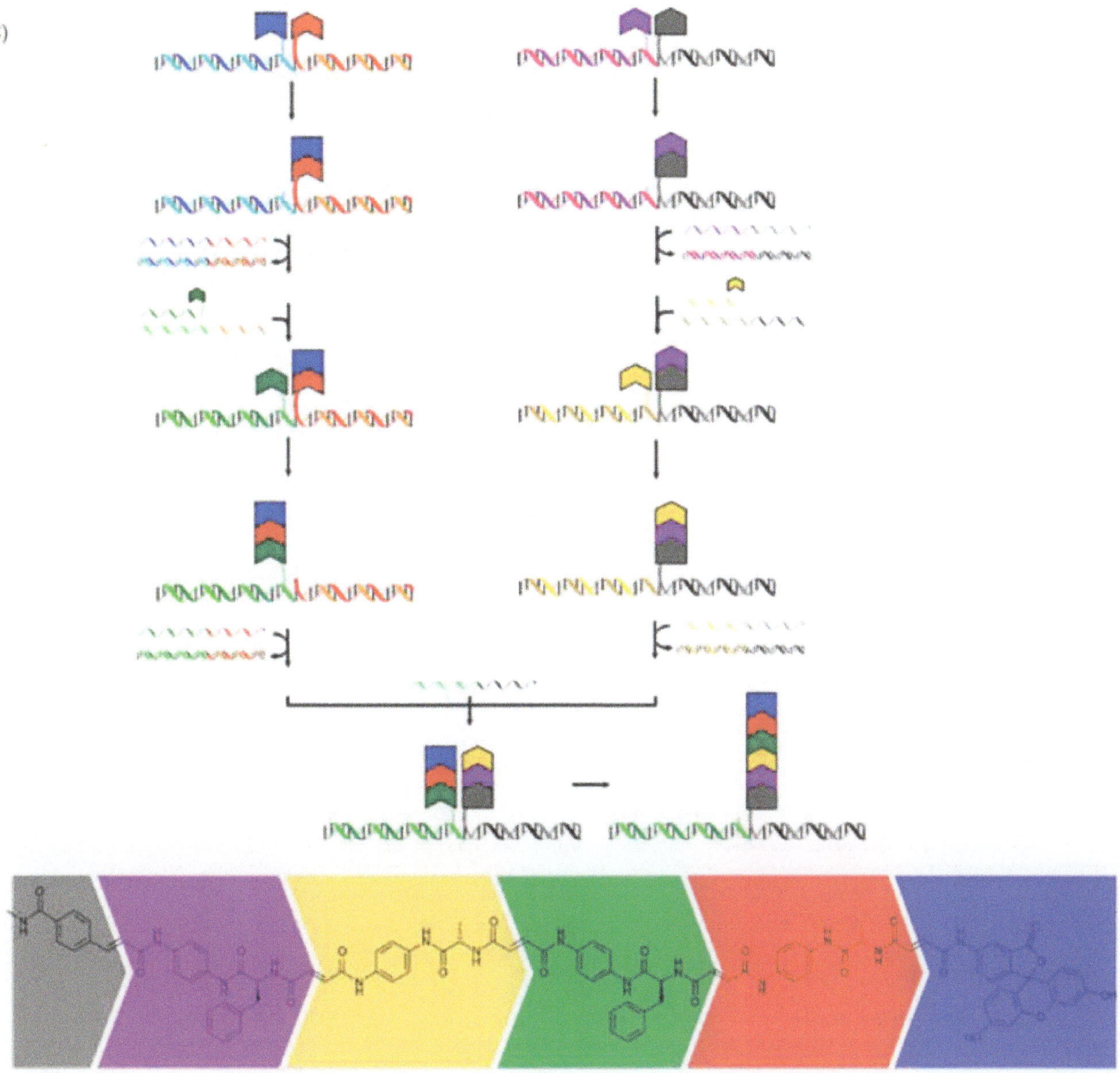

Figure 9: Various stepwise methods using the hybridisation of nucleotides for the synthesis of sequence-defined oligomers. (A) The use of ssDNAs of increasing lengths can be used to prepare hexamers.[68] (B) The use of partially complementary ssDNA strands for the subsequent addition of monomers in Wittig reactions and removal of the empty strands using fully complementary ssDNA strands.[69,70] (C) Binding of two ssDNA strands with monomer residues to a series of template ssDNA strands.[71]

The groups of Turberfield and O'Reilly used a strategy where two end-group functionalised ssDNA fragments are hybridised leaving a non-complementary section.[69,70] After reaction of the monomers in a Wittig reaction the spent ssDNA is hybridised with a (non-functionalised) fully complementary ssDNA strand, thereby effectively removing it from the mixture, leaving the remaining ssDNA free to again be hybridised with an end-group functionalised ssDNA strand which is only partially complementary. Through repetition of such cycles a sequence defined polymer (up to decamers were shown) is formed (Fig. 9b).

The same groups also used the hybridisation between template ssDNA with two shorter, complementary end-group functionalised ssDNA strands in a stepwise DNA template coupling.[71]One of the shorter strands carried a monomer residue at the 5′-terminus, while the other carried another monomer at the 3′-terminus. The addition of the template ssDNA brings the monomers (or, in the first step, a monomer and a monofunctional fluorescein end-group, FAM) close together, allowing the transfer of one of the monomers to the other chain by the Wittig reaction between an aldehyde and ylide functional group. Removal of the template and empty ssDNA, followed by addition of another template and ssDNA with terminal monomer group, results in the addition of another monomer. Such a stepwise procedure can be sped up by coupling oligomers (Fig. 9c).

The system currently closest to the strategy that nature uses in ribosomal translation was presented by the group of Liu, who used a single DNA strand as a template in a system that mimics the process that takes place in the ribosome.[72] As tRNA mimics they used several single stranded RNAs 5′-functionalised with the NHS ester of an amino acid or another small molecules (which will later be the end groups). A section of the tRNA mimic is complementary to a codon section on the template (which contains four codons) while the rest contains a central recognition site for an RNA-cleaving DNAzyme. One of the tRNA mimics (initiator) serves to recognise the start codon, is shorter and does not include a cleavable site or amino acid. The final component is a DNA walker with a 3′-terminal amine group. The 5′-section of the DNA walker is complementary to the RNA of the tRNA mimic closest to the recognition section, followed by a DNAzyme sequence and another section complementary to the final section (closest to the amino acids) of the tRNA mimic. The formation of the oligomers using this procedure is shown in Fig. 10. The template and tRNA mimics self-assemble by hybridisation. The DNA walker and the initiator are added after self-assembly of the other components, ensuring translation starts at the right position. The 3′-section of the walker DNA remains free to hybridise with the 5′-side of the first tRNA mimic. The amine end-group on the walker DNA is at that stage in close proximity to the first activated amino acid group and the first amide bond is formed. The 5′-section binds equally strongly to a section of the first tRNA mimic and moves to that location, bringing the DNAzyme in the DNA walker close to the cleavable site in the tRNA mimic. The cleavage frees up the 3′-section of the DNA walker, which can again bind to the next tRNA mimic. This cycle repeats itself several times.

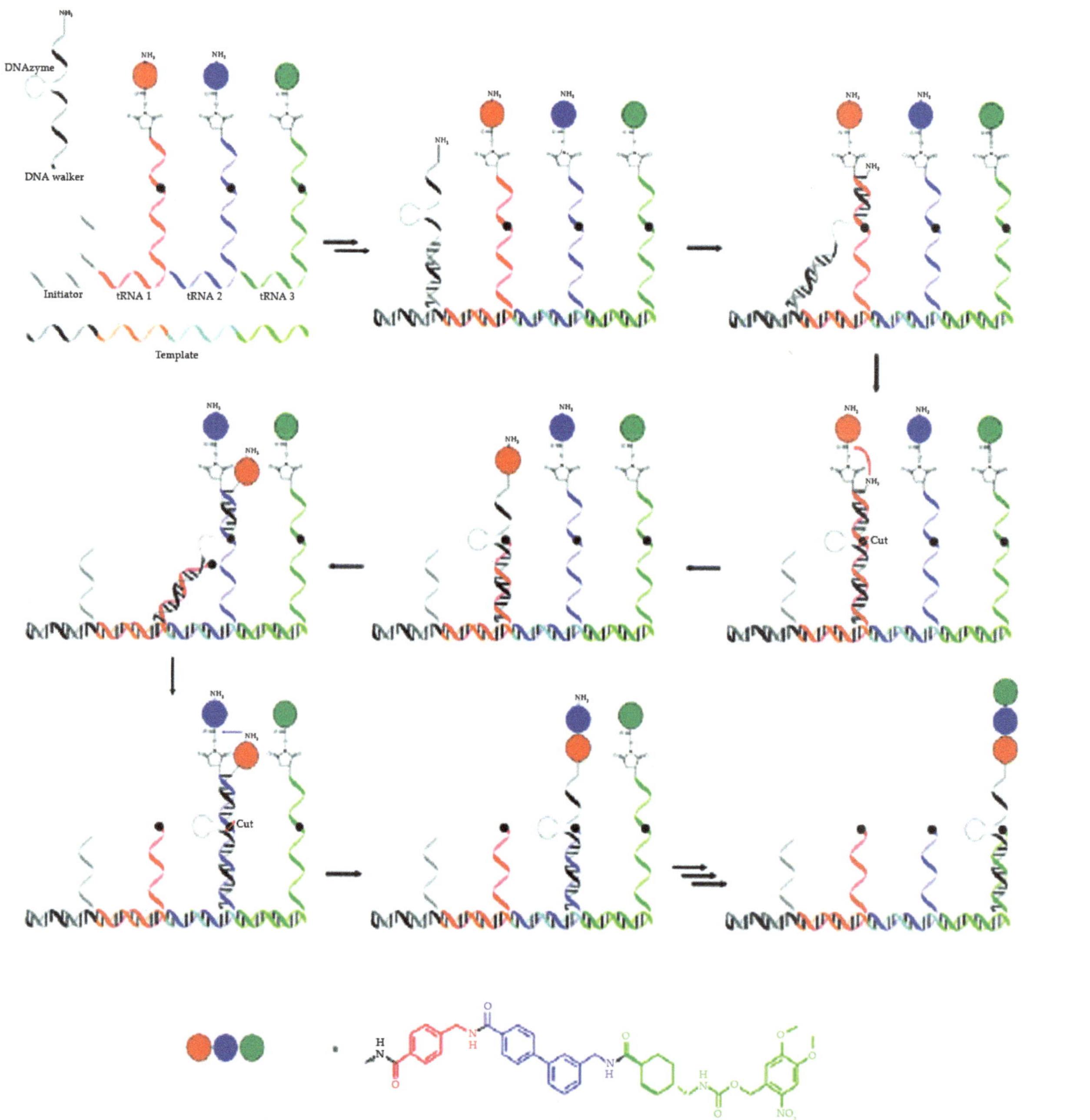

Figure 10: The strategy developed by Liu and co-workers mimicking the process of translation in ribosomes.[72]

Though the procedures described above only yield oligomers rather than true, high-molecular weight polymers, and similar oligomers can be prepared in much more straightforward procedures than the ones described above, *e.g.* through solid-phase synthesis methods, the work by the groups of Liu and the two groups of Turberfield and O›Reilly provide first steps toward novel methods for the synthesis of sequence defined polymers.

Schuster and co-workers used the hybridisation of (base modified) DNA strands for the preparation of sequence-defined linear and cyclic conducting oligomers,[73] continuing earlier studies where conducting oligomers were prepared using similar strategies without introducing sequence control.[74–76]

Two different strategies are presented. The first uses two complementary DNA strands in which some of the cytosine bases are modified with aniline (ANi) or 2,5-bis(2-thienyl)-pyrrole (SNS) moieties. Hybridisation of the strands yields complexes where ANi and SNS moieties are in appropriate distances to each other to allow for the oxidative polymerisation of the moieties using horseradish peroxidase (HRP) and hydrogen peroxide (Fig. 11a).[71] The relatively large size of the SNS groups requires their separation by one base,[73] while ANi moieties are located on neighbouring bases.[72] This strategy could be used to prepare block copolymers (or oligomers) and AABB-periodic copolymers of ANi and SNS residues.

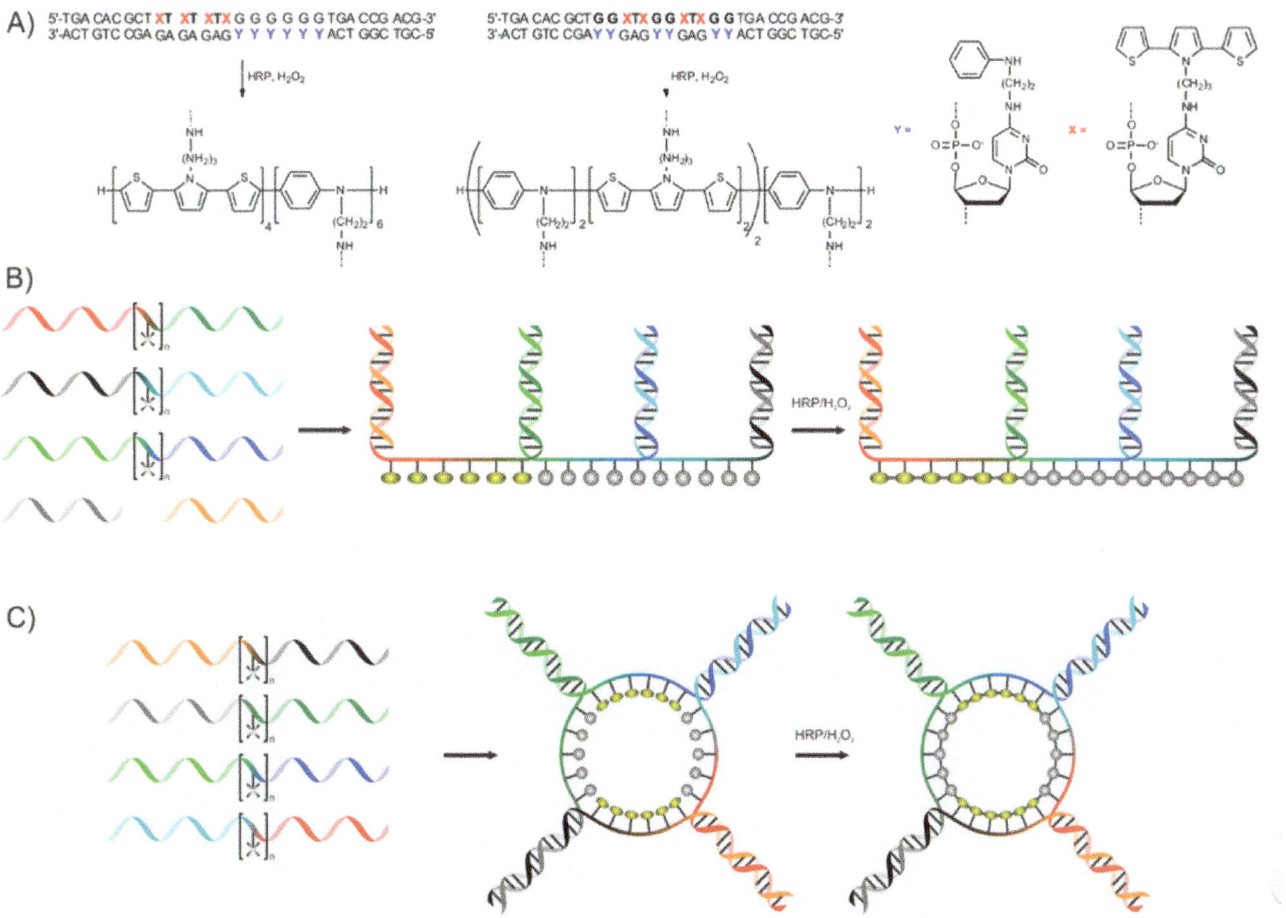

Figure 11: (A) Preparation of sequence-defined conducting polymers using the hybridisation of modified cytosine bases presented by Schuster and co-workers. X and Y represents 2,5-bis(2-thienyl)-pyrrole (SNS) and aniline (ANi) modified cytosine bases respectively. (B) Strategy for the preparation of linear and (C) cyclic sequence-defined conducting copolymers of SNS and ANi.[73]

A more modular strategy, based on the orthogonal hybridisation of flanking stretches of ssDNA on both sides of a central segment bearing SNS or ANi moieties, is also presented.[71] Linear polymers become accessible by using three ssDNA building blocks together with two 'caps' (Fig. 11b). The central segment contains five ANi or six SNS moieties. By using different ssDNA building blocks a range of possible conducting homo- and copolymers could be prepared (ANi15, ANi_{10}-*b*-SNS_6, ANi_5-*b*-SNS_6-*b*-ANi_5, SNS_6-*b*-ANi_5-*b*-SNS_6, SNS_{12}-*b*-ANi_5 and SNS_{18}) in good yields. Cyclic polymers are prepared using either three or four ssDNA building blocks which form cyclic assemblies (Fig. 11c shows the assemblies formed from four ssDNA building blocks). A range of cyclic polymers could be prepared by the oxidative polymerisation using HRP and H_2O_2, with different compositions and block sequences. Finally, it was shown that a longer central segment (bearing ten instead of five ANi moieties) could also be used to prepare cyclic copolymers. The spectroscopic properties of both linear and cyclic (co)polymers was shown to be dependent on the monomer sequence, and experiments using longer central segments bearing 10 ANi moieties showed that the spectroscopic properties of the conducting polymers were not influenced by the presence of DNA. A final step in which the DNA template is removed has not yet been reported by the authors, and the absence of hydrolysis sites might make this difficult. Nevertheless, the template-directed synthesis of conducting copolymers presents a novel method of creating well-defined materials for electronic- and optical applications.

Synthetic Templates

The number of purely synthetic systems where template polymerisation is used to control monomer sequences is still severely limited. The earliest examples where a certain influence over the monomer sequence distribution can be obtained rely on the presence of homopolymers that interact specifically with one of the monomers.[14,16] One example is from the work of Bamford *et al.*, who investigated the *N*-carboxyanhydride (NCA) copolymerisation of the NCAs of L-glutamic acid ethyl ester or phenylalanine with the NCA of sarcosine.[77] If a poly(sarcosine) homopolymer was used as the initiator, hydrogen bond interaction between poly(sarcosine) and the NH-groups in the NCAs of L-glutamic acid ethyl ester or phenylalanine led to the preferred incorporation of these monomers over the monomers of sarcosine, which does not interact with its homopolymer through H-bonding. Similarly, poly(2-vinylpyridine) as a template was shown to yield small blocks of poly(L-alanine) in copolymers of L-alanine and sarcosine (polymerised by NCA polymerisation).[78] The same effect was observed during the copolymerisation of styrene and methacrylic acid in the presence of a poly(2-vinylpyridine) template in radical polymerisations.[79]

The group of Sawamoto has more recently started exploring the use of template polymerisation for the selective incorporation of monomers as a route towards sequence-defined polymers in chain-growth polymerisations. Though their efforts have thus far focused on the specific incorporation of only a single monomer[80,81] or on templated homopolymerisation,[82] a number of different interactions have been shown to successfully yield selective addition of the complementary monomer.

The systems investigated by Sawamoto and co-workers are based on a bifunctional initiator capable of initiating cationic polymerisation and atom-transfer radical polymerisation. The two functional groups are placed on the *ortho*-positions of a phenyl group, thereby ensuring their proximity. A templating group (monomer or homopolymer) is added to the site designated for cationic polymerisation, whereas metal-catalysed atom-transfer radical addition is used to selectively add a complementary monomer. The authors have thus far shown the incorporation of methacrylic acid monomers in the presence of primary amine groups and the incorporation of sodium methacrylate using a 15-crown-5-ether template. By comparison, the addition of methyl methacrylate, benzyl methacrylate (in case of the amine template) or methacryloyloxyethyltrimethylammonium chloride (when using a crown ether template) did not proceed efficiently because of the absence of (or far weaker) interactions between the monomers and the template (Fig. 12).

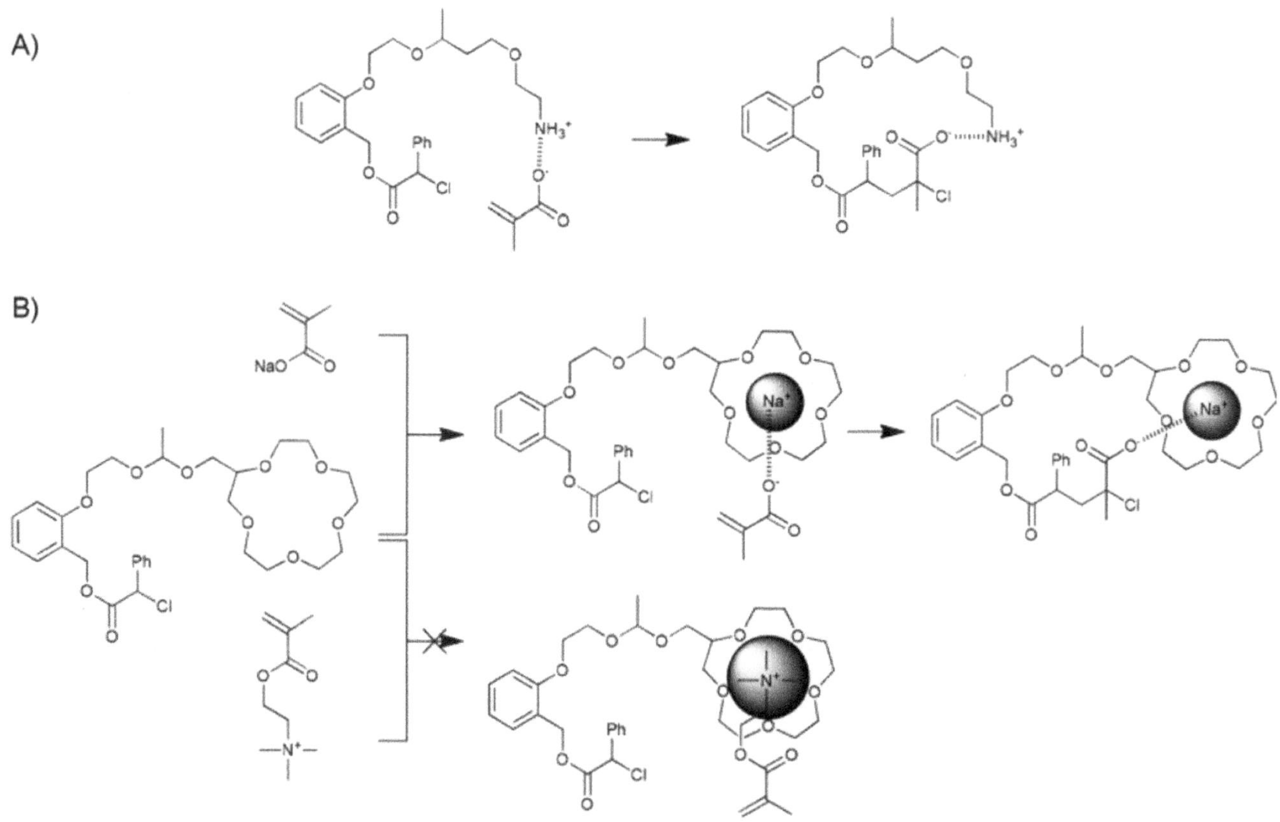

Figure 12: Templated addition of single monomers presented by Sawamoto and co-workers.[80,81]

Though it is debatable whether it should be considered a template polymerisation, supramolecular interactions between monomers can be used to influence monomer sequence. Early examples by Salamone *et al. e.g.* showed that attractive Coulombic forces between monomers can result in an increased degree of alternation in copolymers.[83–86] Combinations of protonated 4-vinyl pyridine and *p*-styrenesulfonate, 2-methacryloyloxyethyltrimethylammonium and 2-methacryloyloxyethanesulfonate or 3-acrylamido-3-methylbutyltrimethylammonium 2-acrylamido-2-methylpropanesulfonate were used (Fig. 13). Similarly, hydrogen bonding between monomers can lead to the formation of highly alternating monomer sequences.[87]

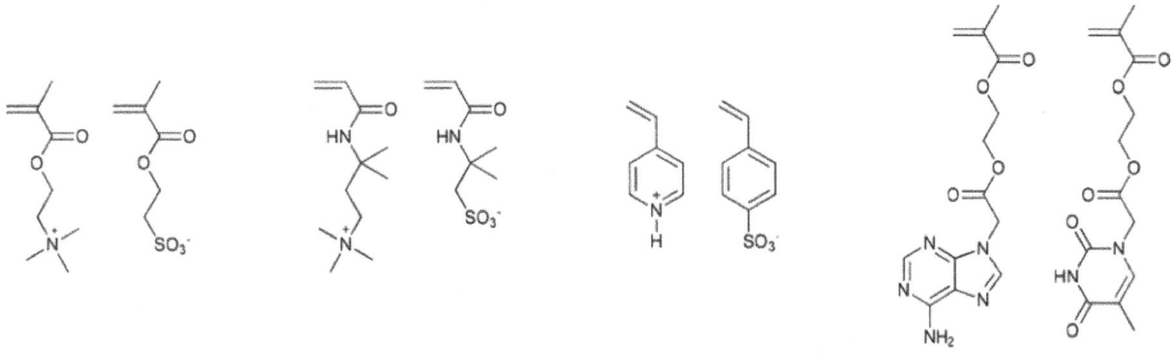

Figure 13: Monomer pairs where supramolecular interactions between the monomers lead to the formation of predominantly alternating monomer sequences.[83–87]

Jana *et al.* used metal–ligand complexation to create chiral metal complexes incorporating two polymerisable moieties ((meth)acrylates), Zn(II) and (−)-sparteine for the preparation of main-chain optically active polymers.[88,89] Racemic diads spaced by a single different monomer are needed to obtain optically active polymers.[90] The coordination of two (meth) acrylic acid monomers and optically active spartein to Zn(II), which is tetracoordinating, yielded a complex, the copolymerisation of which with aryl containing monomers (*e.g.* styrene or 2-vinylnaphthalene) led to the preferential formation of such triads and therefore optically active polymers, though the overall composition of the polymers indicated that the acrylic acid or methacrylic acid diads were mostly separated by more than one aryl containing monomer (Fig. 14a).

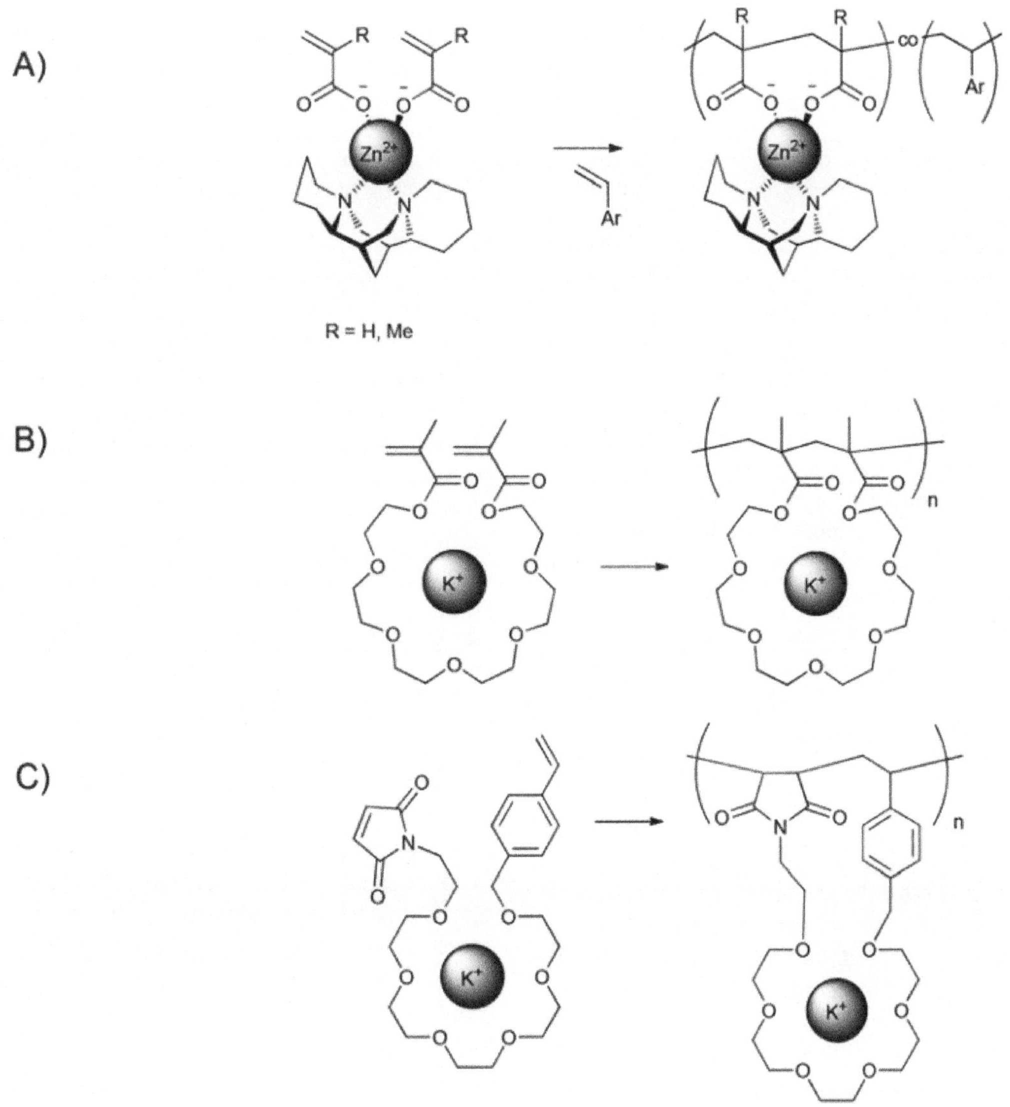

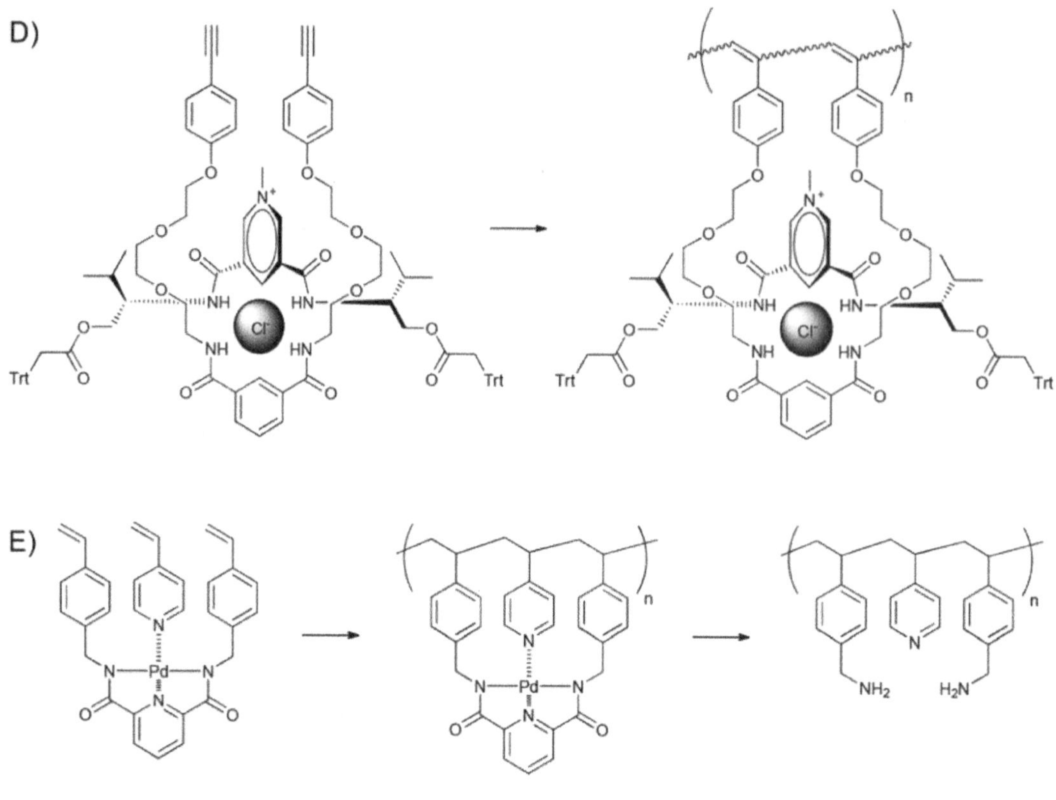

Figure 14: (A) Cyclocopolymerisation of a supramolecular complex of two (meth)acrylic acid monomers with aryl monomers.[88,89] (B) Promotion of cyclopolymerisation of ethylene glycol containing bivalent monomers using potassium-cations.[92] (C) Promotion of the cyclopolymerisation of a heterobifunctional monomer containing both a styrene and a maleimide moiety using a potassium-cation.[93] (D) Promotion of cyclopolymerisation by the binding of chloride in a ternary complex.[94] (E) Preparation of periodic copolymers with ABA-monomer sequences using a supramolecular complex of 4-vinylpyridine, palladium and a bivalent pincer containing monomer.[95]

The work of Jana *et al.* employs a synthetic pathway similar to cyclopolymerisation. In cyclopolymerisations bifunctional monomers are polymerised in such a way that the two polymerisable groups polymerise subsequently (intramolecularly), thereby yielding linear polymer chains with cyclic side-chains.[91] A requirements for bifunctional monomers to polymerise in a cyclopolymerisation is that the polymerisable groups need to be in close proximity to each other. If this requirement is not fulfilled, branching or cross-linking can result. Therefore the linkers between the polymerisable groups are typically fairly short and promote their colocalisation using bulky groups.

A templating approach can be used to induce cyclopolymerisation for monomers with longer linkers though. If the polymerisable groups, *i.e.* methacrylate groups in this case, are linked by oligo-ethylene glycol (EG) linkers, ions can be used as templates.[92] Crown-ether-like complexes are formed between the linkers and metal ions (Fig. 14b). A potassium cation was used to form complexes with dimethacrylate monomers with EG_5, EG_6 and EG_8 monomers, though the highest degree of cyclisation was found for EG_6. By promoting cyclopolymerisation, the intramolecular polymerisation of the two functional groups from one molecule was promoted, and thereby also a control over the monomer sequence is exerted. This same principle was utilised by Zou *et al.* for heterobifunctional monomers that contains both a styrene moiety and a maleimide moiety separated by a EG_6 or EG_7 spacer. Normally styrene and maleimide monomers copolymerise in an alternating fashion, but templating using a potassium-cation ensured that cyclopolymerisation took place and cross-linking was prevented.[93] Isono *et al.* used the ternary complexation of a bivalent monomer, a chloride ion and a cationic, anion coordinating axle compound to promote the cyclopolymerisation of a bivalent monomer (Fig. 14d).[94]

Sawamoto and co-workers used a combination of metal–ligand complexation and cyclopolymerisation to create periodic copolymers (ABA sequences).[95] A tridentate bifunctional monomer was used to complex Pd(II) while a 4-vinylpyridine monomer served as a monodentate ligand. The alignment of the polymerisable groups, which was additionally supported by the H-bonding of bulky fluoroalcohols to the carbonyl groups of the bifunctional monomer, yielded the desired ABA-sequences with a high degree of accuracy. Removal of the templating Pd(II) and hydrolysis of the bifunctional monomer yielded primary amine and pyridine side groups (Fig. 14e).

CONCLUSIONS AND OUTLOOK

Template polymerisation as a method towards sequence-controlled synthetic polymers is a very useful pathway, though at present predominantly for the synthesis of structures closely related to biomacromolecules. Accordingly, the biggest successes have thus far been achieved using the machinery that nature has given rise to. The transcription and translation processes that biology uses could be employed for the incorporation of non-natural monomers with high accuracy by utilising polymerases or ribosomes, for example by using non-natural nucleic acids and non-canonical amino acids (or α-hydroxy acids). The accuracy achieved using nature's machinery also limits the scope of this approach, since only a limited number of monomers can be incorporated in this way, and modifications of the polymerisation systems remains highly challenging, while often selectivity is traded against accuracy. A wider variety of monomers can be incorporated using enzyme-free template polymerisations, even though the accuracy is a far cry from that achieved using specialised proteins, even when oligomers (rather than single monomers) are linked together. Nevertheless the formation of stable double stranded helices between the template and the daughter polymer is still required, a limitation that can be circumvented using various stepwise processes based on the hybridisation of nucleic acid strands, using which in theory any type of step-growth polymerisation can be developed into a sequence-defined polymerisation. Realistically, these methods are only suitable for the preparation of sequence-controlled oligomers though, and the synthesis is still very tedious and inefficient.

The potential of template polymerisations using templates unrelated to nucleic acids is illustrated by the influence that can already be exerted over other properties of polymers, such as the molecular weight, dispersity and tacticity. Though some simple systems have been described where synthetic templates are used to control monomer sequence, this area is still in its infancy. One of the challenges has been that chain-growth polymerisations are often targeted, which are far harder to control than step-growth polymerisations due to the reactive nature of the ions and radicals that are used. Nevertheless promising results have already been obtained which no doubt will lead the way for novel strategies to also control the monomer sequence in polymers prepared through chain-growth polymerisations.

A second challenge is the availability of suitable templates. Whereas the synthesis of nucleic acids has been developed to a high degree, thanks in no small amount to the relevance of these biomacromolecules for biological and biochemical research, the synthesis of templates for other types of polymers still requires a lot of research since nucleic acids present their binding sites at regular distances not necessarily suitable for other polymerisations. Additionally, also the interactions between the template and monomers or the daughter polymer need to be suitable in order to align monomers and increase their effective concentration.

The effort still required to make control over monomer sequence using template polymerisations feasible, especially for polymers unrelated to biopolymers, is still substantial. In the end the benefits will likely outweigh the costs though, as a plethora of materials with novel properties could come within reach.

ACKNOWLEDGEMENTS

The author would like to thank the Chemical Industry Fund (FCI Liebig-fellowship) and the German Research Foundation (DFG, project number BR 4363/3-1) for financial support.

REFERENCES

1. S. D. McCulloch and T. A. Kunkel, *Cell Res.*, 2008, 18, 148–161 .
2. J. M. Berg, J. L. Tymoczko and L. Stryer, *Biochemistry*, W. H. Freeman and Company, New York, 5th edn, 2002 .
3. W. Gilbert, *Nature*, 1986, 319, 618 .
4. A. Eschenmoser, *Science*, 1999, 284, 2118–2124 .
5. G. F. Joyce, *Nature*, 2002, 418, 214–221 .
6. R. Jones, *Nat. Nanotechnol.*, 2008, 3, 699–700 .
7. N. Badi and J.-F. Lutz, *Chem. Soc. Rev.*, 2009, 38, 3383–3390 .
8. J.-F. Lutz, *Polym. Chem.*, 2010, 1, 55–62 .
9. M. Ouchi, N. Badi, J.-F. Lutz and M. Sawamoto, *Nat. Chem.*, 2011, 3, 917–924 .
10. N. Badi, D. Chan-Seng and J.-F. Lutz, *Macromol. Chem. Phys.*, 2013, 214, 135–142 .
11. J.-F. Lutz, M. Ouchi, D. R. Liu and M. Sawamoto, *Science*, 2013, 341, 1238149 .
12. J.-F. Lutz, in *Sequence-Controlled Polymers: Synthesis, Self-Assembly, and Properties*, ed. J.-F. Lutz, T. Y. Meyer, M. Ouchi and M. Sawamoto, ACS Symposium Series 1170, American Chemical Society, Washington, DC, 2014, ch. 1, pp. 1–11 .
13. H. Colquhoun and J.-F. Lutz, *Nat. Chem.*, 2014, 6, 455–456 .
14. Y. Y. Tan, *Prog. Polym. Sci.*, 1994, 19, 561–588 .
15. S. Połowiński, *Prog. Polym. Sci.*, 2002, 27, 537–577 .
16. S. Połowiński, *Template Polymerization*, ChemTec Publishing, Ontario, 1997 .
17. K. Satoh and M. Kamigaito, *Chem. Rev.*, 2009, 109, 5120–5156 .
18. P. K. Lo and H. F. Sleiman, *J. Am. Chem. Soc.*, 2009, 131, 4182–4183 .
19. L. E. Orgel, *Crit. Rev. Biochem. Mol. Biol.*, 2004, 39, 99–123 .
20. J. P. Ferris and G. Ertem, *Science*, 1992, 257, 1387–1389 CAS.
21. S. Miyakawa and J. P. Ferris, *J. Am. Chem. Soc.*, 2003, 125, 8202–8208 .
22. F. Olasagasti, H. J. Kim, N. Pourmand and D. W. Deamer, *Biochimie*, 2011, 93, 556–561 .
23. X. Li and D. R. Liu, *Angew. Chem., Int. Ed.*, 2004, 43, 4848–4870 .
24. Y. Brudno and D. R. Liu, *Chem. Biol.*, 2009, 16, 265–276 .
25. V. B. Pinheiro, A. I. Taylor, C. Cozens, M. Abramov, M. Renders, S. Zhang, J. C. Chaput, J. Wengel, S.-Y. Peak-Chew, S. H. McLaughlin, P. Herdewijn and P. Holliger, *Science*, 2012,336, 341–344 .
26. A. A. Henry and F. E. Romesberg, *Curr. Opin. Biotechnol.*, 2005, 16, 370–377 .
27. R. C. Holmberg, A. A. Henry and F. E. Romesberg, *Biomol. Eng.*, 2005, 22, 39–49 .
28. S. Zhang, J. C. Blain, D. Zielinska, S. M. Gryaznov and J. W. Szostak, *Proc. Natl. Acad. Sci. U. S. A.*, 2013, 110, 17732–17737 .
29. I. Hirao, M. Kimoto and R. Yamashige, *Acc. Chem. Res.*, 2012, 45, 2055–2065 .
30. R. Hili, J. Niu and D. R. Liu, *J. Am. Chem. Soc.*, 2013, 135, 98–101 .

31. O. L. Acevedo and L. E. Orgel, *J. Mol. Biol.*, 1987, 197, 187–193 .
32. L. E. Orgel, *Acc. Chem. Res.*, 1995, 28, 109–118 .
33. J. G. Schmidt, L. Christensen, P. E. Nielsen and L. E. Orgel, *Nucleic Acids Res.*, 1997, 25, 4792–4796 .
34. J. G. Schmidt, P. E. Nielsen and L. E. Orgel, *Nucleic Acids Res.*, 1997, 25, 4797–4802 .
35. I. A. Kozlov, S. Pitsch and L. E. Orgel, *Proc. Natl. Acad. Sci. U. S. A.*, 1998, 95, 13448–13452 .
36. I. A. Kozlov, P. K. Politis, A. van Aehot, R. Busson, P. Herdewijn and L. E. Orgel, *J. Am. Chem. Soc.*, 1999, 121, 2653–2656 .
37. I. A. Kozlov, B. de Bouvere, A. van Aehot, P. Herdewijn and L. E. Orgel, *J. Am. Chem. Soc.*, 1999, 121, 5856–5859 .
38. I. A. Kozlov, L. E. Orgel and P. E. Nielson, *Angew. Chem., Int. Ed.*, 2000, 39, 4292–4295 .
39. I. A. Kozlov, M. Zielinski, B. Allart, L. Kerremans, A. van Aehot, R. Busson, P. Herdewijn and L. E. Orgel, *Chem. – Eur. J.*, 2000, 6, 151–155 .
40. J. A. R. Stutz and C. Richert, *J. Am. Chem. Soc.*, 2001, 123, 12718–12719 .
41. J. C. Leitzel and D. G. Lynn, *Chem. Rec.*, 2001, 1, 53–62 .
42. A. H. El-Sagheer and T. Brown, *Acc. Chem. Res.*, 2012, 45, 1258–1267 .
43. D. M. Rosenbaum and D. R. Liu, *J. Am. Chem. Soc.*, 2003, 125, 13924–13925 .
44. R. E. Kleiner, Y. Brudno, M. E. Birnbaum and D. R. Liu, *J. Am. Chem. Soc.*, 2008, 130, 4646–4659 .
45. Y. Brudno, M. E. Birnbaum, R. E. Kleiner and D. R. Liu, *Nat. Chem. Biol.*, 2010, 6, 148–155 .
46. Y. Ura, J. M. Beierle, L. J. Leman, L. E. Orgel and M. R. Ghadiri, *Science*, 2009, 325, 73–77 .
47. L. Wang, J. Xie and P. G. Schultz, *Annu. Rev. Biophys. Biomol. Struct.*, 2006, 35, 225–249 .
48. J. Xie and P. G. Schultz, *Methods*, 2005, 36, 227–238 .
49. A. Ohta, Y. Yamagishi and H. Suga, *Curr. Opin. Chem. Biol.*, 2008, 12, 159–167 .
50. Z. Tan, A. C. Forster, S. C. Blacklow and V. W. Cornish, *J. Am. Chem. Soc.*, 2004, 126, 12752–12753 .
51. T. Passioura and H. Suga, *Chem. – Eur. J.*, 2013, 19, 6530–6536 .
52. T. Kawakami, H. Murakami and H. Suga, *Nucleic Acids Symp. Ser.*, 2007, 51, 361–362 .
53. T. Kawakami, H. Murakami and H. Suga, *Chem. Biol.*, 2008, 15, 32–42 .
54. A. O. Subtelny, M. C. T. Hartman and J. W. Szostak, *J. Am. Chem. Soc.*, 2008, 130, 6131–6136 .
55. B. Zhang, Z. Tan, L. Gartenmann Dickson, M. N. L. Nalam, V. W. Cornish and A. C. Forster, *J. Am. Chem. Soc.*, 2007, 129, 11316–11317 .
56. A. Frankel, S. W. Millward and R. W. Roberts, *Chem. Biol.*, 2003, 10, 1043–1050 .
57. L. M. Dedkova, N. E. Fahmi, S. Y. Golovine and S. M. Hecht, *J. Am. Chem. Soc.*, 2003, 125, 6616–6617 .
58. L. M. Dedkova, N. E. Fahmi, S. Y. Golovine and S. M. Hecht, *Biochemistry*, 2006, 45, 15541–15551 .
59. A. Ohta, H. Murakami, E. Higashimura and H. Suga, *Chem. Biol.*, 2007, 14, 1315–1322 .
60. A. Ohta, H. Murakami and H. Suga, *ChemBioChem*, 2008, 9, 2773–2778 .
61. S. Fahnestock and A. Rich, *Science*, 1971, 173, 340–343 CAS.
62. Y. Goto, T. Katoh and H. Suga, *Nat. Protoc.*, 2011, 6, 779–790 .
63. H. Kries, R. Wachtel, A. Pabst, B. Wanner, D. Niquille and D. Hilvert, *Angew. Chem., Int. Ed.*, 2014, 53, 10105–10108 .
64. J. Niu, R. Hili and D. R. Liu, *Nat. Chem.*, 2013, 5, 282–292 .
65. Z. J. Gartner, M. W. Kanan and D. R. Liu, *J. Am. Chem. Soc.*, 2002, 124, 10304–10306 .
66. Z. J. Gartner, B. N. Tse, R. Grubina, J. B. Doyon, T. M. Snyder and D. R. Liu, *Science*, 2004, 305, 1601–1605 .
67. T. M. Snyder and D. R. Liu, *Angew. Chem., Int. Ed.*, 2005, 44, 7379–7382 .
68. Y. He and D. R. Liu, *J. Am. Chem. Soc.*, 2011, 133, 9972–9975 .
69. M. L. McKee, P. J. Milnes, J. Bath, E. Stulz, A. J. Turberfield and R. K. O'Reilly, *Angew. Chem., Int. Ed.*, 2010, 49, 7948–7951 .
70. P. J. Milnes, M. L. McKee, J. Bath, L. Song, E. Stulz, A. J. Turberfield and R. K. O'Reilly, *Chem. Commun.*, 2012, 48, 5614–5616 .
71. M. L. McKee, P. J. Milnes, J. Bath, E. Stulz, R. K. O'Reilly and A. J. Turberfield, *J. Am. Chem. Soc.*, 2012, 134, 1446–1449 .
72. Y. He and D. R. Liu, *Nat. Nanotechnol.*, 2010, 5, 778–782 .
73. W. Chen and G. B. Schuster, *J. Am. Chem. Soc.*, 2013, 135, 4438–4449 .
74. B. Datta, G. B. Schuster, A. McCook, S. C. Harvey and K. Zakrzewska, *J. Am. Chem. Soc.*, 2006, 128, 14428–14429 .
75. B. Datta and G. B. Schuster, *J. Am. Chem. Soc.*, 2008, 130, 2965–2973 .
76. W. Chen and G. B. Schuster, *J. Am. Chem. Soc.*, 2012, 134, 840–843 .
77. C. H. Bamford, H. Block and Y. Imanishi, *Biopolymers*, 1966, 4, 1067–1072 .
78. R. A. Volpe and H. L. Frisch, *Macromolecules*, 1987, 20, 1747–1752 .
79. H. L. Frisch and Q. Xu, *Macromolecules*, 1992, 25, 5145–5149 .
80. S. Ida, T. Terashima, M. Ouchi and M. Sawamoto, *J. Am. Chem. Soc.*, 2009, 131, 10808–10809 .
81. S. Ida, M. Ouchi and M. Sawamoto, *J. Am. Chem. Soc.*, 2010, 132, 14748–14750 .
82. S. Ida, M. Ouchi and M. Sawamoto, *Macromol. Rapid Commun.*, 2011, 32, 209–214 .

83. J. C. Salamone, A. C. Watterson, T. D. Hsu, C. C. Tsai, M. U. Mahmud, A. W. Wisniewski and S. C. Israel, *J. Polym. Sci., Polym. Symp.*, 1978, 64, 229–213 .

84. J. C. Salamone, C.-C. Tsai and A. C. Watterson, *J. Macromol. Sci., Chem.*, 1979, 13, 665–672 .

85. J. C. Salamone, L. Quach, A. C. Watterson, S. Krauser and M. U. Mahmud, *J. Macromol. Sci., Chem.*, 1985, 22, 653–664 .

86. J. C. Salamone, M. K. Raheja, Q. Anwaruddin and A. C. Watterson, *J. Polym. Sci., Polym. Lett. Ed.*, 1985, 23, 655–659 .

87. Y. Kang, A. Lu, A. Ellington, M. C. Jewett and R. K. O'Reilly, *ACS Macro Lett.*, 2013, 2, 581–586 .

88. S. Jana and D. C. Sherrington, *Angew. Chem., Int. Ed.*, 2005, 44, 4804–4808 .

89. S. Jana, P. A. G. Cormack, A. R. Kennedya and D. C. Sherrington, *J. Mater. Chem.*, 2009, 19, 3427–3442 .

90. G. Wulff, *Angew. Chem., Int. Ed. Engl.*, 1989, 28, 21–37 .

91. G. B. Butler, *Cyclopolymerization and cyclocopolymerization*, Marcel Dekker, Inc., New York, 1992 .

92. T. Terashima, M. Kawabe, Y. Miyabara, H. Yoda and M. Sawamoto, *Nat. Commun.*, 2013, 4, 2321 .

93. L. Zou, J. Liu, K. Zhang, Y. Chen and F. Xi, *J. Polym. Sci., Part A: Polym. Chem.*, 2014, 52, 330–338 .

94. T. Isono, T. Satoh and T. Kakuchi, *J. Polym. Sci., Part A: Polym. Chem.*, 2011, 49, 3184–3192 .

95. Y. Hibi, M. Ouchi and M. Sawamoto, *Angew. Chem., Int. Ed.*, 2011, 50, 7434–7437

CHAPTER 2 SYNTHESIS AND POST-POLYMERISATION MODIFICATIONS OF ALIPHATIC POLY (CARBONATE) S PREPARED BY RING-OPENING POLYMERISATION

Sarah Tempelaar[a], Laetitia Mespouille[b], Olivier Coulembier[b], Philippe Dubois[b], and Andrew P. Dove[a]

[a]Department of Chemistry, University of Warwick, Coventry, CV4 7AL, UK

[b]Center of Innovation and Research in Materials and Polymers (CIRMAP), Laboratory of Polymeric and Composite Materials, University of Mons, Place du Parc 23, 7000 Mons, Belgium

ABSTRACT

Owing to their low toxicity, biocompatibility and biodegradability, aliphatic poly(carbonate)s have been widely studied as materials for biomedical application. Furthermore, the synthetic versatility of the sixmembered cyclic carbonates for the realization of functional degradable polymers by ring-opening polymerisation has driven wider interest in this area. In this review, the synthesis and ring-opening polymerisation of functional cyclic carbonates that have been reported in the literature in the past decade are discussed. Finally, the post-polymerisation modification methods that have been applied to the resulting homopolymers and copolymers and the application of the materials are also discussed.

INTRODUCTION

Alongside aliphatic poly(ester)s, aliphatic poly(carbonate) s are important biodegradable materials that have received considerable attention.[1–3] These materials are excellent candidates forpharmaceutical applications as a consequence of their low toxicity, biocompatibility, and biodegradability. Copolymers of carbonates such as trimethylene carbonate (TMC) with other cyclic monomers such as lactide and glycolide have already found application as sutures and in other biomedical fields such as controlled drug delivery on account of the favourable materials properties that they bestow into the copolymers.[4,5] In addition, materials prepared from poly(trimethylene carbonate) (PTMC) are known to degrade *in vivo* by surface erosion in contrast to the bulk degradation behaviour shown by poly(ester)s. Furthermore, poly(ester) degradation products can lead to increased levels of acidity, which are not present during poly(carbonate) degradation, that may be hazardous to loaded drugs or living tissues.[6–9]

Poly(carbonate) synthesis is commonly realised either by the copolymerisation of epoxides with carbon dioxide or by the ring-opening polymerisation (ROP) of cyclic carbonate monomers.[1,3,10–13] The synthesis of poly(carbonate)s *via* CO_2/epoxide copolymerisation is a potentially useful method of utilising CO_2; however, the formation of five-membered cyclic carbonate by-products, the presence of ether linkages, and the use of air-sensitive coordination compounds make its application synthetically challenging. The ROP of cyclic carbonates has received significantly more attention and consequently provides an attractive and versatile methodology for the synthesis of functionalised poly(carbonate)s, despite the requirement to synthesise many of the monomers, and has been studied by many groups.[1,3,14] The ROP of cyclic carbonates can be realised using cationic, anionic, coordination-insertion, organocatalytic, and enzymatic methods and has resulted

Citation: Sarah Tempelaar, Laetitia Mespouille, Olivier Coulembier, Philippe Dubois and Andrew P. Dove, Synthesis and post-polymerisation modifications of aliphatic poly(carbonate)s prepared by ring-opening polymerization, DOI: 10.1039/C2CS35268K.

in high levels of control over polymer molecular weight, dispersity and end-group fidelity.[15–24]

For their use in biomedical applications, such as targeted drug delivery and tissue engineering, it is advantageous to be able to tailor the poly(carbonate) structure to enable specific interactionswithcellsororgans.Thepropertiesofpoly(carbonate)s derived from TMC can be tuned by copolymerisation and blending with other biodegradable polymers (*i.e.*poly(lactide), poly(glycolide), poly(caprolactone)) and the introduction of functional end-groups.[25–32] Precise control over the physical properties or the addition of biologically active molecules to the poly(carbonate) structure however, is best achieved by post-derivatisation of pendant functionalities in the polymer backbone. Such pendant functionalities may be introduced *via* the ROP of cyclic carbonate monomers bearing the desired functionality however, not all functional groups are compatible with the ring-opening polymerisation process and in some es the added functionality limits the polymerisation efficiency.[33] As a result, further modification of the polymer backbone following polymerisation is often required. While many further functionalisations of the pendant moieties in the poly(carbonate) backbone have been reported, modifications commonly involve many steps and typically only partial functionalisation is achieved. In recent years, however, post-polymerisation modifications of functional poly(carbonate)s using more efficient chemistries have been reported. In addition, many copolymerisations of functional cyclic carbonate monomers with simple cyclic esters (*i.e.* lactide or ε-caprolactone) have been reported to introduce functionality in the parent poly(ester) backbone in order to overcome the challenging nature of functional cyclic ester synthesis.[34–36]

Herein the synthesis and polymerisation of cyclic carbonates reported in the literature including post-polymerisation modifications of the resulting homo- and co-polymers are discussed with a focus on progress made in the area in the past decade.

SYNTHETIC ROUTES TO FUNCTIONAL POLY(CARBONATE)S

Synthesis of cyclic carbonates can be achieved through a number of routes, commonly involving multiple steps using toxic phosgene derivatives for carbonate formation from a 1,3-diol (Fig. 1).[1,3] Alternative methods that do not involve the use of phosgene include the reaction of oxetane with CO_2, a two-step procedure using a pentafluorophenyl ester intermediate and the use of 1,1′-carbonyldiimidazole (CDI).[37,38] This section of this review article will focus on the specific methods for monomer synthesis centred on the most common feedstocks.

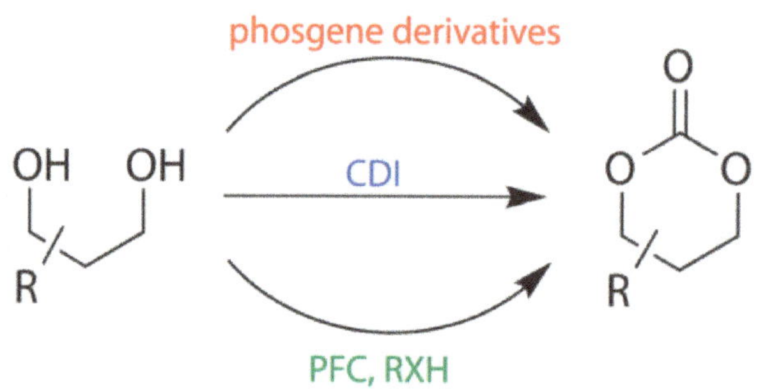

Figure 1: Representative scheme for the synthesis of functional six-membered cyclic carbonatesfrom 1,3-diols using phosgene derivatives, 1,1′-carbonyldiimidazole (CDI) orbis(pentafluorophenyl)carbonate (PFC) with an alcohol or amine (RXH).

Functional Cyclic Carbonate Monomers Derived from 2,2-Bishydroxy(methyl)Propionic Acid (bis-MPA)

2,2-Bishydroxy(methyl)propionic acid (bis-MPA), also widely used in the preparation of biocompatible dendrimers,[39,40] and hydrogel materials,[41] was first used for the synthesis of cycliccarbonates by Bisht and coworkers in 1999.[42] The reported cyclic carbonate could be prepared by direct functionalisation of the acid group in bis-MPA under either acidic or basic conditions (Scheme 1A), before cyclisation to the carbonate using phosgene or a phosgene derivative such as ethyl chloroformate, di- or triphosgene.[42,43] Other possible synthetic pathways from the bis-MPA scaffold to provide functional cyclic carbonates have since been reported. Alternative pathways (Scheme 1, route B and C) are more suitable for the synthesis of monomers that have more sensitive substituents. In one example (Scheme 1, route B), cyclic carbonate synthesis involves protection of the acid functionality by conversion to a benzyl ester followed by the synthesis of benzyl functional cyclic carbonate, 5-methyl-5-benzyloxycarbonyl-1,3-dioxan-2-one (MBC, **15**). Subsequent deprotection of the acid functionality followed byactivation of the free carboxylic acid, typically by the use of dicyclohexylcarbodiimide (DCC) or by conversion to the acyl chloride, allow reaction with an alcohol or an amine.[44] In a further alternative (Scheme 1, route C), the 1,3-diol is acetonide-protected before functionalisation of the carboxylic acid is carried out. Subsequent removal of the acetonide protecting group and formation of the functional cyclic carbonate are realised in the last step. Finally, the use of bis(pentafluorophenyl)carbonate, as demonstrated recently by Hedrick and coworkers, provides an excellent alternative to the use of phosgene derivatives and multi-step procedures (Scheme 1, route D).[37] Here, the synthesis of a pentafluorophenyl-functionalised cyclic carbonate, 5-methyl-5-pentafluorophenyloxycarbonyl-1,3-dioxan-2-one, from bis-MPA is achieved in one step using bis(pentafluorophenyl) carbonate in the presence of caesium fluoridewhich followed by reaction of the pentafluorophenylester with amines or alcohols yields the desired monomers.[37] In summary, 2,2-bishydroxy(methyl)propionic acid (bis-MPA) is an excellent choice as biocompatible and versatile scaffold for the preparation of cyclic carbonate monomers with pendant ester and amide groups.

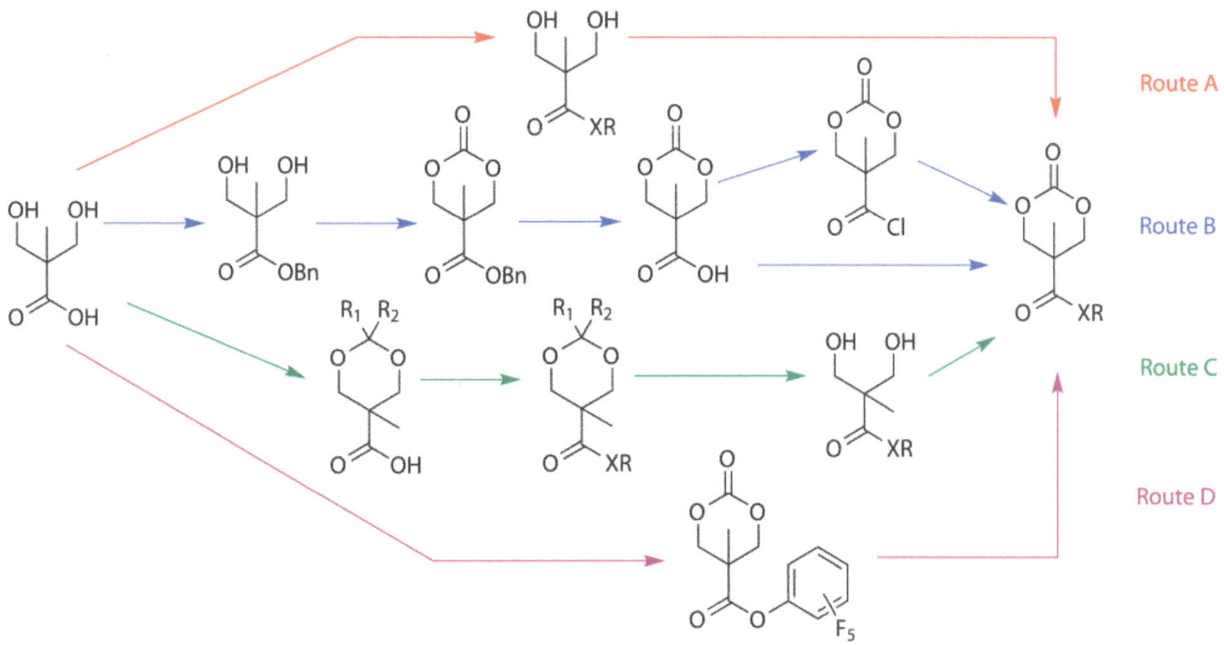

Scheme 1: Synthesis of functional cyclic carbonates derived from 2,2-bishydroxy(methyl)propionic acid (bis-MPA).

Functional cyclic carbonates derived from pentaerythritol, glycerol and trimethylolalkanes

The preparation of functional cyclic carbonates has also been reported in many es using 1,3-diol containing compounds that have three or four hydroxyl groups in their molecular structure. Common precursors are pentaerythritol, glycerol and trimethylolalkanes which, apart from benzyl protection of the hydroxyl functionality (or functionalities) can also be used in the synthesis of other functional cyclic carbonates (Scheme 2).[45–50] A number of synthetic pathways are described in the literature that use glycerol or trimethylolalkanes as a scaffold to provide cyclic carbonates with a pendant functionality. In one example (Scheme 2, route A), two of the available hydroxyl groups are protected with acetonide or benzylidene groups before further reaction of the residual alcohol with an acid chloride or *via* DCC-coupling with a carboxylic acid-functional molecule. The deprotection of the diol and conversion to the cycliccarbonate yields the desired monomers. Alternatively, the precursor has been reacted with a dialkyl carbonate, mediated by immobilised *Candida antarctica lipase B*, to yield a hydroxyl-functional cyclic carbonate intermediate (Scheme 2, route B). The hydroxyl group was then further functionalised using the abovementioned chemistries. In a report by Hatti-Kaul and coworkers, the dialkyl carbonates were also used in the functionalisation of the hydroxyl group (Scheme 2, route B).[51] In a third pathway, under basic conditions, a dialkyl carbonate is used to form an oxetane by transesterification and subsequent decarboxylation. The oxetane can then be ring-opened by HX to give a 1,3-diol that can be converted to the corresponding halogen-containing carbonate (Scheme 2, route C).[37,38]

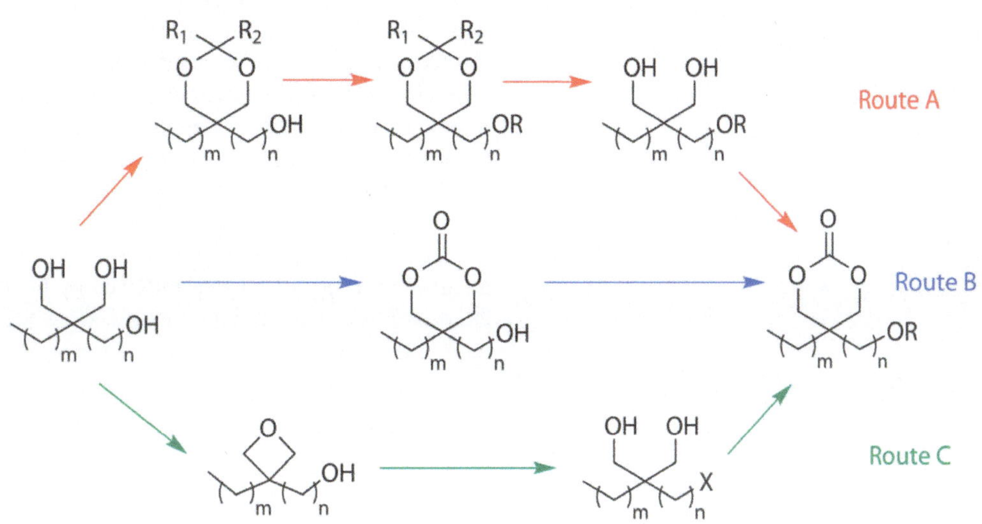

Scheme 2: Synthesis of functional cyclic carbonates derived from glycerol and trimethylolalkanes.

Synthesis of functional monomers derived from pentaerythritol is typically realised by partial functionalisation/protection of two of the hydroxyl functionalities leaving a 1,3-diol that can be cyclised using a phosgene derivative (Scheme 3).[50] In conclusion, glycerol and trimethylolalkanes are good alternatives to bis-MPA in the synthesis of functional cyclic carbonateswith ether and halogen substituents, whilst pentaerythritol provides an excellent opportunity to introduce two hydroxyl groups per monomer unit into the poly(carbonate) backbone.

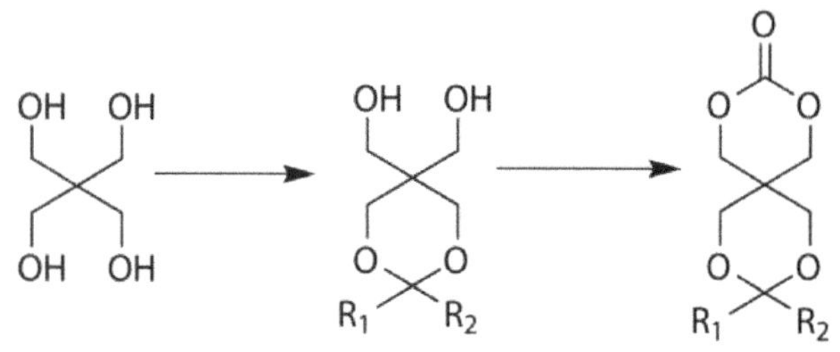

Scheme 3: Synthesis of functional cyclic carbonates derived from pentaerythritol.

Functional cyclic carbonates derived from amino acids or sugars

Amino acids and sugars are natural products that are indispensable for life processes, hence poly(amino acid)s and poly(saccharide)s or sugar-based polymers are highly attractive biocompatible materials. In view of the biocompatibility of the resulting polymers and their degradation products, cyclic carbonates derived from amino acids and sugars have been reported by Sanda *et al.* and by Gross and coworkers, respectively.[52–54] The synthesis of cyclic carbonate monomers derived from L-serine and L-threonine was realised by protection of these amino acids *via* conversion of the amino groups to a carbamic acid benzyl esters (*Z*) or carbamic acid *tert*-butyl esters (Boc) followed by reduction and carbonate formation with triphosgene(Scheme 4).[52] In an alternative report, the serine-derived cyclic carbonate was prepared from the alcohol analogue of serine, serinol, instead.[55]

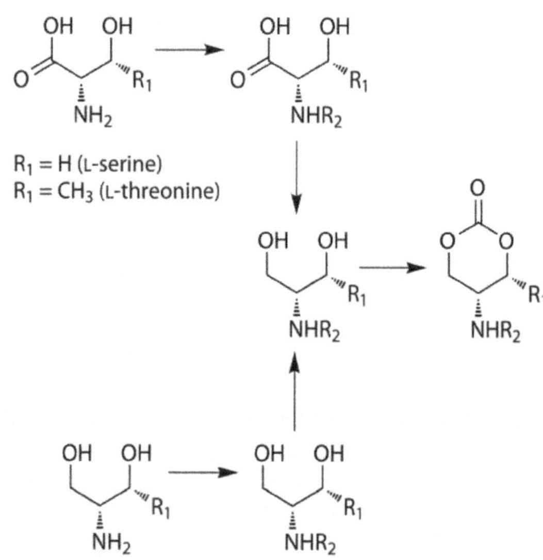

Scheme 4: Preparation of amino-acid derived cyclic carbonate monomers.

Sugar-derived cyclic carbonate monomers have been prepared from the furanose forms of D-xylose and D-glucose.[53,54,56] Synthesis from D-xylofuranose involves the acetonide protection of the two hydroxyl groups to form a 1,3-diol, which is commercially available (Scheme 5A). This 1,3-diol was converted to the cyclic carbonate using ethyl chloroformate. Monomer synthesis from D-glucofuranose involved protection (acetonide and benzyl ester) of three of the hydroxyl groups to give a 1,2-diol which was converted to a 1,3-diol (Scheme 5B) through a multi-step procedure. Carbonate formation was realised using a phosgene-derivative yielding the D-glucose-derived cyclic carbonate.

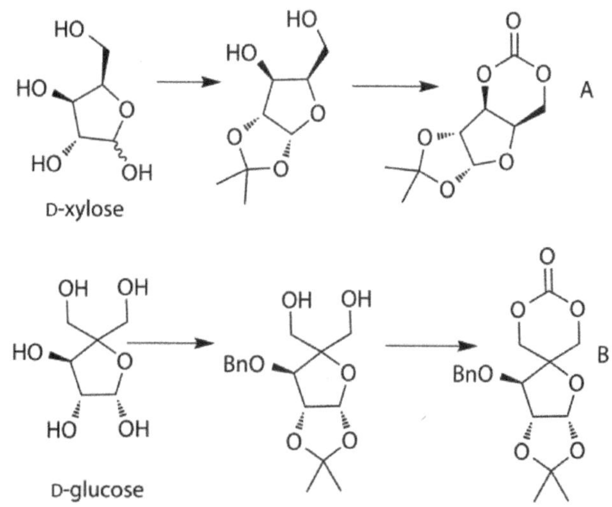

Scheme 5: Synthesis of cyclic carbonates from (A) D-xylofuranose and (B) D-glucofuranose.

Synthesis of Functional Cyclic Carbonates from other Resources

Examples of other reported cyclic carbonate syntheses include those derived from commercially available diols, oxetanes, a malonic acid diethyl ester derivative, dihydroxy acetone and *N*-methyl diethanolamine (Scheme 6).[57–68] Direct synthesis of six-membered cyclic carbonates from oxetanes using CO_2 has been reported.[61] Alternatively ring-opening of 3-methyl-3-oxetanemethanol with HBr has also been reported and resulted in a bromo-functional diol. The bromide was then replaced with a thiol by reaction with NaSH, which was subsequently reacted with an alkene to generate the desired functional 1,3-diol. This diol was converted to the functional cyclic carbonate monomer by reaction with ethyl chloroformate.[64] A monomer prepared from dihydroxy acetone, a glycolytic metabolite, is converted to the dimethoxy acetal before ring-closure mediated with ethyl chloroformate.[62,63] In a final example, a functional malonic acid diethyl ester was functionalised with an alkyl halide in a first step followed by the reduction of the ethylesters to alcohols that enabled subsequent ring closure withdiphosgene.[65]

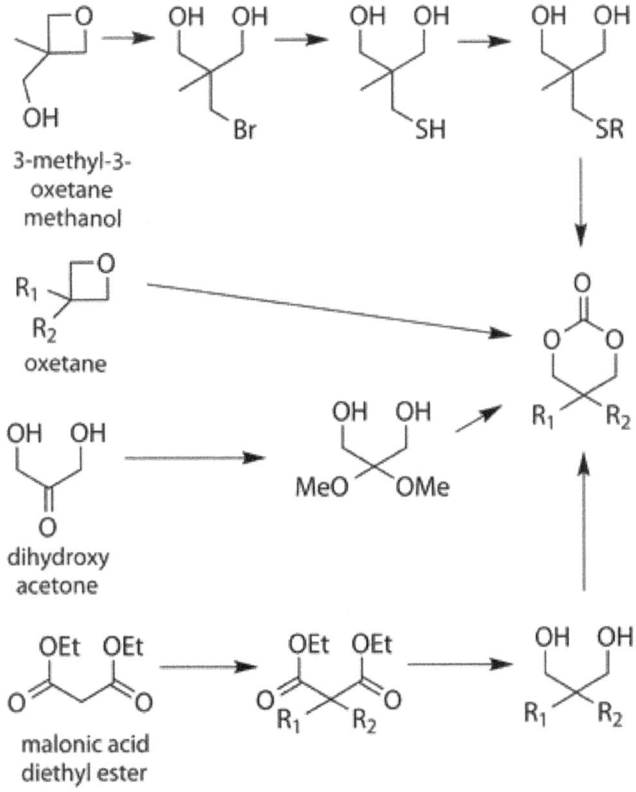

Scheme 6: Synthesis of functional six-membered cyclic carbonates derived from oxetanes, malonic acid diethyl ester and dihydroxy acetone.

RING-OPENING POLYMERISATION OF FUNCTIONAL CYCLIC CARBONATES

Cyclic Carbonates with Alkyl-Containing Functionalities

The introduction of a pendant functional group on a poly(carbonate) is desirable, but homopolymerisation of functional cyclic carbonates is not always necessary. Indeed, for some applications it is even advantageous to have a polymer backbone in which only a percentage of monomer units have a pendant functionality.[66,67] Furthermore, in es in which the polymerisability of a functional monomer is not optimal, copolymerisation often enables a solution to be achieved.[33] In the preparation of functional poly(carbonate) s throughcopolymerisation, trimethylene carbonate (TMC) is often used as co-monomer, however other simple alkyl functional monomers have also been reported.[44,68] Hedrick and coworkers have recently reported copolymerisations of functional monomers with a simple ethyl functional monomer derived from bis-MPA, 5-methyl-5-ethyloxycarbonyl-1,3-dioxan-2-one (MEC, 7), while Zhuo and coworkers report the use of dimethyl trimethylene carbonate, DTC, **1** (Table 1).[37,69-78] Only two examples of MEC homopolymerisations have been reported in the literature.[37,80] Using the sparteine/1-(3,5-bis(trifluoromethyl)-phenyl)-3-cyclohexyl-2-thiourea (TU) catalyst system in methylene chloride at room temperature, the polymerisation ([M]/[I] = 108) yielded a polymer with a molecular weight of 17 100 g mol^{-1} and a dispersity of 1.27, however application of a hexafluoroalcohol as a cocatalyst in place of the thiourea under comparable conditions resulted in polymers with decreased dispersity (1.13) and a lower molecular weight (3200 g mol^{-1}; ([M]/[I] = 88). In contrast, homopolymerisation of DTC has been reported using a number of catalysts.[79-84] The cationic ROP of DTC was reported in methylene chloride at room temperature using methyl trifluoromethane sulfonate (TfOMe) (<30% monomer conversion; M_n = 1800–2100 g mol^{-1}; $Đ_M$ = 1.30–1.38) or trifluoromethane sulfonic acid (<50% monomer conversion; M_n = 3000–3500 g mol^{-1}; $Đ_M$ = 1.10–1.11) or innitrobenzene at 120 °C using methyl iodide (50% monomer conversion; M_n = 1100 g mol^{-1}; $Đ_M$ = 1.52). Homopolymerisation of DTC has also been reported, catalysed by tin octanoateunder bulk conditions at 120 °C, and resulted in polymers with molecular weights ranging from 2900 to 8300 g mol^{-1} ($Đ_M$ = 1.52–1.58) or using rare earth catalysts at temperatures between 0 and 45 °C (M_n = 116 000–171 000 g mol^{-1}; $Đ_M$ = 2.28–2.79). Poly(DTC) is a hydrophobic, semi-crystalline material with reported melting points of 90–93 °C[78] and 122 °C[82] which consequently leads to decreased degradation rates of the homopolymer compared to poly(TMC) as well as for some of its copolymers.[85,79-83]

Table 1: Cyclic Carbonates with alkyl-containing functionalities

Monomer	Number	Ref.	Monomer	Number	Ref.
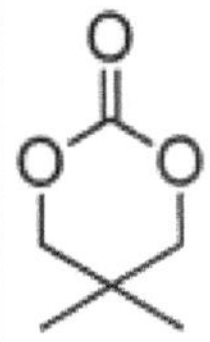	**1 (DTC)**	75–85	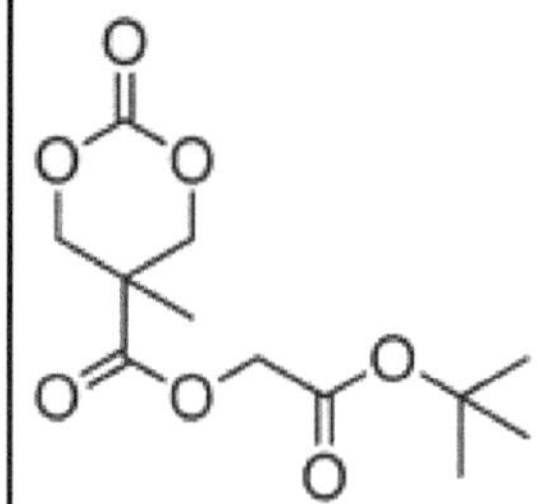	**9**	87
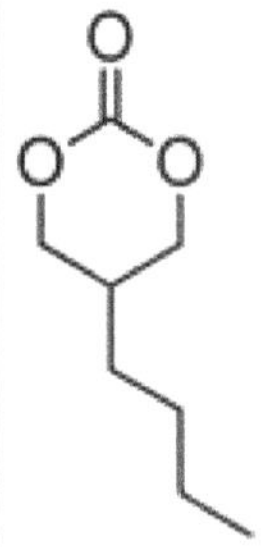	**2**	81	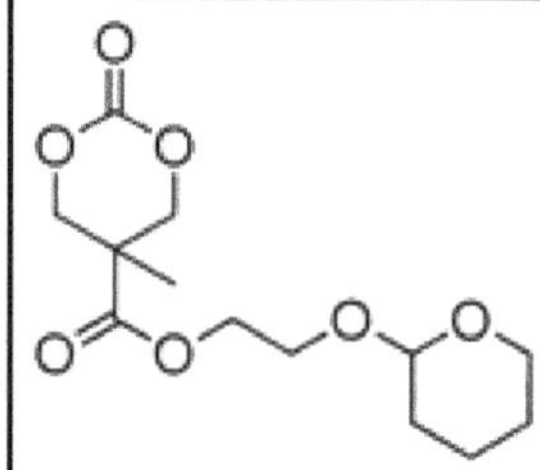	**10**	37 and 74
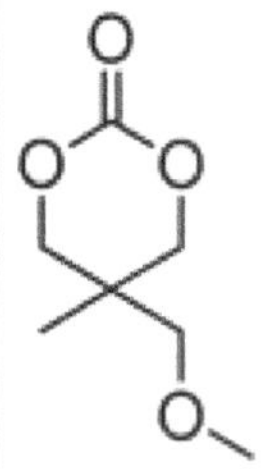	**3**	61	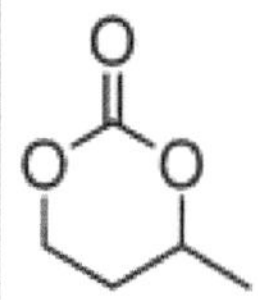	**11**	49, 79, 85, 90 and 91
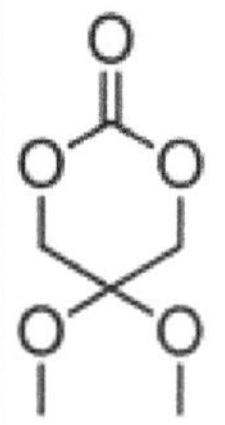	**4**	62, 63 and 86	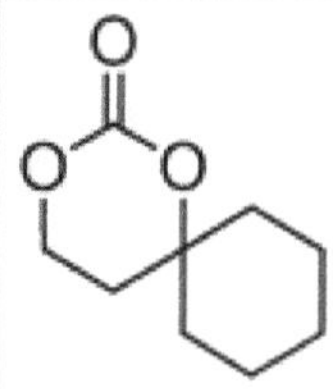	**12**	81
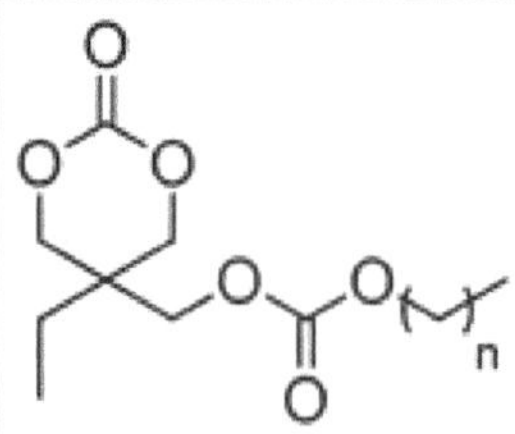	**5** ($n = 0$) **6** ($n = 1$)	51	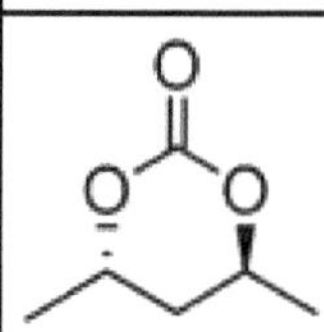	**13**	68
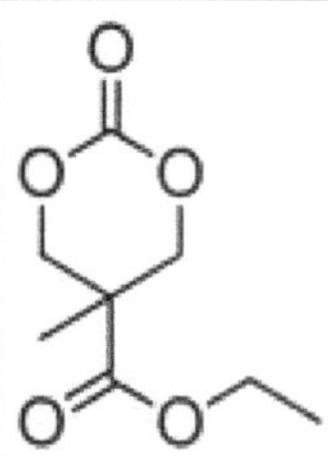	**7 (MEC)**	37, 69–74 and 80	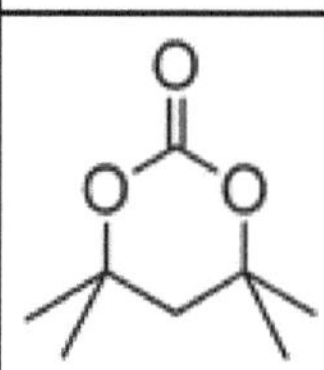	**14**	68
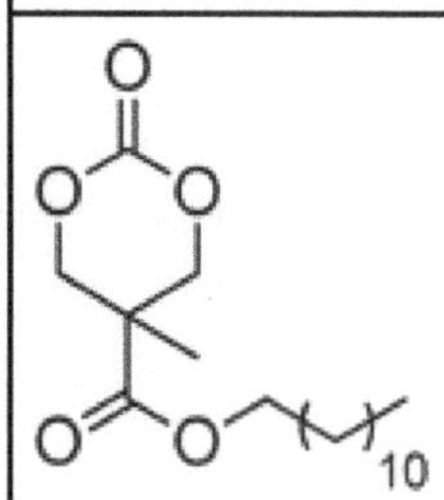	**8**	70			

Many other alkyl-functional monomers have been reported (Table 1; 1–14), however there are few reports on the properties of their resultant homopolymers.[37,51,61–63,70,74,81,86,87]Homopolymerisation of 2 in bulk using TfOMe as a catalyst (60–100 °C; M_n = 3500–5000 g mol^{-1}; $Đ_M$ = 1.67–2.58) or in chloro- or nitrobenzene at 120 °C initiated by methyl iodide (M_n= 600–800 g mol^{-1}; $Đ_M$ = 1.40–1.47)[81] has been reported whilst one example of the homopolymerisation of 10 using the organic catalyst 1,8-diazabicyclo[5.4.0]undec-7-ene (DBU) in combination with the TU cocatalyst is reported in dichloromethane at room temperature ([M]/[I] = 99; M_n = 6600 g mol^{-1}; $Đ_M$ = 1.24).[37] Ring-opening of 4 using a range of metal-based and organic catalysts has recently received a lot of attention. Reported polymerisations were either carried out in bulk at elevated temperature (130 °C), or at temperatures between 60 and 110 °C in toluene.[76,86] ROP of 4 in bulk resulted in a polymer with M_n = 56 400 g mol^{-1} ($Đ_M$ = 1.44), whilst good results were obtained for both metal-based and the organic catalyst 4-dimethylaminopyridine (DMAP) in the solution polymerisation of 4 ([M]/[I] = 100; M_n = 6800–17 000 g mol^{-1}; $Đ_M$ = 1.18–1.25) as well. Poly(4) is a rigid, brittle, amorphous material with a glass transition temperature at 39–45 °C and a degradation temperature (225 °C) that is slightly lower than PTMC (230 °C).[62,86,88,89]

Other alkyl functional monomers that have a group at the alpha position on the carbonate ring are 1,3-dioxaspiro[5.5] undecan-2-one, 12[81] and α-methyl trimethylene carbonate,[49,79,85,90,91] 11. Endo and coworkers reported the cationic homopolymerisation of 11 in 2001, however only very low monomer conversions were observed and the resulting polymerwas not isolated.[79] Higher monomer conversions have since been reported using rare earth initiators, aluminium triflate, β-diiminate zinc complexes and organic bases.[90,91] Good control was obtained using Al(OTf)$_3$ at 110 °C and 2-*tert*-butylimino-2-diethylamino-1,3-dimethylperhydro-1,3,2-diazaphosphorine (BEMP) and 1,4,7-triazabicyclodecene (TBD) resulting in poly(carbonate)s with molecular weights between 5700 and 10 000 g mol^{-1} and low dispersities. The substitution pattern of the carbonate ring can lead to two different propagating alcoholsupon ring-opening which, in turn means that different microstructures can be expected depending on the regioselectivity of the catalysts applied in the ROP. Guillaume and coworkers investigated the regioselectivity of the resulting polymers using a range of catalytic systems.[90] Polymers with a high degree of regioregularity were obtained using β-diiminate zinc complexes, whereas organocatalysts showed poor regioselectivity. It was rationalised that this difference in regioselectivity was caused by the bulky nature of the β-diiminate ligand present in the zinc-catalysed reactions, but not in those catalysed by aluminium triflate or organic bases. The thermal properties of the poly(11) homopolymers reveal amorphous polymers with a glass transition temperature (T_g) between −18 and −10 °C, which is very similar to that of PTMC (−15 °C).[71] In addition, 11 was copolymerised with ε-caprolactone and lactide to form highly hydrophobic copolymers and was used in the synthesis of poly(butylene succinate-*co*-11) by a combination of ROP and condensation polymerisation.[49,91]

Cyclic Carbonates with Aryl-Containing Functionalities

To date, many aryl-functional monomers and their respective polymers have been reported. Most commonly, those containing a benzyl group have been described. Monomers such as 5-methyl-5-benzyloxycarbonyl-1,3-dioxan-2-one, MBC 15 (Table 2) are readily polymerised and, moreover, the benzyl functionality can often be removed easily to yield hydrophilic materials or intermediates in the synthesis of carbonate monomers with more sensitive functionalities. For these reasons, benzyl-functional monomers have been used extensively in the preparation of functional aliphatic poly(carbonates) and poly(ester carbonate)s in recent years. MBC (15) and 5-benzyloxy-trimethylene carbonate (17), in particular have received much attention.[42,71,75,78,86,91–102] Homopolymerisation of MBC was first reported by Bisht and coworkers in 1999 using enzymatic methods.[42] Polymerisation was studied using seven different commercially available lipases at 80 °C in bulk among which lipase AK (from *Pseudomonas fluorescens*) gave the highest monomer conversion (97%) and molecular weight (M_n) 6100 g mol^{-1} with no decarboxylation observed. Molecular weights remained low due to competitive initiation from water however higher molecular weights could be obtained when the lipase was dried extensively. Increasing molecular weight also led to significant broadening of the dispersities ($Đ_M \leq 7.5$). In order to control the incorporation of benzyl groups in the poly(carbonate) structures, copolymerisation with TMC was conducted under the same conditions and led to statistical copolymers being isolated.[92] Organocatalytic ring-opening of MBC performed at room temperature with DBU as organocatalyst has resulted in well-defined homopolymers with controlled molecular weights (M_n = 4500–11 400 g mol^{-1}) and low dispersities ($Đ_M$ = 1.18–1.19).[99] The resulting poly(MBC) homopolymers are amorphous sticky polymers with a glass transition temperature (T_g) between 3 and 8 °C.[92,99] MBC has furthermore been reported in copolymerisations with lactide using tin- and zinc-based catalysts, mostly at elevated temperatures,[94–96] and in copolymerisations with MEC or lactide (LA) using organiccatalysts (DBU or TU-amine) at room temperature.[71,98] In the latter e, reactivity ratios in the copolymerisation of lactide and MBC were found to be 1.1 for lactide and 0.072 for MBC. Micelles obtained from block copolymers prepared by the statistical copolymerisation of MBC with lactide, initiated from a poly(ethylene oxide), PEO, macroinitiator, demonstrated that the introduction of carbonate moieties did not only enhance solubilisation and drug loading of the hydrophobic drug bicalutamide, but also resulted in significant reductions in the value of critical micelle concentration (CMC) compared to micelles without MBC incorporated.[96]

Table 2: Cyclic Carbonates with aryl-containing functionalities

Monomer	Number	Ref.
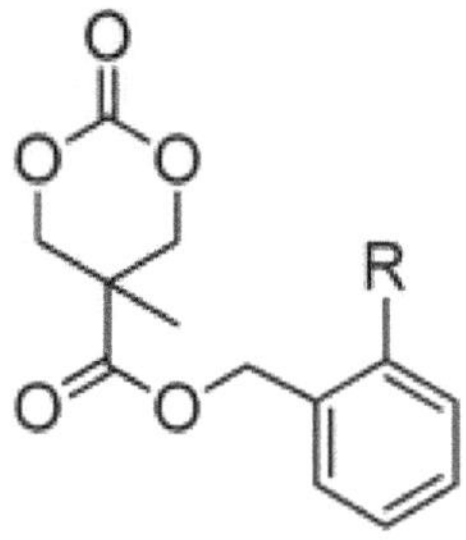	**15 (MBC)** (R = H) **16** (R = NO_2)	42, 71 and 92–99
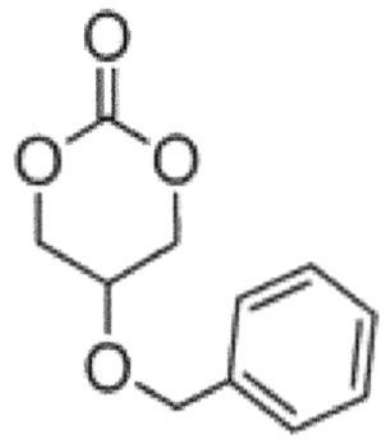	**17**	49, 75, 78, 86 and 100–102
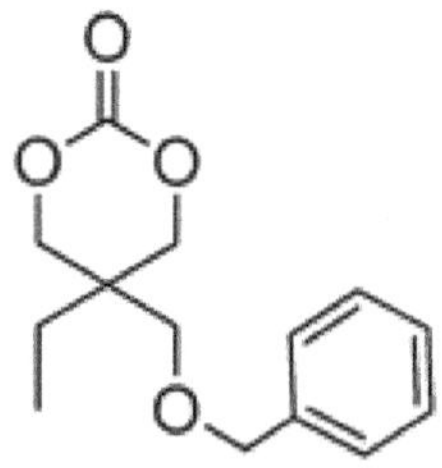	**18**	49
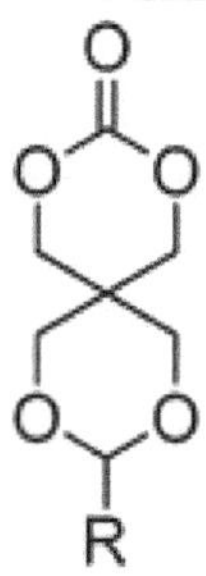	**19** (R = Ph) **20** (R = *p*OMePh) **21** (R = 2,4,6-OMe_3Ph)	50 and 103–105
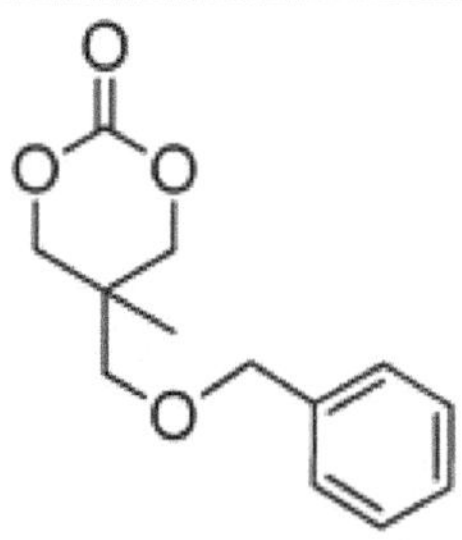	**22**	61
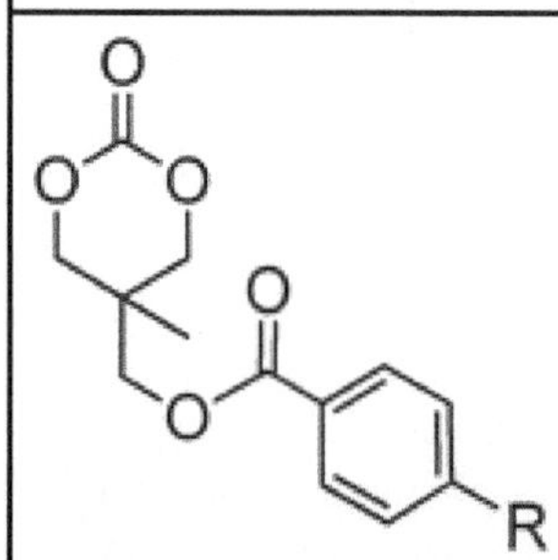	**23** (R = H) **24** (R = OMe) **25** (R = NO_2)	79
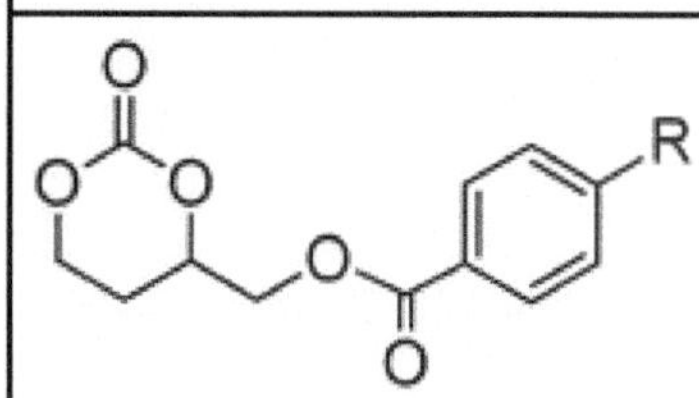	**26** (R = H) **27** (R = OMe) **28** (R = NO_2)	79

Another frequently reported benzyl functional monomer is 5-benzyloxy-trimethylene carbonate (17). This monomer has been polymerised under bulk conditions at elevated temperatures using aluminium- and tin-based catalysts, a range of organic catalysts, porcine pancreas lipase immobilised on silica particles and under reduced pressure without added ca talyst.[75,78,86,101,102]Reported poly(17) homopolymers displayed medium to large dispersities ($Đ_M$ = 1.41–1.92) and polymer molecular weights (M_n = 8800–22 400 g mol^{-1}) that are much lower than those expected for polymerisations with [M]/[I] = 1000. The resultant materials are rubbery at room temperature with T_g = 0–6 °C and no observed melting point.[100,101] Reported copolymerisations include random and block copolymers with DTC, lactide and PEO.[75,100,102] Another example is the combined ROP and polycondensation reaction of 17 or 18 with poly(butylene succinate) macromers.[49]

Other reported aryl-functional monomers, 19, 22, 23 and 26 have been studied to a lesser extent.[50,61,79,103,104] ROP of benzyl-functional monomer 9-phenyl-2,4,8,10-tetraoxaspiro-[5,5]-undecan-3-one (19) was investigated using diethyl zinc as initiator at 130 °C under bulk conditions for its homopolymerisation and its copolymerisation with both lactide and ε-caprolactone.[50,104] The resulting poly(19) homopolymers displayed a T_g = 124 °C, which is high compared to other poly(carbonate) homopolymers. Cationic ROP of 23 and 26 to form the respective homopolymers was reported at room temperature using a range of catalysts, however polymerisations were plagued by side reactions including backbiting and decarboxylation leading to poor control and low molecular weight products.[79] Suppression of decarboxylation was achieved using MeI as initiator, but increased reaction times and temperatures (120 °C) were required for the polymerisation to proceed.

Beyond simple benzyl-functional monomers, a number of monomers with methoxy- and nitro-substituted phenyl groups have been reported (Table 2).[79,105–107] The cationic ROP of 24,25, 27 and 28 was reported by Endo and coworkers.[79] All four monomers were polymerised using methyl iodide at 120 °C to yield the respective homopolymers. Homopolymers were obtained with M_n = 10 000 g mol^{-1} ($Đ_M$ = 1.22) for poly(24), M_n = 3400 g mol^{-1} ($Đ_M$ = 1.88) for poly(25), M_n = 4400 g mol^{-1} ($Đ_M$ = 1.58) for poly(27) and M_n = 7100 g mol^{-1} ($Đ_M$ = 1.39) for poly(28). Polymerisation was also successfully conducted at room temperature in dichloromethane in e of 24 and 25 using TfOMe or TfOH as catalysts yielding polymers with M_n = 1300–3500 g mol^{-1} ($Đ_M$ = 1.13–1.24) and M_n = 3000 g mol^{-1} ($Đ_M$ = 1.16), respectively. Nitrobenzyl-functional monomer 5-methyl-5-(2-nitro-benzoxycarbonyl)-1,3-dioxan-2-one (16), methoxy functional monomers mono-2,4,6-trimethoxybenzylidene-pentaerythritol carbonate (21) and mono-4-methoxybenzylidene-pentaerythritol carbonate (20) have all been used in the synthesis of PEO block copolymers prepared by the homopolymerisation of the monomers or copolymerisation with lactide from methoxy-PEO macroinitiators.[105] Polymerisations were performed using tin octanoate as catalyst under bulk conditions at 110–120 °C (16) or using zinc bis[bis(trimethylsilyl)amide] in methylene chloride at 50 °C (20 and 21).[105–107]

Cyclic Carbonates with Alkene- and Alkyne-Containing Functionalities

Alkene and alkyne functional groups participate in many organic reactions, including highly efficient and orthogonal reactions, as a consequence of their (double) unsaturation such asMichael addition, radical thylation, Huisgen 1,3-dipolar cycloaddition, epoxidation[120] and UV-cross-linking.[124,125] The introduction of such functionalities to the poly(carbonate) backbone leaves the resulting polymers open for further functionalisation.[108–112] Notably, there have been several reports of the synthesis and subsequent modification of (pendant) alkenes and alkyne groups in poly(ester)s.[113,114]

The introduction of pendant unsaturated groups in poly(carbonate)s prepared by the ring-opening of cyclic carbonate monomers has received increased attention in the past few years. The ROP of allyl ester-functional cyclic carbonate 5-methyl-5-allyloxycarbonyl-1,3-dioxan-2-one (MAC, **29**), for example, has been described in the preparation of a range of (co)polymers by Storey, Jing, Zhuo, Dove, Cui and their respective coworkers (Table 3).[76,115–122] Reported homopolymerisations of MAC were either performed without added catalyst at elevated temperature (at 115 °C; M_n = 13 700 g mol^{-1}; $Đ_M$ = 1.8), or in toluene at 95 °C using tin(II)-based poly(lactide) macroinitiator ([M]/[I] = 40; M_n = 9200 g mol^{-1}; $Đ_M$ = 2.5) and at 110 °C catalyzed by diethyl zinc ([M]/[I] = 200; M_n = 19 000 g mol^{-1}; $Đ_M$ = 1.6).[116,120] However, in all es polymerisations showed relatively poor control and resulted in branched polymers in some es. More recently Dove and coworkers demonstrated that the ROP of MAC using efficient organocatalysts led to a high level of control over the resulting polymer such that the application of the sparteine/TU catalyst system resulted in the isolation of PMAC with M_n up to *ca.* 13 000 g mol^{-1} ([M]/[I] = 100) and low dispersities ($Đ_M$ = 1.12).[121] Notably in this study, the polymer T_g was reported to be −26.4 °C, lower than that previously reported (T_g = −12 °C),[116,120] attributed to an absence of chain branching as a consequence of very low levels of transesterification. The ROP of MAC has also been reported in the synthesis of 3-armed star polymers initiated from triethanolamine using a *O,N,N,O*-tetradentate salan lutetium alkylcomplex.[122] The resulting polymers displayed low dispersities ($Đ_M$ < 1.10) and molecular weights close to the predicted values from the monomer-to-initiator ratio. Furthermore, the polymerisation proceeded well with low catalyst loadings (compared to initiator) thus demonstrating its 'immortal' nature. Other studies involving MAC have mostly focused on itscopolymerization with lactide,[116] including the formation of amphiphilic block copolymers composed of PMAC-*co*-PLA and poly(ethylene oxide) to form micelles[118] and polycations from poly(ethylenimine)-grafted polycarbonate.[119] While ROP of the related allyl-functional monomer 5-allyloxy-1,3-dioxan-2-one (30) has been reported, only the synthesis of 5-(2-propenyl)-5-propyl-1,3-dioxan-2-one (33) is known.[45,46,65] Homopolymerisation of 30 was reported by Parzuchowski *et al.*[45] and He *et al.*[46] Polymerisations were carried out in ethyl acetate at 65 °C using DMAP as catalyst or under bulk conditions at 120 °C using immobilised *porcine pancreas lipase* on silica or tin octanoate as catalyst. No evidence of decarboxylation was observed during the polymerisation and resulted in a semi-crystalline poly(30) that was characterized by a melting temperature around room temperature (T_m = 24 °C) and a low glass transition temperature (T_g = −40 °C). In an effort to improve the thermal properties of the poly(carbonate)s, while taking advantage of this lateral reactive groups, Gross and coworkers prepared the cyclohexene functional monomer 2,2-[2-pentene-1,5-diyl]-trimethylene carbonate (31).[60,68,123] It was hypothesised that the introduction of the cyclohexene substituent would yield poly(carbonate)s with a more rigid backbone and, as a consequence of this rigidity, a higher glass transition temperature. Indeed, polymers obtained by the metal-mediated (Al, Sn, Zn) ROP of 31 in bulk at 120 °C exhibited a T_g = 30 °C ([M]/[I] = 200; M_n = 6000–255 000 g mol^{-1}; $Đ_M$ = 1.3–3.4) which is significantly higher than those reported for allyl-functional poly(carbonate)s and TMC. In contrast, the homopolymer obtained from ROP of cinnamate-functional 5-methyl-5-cinnamoyloxymethyl-1,3-dioxan-2-one (32) with a molecular weight of 6000 g mol^{-1} ($Đ_M$ = 1.87) had a T_g of 42 °C.[124] Both its homopolymerisation and copolymerisation with L-lactide were performed in bulk at 120 °C using diethyl zinc as catalyst.

Table 3: Cyclic carbonates with alkene- and alkyne-containing functionalities

Monomer	Number	Ref.	Monomer	Number	Ref.
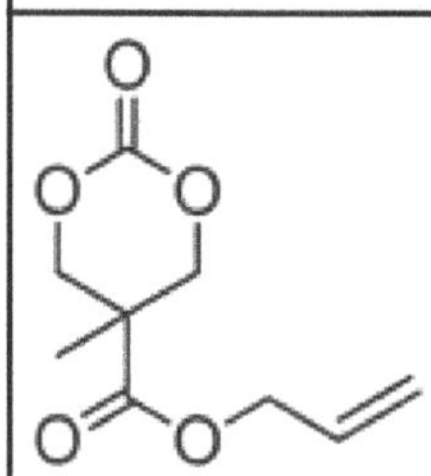	**29 (MAC)**	76 and 115–122	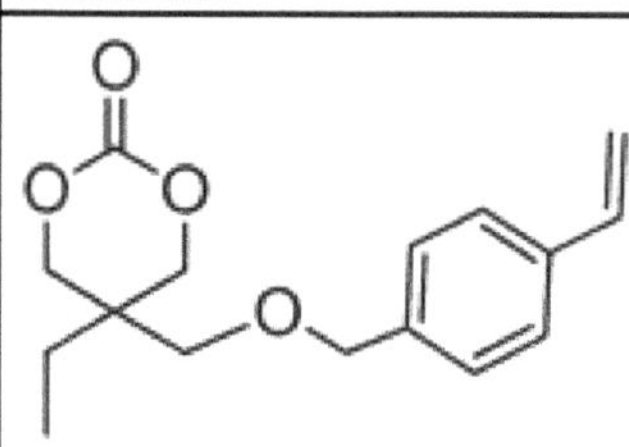	**36**	48
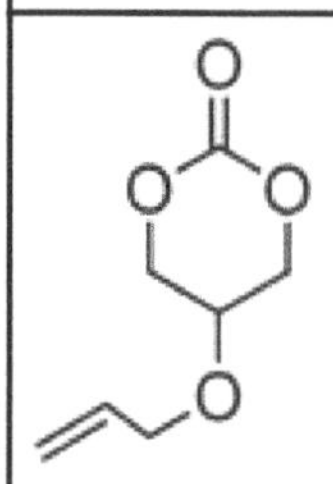	**30**	45 and 46	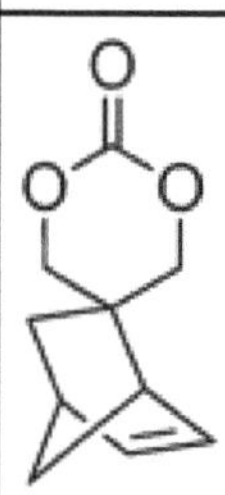	**37**	128
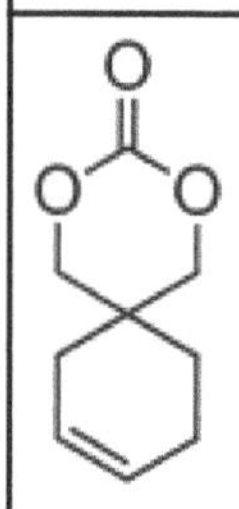	**31**	60, 123 and 68	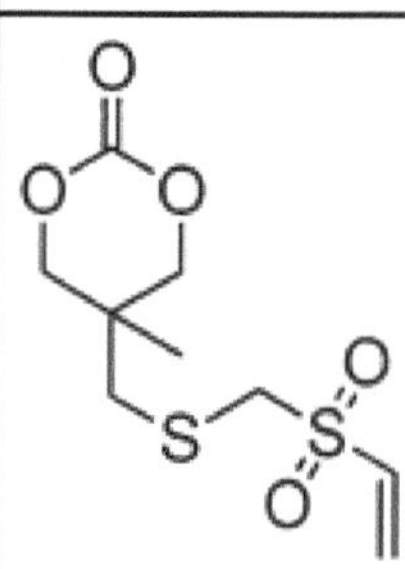	**38**	64
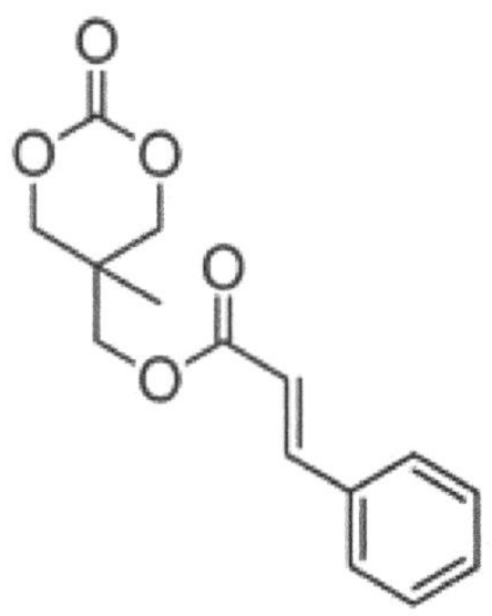	**32**	124	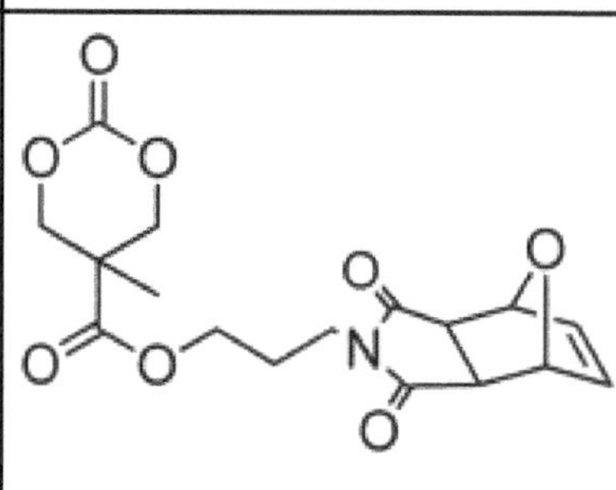	**39**	129
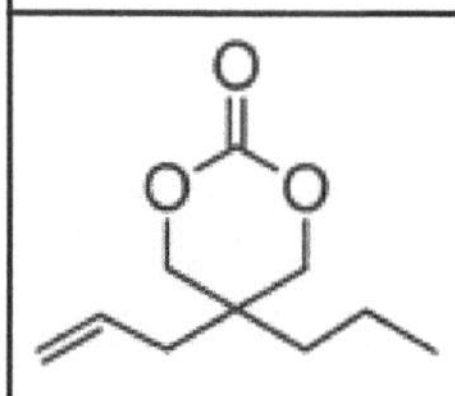	**33**	65	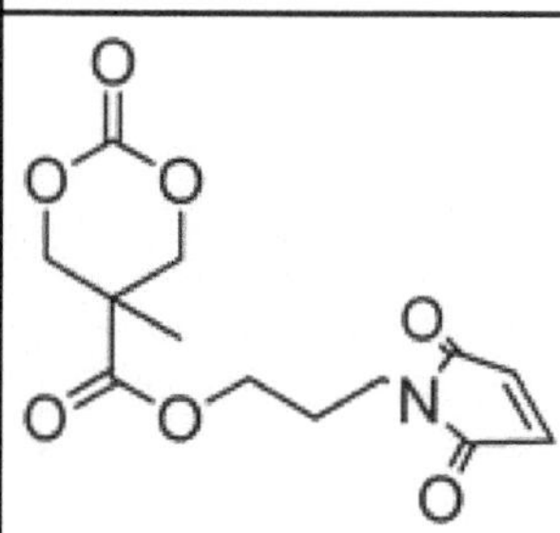	**40**	129

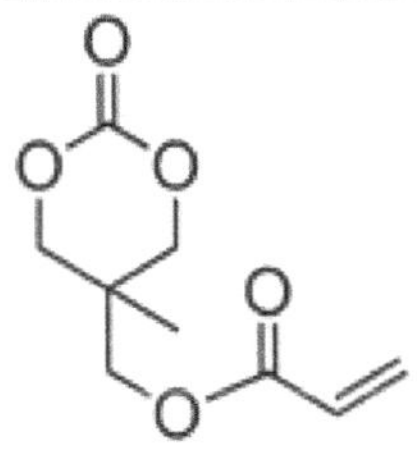	**34**	47, 126 and 127	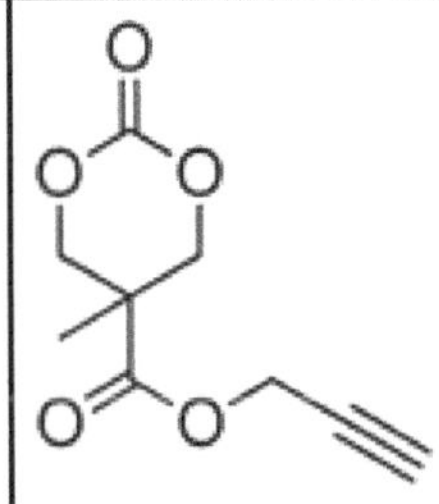	**41 (MPC)**	44 and 130–134
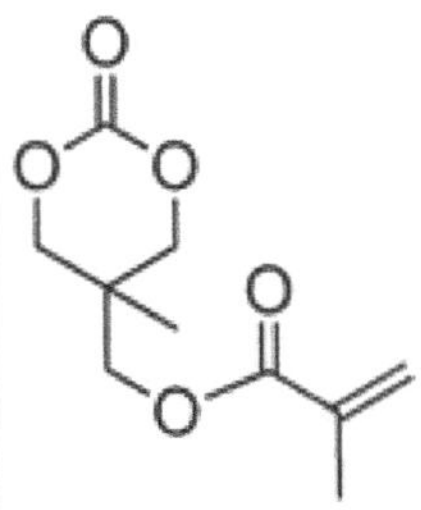	**35**	47			

Acrylate-functional monomer 34 and methacrylate-functional monomer 35 were reported by Zhong and coworkers in copolymerisations with ε-caprolactone and lactide (incorporations lower than 33%) initiated from PEO to result in amphiphilic block copolymers.[47,126,127] The monomers were polymerised in toluene using tin octanoate at 110 °C or in methylene chloride at room temperature using bis[bis(trimethylsilyl)amido] zinc as a catalyst. Styrene-functional 5-ethyl-5-[(4-vinylphenyl)methoxymethyl]-1,3-dioxan-2-one (36), reported by Miyagawa *et al.*, was polymerised by anionic ROP in THF at 0 °C in the presence of potassium *tert*-butoxide (*t*-BuOK).[48] The resulting poly(36) homopolymer was characterised by a M_n = 14 000 g mol^{-1}($Đ_M$ = 1.44) without polymerization of the pendant styrene groups.

Cyclic carbonate, 5,5-(bicyclo[2.2.1]hept-2-en-5,5-ylidene)-1,3-dioxan-2-one (37), with a pendant norbornene functionality was polymerised in bulk with a range of amine catalysts at 100–140 °C among which DBU proved to be the most active.[128] The resulting poly(37)s displayed high glass transition temperatures (T_g = 108 °C) as a consequence of the backbone rigidity resulting from the norbornene functionality. Copolymerisation of a vinyl sulfone-functional carbonate 38 was reported with TMC, caprolactone or lactide in toluene at 110 °C, usingisopropanol as initiator and tin octanoate as catalyst has been reported.[64] The pendant vinyl sulfone remained intact during the polymerisation allowing subsequent modification of thepolymer surface with thiol derivatives.[64] Onbulak *et al.* recently reported the synthesis and polymerisation of cyclic carbonate 39 which has a furan-protected maleimide as a pendant functionality.[129] The possibility to expose the maleimide functionality after polymerisation was demonstrated by the synthesis of 40*via* deprotection in toluene at 110 °C.Homopolymerisation of 39 was carried out at room temperature using DBU as catalyst and benzyl alcohol as initiator. As polymerisations became severely retarded after 60% monomer conversion, polymerisations were stopped after 50% monomer conversion to avoid transesterification reactions resulting in poly(39)s with low dispersities ($Đ_M$ = 1.1–1.2). In order to overcome poor polymerisability of 39, its copolymerisation with LA was carried out, yielding well-defined copolymers with pendant furan-protected maleimides. Surprisingly, the only propargyl functional cyclic carbonate reported is 5-methyl-5-propargyloxycarbonyl-1,3-dioxan-2-one (MPC, 41) and to date this monomer has only been copolymerised with lactide or trimethylene carbonate.[44,130–134] Copolymerisations were carried out in toluene or in bulk at 100–110 °C using diethyl zinc as catalyst/initiator[130–134] or in methylene chloride at room temperature using DBU/TU co-catalysts[44] resulting in copolymers with composition in MPC in agreement with the feed and ranging from 7 to 20%. In the latter study, it was found that MPC was incorporated preferentially to TMC at the beginning of the polymerisation.

Cyclic Carbonates with Halogen- and Azide-Containing Functionalities

Cyclic carbonates carrying halogen groups are an attractive class of monomers as it allows subsequent nucleophilic substitution and introduction of novel functionalities even afterpolymerisation. Mindemark *et al.* recently reported the synthesis and polymerisation of four halide-containing monomers, 42–45 (Table 4).[38] The resulting homopolymers of these chloro- and bromo-functional monomers, obtained by ROP at elevated temperature (110 °C) in bulk using tin octanoate as a catalyst, yielded chloride-functional polymers with M_n = 16 000–38 500 g mol^{-1} and $Đ_M$ = 1.46–1.81 and bromide functional polymers with M_n = 8500–31 900 g mol^{-1} and $Đ_M$ = 1.41–1.63, respectively. All polymers were semi-crystalline in nature with small variations of the measured melting points and glass transitions (P42: T_g = 15 °C, T_m = 105 °C; P43: T_g = 10–13 °C, T_m = 131–152 °C; P44: T_g = 18–23 °C, T_m = 100 °C; P45: T_g = 15–18 °C, T_m = 121–153 °C) However, a significant difference in solubility was observed between halide-containing polymers with a pendant methyl and an ethyl substituent. Indeed, whereas poly(43) and poly(45) are insoluble in most common organic solvents, poly(42) and poly(44) were found to be readily soluble in methylene chloride and tetrahydrofuran. Interestingly,copolymerisation of these monomers with TMC also resulted in semicrystalline poly(carbonate)s. A similar cyclic carbonate monomer with two pendant bromide functionalities, 2,2-bis(bromomethyl) trimethylene carbonate (46), has also been reported.[57,58,135] Its homopolymerisation and its copolymerisation with ε-caprolactone were reported in bulk at 120 °C using tin octanoate as catalyst and the resulting polymers were subsequently modified with NaN_3 and used for the preparation of poly(carbonate) nanoparticles by the Huisgen 1,3-dipolar cycloaddition. The synthesis and polymerisation of halogen-functional cyclic carbonates derived from bis-MPA were also reported by Hedrick and coworkers (Table 4).[37,44,69,72,138,139] In a recent study, the pentafluorophenol pathway was exploited to form the cyclic carbonate 53 intermediate as outlined previously.

Table 4: Cyclic carbonates with halogen- and azide-containing functionalities

Monomer	Number	Ref.	Monomer	Number	Ref.
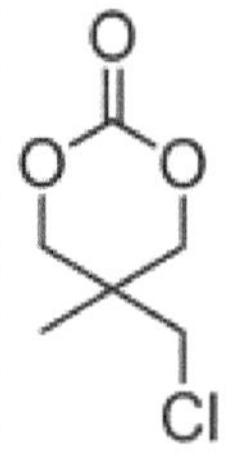	**42**	38	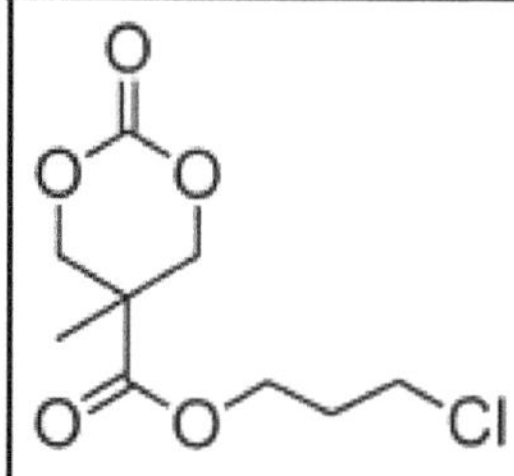	**48**	37, 44, 69, 72 and 139
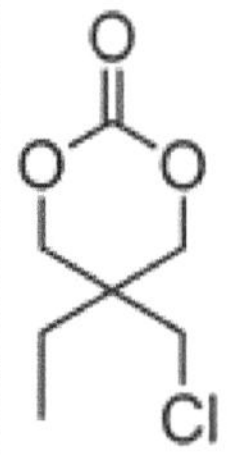	**43**	38	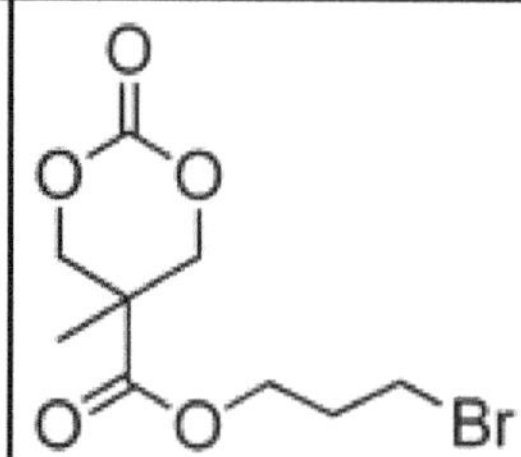	**49**	37, 72 and 138
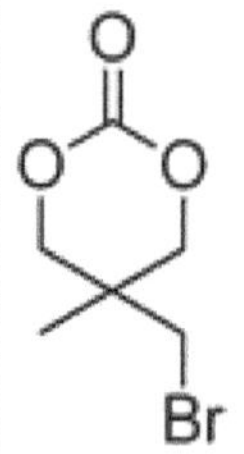	**44**	38	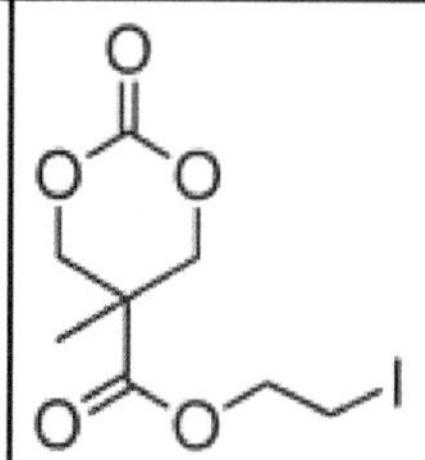	**50**	72
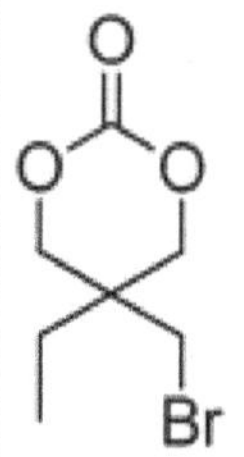	**45**	38	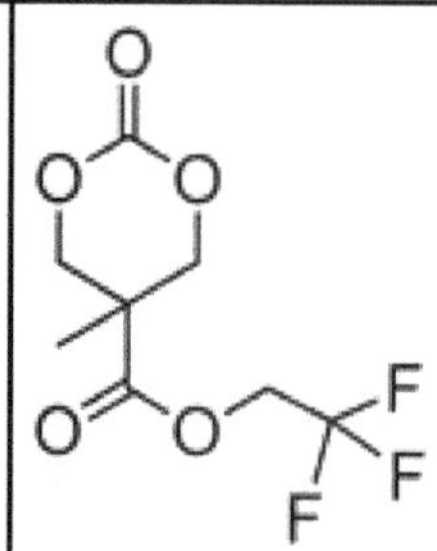	**51**	44
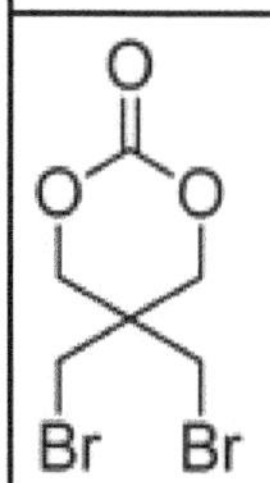	**46**	57 and 58	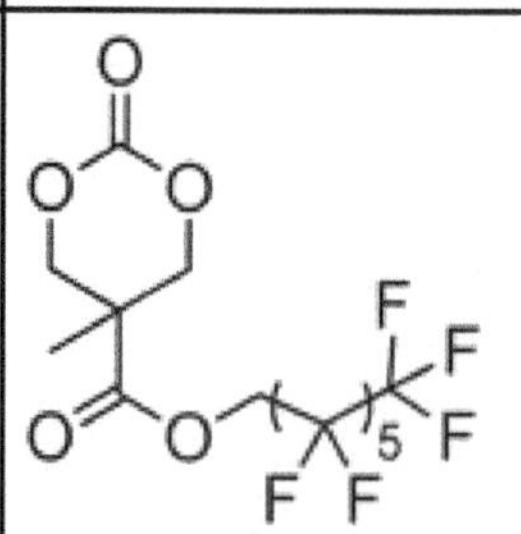	**52**	37

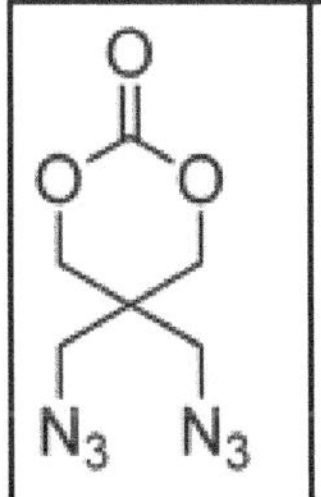	**47**	57, 58, 136 and 137	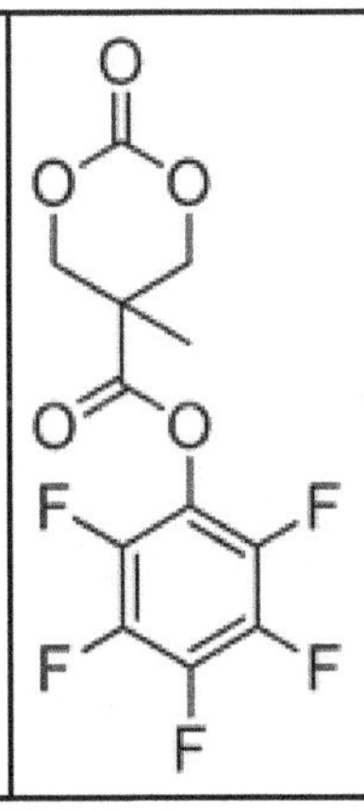	**53**	37

The synthesis of halogen-containing poly(48) ([M]/[I] = 55; M_n = 12 200 g mol^{-1}; $Đ_M$ = 1.17), poly(49) ([M]/[I] = 52; M_n = 11 700 g mol^{-1}; $Đ_M$ = 1.11), poly(50) ([M]/[I] = 51; M_n = 10 500 g mol^{-1}; $Đ_M$ = 1.22) and poly(51) ([M]/[I] = 54; M_n = 11 500 g mol^{-1}; $Đ_M$ = 1.26) was achieved *via* organocatalytic ROP at room temperature using the TU/DBU catalyst system.[37,72]Distribution of the pendant halogen functionalities in the poly(carbonate) backbone was achieved under these conditions in copolymerisations with MEC (with 48, 49, 50) or TMC (with 52) and resulted in polymers with controlled molecular weights and low dispersities.[37,44,69,72,138,139]

Cyclic Carbonates with Lateral Nitrogen-Containing Functionalities: Amino-, Amido, Carbamate and Urea Functional Cyclic Carbonates

Nitrogen-containing cyclic carbonates include those with pendant amides and pendant amino, amido and urea functionalities.[37,52,55,59,87,140,141] Cyclic carbonates with pendant amides are those with pendant carbamic acid benzyl ester (Z) groups and carbamic acid *tert*-butyl ester (Boc) functional groups, which are used to protect functional groups that have primary aminesduring the polymerisation. Endo and coworkers reported the anionic ROP of 55, 56 and 59 (Table 5) which was carried out in THF at temperatures between −78 and −30 °C using a range of initiators.[52] Molecular weights of the resulting poly(55) homopolymers ranged from 5600 g mol^{-1} to 20 000 g mol^{-1} with moderate dispersities being observed ($Đ_M$ = 1.20–1.46). Interestingly, poly(56) (M_n = 13 000–24 000 g mol^{-1}) and poly(59) (M_n = 5700–21 000 g mol^{-1}) showed inverse optical rotations compared to the respective cyclic carbonates and lower dispersities ($Đ_M$ = 1.18–1.24 and 1.11–1.20, respectively) than poly(55). Additionally, ROP of 56 and 59 showed a preference (9 : 1) for ring-opening at the least-hindered acyl bond. The bulk copolymerisation of 55 with lactide has also been reported at 120 °C using diethylzinc as catalyst and yielded poly(55) with comparable molecular weight and dispersity (M_n = 24 100 g mol^{-1}; $Đ_M$ = 1.51).[55] While the polymerisation of the Boc-protected-amine cyclic carbonate, 54, has not been reported, the organocatalytic ROP of related monomer 57 has been reported with TMC.[37,44] The copolymerisation was performed using the TU/DBU catalytic system at room temperature in methylene chloride and resulted in copolymers of 57 and TMC with low dispersities ($Đ_M$ = 1.07). Synthesis of poly(58) homopolymers with low degrees of polymerisation (DP = 8–22) and copolymers of 58 with other cyclic carbonate monomers was realised using the TU/DBU catalyst system in CH_2Cl_2 at room temperature.[87,140] Homopolymers obtained from the organocatalytic ROP of 58 exhibited low dispersities ($Đ_M$ = 1.11–1.16) for molecular weights M_n ranging from 3800 to 10 000 g mol^{-1} and displayed excellent cellular uptake with longer oligomers being observed to mediate cell adhesion.[140] Furthermore, thesehomopolymers were also applied as components of non-covalent, degradable hydrogels.[87]

Table 5: Cyclic carbonates with lateral nitrogen-containing functionalities: amino-, amido, carbamate and urea functional cyclic carbonates

Monomer	Number	Ref.	Monomer	Number	Ref.
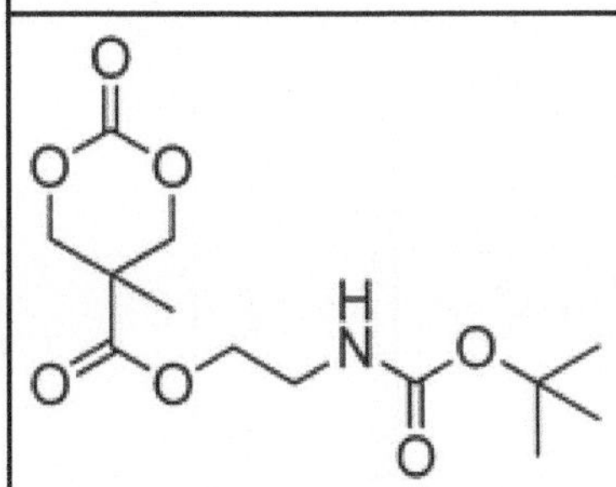	**54**	37 and 44	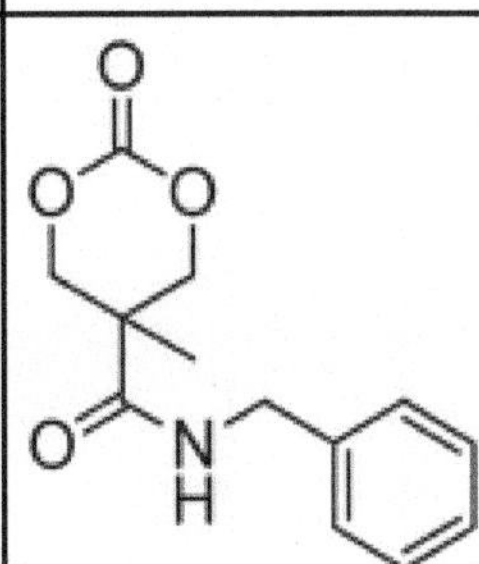	**61**	44

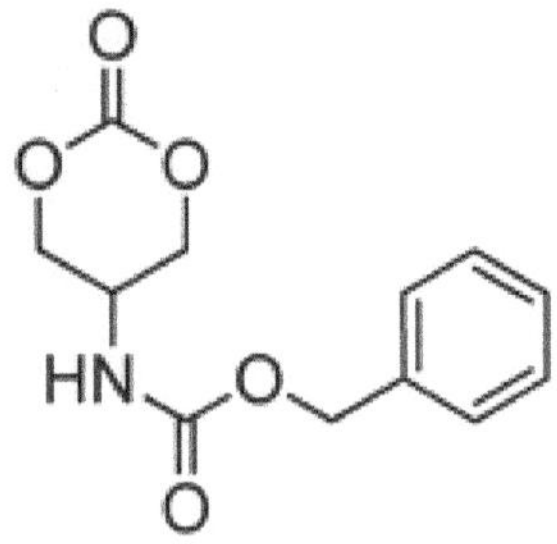	**55**	52 and 55	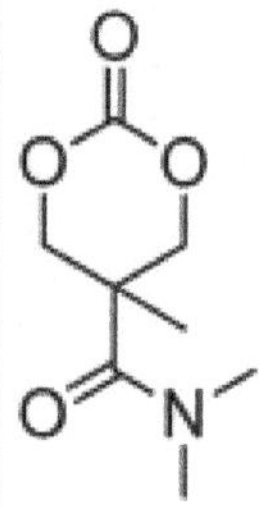	**62**	37
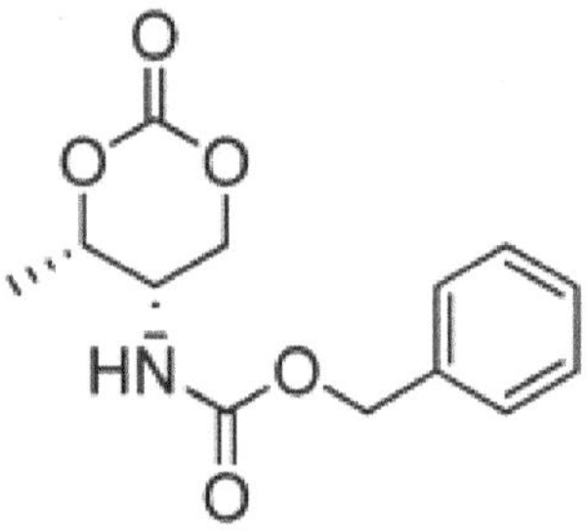	**56**	52	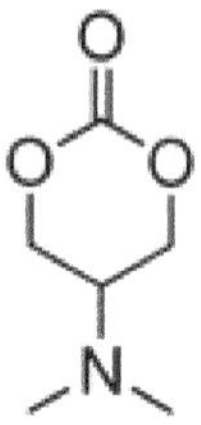	**63**	141
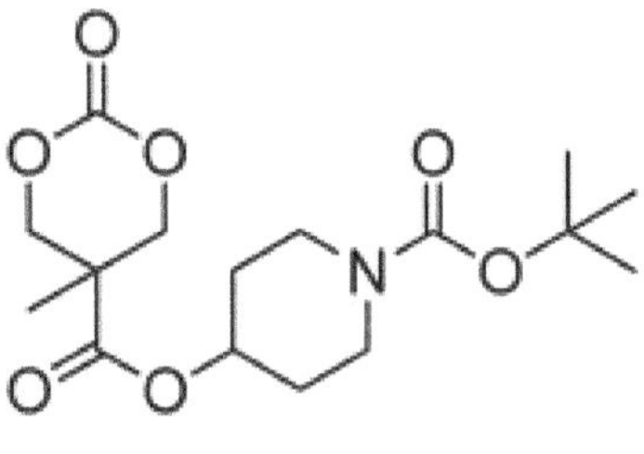	**57**	44	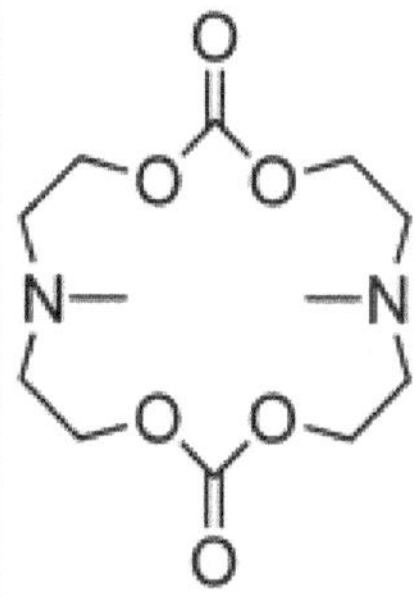	**64**	59
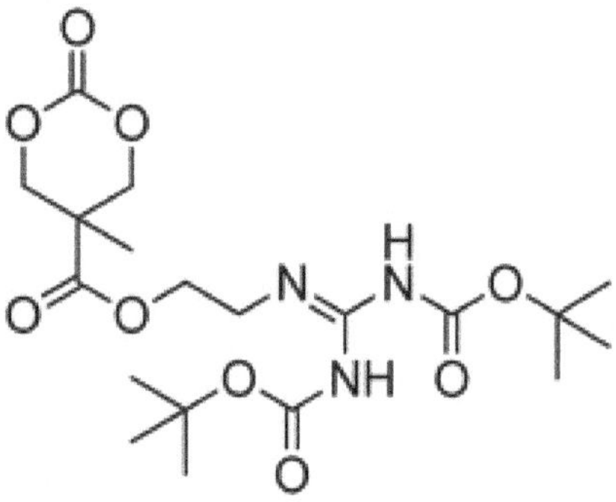	**58**	87 and 140	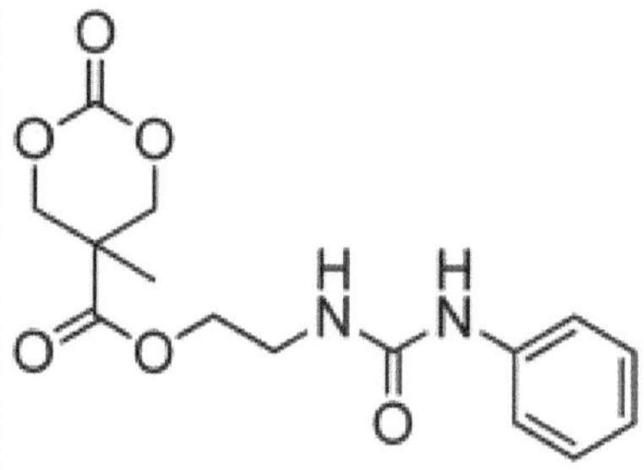	**65**	71, 73 and 142
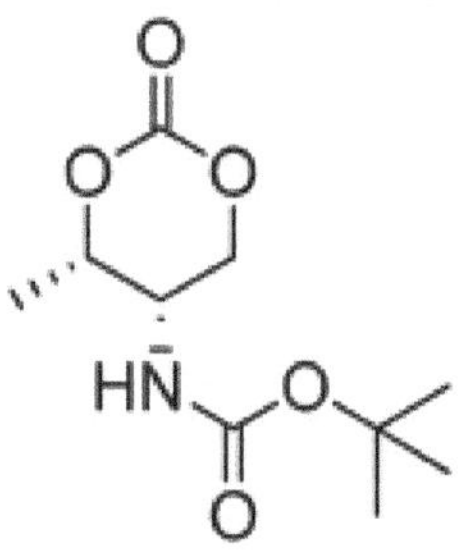	**59**	52	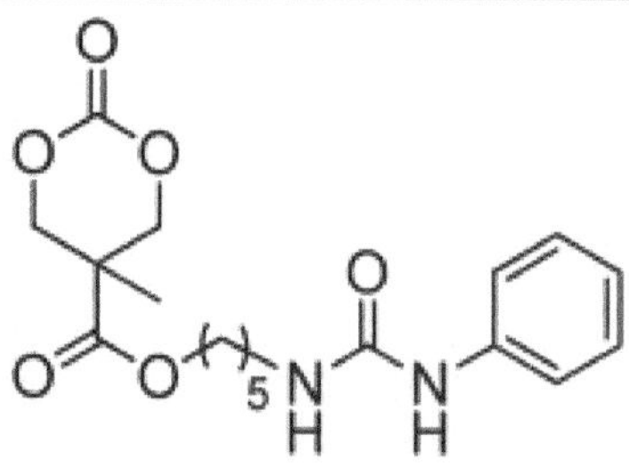	**66**	71, 73 and 142
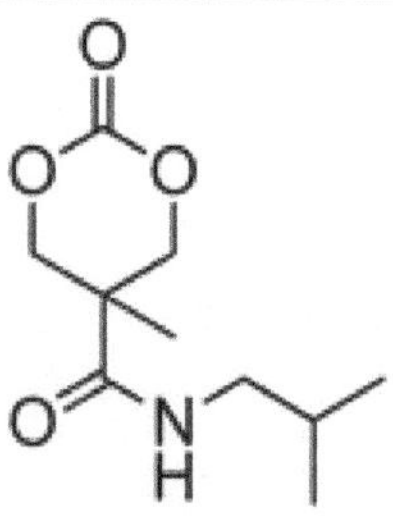	**60**	44			

Although the synthesis of amido-functional cyclic carbonates 60 and 61 has been reported (Table 5),[44] no report was found describing their polymerisation. No other reports of the ROP of this type of amido-functional cyclic carbonate monomers were found, possibly as polymerisation could be complicated by the presence of the secondary amine. In contrast,homopolymerisation of 62, with a tertiary amide pendant group, has been reported by Sanders *et al.* using the TU/DBU catalyst system at room temperature to yield poly(62) with M_n = 10 500 g mol^{-1} and $Đ_M$ = 1.32 ([M]/[I] = 63).[37] The homopolymerisations of the tertiary amine-containing 16-membered 6,14-dimethyl-1,3,9,11-tetraoxa-6,14-diazacyclohexadecane-2,10-dione (64), which contains two carbonate groups and two tertiary amines, was reported by Wang *et al.* in 2010.[59] ROP was attempted using both enzyme- and metal-based catalyst systems. Polymerisations performed in bulk at elevated temperatures (120 °C) using tin octanoate resulted in decarboxylation of the monomer whilst polymerisations performed in toluene at temperatures between 70 and 100 °C did not result in any polymerisation being observed. In contrast, the enzymatic ROP of 64 in toluene using Novozym-435 resulted in the formation of poly(64) both at 70 °C and at room temperature and displayed living characteristics. Poly(64) with a molecular weight of 8000 g mol^{-1} and a $Đ_M$ = 1.34 was obtained by the enzymatic process and were amorphous materials presenting excellent biocompatibility and fast degradability. Likewise, homopolymers obtained from the enzymatic ROP (60 °C; M_n = 4900 g mol^{-1},$Đ_M$ = 1.40) of six-membered dimethylamine-functional cyclic carbonate, 2-dimethylaminotrimethylene carbonate (63), showed low toxicity and good degradability.[141] Finally,polymerisation of urea-functional monomers 65 and 66 were reported using the TU/(−)-sparteine catalyst system at room temperature in CH_2Cl_2. Poly(66) was obtained with M_n = 10 500 g mol^{-1} ([M]/[I] = 32; $Đ_M$ = 1.14) and displayed a T_g at approximately 15 °C.[142] The synthesis of block copolymers with a PEO block, a poly(TMC) and a poly(66) or poly(65) block have also been reported, as well as a block copolymer with a PEO block and a poly(MEC-*co*-65) sequence, was reported using the same mild conditions. These block copolymers were used for the formation of micelles that displayed low CMC's and increased stability of drug-loaded micelles due to the hydrogen-bonding interactions between the urea groups.[71,73] Strong H-bonding interaction arising from the urea-groups was also highlighted in semi-interpenetrating networks by a decrease of the overall degree of swelling.[142,143]

Cyclic Carbonates with Protected Sugar Functionalities

Beyond their degradable and biocompatible properties, carbohydrates display specific interactions with proteins (lectins) making them very desirable for targeted drug delivery systems.[144]Introduction of these groups on a poly(carbonate) backbone presents an attractive, non-toxic and biodegradable method to targeted drug delivery. As the hydroxyl groups in sugar-functional cyclic carbonates are incompatible with the ring-opening process, the hydroxyl groups require protection prior to the polymerisation of the cyclic carbonate. Gross and coworkers first prepared a cyclic carbonate monomer directly from xylose saccharide, 1,2-*O*-isopropylidene-D-xylofuranose-3,5-cyclic carbonate (71, Table 6).[54,145,146] This work was extended with the synthesis and polymerisation of glucose based 1,2-*O*-isopropylidene-3-benzyloxy-pentofuranose-4,4′-cyclic carbonate (70).[53] Monomers were homopolymerised as well as copolymerised with L-lactide under bulk conditions at elevated temperatures (120–180 °C) or in dioxane at 25–60 °C. The resulting poly(70) and poly(71) homopolymers were obtained with number-average molecular weights of 7900 g mol^{-1} ($Đ_M$ = 1.47),[145] and 2100–13 200 g mol^{-1} ($Đ_M$ = 1.22–1.92), respectively.[146] Thermal analysis of the polymers revealed that poly(70) displayed aT_g = 128 °C[145] whereas poly(71) was also found to be semicrystalline and displayed a T_g = 128 °C.[146] Deprotection of ketal group of polymers containing monomer 71 was successfully achieved by using trifluoroacetic acid. Unfortunately however, deprotection resulted in some degradation due to the hydrolysis of the pendant ester bonds that link the carbohydrate to thecarbonate. While deprotection of the homopolymer of 70 has not been reported to date, conversely, removal of a benzyl protecting group in copolymers containing monomer 70 can be achieved selectively by hydrogenation over palladium/carbon (90% removal) however again, acetonide deprotection resulted in some degradation. Deprotection of its copolymer withlactide showed a lowering in glass transition after both deprotection steps.

Table 6: Cyclic carbonates with protected sugar functionalities

Monomer	Number	Ref.
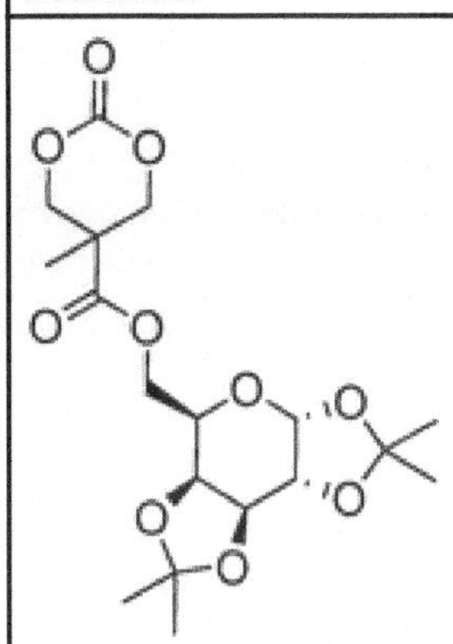	**67**	147

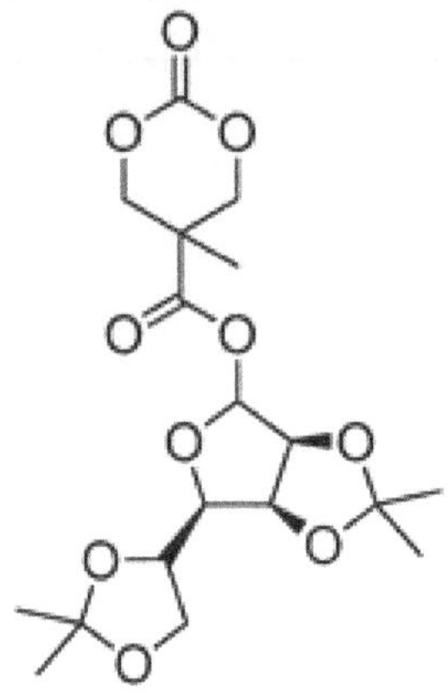	**68**	147
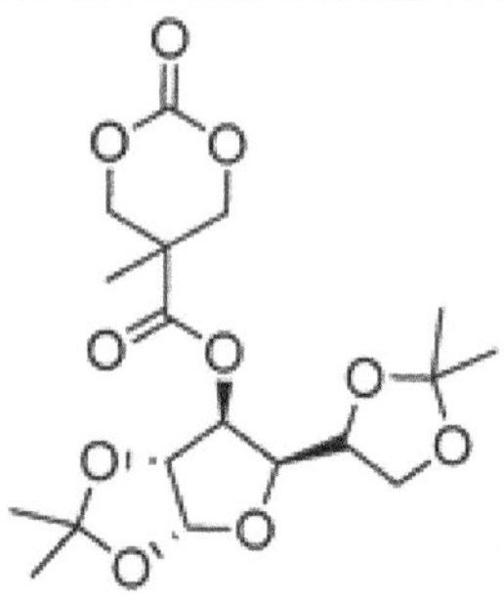	**69**	147
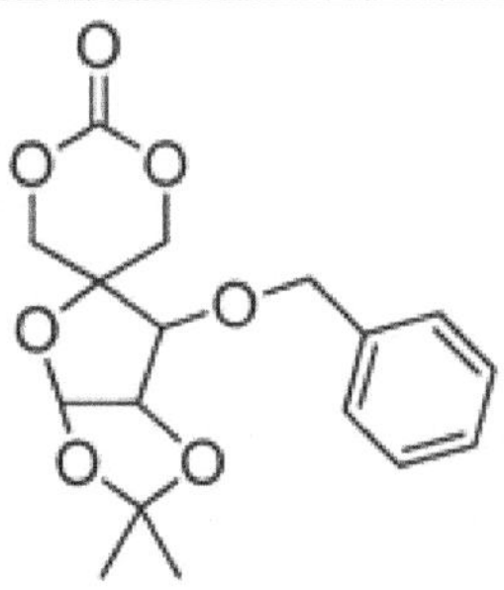	**70**	53 and 146
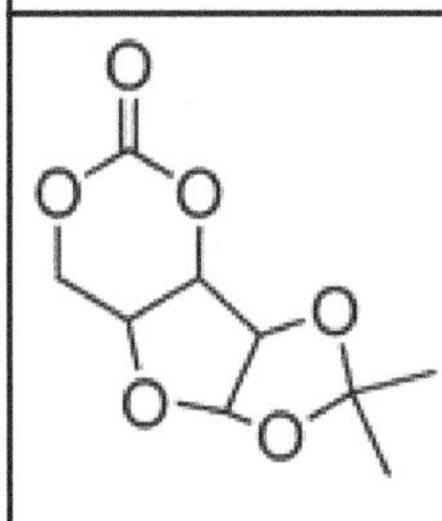	**71**	54, 145 and 146

Homopolymerisation of cyclic carbonates 67, 68 and 69 with pendant protected saccharides was carried out at room temperature using the TU/DBU organocatalytic system.[147]Homopolymers were obtained with molecular weights between 17 000 and 19 000 g mol^{-1} and dispersities of 1.20–1.28. Recovery of the hydroxyl groups was carried out by deprotection of the acetonide groups using formic acid solution. Interestingly, full deprotection was achieved and no degradation of the polycarbonate backbone was observed. Block copolymers of 67–69with TMC were reported for the preparation of micelles in aqueous solutions, and were prepared by sequential addition of TMC to poly(67), poly(68) and poly(69).

Macromonomers, bis- and tris-Carbonates

A number of cyclic carbonate-containing macromonomers and cross-linkers have been reported.[33,37,44,70,77,87,142] In particular those containing ethylene glycol moieties (72–74) have received increased interest in recent years (Table 7). PEO-functional cyclic carbonate macromonomer (72), prepared from methoxy poly(ethylene oxide)s, have been reported with molecular weights ranging from 350 g mol^{-1} to 1900 g mol^{-1} ($Đ_M$ = 1.03–1.14). While homopolymerisation of 72 in methylene chloride solution at room temperature using DBU as a catalyst was possible (with M_n = 1900 g mol^{-1} and $Đ_M$ = 1.03), the reaction showed a maximum at 55% monomer conversion.[33] The limited, low concentration of the cyclic carbonate (0.2 M), due to the large PEO pendant chain, was thought to cause the ring-chain equilibrium to dominate such that a high molecular weight polymers could not be obtained. To obtain higher molecular weightpolymers, the cyclic carbonate concentration was increased by copolymerisation of 72 with TMC.[33,44] Random copolymerisation of 72 with dimethyltrimethylene carbonate (DTC) from a poly(ε-caprolactone) macroinitiator at

130 °C in bulk using tin octanoate as catalyst or with dodecyl-functional monomer 8 in CH_2Cl_2 at room temperature using organic catalysts were further reported.[70,77] The resulting block/graft copolymers could be self-assembled with the latter example displaying thermoresponsive behaviour owing to the combination of hydrophilic/hydrophobic groups in the second block. An organocatalytic approach to hydrogels was presented by Nederberg *et al.* by the synthesis of a bifunctional cyclic carbonate macromonomer from poly(ethylene oxide) diols.[33] Poly(ethylene oxide)-α,ω-methylcarboxytrimethylene carbonate crosslinkers, 74, were prepared with M_n = 3400 g mol^{-1}, 8000 g mol^{-1}and 18 500 g mol^{-1} and copolymerised with TMC in CH_2Cl_2 at room temperature in the presence of DBU to form a poly(carbonate)/poly(ethylene oxide) cross-linked hydrogel. Hydrogels prepared with increased amounts of TMC showed hydrophobic PTMC domains which contribute to an increase of the toughness of the resulting hydrogels.[33,142] This led to further investigation by synthesis of telechelic, carbonate functional poly(ε-caprolactone), 75, with a molecular weight of 8000 g mol^{-1} ($Đ_M$ = 1.40; DP = 70). Copolymerisation of this hydrophobic crosslinker with 74 (8000 g mol^{-1}) and TMC to introduce hydrophobic segments resulted in particularly tough materials. Furthermore hydrogels prepared by copolymerisation from a poly(carbonate) macroinitiator bearing pendant urea functionalities (*i.e.* poly(66)) are presented as another method to strengthen the resulting network *via* hydrogen bonding with the poly(ethylene oxide) crosslinkers.[33,87,142] Very recently, Akashi *et al.* developed macromonomers based on oligo poly(ethylene oxide) connected to the cyclic carbonate by an ether linker (73).[148] Interestingly, the resulting homopolymers (M_n = 960–11 000 g mol^{-1}, $Đ_M$ = 1.06–1.42) displayed tunable thermoresponsive properties around body temperature.

Table 7: Macromonomers, bis- and tris-carbonates

Monomer	Number	Ref.	Monomer	Number	Ref.
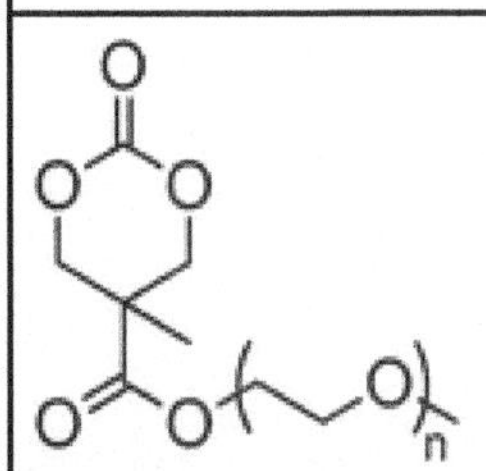	**72**	33, 44, 70 and 77	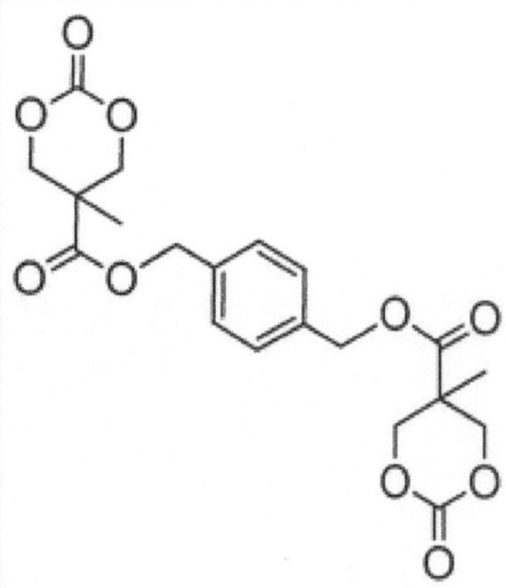	**79**	44
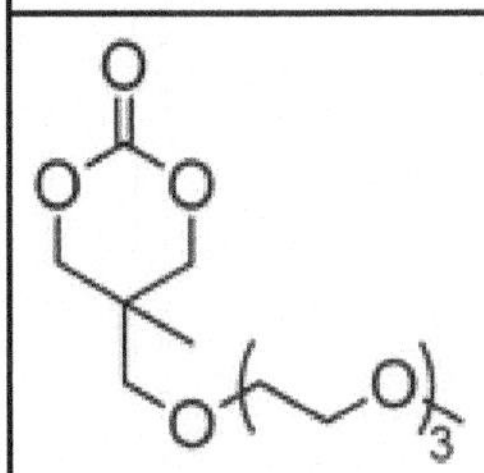	**73**	148	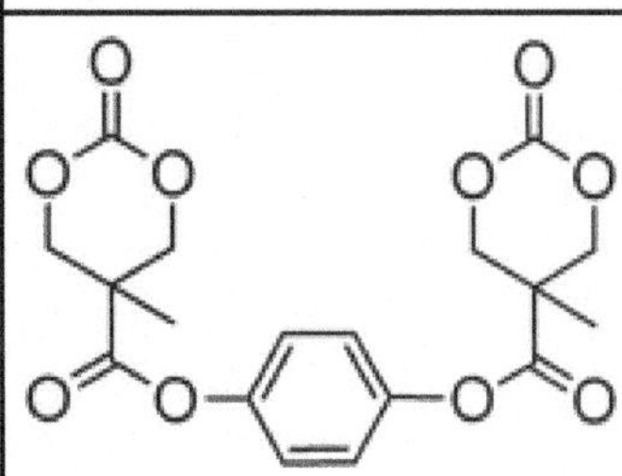	**80**	150
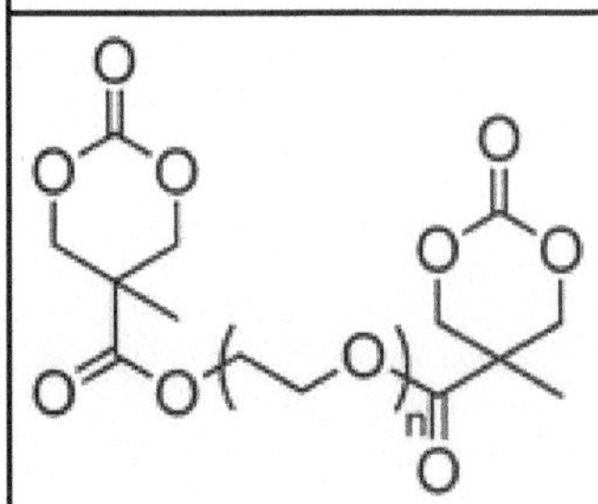	**74**	150	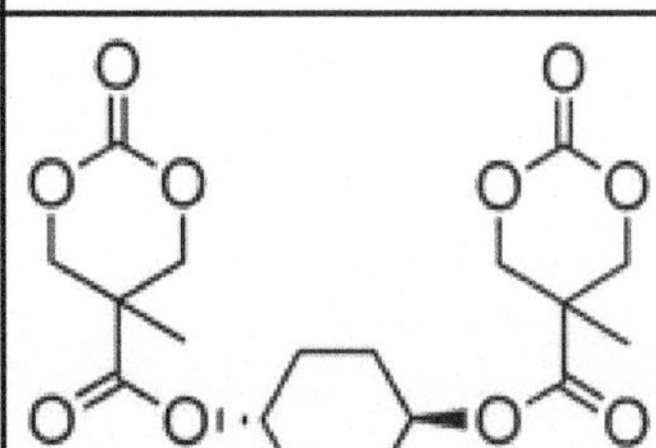	**81**	150

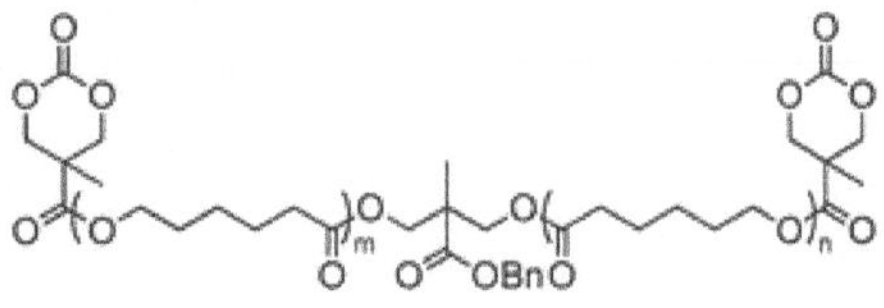	**75**	150	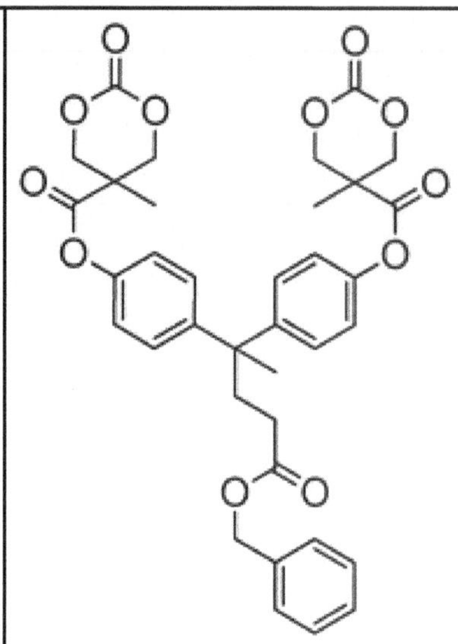	**82**	150
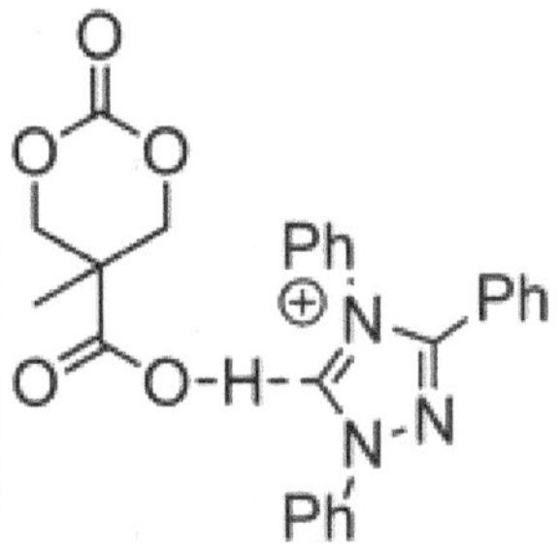	**76**	149	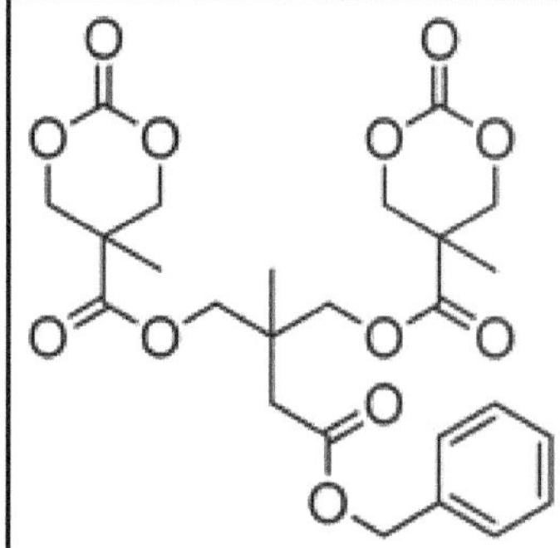	**83**	150
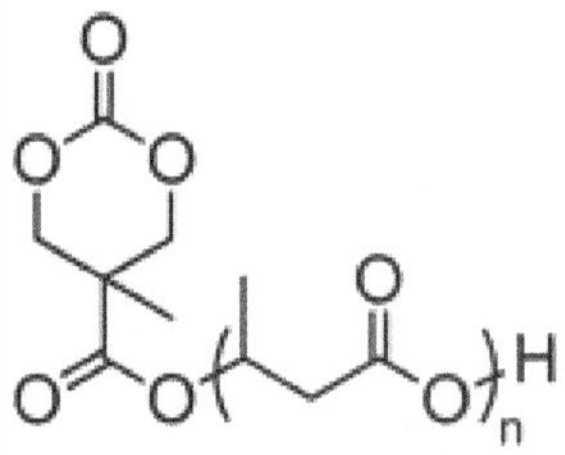	**77**	149	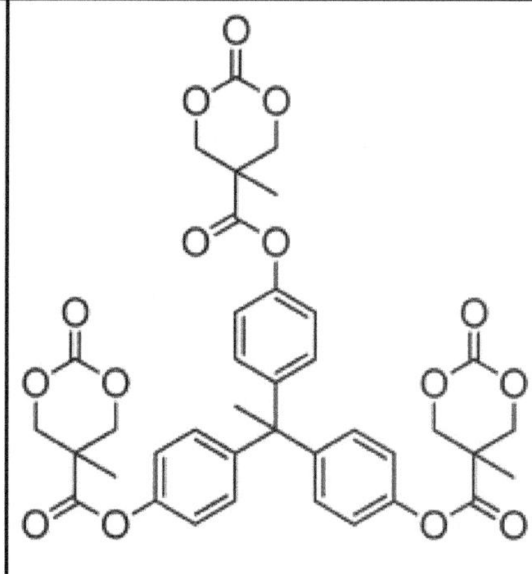	**84**	150
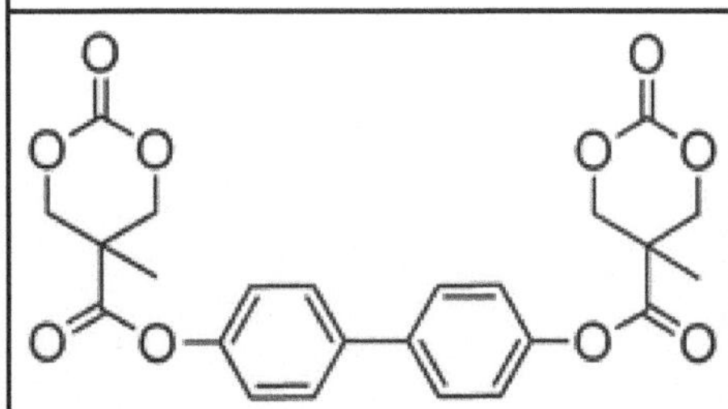	**78**	150			

The synthesis of a triazolium-based cyclic carbonate, 76, that can act as an initiator for the anionic ROP of β-butyrolactone (BL) at elevated temperature was reported by Dubois and coworkers.[149] To prove that the carbene catalyst was selective for the polymerisation of BL, its polymerisation from 76 was carried out at 90 °C in bulk. The resulting poly(β-butyrolactone) PBL cyclic carbonate macromonomer (77) was characterised by low dispersity ($Đ_M$ = 1.21) and preservation of the cyclic carbonate end-group as attested by ^{1}H NMR spectroscopy. As an alternative, 76 was copolymerised with TMC at room temperature with a (−)-sparteine/4-chlorophenol cocatalyst system to give a co-poly(carbonate) with pendant triazolium groups (M_n = 4100 g mol^{-1}; $Đ_M$ = 1.19) which, in a subsequent step could then be used as initiator for the ROP of BL in the preparation of graft copolymers.

A range of other bis-cyclic carbonate cross-linkers and one tris-cyclic carbonate cross-linker have also been reported (78–84; Table 7) for the preparation of biodegradable cross-linked networks.[44,150] Monomer 81 exists in both the *trans*- and *cis*-isomers which could be separated by column chromatography and isolated in a 1 : 1 ratio. Cross-linked networks were obtained by copolymerisation of the bifunctional monomers with TMC and ε-caprolactone at 140 °C in bulk in the presence of tin octanoate. As expected and in agreement with Flory's theory,[151] the as-obtained cross-linked networks displayed increased glass transition temperatures with increased amount of cross-linker.

Other Functional Monomers

Several other six-membered cyclic carbonate monomers have been reported. Azidation of the 1,3-diol precursor to 46, followed by carbonate formation using ethyl chloroformate yielded diazido-functional 85. Poly(85) could be obtained at elevated temperatures under bulk conditions with tin-based catalysts (M_n = 10 700–57 900 g mol^{-1}; $Đ_M$ = 1.20–1.41) or using DBU inmethylene chloride at room temperature resulting in polymers with low dispersities ($Đ_M$ < 1.1). Poly(85) was also shown to be a semicrystalline material with a melting temperature ranging from 50 to 100 °C and a glass transition temperature that is slightly below 0 °C.[136,137] Furthermore, 85 was reported in copolymerisations with DTC, lactide and in the preparation of block copolymers, by initiation from PEO macroinitiators. These copolymers were subsequently studied in hydrogel formation by "Click reaction".[57,58,136,137] In a comparable manner, Mindemark and Bowden described the azidation of 44 and 45 to yield the azide-functional analogues 86 and 87. While the ROP of these monomers was not reported, a range of triazole-functional monomers (88–97; Table 8) were prepared by Cu(I)-catalysed Huisgen dipolar cycloaddition with terminal alkynes. ROP of 89, 91, 93 and 95 were reported (target DP = 50) using the TU/DBU organocatalyst system to yield polymers with low dispersities (1.13–1.25) and M_n between 2300 and 5600 g mol^{-1}.[152]

Table 8: Miscellaneous functional cyclic carbonates

Monomer	Number	Ref.	Monomer	Number	Ref.
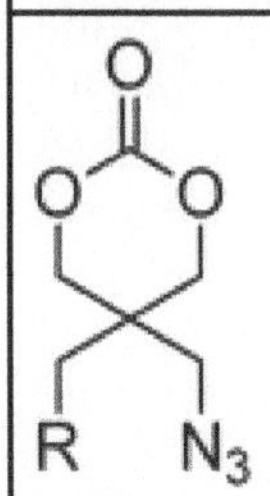	**85** (R = N_3) **86** (R = H) **87** (R = Me)	57, 58, 136, 137 and 152	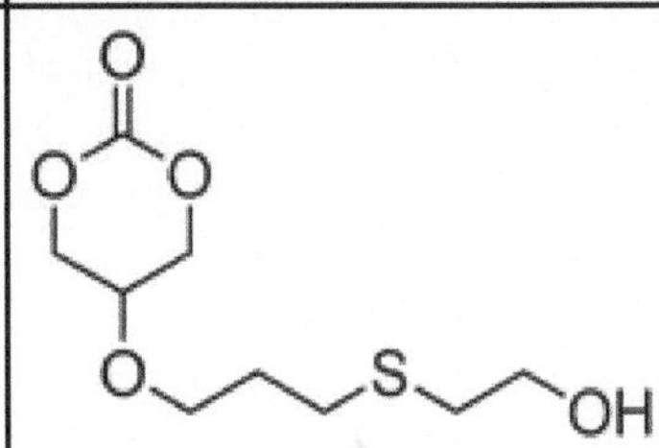	**102**	45
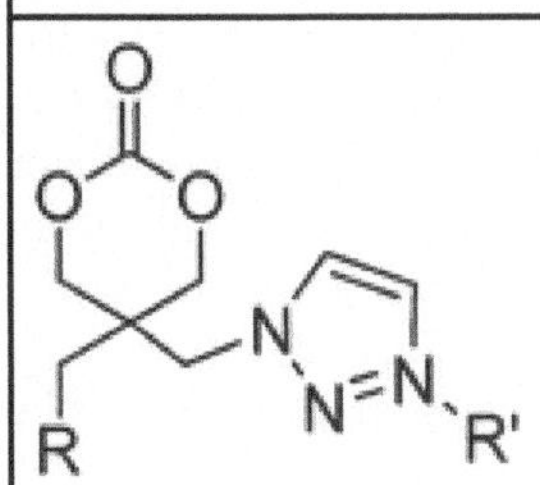	**88** (R = H, R′ = Ph) **89** (R = Me, R′ = Ph) **90** (R = H, R′ = C_3H_5) **91** (R = Me, R′ = C_3H_5) **92** (R = H, R′ = C_3H_7) **93** (R = Me, R′ = C_3H_7) **94** (R = H, R′ = C(= CH_2) CH_3) **95** (R = Me, R′ = C(= CH_2) CH_3) **96** (R = H, R′ = C(OH)Me_2) **97** (R = Me, R′ = C(OH)Me_2)	152	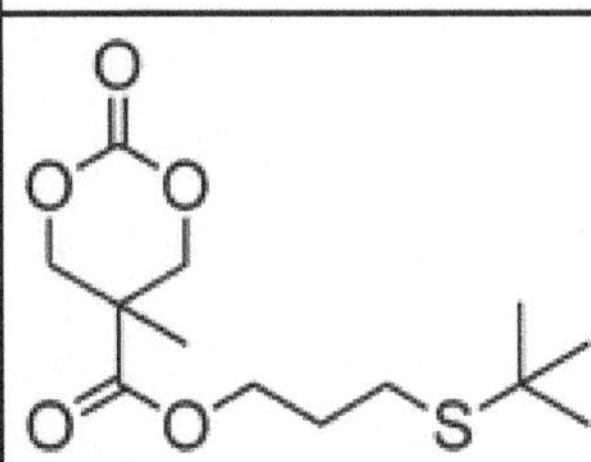	**103**	44
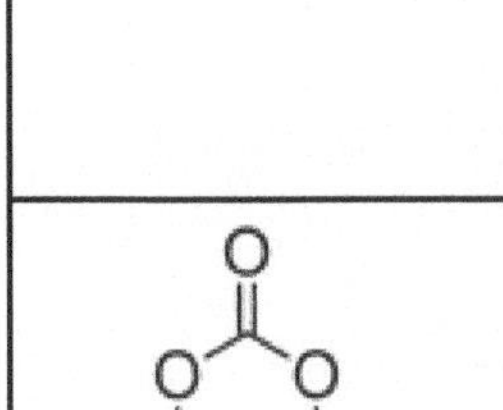	**98**	65	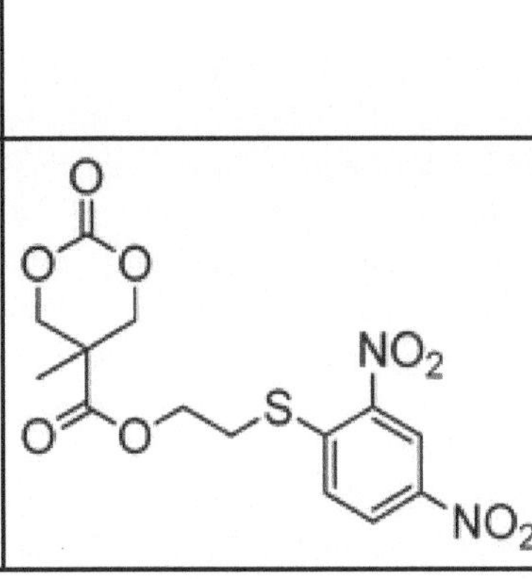	**104**	37

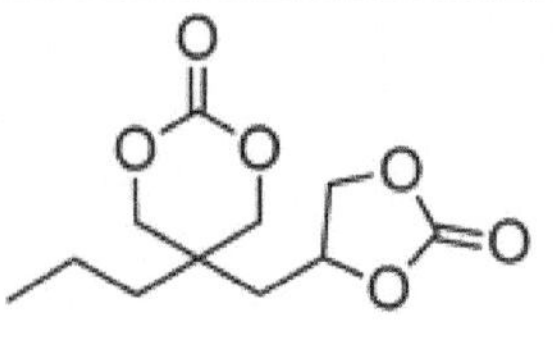	**99**	65	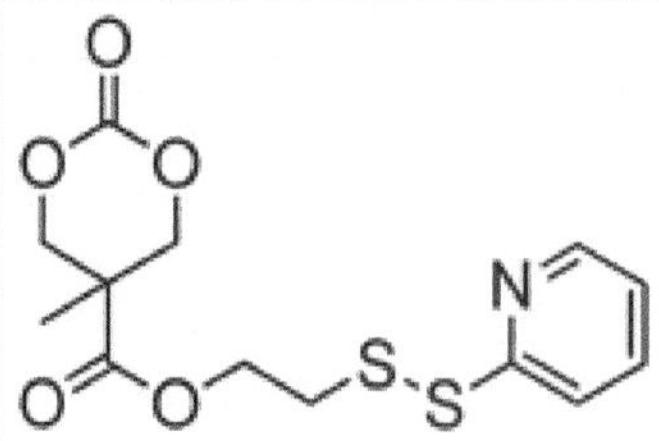	**105**	37 and 44
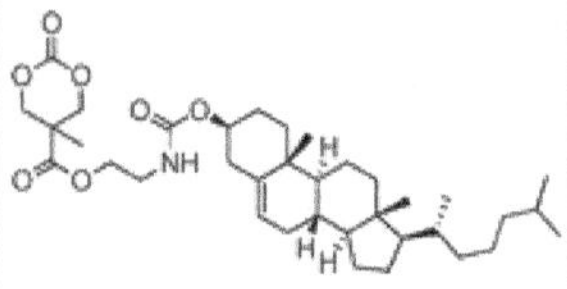	**100**	160	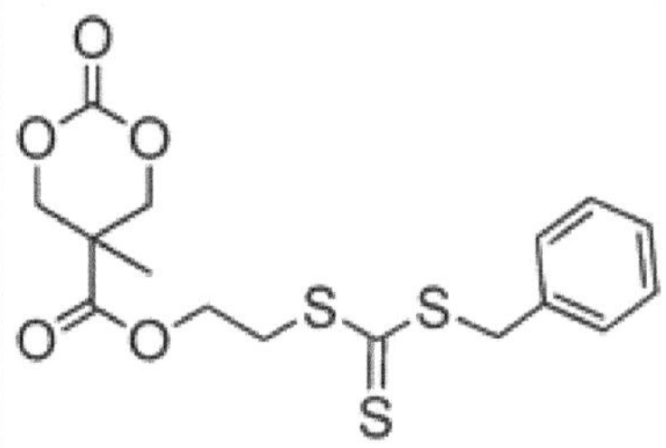	**106**	155
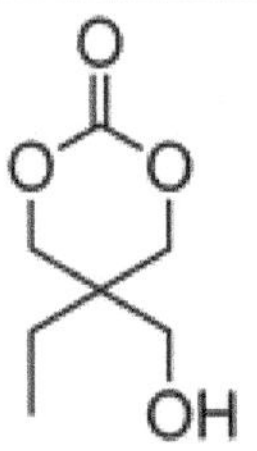	**101**	153 and 154	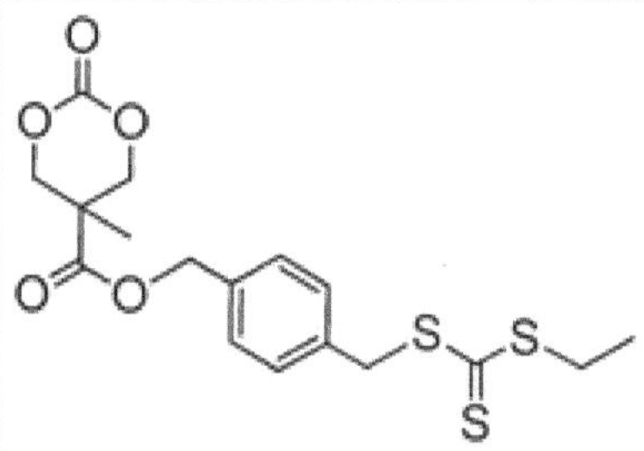	**107**	156

Other functional monomers reported include epoxide-functional cyclic carbonate, 98, which was converted to bifunctional cyclic carbonate 5-(2-oxo-1,3-dioxolan-4-yl) methyl-5-propyl-1,3-dioxan-2-one (99, Table 8).[65] The selective polymerisation of the six-membered cyclic carbonate at 60 °C with DBU was achieved and the resulting homopolymer, with the pendant five-membered cyclic carbonate intact, was obtained in 50% yield (M_n = 14 900 g mol^{-1}; $Đ_M$ = 1.41). Hydroxyl-functional monomers 101 and 102 are incompatible for the controlled ROP process as the hydroxyl functionality acts as an initiating group. However, advantage was taken of this fact to prepare biodegradable, hydrophilic, hyperbranched hydroxyl-enriched aliphatic polycarbonates.[45,153,154] Attaching PEG arms onto the hyperbranched structure led to self-assembly into supramolecular micelles with excellent stability and good cell biocompatibility.[117]

In a recent report by Lee *et al.*, a cyclic carbonate with a pendant cholesterol group, cholesteryl-2-(5-methyl-2-oxo-1,3-dioxane-5-carboxyloyloxy) ethyl carbamate (100) was homopolymerised (DP 11) and copolymerised with TMC from a PEO macroinitiator using organic catalysts DBU and TU.[160] Micelles prepared from the abovementioned block copolymersshowed increased loading of the anticancer drug paclitaxel. It was hypothesised that this excellent compatibility is a result of the rigid aromatic structure of cholesterol and its available stereocenters that may interact with those in paclitaxel. Sulphide and disulphide functional cyclic carbonates 103, 104 and 105 (Table 8) have been reported, but their polymerisation has not yet been reported.[37,44] Other sulfur-containing monomers have been recently reported, including those containing pendant groups capable of acting as chain-transfer agents (CTAs) in reverse addition-fragmentation chain transfer (RAFT) polymerisations.[155,156] Mespouille *et al.* reported the synthesis of a bis-MPA-based cyclic carbonate bearing a 2-(benzylsulfanylthiocarbonylsulfanyl)ethyl ester, 106.[155] ROP of 106 was investigated using both DBU and TU/(−)-sparteine catalysts with the latter system yielding polymers with lower dispersities ($Đ_M$ = 1.11–1.78). The polymerisation of *N*-isopropylacrylamide (NiPAAM) from the resulting polymers in a Z-group approach (in which the CTA remains bound to the backbone) in DMF at 70 °C for 8 h using AIBN as coinitiator resulted in successful chain grafting from the poly(carbonate) backbone. More recently, Williams, O'Reilly and Dove reported the synthesis of poly(carbonate) graft copolymers using ROP and an R-group RAFT approach in order that the CTA moiety be located at the graft chain termini.[156] Here, the ROP of an analogous dual-functional monomer, 107, was undertaken simply using DBU as a catalyst to yield polymers with predictable molecular weights (in comparison to [M]/[I]) and low dispersities ($Đ_M$ < 1.16). Graft copolymers with NiPAAM, styrene and methyl acrylate were prepared in $CHCl_3$ at 65 °C using AIBN as a coinitiator. Notably, the NiPAAM-grafted poly(carbonate)s displayed altered T_g's and lower critical solution temperatures that were in line with those expected based on their structure. Additionally, direct dissolution of thesepolymers in nanopure water at 3 °C resulted in the isolation of degradable nanoparticles.

POST-POLYMERISATION MODIFICATION OF FUNCTIONAL POLY(CARBONATE)S

Deprotection of Aryl- and Alkyl-Functional Poly(Carbonate)s

Alkyl- and aryl-functional groups are inert during most polymerisation conditions and are therefore often used to protect functional groups such as hydroxyl and carboxylic acid groups which may be incompatible in a ROP process. Aliphatic poly(carbonate)s containing pendant hydroxyl and carboxylic acid functions can be obtained by introducing benzyl ether or ester protecting groups respectively such that deprotection can be performed directly following the polymerisation process. Deprotection of the benzyl groups of poly(carbonate)s such as poly(MBC) and poly(**17**) has been reported by catalytic hydrogenolysis over palladium on carbon or palladium hydroxide on carbon to lead to carboxyl- and hydroxyl-functional poly(carbonate)s.[42,78,90,92,93,95–102,104,157–159] In both es the reduction in the molecular weight was confirmed by a shift to higher retention time observed by GPC analysis. Poly(17) and poly(MBC) are highly hydrophobic materials before deprotection, but the resulting hydroxyl and carboxylic acid functional polymers are hydrophilic materials that show enhanced biocompatibility and hydrosolubility compared to the benzyl-containing material.[95,102] Homopolymers and copolymers of MBC also reveal a change in T_g after deprotection with poly(MBC) being isolated as a sticky amorphous polymer with T_g = *ca.* 3 °C. After deprotection, the polymer was isolated as a semicrystalline white powder with a higher T_g (47–60 °C) and a T_m = *ca.* 170 °C. Poly(17) homopolymers displayed an impressive 64-fold increase in water absorption after debenzylation. The bulk autodegradation process of poly(lactide) was increased in copolymers with 17 after benzyl deprotection due to the presence of the hydroxyl groups. The rate of degradation was found to be even faster under acidic conditions (pH = 4.4) than in neutral conditions (pH = 7.4).[100] In a similar manner, hydroxyl-functional polymers were obtained by removal of the benzylidene group in polymers prepared from 19.[50] In contrast to poly(17), poly(19) reveals two hydroxyl groups per repeat unit due to the double hydroxyl protection of the benzylidene group. Although to date no thermal properties are reported for poly(19), copolymers of lactide with 10% incorporation of deprotected 19 already show a decrease in T_g of 14 °C due to the increased flexibility of the polymer chains after removal of the rigid benzylidene groups. Interestingly, these lactide copolymers showed surface degradation in contrast to pure poly(lactide) which usually displays a bulk degradation profile. Facile deprotection of the benzylidene group could be achieved under acidic conditions in e of block copolymers of PEO-*b*-P(20) or PEO-*b*-P(21). In poly(20) and poly(21) the benzylidene groups are substituted with methoxy groups at the *para*-position and the *ortho*- and *para*-positions, respectively.[105] Despite the high number of hydroxyl groups on the polymer, the deprotected poly(20/21) block is still insoluble in water and hence the block copolymers self-assemble into stable micelles at pH 7.4 and also show altered degradation profiles in slightly acidic conditions (pH 4–5). This polymer was successfully employed for the encapsulation of anti-cancer drugs and their subsequent pH-controlled release.[105,161,162]

Fukushima *et al.*[74] reported another example of a poly(carbonate) that deprotects under acidic conditions. In this contribution, deprotection of tetrahydropyranyloxy protecting groups in poly(10) was carried out at 50 °C using an ion-exchange resin to reveal hydroxyl functional polymers. The hydroxyl functionalities then acted as initiating sites in further ROP of stereopure poly(lactide) to form graft-copolymers with a poly(carbonate) backbone and stereopure poly(lactide) grafts. After mixing these PEO-poly(carbonate) block-copolymers grafted with PLA of opposite chirality (PLLA and PDLA), micelles could be formed at extremely low concentrations owing to the stereocomplexation driving force between PDLA and PLLA in the core of the micelles.

Jing *et al.*[106,107] reported recently that the nitrobenzyl lateral groups in PEO-*b*-P(16) block copolymers could be deprotected under UV irradiation at 365 nm to convert the poly(carbonate) block from a hydrophobic block to a hydrophilic carboxylic acid-functional block. Micelles prepared from these PEO-*b*-P(16) block copolymers were shown to dissociate readily when irradiated due to the conversion of the amphiphilic block copolymers into hydrophilic copolymers.[106,107]

The newly available carboxylic acid and hydroxyl groups after deprotection are also open to further reactions. Further modification of carboxylic acid functional polymers were reportedvia *N,N'*-diisopropylcarbodiimide (DIC), 1-ethyl-3-(3-dimethylaminopropyl) carbodiimide (EDC) or DCC couplings with free alcohol groups present in biologically active molecules or with *N*-hydroxysuccinimide (NHS) or hydroxybenzotriazole (HOBt) for further reaction with amines.[95,96,98–100] Conjugation using these methods is limited since quantitative conversion of the acid groups into esters is rarely observed and can be as low as 20% in some reported es. Examples of post-polymerisation modifications of acid-functional polymers obtained after deprotection included conjugation with aliphatic amines in the formation of DNA/polymer nanoparticles, the grafting of PEO-amines[98] to form amphiphilic copolymers with a hydrophobic backbone and hydrophilic grafts that self-assemble in aqueous environments and conjugation with peptides[95] or drugs such as heparin,[93] docetaxel[95] and paclitaxel.[96] Further modifications of hydroxyl-functional monomers include reaction of the hydroxyl groups with acid chlorides (*e.g.* 4-isobutylmethylphenyl acetic acid).[100]

Post-Polymerisation Modifications of Alkene-Functional Poly(Carbonate)s

Alkenes participate in many reactions and are for that reason utilised extensively. In the post-polymerisation functionalisation of biodegradable polymers, the presence of carbon–carbon double bonds in the polymer chain[113] or as lateral group have been utilised in cross-linking reactions and in the addition of new pendant functional moieties.[44–48,60,65,68,71,76,115–134] Reported post-polymerisation functionalisations of alkenyl-functional poly(carbonate)s can divided into three general categories: cross-linking, thiol additions and epoxidations. Cross-linking of the hydrophobic core of micelles prepared from block copolymers with hydrophilic PEO blocks and hydrophobic poly(ester-carbonate) blocks with pendant allyl- or acryloyl-groups was realised by heating in the presence of azobisisobutyronitrile (AIBN) at 60 °C and by UV irradiation of the micelles in the presence of a biocompatible photoinitiator respectively.[115,116,126]Micelles with cross-linked poly(ester-carbonate) cores, loaded with anti-cancer drug paclitaxel and with galactose groups at the micelle surface for specific targeting displayed enhanced antitumor efficacy in *in vivo* studies. The acroyl-functional monomer, 34, has also been copolymerised to form triblock PEG-*b*-Poly(34)-*b*-PLA copolymers.[163] Here, the micellization in the presence of paclitaxel was followed by irradiation with UV light to crosslink the acroyl groups and stabilize the particle to prevent premature release of drug. In an extension to this work, synthesis of PEG-initiated statistical copolymers of 34 with 21 enabled the synthesis of crosslinked nanoparticles that were able to be deprotected upon lowering of the pH to enable a triggered release of the cargo.[164] In another example, the radical cross-linking of a poly(carbonate) homopolymer with pendant styrene groups (poly(36)) was realised in the presence of the radical initiator benzoyl peroxide.[48] The cross-linking was carried out in the presence of styrene to yield a poly(carbonate)/poly(styrene) network. Notably, the resulting cross-linked networks could be unlinked in the presence of *t*-BuOK, resulting in poly(styrene) with pendant six-membered cyclic carbonates. Photo-cross-linking of poly(ester-carbonates) (copolymersof L-lactide and 32) with pendant cinnamate functionalities could be cross-linked by UV irradiation at 365 nm providing biodegradable networks.[124] This type of cross-linking occurs *via*dimerisation through the [2+2] cycloaddition reaction of two cinnamate moieties upon irradiation with UV light.

The reaction of thiols and alkenes ('thiol–ene') has been utilised extensively as an efficient method in the preparation of functional polymers. Reactions proceed either *via* the free-radical addition of thiols to carbon–carbon double bonds or *via* the Michael-addition of thiols to electron-deficient carbon–carbon double bonds.[111] Radical-type addition of thiols to the poly(carbonate) backbone has been reported for polymers containing pendant allyl functionalities.[45,116,118,121] Homopolymers of MAC (29) were readily functionalized under radical 'thiol–ene' conditions (>2 eq. thiol, 90 °C, 24 h) in 1,4-dioxane with a range of thiol-containing species including some with functional groups that are not compatible with ring-opening polymerization such as alcohol and carboxylic acid groups.[121] As expected the changes to the side-chain led to changes in the polymer T_g being observed, although the thermal degradation behaviour was largely unchanged. Copolymers prepared *via* the ROP of lactide and allyl-functional monomers (MAC, 29) could be functionalised with a number of thiols by UV irradiation at 254 nm in the presence of 1-hydroxycyclohexylphenyl ketone as radical initiator or without radical initiator.[45,118] Although the reaction proceeds to high conversion at equimolar ene-thiolratios, again an excess was necessary to achieve full conversion (5 : 1). Functionalisation of poly(ester-carbonates) with folic acid *via* radical thiol addition resulted in polymers which displayed increased cell affinity.[116] Similarly, a range of alkene functional poly(carbonate)s could be modified with thiols *via* Michael-addition.[47] PLA- or PCL-based polymers with low contents of acryloyl or methacryloyl functional cyclic carbonate monomers (34 and 35), were further functionalised in dimethylformamide at room temperature with a number of thiols.[47]Quantitative conversions (using 10 equivalents of thiol) of the pendant (meth)acrylates was obtained for functionalisation with mercaptoethanol, cysteine and 2-mercaptoethylamine, but conjugation with a peptide only proceeded to 58% conversion whilst mercaptopropanoic acid only displayed a conversion of 30%. Conjugation of block copolymers composed of PEO and a small acryloyl-functional poly(carbonate) blocks (DP < 20) was performed with thiolated chitosan that led to *in situ* hydrogel formation.[127] Such methodology was also extended to functionalise statistical copolymers of 34 with ε-caprolactone with cysteamine that was subsequently converted into a CTA for the RAFT polymerisation of hydroxyethylmethacrylate. In turn, these copolymers were assembled into graft copolymer micelles that showed pH-responsive release of doxorubicin.[165]

Vinyl sulfone-functional copoly(carbonate)s and poly(ester-carbonate)s, obtained from the copolymerisation of 38 with other cyclic monomers, could also be quantitatively functionalised *via* the Michael addition of thiols to the vinyl group withoutcatalyst and under aqueous conditions.[64] Another example of functionalisation of poly(carbonate)s and poly(ester-carbonate)s *via* Michael-addition of thiols was presented by Onbulak *et al.*[129] Polymers with pendant furan-protected maleimides were deprotected *via* thermal *retro*-Diels Alder to give maleimide functional polymers that were further functionalised with thethiol-containing molecules, 6-(ferrocenyl)hexanethiol and 1-hexanethiol, in the presence of triethylamine. Notably, both deprotection and thiol addition proceeded without degradation of the polymer backbone.

Finally, alkene functional poly(carbonate)s could also be epoxidised using *meta*-chloroperoxybenzoic acid (*m*-CPBA). Epoxidation of functional aliphatic poly(carbonate)s and poly(ester-carbonate)s was reported for pendant allyl groups and pendant cyclohexenyl groups with efficiencies ranging from 35 to 99%. Hydrolysis of the epoxide groups was noted to occur in some es.[46,76,119–123] In e of epoxidation of PMAC homopolymers an increase in glass transition temperature from −12 °C to 14 °C was observed. The introduction of this new reactive functionality was applied in the grafting of poly(ethylene imine) to the poly(carbonate) backbone in the formation of biodegradable polycations and polycation/DNA polyplexes by Zhuo and coworkers.[119]

Post-Polymerisation Modifications of Poly(carbonate)s *via* Dipolar Cycloadditions

Alkyne- and azide-functional poly(carbonate)s or poly(ester-carbonate)s were successfully functionalised *via* 1,3-dipolar cycloaddition reactions.[57,58,69,73,150,126–129] Modifications of poly(ester-carbonates) obtained from copolymerisation of MPC (41) with lactide were successfully functionalised with azido-containing molecules using copper(II) sulfate pentahydrate andsodium ascorbate in water–DMSO mixtures.[130,133,134] Modifications with glucose and galactose azides were shown to lead to increased biocompatibility and enhanced hydrophilicity and cellular affinity of the resulting polymers.[134] Similarly, some polymers were also modified with CO-protected hemoglobin and proteins including Testis-specific protease 50 (TSP50), a breast cancer sensitive protease, and bovine serum albumin (BSA). The latter achieved only low grafting efficiencies of 0.7–1.6 mol%.[130,133] Alternatively, azide-functional poly(carbonate)s could be obtained *via* polymerisation of azide functional cyclic carbonates or *via* modification of bromide functional poly(carbonate)s with sodium azide in DMF at various temperatures.[57,58] The latter method, however, has limitations as azidation of bromide-functional polymers did not exceed 25% for homopolymers and degradation of the poly(carbonate) backbone was observed in both bromide functional homopolymers and copolymers.[57,58] Successful cycloaddition of alkyne-functional molecules to azido-functional poly(carbonate)s *via* copper catalysed methods (CuAAC) and *via* strain-promoted methods (SPAAC) was reported by Song and coworkers and was used in the preparation of hydrogels by cross-linking of poly(carbonate)s with DABCO-terminated poly(ethylene oxide)s.[136,137]

Quaternisation of Halogen-Functional Poly(Carbonate)s

A range of polymers containing quaternary amines have shown to exhibit antimicrobial activity. These polymers are, however, often not degradable which limits their use. The quaternisation of amines by reaction with halogen-containing compounds enables poly(carbonate) homopolymers and copolymers bearing pendant halogen functionalities to be reacted withtertiary amines such as trimethylamine or *N,N,N′,N′*-tetramethylethylenediamine. Recently, Yang and coworkers reported the first example of quaternised poly(carbonate) nanoparticlesprepared from block copolymers of PTMC and a chloro-functional poly(carbonate) prepared from monomer 48.[139] Investigation of quaternisation conditions revealed a dependence on the halogen substituent on the pendant chain. While chloro-functional poly(carbonate)s required heating at 90 °C, bromo- and iodo-functional poly(carbonate)s could be fully converted at room temperature. Another application for these cationic poly(carbonate)s was found in the synthesis of poly(carbonate)/DNA complexes.[69,72,138,139]

Deprotection of Carbamate-Functional Poly(Carbonate)s

Poly(carbonate)s with pendant primary amines can be prepared *via* the ROP of monomers in which the primary amines are protected by conversion to benzyl- or *tert*-butyl-carbamate groups. Deprotection of the benzyl carbamate (Z) groups is carried out *via* hydrogenation catalysed by palladium while the *tert*-butyl-carbamate (Boc) groups can easily be removed usingtrifluoroacetic acid.[52,55] Resulting primary amine-functional poly(carbonate)s and poly(ester-*co*-carbonate) s have been obtained whilst retaining low dispersities. As expected, a reduction in molecular weight was observed after deprotection and polymers were highly hydrophilic in contrast to the protected polymers. Primary amine functional poly(ester-carbonate)s could be further modified by reaction with peptides in the presence of CDI. Films of the resulting polymers displayed increased cell adhesion and multiplication.[55] In another report, two Boc-protecting groups were removed to reveal pendant guanidinium groups. Hydrogels prepared from bis(carbonate) poly(ethylene oxide) cross-linker, Boc protected guanidinium functional cyclic carbonate and *tert*-butoxy protected carboxylic acid functional cyclic carbonate revealed additional stabilisation of the hydrogel network after removal of the protecting groups withtrifluoroacetic acid as a result of the hydrogen bonding between the free guanidinium and carboxylic acid groups.[87]

Other Post-Polymerisation Modifications of Functional Poly (Carbonate)s

Putnam and coworkers has reported a post-polymerisation conversion of dimethoxy-functional poly(carbonate), poly(4), and its copolymers into dihydroxy acetone-based poly(carbonate)s.[63] The conversion to dihydroxyacetone functional poly(carbonate)s was accomplished by refluxing poly(4) with iodine in acetone. The resulting polymer presents an increase in glass transition temperature compared to the precursor from 45 to 68 °C. In addition, the resulting polymers showed very low solubility in common organic solvents. Moreover,copolymers composed of lactide and dihydroxyacetone-functional carbonate showed increasing degradation kinetics in copolymers enriched in functional carbonate.

CONCLUSIONS

The combination of the many versatile methods reported for the synthesis of functional cyclic carbonate monomers from a range of biocompatible scaffolds and the excellent control in theirring-opening polymerisation, have led to preparation of a multitude of functional poly(carbonate)s with controlled molecular weights, low dispersities and high end-group fidelity. Despite the development of the large range of functional aliphatic poly(carbonate)s, their specific design to achieve better-controlled drug release and advanced tissue engineering scaffolds is very rare. Further development of controlled poly(carbonate) synthesis in combination with efficient post-polymerisation modifications with a specific focus on application could open up new ways to achieve their widespread biomedical application.

ACKNOWLEDGEMENTS

The Research Councils UK (RCUK) are acknowledged for funding a fellowship to A.P.D. We gratefully acknowledge financial support from EPSRC (EP/F068808/1) for funding a studentship (S.T.). This work was supported by the European Commission and Région Wallonne FEDER program (Materia Nova) and OPTI²MAT program of excellence, by the Interuniversity Attraction Pole program of the Belgian Federal Science Policy Office (PAI 6/27) and by FNRS-FRFC. O. C. is a FNRS Research Associate.

NOTES AND REFERENCES

1. G. Rokicki, *Prog. Polym. Sci.*, 2000, 25, 259–342 .
2. L. S. Nair and C. T. Laurencin, *Prog. Polym. Sci.*, 2007, 32, 762–798 .
3. H. Keul, *Polycarbonates*, Wiley-VCH Verlag GmbH & Co. KGaA, New York, 2009, pp. 307–327 .
4. Z. Zhang, D. W. Grijpma and J. Feijen, *J. Controlled Release*, 2006, 111, 263–270 .
5. H. Seyednejad, A. H. Ghassemi, C. F. van Nostrum, T. Vermonden and W. E. Hennink, *J. Controlled Release*, 2011, 152, 168–176 .
6. C. G. Pitt and Z. W. Gu, *J. Controlled Release*, 1987, 4, 283–292 .
7. K. Fu, D. W. Pack, A. M. Klibanov and R. Langer, *Pharm. Res.*, 2000, 17, 100–106 .
8. A. P. Pêgo, M. J. A. Van Luyn, L. A. Brouwer, P. B. Van Wachem, A. A. Poot, D. W. Grijpma and J. Feijen, *J. Biomed. Mater. Res., Part A*, 2003, 67A, 1044–1054 .
9. Z. Zhang, R. Kuijer, S. K. Bulstra, D. W. Grijpma and J. Feijen, *Biomaterials*, 2006, 27, 1741–1748 .
10. B. Ochiai and T. Endo, *Prog. Polym. Sci.*, 2005, 30, 183–215 .
11. G. W. Coates and D. R. Moore, *Angew. Chem., Int. Ed.*, 2004, 43, 6618–6639 .
12. D. J. Darensbourg, *Chem. Rev.*, 2007, 107, 2388–2410 .

13. D. J. Darensbourg, R. M. Mackiewicz, A. L. Phelps and D. R. Billodeaux, *Acc. Chem. Res.*, 2004, 37, 836–844 .

14. J. Feng, R. X. Zhuo and X. Z. Zhang, *Prog. Polym. Sci.*, 2012, 37, 211–236 .

15. S. Kobayashi, *Macromol. Rapid Commun.*, 2009, 30, 237–266 .

16. A. P. Dove, *Chem. Commun.*, 2008, 6446–6470 .

17. S. Matsumura, K. Tsukada and K. Toshima, *Macromolecules*, 1997, 30, 3122–3124 .

18. R. F. Storey, B. D. Mullen, G. S. Desai, J. W. Sherman and C. N. Tang, *J. Polym. Sci., Part A: Polym. Chem.*, 2002, 40, 3434–3442 .

19. N. E. Kamber, W. Jeong, R. M. Waymouth, R. C. Pratt, B. G. G. Lohmeijer and J. L. Hedrick, *Chem. Rev.*, 2007, 107, 5813–5840 .

20. F. Nederberg, B. G. G. Lohmeijer, F. Leibfarth, R. C. Pratt, J. Choi, A. P. Dove, R. M. Waymouth and J. L. Hedrick, *Biomacromolecules*, 2007, 8, 153–160 .

21. O. Coulembier, D. P. Sanders, A. Nelson, A. N. Hollenbeck, H. W. Horn, J. E. Rice, M. Fujiwara, P. Dubois and J. L. Hedrick, *Angew. Chem., Int. Ed.*, 2009, 48, 5170–5173 .

22. M. Pastusiak, P. Dobrzynski, B. Kaczmarczyk and J. Kasperczyk, *J. Polym. Sci., Part A: Polym. Chem.*, 2011, 49, 2504–2512 .

23. Y. Yamamoto, S. Kaihara, K. Toshima and S. Matsumura, *Macromol. Biosci.*, 2009, 9, 968–978 .

24. S. Matsumura, K. Tsukada and K. Toshima, *Macromolecules*, 1997, 30, 3122–3124 .

25. K. Odelius and A.-C. Albertsson, *J. Polym. Sci., Part A: Polym. Chem.*, 2008, 46, 1249–1264 .

26. B. L. Dargaville, C. Vaquette, H. Peng, F. Rasoul, Y. Q. Chau, J. J. Cooper-White, J. H. Campbell and A. K. Whittaker, *Biomacromolecules*, 2011, 12, 3856–3869 .

27. S. Dai, L. Xue and Z. Li, *ACS Catal.*, 2011, 1, 1421–1429 .

28. M. Normand, E. Kirillov, J. F. Carpentier and S. M. Guillaume, *Macromolecules*, 2012, 45, 1122–1130 .

29. J. Liu, C. Zhang and L. Liu, *J. Appl. Polym. Sci.*, 2008, 107, 3275–3279 .

30. R. Chapanian, M. Y. Tse, S. C. Pang and B. G. Amsden, *J. Pharm. Sci.*, 2012, 101, 588–597 .

31. Z. Ma, Y. Hong, D. M. Nelson, J. E. Pichamuthu, C. E. Leeson and W. R. Wagner, *Biomacromolecules*, 2011, 12, 3265–3274 .

32. M. Helou, G. Moriceau, Z. W. Huang, S. Cammas-Marion and S. M. Guillaume, *Polym. Chem.*, 2011, 2, 840–850 .

33. F. Nederberg, V. Trang, R. C. Pratt, A. F. Mason, C. W. Frank, R. M. Waymouth and J. L. Hedrick, *Biomacromolecules*, 2007, 8, 3294–3297 .

34. O. Thillaye du Boullay, N. Saffon, J.-P. Diehl, B. Martin-Vaca and D. Bourissou, *Biomacromolecules*, 2010, 11, 1921–1929 .

35. R. J. Pounder and A. P. Dove, *Biomacromolecules*, 2010, 11, 1930–1939 .

36. R. J. Pounder and A. P. Dove, *Polym. Chem.*, 2010, 1, 260–271 .

37. D. P. Sanders, K. Fukushima, D. J. Coady, A. Nelson, M. Fujiwara, M. Yasumoto and J. L. Hedrick, *J. Am. Chem. Soc.*, 2010, 132, 14724–14726 .

38. J. Mindemark and T. Bowden, *Polymer*, 2011, 52, 5716–5722 .

39. P. Wu, M. Malkoch, J. N. Hunt, R. Vestberg, E. Kaltgrad, M. G. Finn, V. V. Fokin, K. B. Sharpless and C. J. Hawker, *Chem. Commun.*, 2005, 5775–5777 .

40. M. Malkoch, E. Malmstrom and A. Hult, *Macromolecules*, 2002, 35, 8307–8314 .

41. M. Malkoch, R. Vestberg, N. Gupta, L. Mespouille, Ph. Dubois, A. F. Mason, J. L. Hedrick, Q. Liao, C. W. Frank, K. Kingsbury and C. J. Hawker, *Chem. Commun.*, 2006, 2774–2776 .

42. T. F. Al-Azemi and K. S. Bisht, *Macromolecules*, 1999, 32, 6536–6540 .

43. K. D. Weilandt, H. Keul and H. Hocker, *Macromol. Chem. Phys.*, 1996, 197, 3851–3868 .

44. R. C. Pratt, F. Nederberg, R. M. Waymouth and J. L. Hedrick, *Chem. Commun.*, 2008, 114–116 .

45. P. G. Parzuchowski, M. Jaroch, M. Tryznowski and G. Rokicki, *Macromolecules*, 2008, 41, 3859–3865 .

46. F. He, Y.-P. Wang, G. Liu, H.-L. Jia, J. Feng and R.-X. Zhuo, *Polymer*, 2008, 49, 1185–1190 .

47. W. Chen, H. Yang, R. Wang, R. Cheng, F. Meng, W. Wei and Z. Zhong, *Macromolecules*, 2010, 43, 201–207 .

48. T. Miyagawa, M. Shimizu, F. Sanda and T. Endo, *Macromolecules*, 2005, 38, 7944–7949 .

49. J. Yang, Q. Hao, X. Liu, C. Ba and A. Cao, *Biomacromolecules*, 2004, 5, 209–218 .

50. Z. Xie, C. Lu, X. Chen, L. Chen, Y. Wang, X. Hu, Q. Shi and X. Jing, *J. Polym. Sci., Part A: Polym. Chem.*, 2007, 45, 1737–1745 .

51. S.-H. Pyo, P. Persson, S. Lundmark and R. Hatti-Kaul, *Green Chem.*, 2011, 13, 976–982 .

52. F. Sanda, J. Kamatani and T. Endo, *Macromolecules*, 2001, 34, 1564–1569 .

53. R. Kumar, W. Gao and R. A. Gross, *Macromolecules*, 2002, 35, 6835–6844 .

54. X. Chen and R. A. Gross, *Macromolecules*, 1999, 32, 308–314 .

55. X. Hu, X. Chen, Z. Xie, H. Cheng and X. Jing, *J. Polym. Sci., Part A: Polym. Chem.*, 2008, 46, 7022–7032 .

56. R. D. Youssefyeh, J. P. H. Verheyden and J. G. Moffatt, *J. Org. Chem.*, 1979, 44, 1301–1309 .

57. X. Zhang, Z. Zhong and R. Zhuo, *Macromolecules*, 2011, 44, 1755–1759 .

58. W. Zhu, Y. Wang, Q. Zhang and Z. Shen, *J. Polym. Sci., Part A: Polym. Chem.*, 2011, 49, 4886–4893 .

59. H.-F. Wang, W. Su, C. Zhang, X.-H. Luo and J. Feng, *Biomacromolecules*, 2010, 11, 2550–2557 .

60. X. Chen, S. P. McCarthy and R. A. Gross, *Macromolecules*, 1997, 30, 3470–3476 .

61. D. J. Darensbourg, A. I. Moncada and S. H. Wei,

Macromolecules, 2011, 44, 2568–2576 .

62. M. Helou, J.-M. Brusson, J.-F. Carpentier and S. M. Guillaume, *Polym. Chem.*, 2011, 2, 2789–2795 .
63. J. R. Weiser, P. N. Zawaneh and D. Putnam, *Biomacromolecules*, 2011, 12, 977–986 .
64. R. Wang, W. Chen, F. Meng, R. Cheng, C. Deng, J. Feijen and Z. Zhong, *Macromolecules*, 2011, 44, 6009–6016 .
65. T. Endo, K. Kakimoto, B. Ochiai and D. Nagai, *Macromolecules*, 2005, 38, 8177–8182 .
66. A. E. van der Ende, E. J. Kravitz and E. Harth, *J. Am. Chem. Soc.*, 2008, 130, 8706–8713 .
67. A. E. van der Ende, J. Harrell, V. Sathiyakumar, M. Meschievitz, J. Katz, K. Adcock and E. Harth, *Macromolecules*, 2010, 43, 5665–5671 .
68. R. M. Burk and M. B. Roof, *Tetrahedron Lett.*, 1993, 34, 395–398 .
69. Y. Qiao, C. Yang, D. J. Coady, Z. Y. Ong, J. L. Hedrick and Y.-Y. Yang, *Biomaterials*, 2012, 33, 1146–1153 .
70. S. H. Kim, J. P. K. Tan, K. Fukushima, F. Nederberg, Y. Y. Yang, R. M. Waymouth and J. L. Hedrick, *Biomaterials*, 2011, 32, 5505–5514 .
71. S.-H. Kim, J. P. K. Tan, F. Nederberg, K. Fukushima, J. Colson, C. Yang, A. Nelson, Y. Y. Yang and J. L. Hedrick, *Biomaterials*, 2010, 31, 8063–8071 .
72. Z. Y. Ong, K. Fukushima, D. J. Coady, Y. Y. Yang, P. L. R. Ee and J. L. Hedrick, *J. Controlled Release*, 2011, 152, 120–126 .
73. J. P. K. Tan, S. H. Kim, F. Nederberg, K. Fukushima, D. J. Coady, A. Nelson, Y. Y. Yang and J. L. Hedrick, *Macromol. Rapid Commun.*, 2010, 31, 1187–1192 .
74. K. Fukushima, R. C. Pratt, F. Nederberg, J. P. K. Tan, Y. Y. Yang, R. M. Waymouth and J. L. Hedrick, *Biomacromolecules*, 2008, 9, 3051–3056 .
75. F. He, Y. Wang, J. Feng, R. Zhuo and X. Wang, *Polymer*, 2003, 44, 3215–3219 .
76. F. He, C. F. Wang, T. Jiang, B. Han and R. X. Zhuo, *Biomacromolecules*, 2010, 11, 3028–3035 .
77. X. Zhang, F. Chen, Z. Zhong and R. Zhuo, *Macromol. Rapid Commun.*, 2010, 31, 2155–2159 .
78. X. Zhang, H. Mei, C. Hu, Z. Zhong and R. Zhuo, *Macromolecules*, 2009, 42, 1010–1016 .
79. N. Nemoto, F. Sanda and T. Endo, *J. Polym. Sci., Part A: Polym. Chem.*, 2001, 39, 1305–1317 .
80. O. Coulembier, D. P. Sanders, A. Nelson, A. N. Hollenbeck, H. W. Horn, J. E. Rice, M. Fujiwara, P. Dubois and J. L. Hedrick, *Angew. Chem., Int. Ed.*, 2009, 48, 5170–5173 ; O. Coulembier, D. P. Sanders, A. Nelson, A. N. Hollenbeck, H. W. Horn, J. E. Rice, M. Fujiwara, P. Dubois and J. L. Hedrick, *Angew. Chem.*, 2009, 121, 5272–5275 .
81. T. Ariga, T. Takata and T. Endo, *Macromolecules*, 1997, 30, 737–744 .
82. J. Ling, Z. Shen and Q. Huang, *Macromolecules*, 2001, 34, 7613–7616 .
83. H. Peng, J. Ling, J. Liu, N. Zhu, X. Ni and Z. Shen, *Polym. Degrad. Stab.*, 2010, 95, 643–650 .
84. B. Liu and D. Cui, *J. Appl. Polym. Sci.*, 2009, 112, 3110–3118 .
85. J. Cai, K. J. Zhu and S. L. Yang, *Polymer*, 1998, 39, 4409–4415 .
86. M. Helou, O. Miserque, J. M. Brusson, J. F. Carpentier and S. M. Guillaume, *Chem.–Eur. J.*, 2010, 16, 13805–13813 .
87. C. Bartolini, L. Mespouille, I. Verbruggen, R. Willem and Ph. Dubois, *Soft Matter*, 2011, 7, 9628–9637 .
88. A. N. Zelikin, P. N. Zawaneh and D. Putnam, *Biomacromolecules*, 2006, 7, 3239–3244 .
89. P. N. Zawaneh, A. M. Doody, A. N. Zelikin and D. Putnam, *Biomacromolecules*, 2006, 7, 3245–3251 .
90. P. Brignou, J. F. Carpentier and S. M. Guillaume, *Macromolecules*, 2011, 44, 5127–5135 .
91. J. Ling, Y. Dai, Y. Zhu, W. Sun and Z. Shen, *J. Polym. Sci., Part A: Polym. Chem.*, 2010, 48, 3807–3815 .
92. T. F. Al-Azemi, J. P. Harmon and K. S. Bisht, *Biomacromolecules*, 2000, 1, 493–500 .
93. F. Gong, X. Cheng, S. Wang, Y. Zhao, Y. Gao and H. Cai, *Acta Biomater.*, 2010, 6, 534–546 .
94. Y. Yuan, X. Jing, H. Xiao, X. Chen and Y. Huang, *J. Appl. Polym. Sci.*, 2011, 121, 2378–2385 .
95. Z. Xie, X. Hu, X. Chen, T. Lu, S. Liu and X. Jing, *J. Appl. Polym. Sci.*, 2008, 110, 2961–2970 .
96. M. Danquah, T. Fujiwara and R. I. Mahato, *Biomaterials*, 2010, 31, 2358–2370 .
97. W. Lu, F. Li and R. I. Mahato, *J. Pharm. Sci.*, 2011, 100, 2418–2429 .
98. J. Lu and M. S. Shoichet, *Macromolecules*, 2010, 43, 4943–4953 .
99. W. Y. Seow and Y. Y. Yang, *J. Controlled Release*, 2009, 139, 40–47 .
100. W. C. Ray III and M. W. Grinstaff, *Macromolecules*, 2003, 36, 3557–3562 .
101. X. L. Wang, R. X. Zhuo, L. J. Liu, F. He and G. Liu, *J. Polym. Sci., Part A: Polym. Chem.*, 2002, 40, 70–75 .
102. J. Feng, X. L. Wang, F. He and R. X. Zhuo, *Macromol. Rapid Commun.*, 2007, 28, 754–758 .
103. L. L. Mei, G. P. Yan, X. H. Yu, S. X. Cheng and J. Y. Wu, *J. Appl. Polym. Sci.*, 2008, 108, 93–98 .
104. H. Chen, G. P. Yan, L. Li, C. W. Ai and X. H. Yu, *J. Appl. Polym. Sci.*, 2009, 114, 3087–3096 .
105. W. Chen, F. Meng, F. Li, S. J. Ji and Z. Zhong, *Biomacromolecules*, 2009, 10, 1727–1735 .
106. Z. Xie, X. Hu, X. Chen, G. Mo, J. Sun and X. Jing, *Adv. Eng. Mater.*, 2009, 11, B7–B11 .
107. Z. Xie, X. Hu, X. Chen, J. Sun, Q. Shi and X. Jing, *Biomacromolecules*, 2008, 9, 376–380 .
108. M. A. Tasdelen, *Polym. Chem.*, 2011, 2, 2133–2145 .
109. R. K. Iha, K. L. Wooley, A. M. Nyström, D. J. Burke, M. J. Kade and C. J. Hawker, *Chem. Rev.*, 2009, 109, 5620–5686 .
110. M. Meldal and C. W. Tornøe, *Chem. Rev.*, 2008, 108, 2952–3015 .

111. C. E. Hoyle and C. N. Bowman, *Angew. Chem., Int. Ed.*, 2010, 49, 1540–1573 .

112. D. Konkolewicz, A. Gray-Weale and S. Perrier, *J. Am. Chem. Soc.*, 2009, 131, 18075–18077 .

113. Z. Ates, P. D. Thornton and A. Heise, *Polym. Chem.*, 2011, 2, 309–312 .

114. R. Riva, S. Schmeits, F. Stoffelbach, C. Jérôme, R. Jérôme and P. Lecomte, *Chem. Commun.*, 2005, 5334–5336 .

115. X. Hu, X. Chen, Z. Xie, S. Liu and X. Jing, *J. Polym. Sci., Part A: Polym. Chem.*, 2007, 45, 5518–5528 .

116. X. Hu, X. Chen, S. Liu, Q. Shi and X. Jing, *J. Polym. Sci., Part A: Polym. Chem.*, 2008, 46, 1852–1861 .

117. X. Hu, X. Chen, J. Wei, S. Liu and X. Jing, *Macromol. Biosci.*, 2009, 9, 456–463 .

118. J. Yue, X. Li, G. Mo, R. Wang, Y. Huang and X. Jing, *Macromolecules*, 2010, 43, 9645–9654 .

119. C. F. Wang, Y. X. Lin, T. Jiang, F. He and R.-X. Zhuo, *Biomaterials*, 2009, 30, 4824–4832 .

120. B. D. Mullen, C. N. Tang and R. F. Storey, *J. Polym. Sci., Part A: Polym. Chem.*, 2003, 41, 1978–1991 .

121. S. Tempelaar, L. Mespouille, P. Dubois and A. P. Dove, *Macromolecules*, 2011, 44, 2084–2091 .

122. W. Zhao, Y. Wang, X. Liu and D. Cui, *Chem. Commun.*, 2012, 48, 4588–4590 .

123. X. Chen, S. P. McCarthy and R. A. Gross, *Macromolecules*, 1998, 31, 662–668 .

124. X. Hu, X. Chen, H. Cheng and X. Jing, *J. Polym. Sci., Part A: Polym. Chem.*, 2009, 47, 161–169 .

125. P. Lundberg, A. Bruin, J. W. Klijnstra, A. M. Nyström, M. Johansson, M. Malkoch and A. Hult, *ACS Appl. Mater. Interfaces*, 2010, 2, 903–912 .

126. R. Yang, F. Meng, S. Ma, F. Huang, H. Liu and Z. Zhong, *Biomacromolecules*, 2011, 12, 3047–3055 .

127. Y. Yu, C. Deng, F. Meng, Q. Shi, J. Feijen and Z. Zhong, *J. Biomed. Mater. Res., Part A*, 2011, 99A, 316–326 .

128. M. Murayama, F. Sanda and T. Endo, *Macromolecules*, 1998, 31, 919–923 .

129. S. Onbulak, S. Tempelaar, R. J. Pounder, O. Gok, R. Sanyal, A. P. Dove and A. Sanyal, *Macromolecules*, 2012, 45, 1715–1722 .

130. T.-H. Lia, X.-B. Jing and Y.-B. Huang, *Polym. Adv. Technol.*, 2011, 22, 1266–1271 .

131. Y. Han, Q. Shi, J. Hu, Q. Du, X. Chen and X. Jing, *Macromol. Biosci.*, 2008, 8, 638–644 .

132. Q. Shia, X. Chena, T. Lua and X. Jing, *Biomaterials*, 2008, 29, 1118–1126 .

133. Q. Shi, Y. Huang, X. Chen, M. Wu, J. Sun and X. Jing, *Biomaterials*, 2009, 30, 5077–5085 .

134. C. Lu, Q. Shi, X. Chen, T. Lu, Z. Xie, X. Hu, J. Ma and X. Jing, *J. Polym. Sci., Part A: Polym. Chem.*, 2007, 45, 3204–3217 .

135. P. F. Gou, W. P. Zhu and Z. Q. Shen, *Polym. Chem.*, 2010, 1, 1205–1214 .

136. J. Xu, T. M. Filion, F. Prifti and J. Song, *Chem.–Asian J.*, 2011, 6, 2730–2737 .

137. J. Xu, F. Prifti and J. Song, *Macromolecules*, 2011, 44, 2660–2667 .

138. C. Yang, Z. Y. Ong, Y. Y. Yang, P. L. R. Ee and J. L. Hedrick, *Macromol. Rapid Commun.*, 2011, 32, 1826–1833 .

139. F. Nederberg, Y. Zhang, J. P. K. Tan, K. Xu, H. Wang, C. Yang, S. Gao, X. D. Guo, K. Fukushima, L. Li, J. L. Hedrick and Y. Y. Yang, *Nat. Chem.*, 2011, 3, 409–414 .

140. C. B. Cooley, B. M. Trantow, F. Nederberg, M. K. Kiesewetter, J. L. Hedrick, R. M. Waymouth and P. A. Wender, *J. Am. Chem. Soc.*, 2009, 131, 16401–16403 .

141. X. Zhang, M. Cai, Z. Zhong and R. Zhuo, *Macromol. Rapid Commun.*, 2012, 33, 693–697 .

142. F. Nederberg, V. Trang, R. C. Pratt, S. H. Kim, J. Colson, A. Nelson, C. W. Frank, J. L. Hedrick, P. Dubois and L. Mespouille, *Soft Matter*, 2010, 6, 2006–2012 .

143. M. Fujiwara, J. L. Hedrick, D. P. Sanders, Y. Terui and M. Yasumoto, *(International Buisness Machines, USA; Central Glass Co. Ltd. Japan). Polymers bearing pendant pentafluorophenyl ester groups, and methods of synthesis and functionalization thereof. US Pat.*, 8,143,369 B2, March 27, 2012 .

144. S. Slavin, J. Burns, D. M. Haddleton and C. R. Becer, *Eur. Polym. J.*, 2011, 47, 435–446 .

145. Y. Shen, X. Chen and R. A. Gross, *Macromolecules*, 1999, 32, 3891–3897 .

146. Y. Shen, X. Chen and R. A. Gross, *Macromolecules*, 1999, 32, 2799–2802 .

147. F. Suriano, R. Pratt, J. P. K. Tan, N. Wiradharma, A. Nelson, Y. Y. Yang, P. Dubois and J. L. Hedrick, *Biomaterials*, 2010, 31, 2637–2645 .

148. H. Ajiro, Y. Takahashi and M. Akashi, *Macromolecules*, 2012, 45, 2668–2674 .

149. O. Coulembier, S. Moins and P. Dubois, *Macromolecules*, 2011, 44, 7493–7497 .

150. T. F. Al-Azemi and K. S. Bisht, *Polymer*, 2002, 43, 2161–2167 .

151. P. J. Flory and P. Rehner, *J. Chem. Phys.*, 1943, 11, 521–526 .

152. J. Mindemark and T. Bowden, *Polym. Chem.*, 2012, 3, 1399–1401 .

153. X.-H. Luo, F.-W. Huang, S.-Y. Qin, H.-F. Wang, J. Feng, X.-Z. Zhang and R.-X. Zhuo, *Biomaterials*, 2011, 32, 9925–9939 .

154. W. Su, X.-H. Luo, H.-F. Wang, L. Li, J. Feng, X. Z. Zhang and R.-X. Zhuo, *Macromol. Rapid Commun.*, 2011, 32, 390–396 .

155. L. Mespouille, F. Nederberg, J. L. Hedrick and P. Dubois, *Macromolecules*, 2009, 42, 6319–6321 .

156. R. J. Williams, R. K. O'Reilly and A. P. Dove, *Polym. Chem.*, 2012, 3, 2156–2164 .

157. F. Li, M. Danquah and R. I. Mahato, *Biomacromolecules*, 2010, 11, 2610–2620 .

158. J. Lu, M. Shi and M. S. Shoichet, *Bioconjugate Chem.*, 2009, 20, 87–94 .

159. X. Hu, S. Liu, X. Chen, G. Mo, Z. Xie and X. Jing,

Biomacromolecules, 2008, 9, 553–560 .

160. A. L. Z. Lee, S. Venkataraman, S. B. M. Sirat, S. Gao, J. L. Hedrick and Y. Y. Yang, *Biomaterials*, 2012, 33, 1921–1928 .

161. E. J. Vandenberg and D. Tian, *Macromolecules*, 1999, 32, 3613–3619 .

162. W. Chen, F. Meng, R. Cheng and Z. Zhong, *J. Controlled Release*, 2010, 142, 40–46 .

163. J. Xiong, F. Meng, C. Wang, R. Cheng, Z. Liu and Z. Zhong, *J. Mater. Chem.*, 2011, 21, 5786–5794 .

164. Y. Wu, W. Chen, F. Meng, Z. Wang, R. Cheng, C. Deng, H. Liu and Z. Zhong, *J. Controlled Release*, 2012 DOI:10.1016/j.jconrel.2012.07.011.

165. J. Cheng, X. Wang, W. Chen, F. Meng, C. Deng, H. Liu and Z. Zhong, *J. Mater. Chem.*, 2012, 22, 11730–11738

CHAPTER 3 GROWING HYPERBRANCHED POLYMERS USING NATURAL SUNLIGHT

Jun-Jie Yan[1], Jiao-Tong Sun[1], Ye-Zi You[1], De-Cheng Wu[2], and Chun-Yan Hong[1]

[1]CAS Key Laboratory of Soft Matter Chemistry, Department of Polymer Science and Engineering, University of Science and Technology of China, Hefei, 230026, Anhui, P. R. China

[2]Beijing National Laboratory for Molecular Sciences, State Key Laboratory of Polymer Physics & Chemistry, Institute of Chemistry, Chinese Academy of Sciences, Beijing 100190, China

ABSTRACT

In nature, a sapling can grow into a big tree under irradiation of sunlight. In chemistry, a similar concept that a small molecule only exposing to sunlight grows into a hyperbranched macromolecule has not been realized by now. The achievement of the concept will be fascinating and valuable for polymer synthesis wherein sunlight is inexpensive, abundant, renewable, and nonpolluting. Herein, we report a new strategy in which small monomers can directly grow into big hyperbranched macromolecule under irradiation of sunlight without any catalyst.

INTRODUCTION

Natural photosynthesis of converting solar energy to chemical energy has inspired many chemists to devote their efforts to mimicking its behavior for synthesizing organic molecules or macromolecules1·2·3·4·5. As early as 1912, Ciamician recognized that sunlight had the potential to serve as an inexpensive, abundant, renewable, and nonpolluting energy for chemical synthesis5. However, most conventional organic molecules lack the ability to absorb visible light, greatly restricting the application of sunlight. Currently, most developed efficient organic photo-syntheses and photo-polymerizations typically require high-intensity UV light6. Recently, some significant progresses in photosynthesis are to employ photo-catalysts with good abilities of absorbing visible light for sensitizing organic molecules to achieve photochemical reactions or photo-polymerizatio ns7·8·9·10·11·12·13·14·15·16·17·18. Although these as-developed methods are very efficient for photosynthesis of visible light, expensive photosensitive catalysts limit the application on industrial scales. Hyperbranched polymers are one kind of polymers with dendritic structure19·20·21. A sapling can grow into a big tree under irradiation of sunlight. Can small molecules directly grow into hyperbranched polymers when exposed to sunlight without any catalyst?

Recently, it was reported that thiyl radicals can be generated from thiols directly by sunlight shining at low or ambient temperature22·23. It suggests that thiyl radical based thiol-yne and thiol-ene reactions can be triggered by direct sunlight instead of the usual need of UV in the presence of photo-initiator. Aminolysis of *N*-acetylhomocysteine thiolactone is a facile method for the introduction of sulfhydryl groups in natural proteins24·25. The method provides a feasible approach to *in situ* generating active thiol groups. Herein, inspired from nature, we report a novel sunlight-activated thiol-yne and thiol-ene strategy to produce hyperbranched polymers from rationally-designed small molecules. The molecules containing both thiols and alkyne or alkene groups

Citation: Jun-Jie Yan, Jiao-Tong Sun, Ye-Zi You, De-Cheng Wu & Chun-Yan Hong, Growing Hyperbranched Polymers Using Natural Sunlight, doi:10.1038/srep02841.

are *in situ* generated from ring-opening of alkyne or alkene -containing thiolactone monomers by amino monomers. Then the thiols are activated into thiyl radicals under irradiation of sunlight, and the thiyl radicals react with alkyne or alkene groups to obtain the hyperbranched polymers.

RESULTS

The schematic illustration of growing hyperbranched polymer from small molecules under irradiation of sunlight is proposed in Figure 1. An alkyne-containing thiolactone monomer, prop-2-yn-1-yl(2-oxotetrahydrothiophen-3-yl)carbamate (POTC, **1**), is designed and synthesized as a model molecule. A suitable primary amine, *N,N*-dimethylpropane-1,3-diamine (DMPDA), is selected to ring-open thiolactone group of the POTC monomer to yield a AB_2-type intermediate(2) containing one *in situ* generated thiol group and one alkyne group. The thiol can be activated into thiyl radical(3) by sunlight[22,23], and the radical attacks an alkyne to form an intermediate(4) with an alkene group, an alkyene group, and a thiol group. Since alkene possesses much higher reactivity than alkyne[26,27,28], the thiyl radical is easier to react with the formed alkene to get an intermediate(5). The alternate thiol-yne and thiol-ene click reactions should eventually produce a hyperbranched polymer with a larger number of peripheral alkyne groups and 100% degree of branching if the formed alkene groups always have much higher reactivity than the remained alkyne groups regardless of the factors of concentration and steric hindrance. To verify the proposed mechanism, dynamic ^{1}H NMR spectra were *in situ* recorded to monitor polymerization process, and the results were summarized in Figure 2. At an early stage, the formation of the AB_2 intermediate(2) was detected in Figure 2b andFigure 3. At an initial 2 hours, the nearly alternate thiol-yne and thiol-ene click reactions were proven in Figure 2c by decrease of the signal intensity of alkyne protons (δ = 2.89 ppm) and a negligible signal of the formed alkene protons (δ = 5.6 ppm). With increased reaction time, the detectable signals of the alkene protons indicated the polymerization process did not follow the strict alternated thiol-yne and thiol-ene click reactions (see Figure 2d and Figure 3d). Figure 2e showed that 97% of the *in situ* generated thiols were consumed after irradiation of sunlight for cumulative 15 hours, and the as-produced hyperbranched polymers had an average molecular weight to be 15300 with PDI of 1.62. The degree of branching (DB) of the final product was calculated to be ~ 0.92 according to a formula indicated by the corresponding integral ratios of the proton signals of linear alkene units, terminal alkyne units and dendritic units in Figure 1 and Figure 2e[19b].

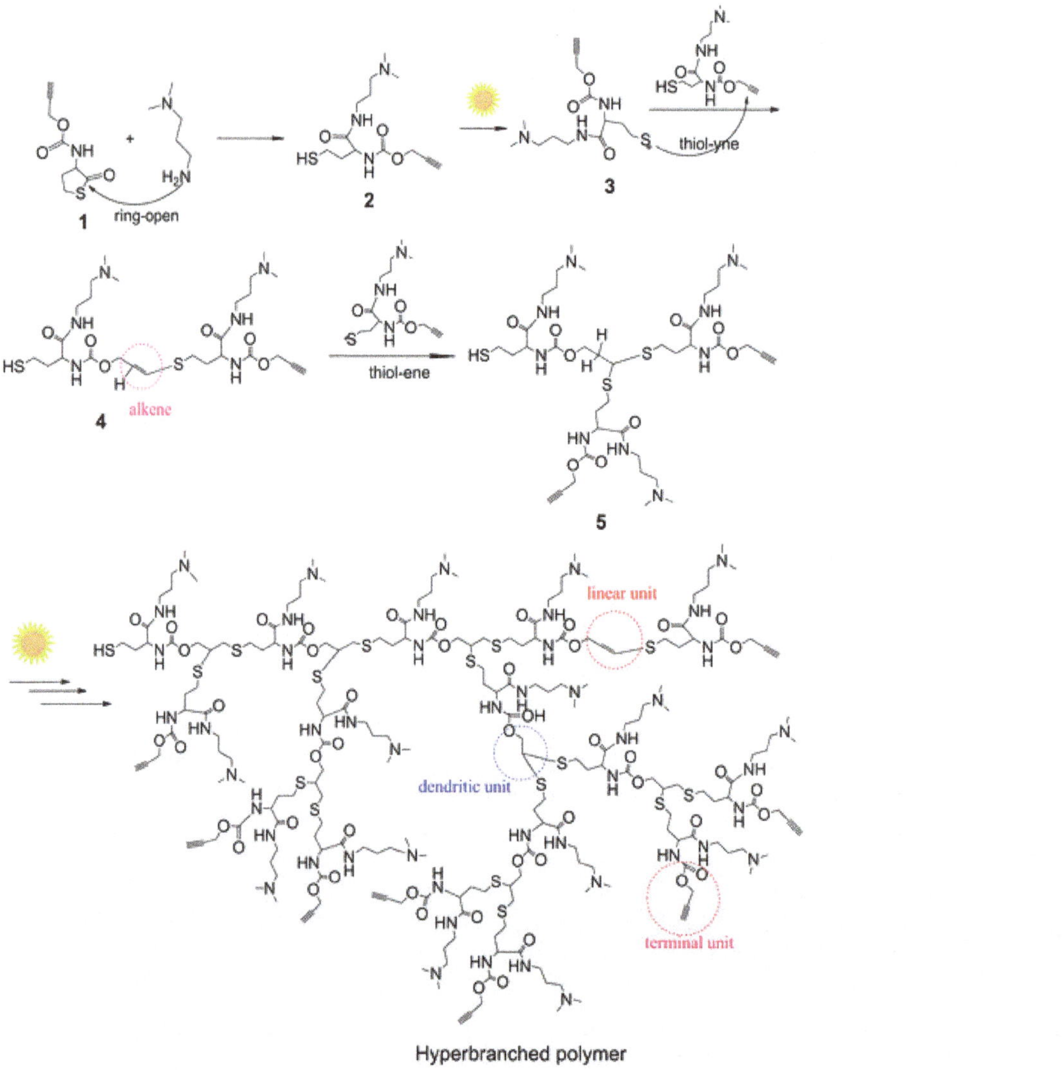

Figure 1: Schematic illustration of growing a hyperbranched polymer via irradiation of the mixture of an alkyne-containing thiolactone and a primary amine by sunlight.

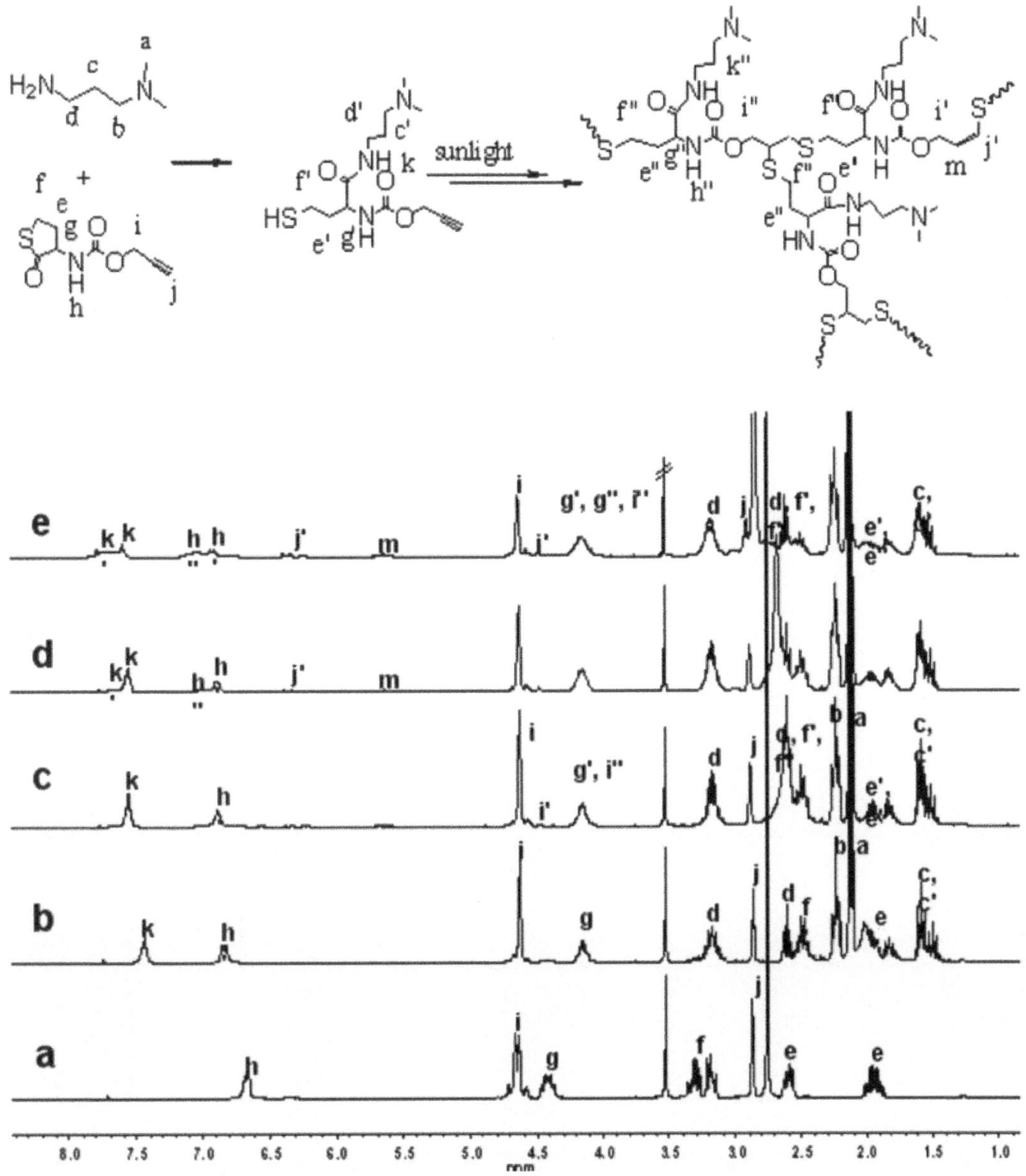

Figure 2: ^{1}H NMR spectra recorded *in situ* **for the reactions of POTC + DMPDA (molar feed ratio is 1:1.25)** and thiol-click in 1,4-dioxane-d_8 **under irradiation of sunlight, the time of sunlight irradiation is 0 min (a), 30 min (b), 110** min (c), 420 min (d) and 900 min (e).

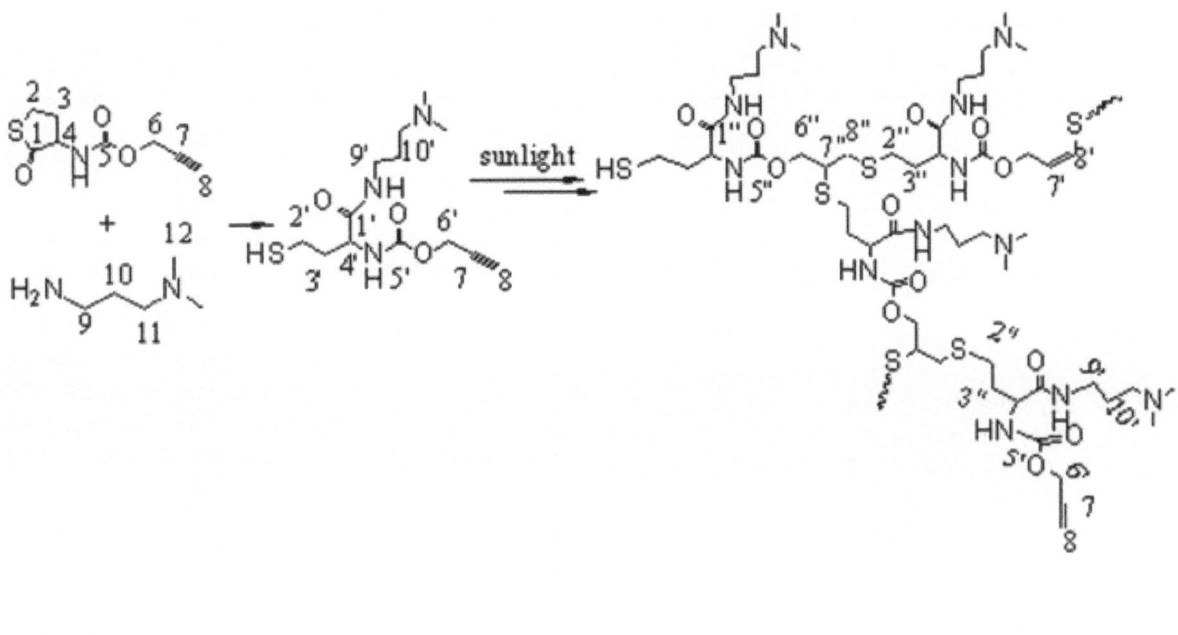

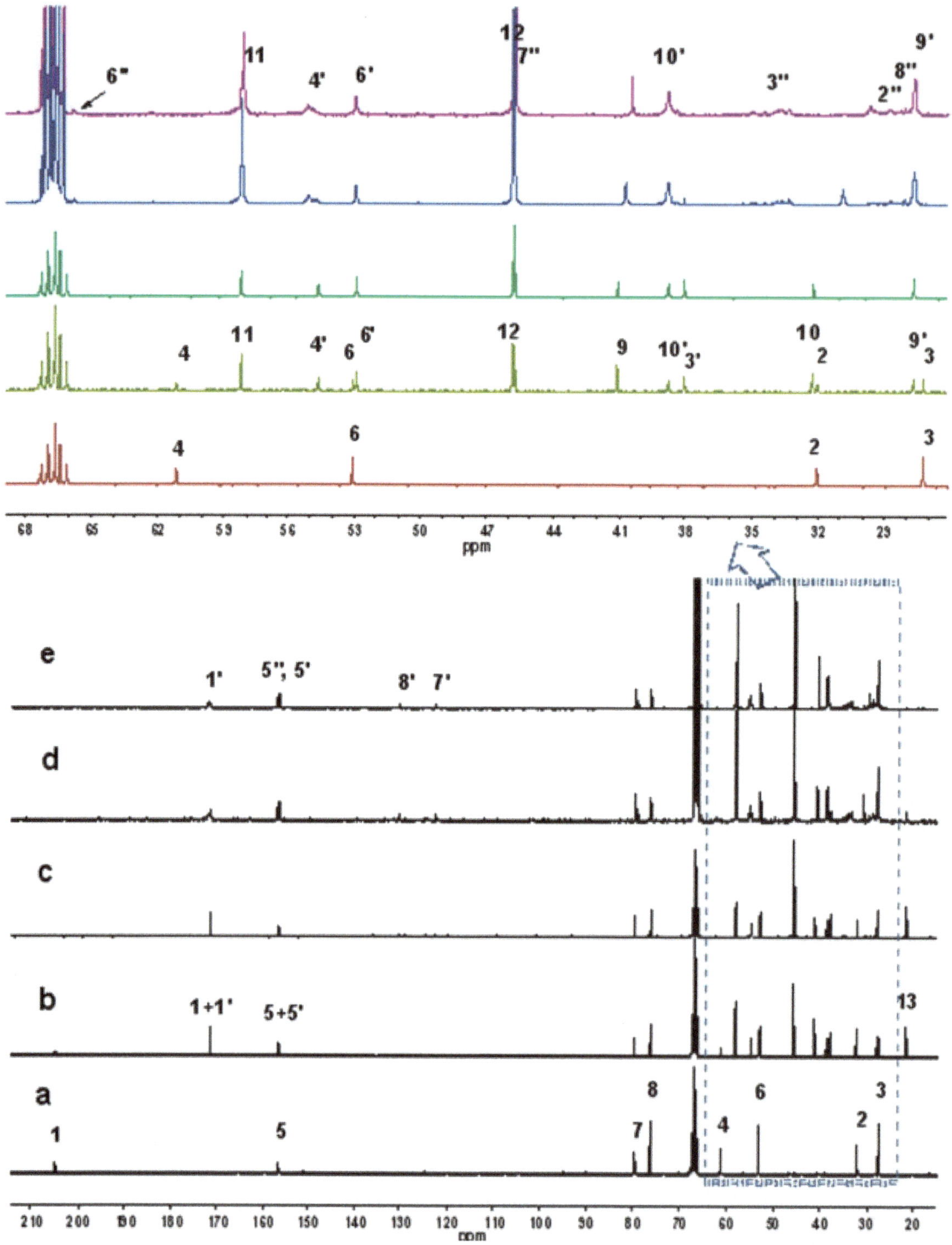

Figure 3: ^{13}C NMR spectra recorded *in situ* **for the reactions of POTC + DMPDA (molar feed ratio is 1:1.25) and** thiol-click in 1,4-dioxane-d_8 **under irradiation of sunlight, the time of sunlight irradiation is 0 min (a), 15 min (b), 110** min (c), 420 min (d) and 900 min (e).

DISCUSSION

In view of the fact that thiols are easily oxidized when exposed to air, we adopt the aminolysis of thiolactone to *in situ* generate thiols. The nucleophilic ring-opening reaction was monitored via dynamic ^{1}H and^{13}C NMR. Figure 2b testified that the aminolysis of thiolactone almost completed within ~ 30 min, due to the corresponding chemical shifts of methine proton of thiolactone (g) from 4.41 ppm to 4.13 ppm (g′), and amine (h) from 6.67 ppm to 6.83 ppm (h′). The result can be further verified by the transformation of the lactone unit (chemical shift at 205 ppm) into amide unit (chemical shift at 172 ppm) in ^{13}C NMR spectra (seeFigure 3) after ring-opening. The aminolysis of thiolactone can be carried out in the presence of primary amines under dark condition, but its rate is a slightly slower than that under irradiation of sunlight. However, secondary and tertiary amines can't ring-open thiolactone even after extended times (e. g., 5 h) (see Supplementary Fig. S10 and S11online).

In this experiment, the generation of thiyl radicals from thiols is of key importance for growing hyperbranched polymers. For identifying whether thiyl radicals are certainly generated from thiols under irradiation of sunlight, some control experiments were carried out. If the radicals can be generated from the employed thiolactone, alkene will form when POTC monomer are directly irradiated by sunlight without adding *N,N*-dimethylpropane-1,3-diamine. However, no reaction was detected for POTC monomer exposing to sunlight for 15 h or heating at 70°C for 10 h without the amine (see Supplementary Fig. S12 and S13 online). Therefore, we deduce that the thiyl radical is produced from *in situ* generated thiols when exposed to sunlight, which coincides with the previous findings[22,23]. Sunlight is a necessary external stimulus to activate thiols into thiyl radicals, so controlling exposure to sunlight irradiation or not provides a facile and efficient way to manipulate the formation of thiyl radicals, resulting in 'on/off' function of thiol-yne reaction. Thereby, the consumption of thiols and the growth of hyperbranched polymers can be stopped and recovered. As shown in Figure 3 and Figure 4a, the amounts of the thiols decreased with the increase of irradiation time while it remained unchanged under dark condition after removal from sunlight, and the consumption of thiols continued when sunlight was resumed. About 97% of thiols were consumed after irradiation of sunlight for cumulative 15 h. Similar result was also obtained from the variation of molecular weight of the hyperbranched polymer. Figure 4b show that the weight average molecular weight (M_w) of the polymer is 3600 with DB of 98% and PDI of 1.32 when the polymerization mixture was irradiated by sunlight for first 7 h, and its M_w remained unchanged under dark condition, indicating growth termination of the hyperbranched polymer. The growth of molecular weight recovered after resumption of sunlight, and its M_w reached 8200 with DB of 94.8% and PDI of 1.51 under irradiation of next 4 h. The growth stopped in the dark condition for another 20 h, and recovered again after resumption of sunlight and produced a final hyperbranched polymer with a M_w to be 15300, DB of 92% and PDI of 1.62 under sunlight irradiation for cumulative 15 h. These results demonstrate that this approach to grow hyperbranched polymers is a controlled process.

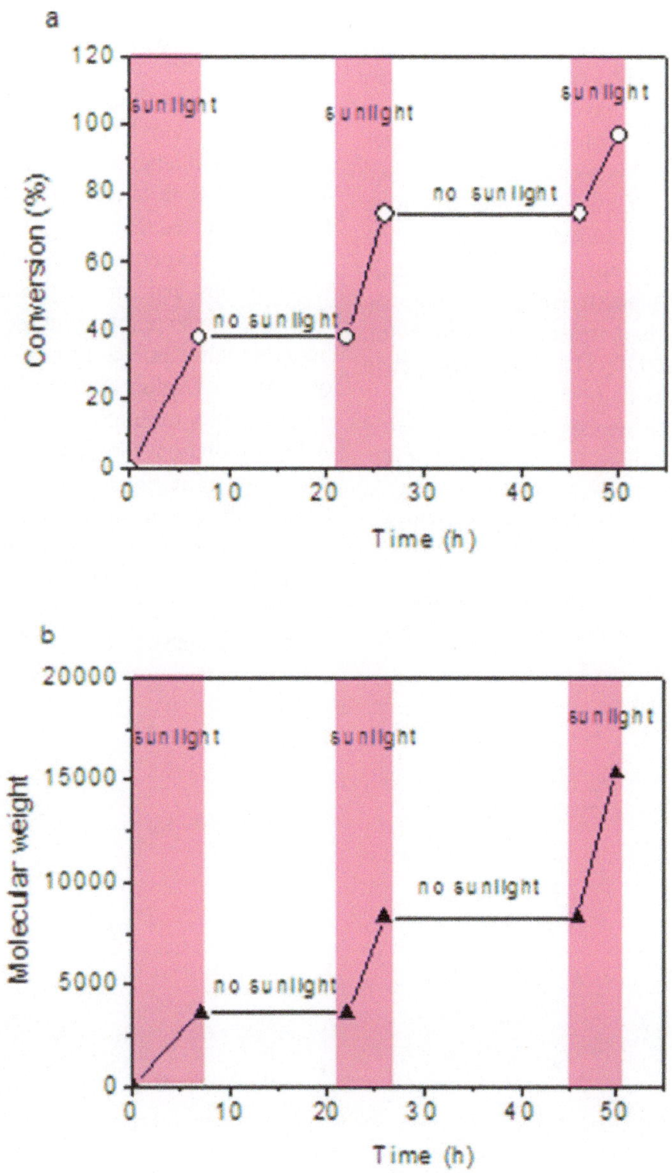

Figure 4: The variations of (A) thiol amount and (B) molecular weight of the formed hyperbranched polymer as a function time with/without irradiation of sunlight.

In a previous report, the polymers obtained via thiol-yne polymerization under UV irradiation does not have high molecular weight26·27, which may result from that there are some side reactions in the presence of photo/thermal initiators25, but the DB could reach 100% based on thiol-yne polymerization under UV irradiation because thiol-ene reaction is much faster than thiol-yne reaction27. In this experiment, there is some residual alkenes in the final hyperbranched polymer (Figure 2e, alkene, δ = 5.0–6.5 ppm), and the DB is determined to be 92 ~ 98% for the obtained polymers. The residual alkenes may result from two factors. The concentration of alkynes is greatly higher than that of alkenes near the end of polymerization as well as higher steric hindrance of the alkenes located inside polymer backbones, reducing differentiation of the reactivity between thiol-yne and thiol-ene reactions.

This method of growing hyperbranched polymers is a versatile approach. General primary amines were suitable for the system, such as 1-(2-aminoethyl)piperazine (AEPZ) and *N'*-(3-aminopropyl)-*N*,*N*-dimethylpropane-1,3-diamine (DMDPTA). The M_w of the corresponding hyperbranched polymers are 22000 and 21000 with the DBs of 0.88 and 0.91, PDIs of 2.5 and 2.17 for the polymerization of the mixtures of POTC with AEPZ and DMDPTA, respectively. The as-produced hyperbranched polymers have excellent aqueous solubility, providing prerequisites of functional candidates for bio-applications. Furthermore, this method can be applied in preparing functionalized biocompatible polymers. Amino ended PEG (PEG-NH_2, M_n = 600) was used as a ring-opening agent for thiolactone, and the results verified that PEG-NH_2 has a similar reactivity as small amino compounds. Viscosity of the reactants obviously increased after irradiation of sunlight for 15 h, and an eventual PEG based hyperbranched polymer had a M_w of 12000, DB of 0.95 and PDI of 1.61. Glucosamine was also employed in synthesis of glucose-based hyperbranched polymers using sunlight. The mixture of POTC and glucosamine in DMF was irradiated by sunlight for 20 h. The glucose-based hyperbranched polymer obtained had a M_w of 16200, DB of 0.98 and PDI of 3.3.

The method of preparing hyperbranched polymers by sunlight is also applicable for the systems only based on thiol-ene reaction. The mixture of *N*-(allyloxy)carbonylhomocysteine thiolactone (NACHT) and PEG with amine and vinyl groups at two ends exposing to sunlight also produces a hyperbranched polymer, and the approach is clearly illustrated in Figure 5. First, amine of PEG ring-opens *N*-(allyloxy) carbonylhomocysteine thiolactone, yielding an intermediate containing two alkene units and one thiol unit (A_2B, Figure 5), which can be verified by the NMR results shown in Figure 6. After the ring-opening, the corresponding chemical shifts of methine proton of thiolactone from 4.41 ppm (e) to 4.13 ppm (e′) (see Figure 6). The result can be further verified by the transformation of the lactone unit (chemical shift at 205 ppm) into amide unit (chemical shift at 172 ppm) in ^{13}C NMR spectra (see Supplementary Fig. S14 online), the chemical shift of carbon (1) next to sulfur in thiolactone at 32 ppm moving to the carbon (26) next to thiol at 20 ppm, the chemical shift of carbon (2) in thiolactone at 28 ppm moving to the carbon (25) at 38 ppm. All these results indicate the formation of A_2B intermediate. Subsequently, the thiols transfer into thiyl radicals by sunlight irradiation, triggering thiol-ene reaction to yield hyperbranched polymer with a M_w of 24500 and PDI of 5.0. It is clear that the signal for carbon (25) next to thiol decreased with the increase of irradiation time (see Supplementary Fig. S14), indicating thiol added onto alkene. On the other hand, the peaks of vinyl unit located at 5.0–6.0 ppm before thiol added onto alkene, some of them moved to the peaks at 1.7–2.0 ppm after thiol added onto alkene (see Figure 6), which further indicated that sunlight triggered thiol-ene reaction. The as-obtained hyperbranched polymer contains many terminal vinyl groups that can be further functionalized via thiol-ene reaction. For example, glucose can be easily grown onto the hyperbranched polymer via sunlight irradiation of the mixture of 1-thio-α-D-glucopyranose (α-GlcSH) and hyperbranched polymer with terminal vinyl groups under argon atmosphere. After irradiation for 12 h, the alkene units are completely consumed. The signals of protons of alkene at 5.9 ppm are completely absent with the appearance of typical proton signals of glucose between 4.1 and 5.6 ppm (see Supplementary Fig. S16 online). After purification via dialysis (MW cut-off: 1000) against water for 2 days, the glucose-functionalized polymer has a molecular weight of 31000.

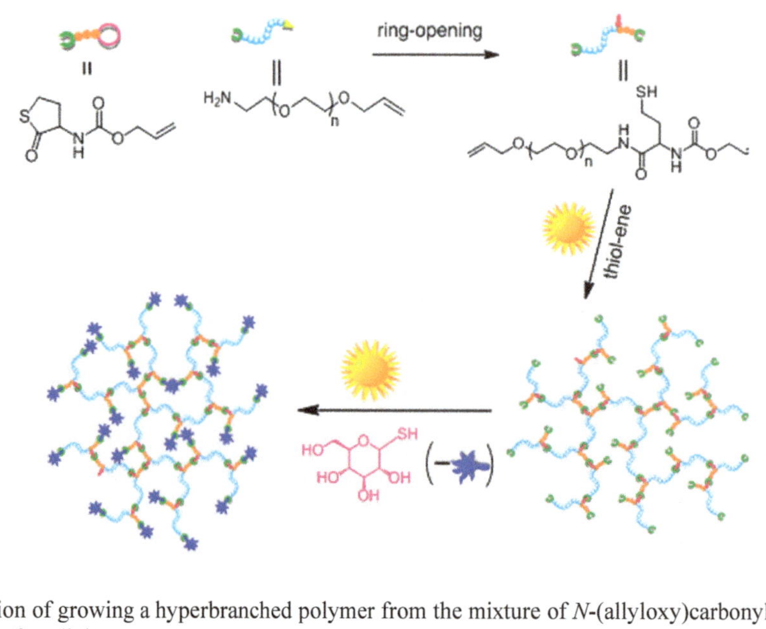

Figure 5: Schematic illustration of growing a hyperbranched polymer from the mixture of *N*-(allyloxy)carbonylhomocysteinethiolactone and amino-PEG under irradiation of sunlight.

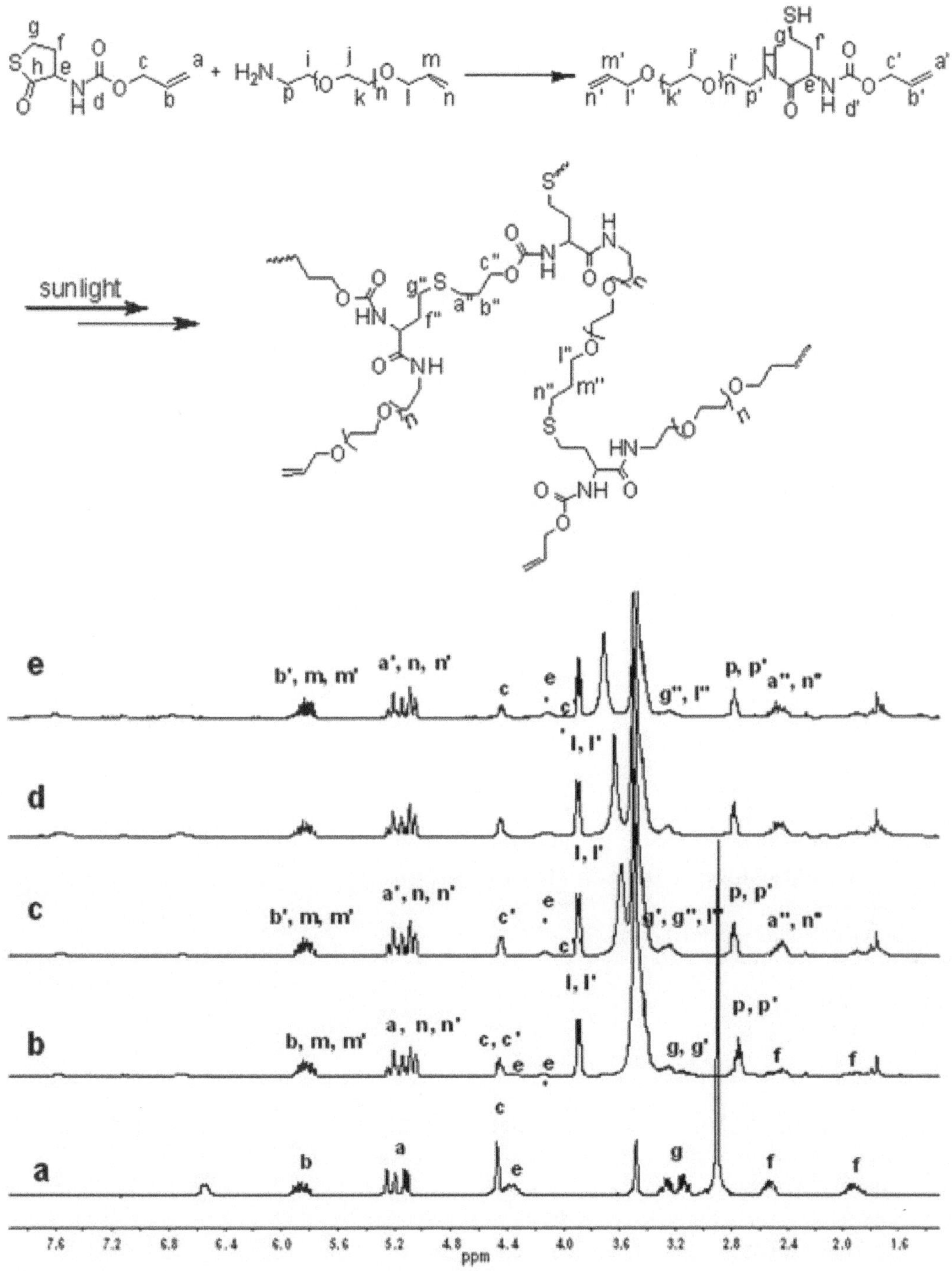

Figure 6: ^{1}H NMR spectra recorded *in situ* **for the reaction of** *N*-(allyloxy)carbonylhomocysteine thiolactone with PEG-amine and thiol-ene reaction in 1,4-dioxane-d_8 **under irradiation of sunlight, the irradiation time is 0 min (a), 30 min (b), 110 min (c), 420 min (d), and 900 min (e).**

In summary, we report a novel strategy to grow hyperbranched polymers from small molecules directly under irradiation of sunlight without any photocatalyst. This facile, innovative strategy offers a versatile platform for synthesizing highly functional hyperbranched polymers for bioscience application, and it is green and friendly to environment in a sustainable manner.

METHODS

The syntheses of monomers are shown in Supplementary Materials. All the polymerizations are carried out under irradiation of natural sunlight (weather conditions: sunny; temperature is 25 ~ 36°C; the highest temperature of the reaction mixture is ~ 55°C). A typical polymerization procedure is as following: Prop-2-yn-1-yl (2-oxo tetrahydro thiophen-3-yl)carbamate (50.2 mg, 0.252 mmol) and *N,N*-dimethyl-1,3-propanediamine (32.1 mg, 0.315 mmol) were dissolved in 1,4-dioxane-d_8(550 μL) in a NMR tube filled with argon atmosphere. The tube was sealed and the reaction was then subjected to natural sunlight. The reaction was monitored via ^{1}H NMR and ^{13}C NMR. The degree of branching (DB) of the as-produced polymer was calculated by the following Formula as originally defined by Fréchet et al.19

$$DB = \frac{I_{dendritic} + I_{terminal}}{I_{dendritic} + I_{terminal} + I_{linear}} \times 100\%$$

(I denotes the integral value of protons).

ACKNOWLEDGEMENTS

Financial support from National Natural Science Foundation of China (51033005, 20974103, 21074121 and 21090354), the National 'Young Thousand Talents Program', and the Program for New Century Excellent Talents in Universities (NCET-08-0520) is gratefully acknowledged.

REFERENCES

1. Kim, J. H., Lee, M., Lee, J. S. & Park, C. B. Self-assembled light-harvesting peptide nanotubes for mimicking natural photosynthesis. *Angew. Chem. Int. Ed.* 51, 517–520 (2012).
2. Freemantle, M. Mimicking natural photosynthesis. *Chem. Eng. News* 76, 37–46 (1998).
3. Zhou, H. *et al.* Artificial inorganic leafs for efficient photochemical hydrogen production inspired by natural photosynthesis. *Adv. Mater.* 22, 951–956 (2010).
4. Gust, D., Moore, T. A. & Moore, A. L. Mimicking photosynthetic solar energy transduction. *Acc. Chem. Res* 34, 40–48 (2001).
5. Ciamician, G. The photochemistry of the future. *Science* 36, 385–394 (1912).
6. Nicewicz, D. A. & MacMillan, D. W. C. Merging photoredox catalysis with organocatalysis: The direct asymmetric alkylation of aldehydes. *Science* 322, 77–80 (2008).
7. Ischay, M. A., Anzovino, M. E., Du, J. N. & Yoon, T. P. Efficient visible light photocatalysis of [2 + 2] enone cycloadditions. *J. Am. Chem. Soc.* 130, 12886–12887 (2008).
8. Narayanam, J. M. R. & Stephenson, C. R. J. Visible light photoredox catalysis: applications in organic synthesis. *Chem. Soc. Rev.* 40, 102–112 (2011).
9. Zhang, G., Song, I. Y., Ahn, K. H., Park, T. & Choi, W. Free radical polymerization initiated and controlled by visible light photocatalysis at ambient temperature. *Macromolecules* 44, 7594–7599 (2011).
10. Wang, X. C. *et al.* A metal-free polymeric photocatalyst for hydrogen production from water under visible light. *Nature Materials* 8, 76–80 (2009).
11. Yoon, T. P., Ischay, M. A. & Du, J. N. Visible light photocatalysis as a greener approach to photochemical synthesis. *Nature Chemistry* 2, 527–532 (2010).
12. Asahi, R., Morikawa, T., Ohwaki, T., Aoki, K. & Taga, Y. Visible-light photocatalysis in nitrogen-doped titanium oxides. *Science* 293, 269–271 (2001).
13. Zhang, G., Song, I. Y., Ahn, K. H., Park, T. & Choi, W. Free radical polymerization initiated and controlled by visible light photocatalysis at ambient temperature. *Macromolecules* 44, 7594–7599 (2011).
14. Wang, Y., Li, L., Yang, K., Samuelson, L. A. & Kumar, J.Nanocrystalline TiO_2-catalyzed solid-state polymerization of diacetylene in the visible region. *J. Am. Chem. Soc.* 129, 7238–7239 (2007).
15. Shi, Y., Liu, G., Gao, H., Lu, L. & Cai, Y. Effect of Mild visible light on rapid aqueous RAFT polymerization of water-soluble acrylic monomers at ambient temperature: initiation and activation.*Macromolecules* 42, 3917–3926 (2009).
16. Yilmaz, G. *et al.* Thioxanthone-fluorenes as visible light photoinitiators for free radical polymerization. *Macromolecules*43, 4520–4526 (2010).
17. Tehfe, M.-A., Lalevée, J., Gigmes, D. & Fouassier, J. P. Green chemistry: sunlight-induced cationic polymerization of renewable epoxy monomers under air. *Macromolecules* 43, 1364–1370 (2010).
18. Fors, B. P. & Hawker, C. J. Control of a living radical polymerization of methacrylates by light. *Angew. Chem. Int. Ed.* 51, 8550–8553 (2012).
19. Hawker, C. J., Lee, R. & Fréchet, J. M. J. One-step synthesis of hyperbranched dendritic polyesters. *J. Am. Chem. Soc.* 113, 4583–4588 (1991).
20. Kim, Y. H. & Webster, O. W. Hyperbranched polyphenylenes.*Polym Prepr* 29, 310–311 (1988).
21. Voit, B. I. Dendritic polymers: hyperbranched polymers: a chance and a challenge. *C. R. Chimie* 6, 821–832 (2003).
22. Ten Brummelhuis, N., Diehl, C. & Schlaad, H. Thiol-ene modification of 1,2-polybutadiene using UV light or sunlight.*Macromolecules* 41, 9946–9947 (2008).
23. Bowman, C. N. & Hoyle, C. E. Thiol-ene click chemistry. *Angew. Chem. Int. Ed.* 49, 1540–1573 (2010).
24. Benesch, R. & Benesch, R. E. Thiolation of protein. *Proc. Natl. Acad. Sci. U.S.A.* 44, 848–853 (1958).
25. Espeel, P., Goethals, F. & Du Prez, F. E. One-pot multistep reactions based on thiolactones: extending the realm of thiol-ene chemistry in polymer synthesis. *J. Am. Chem. Soc.* 133, 1678–1681 (2011).
26. Han, J., Zhao, B., Tang, A. J. & Gao, C. Fast and scalable production of hyperbranched polythioether-ynes by a combination of thiol-halogen click-like coupling and thiol-yne click polymerization.*Polym. Chem.* 3, 1918–1925 (2012).
27. Konkolewicz, D., Gray-Weale, A. & Perrier, S. Hyperbranched polymers by thiol-yne chemistry: from small molecules to functional polymers. *J. Am. Chem. Soc.* 131, 18075–18076 (2009).
28. Rahane, S. B., Hensarling, R. M., Sparks, B. J., Stafford, C. M. &Patton, D. L. Synthesis of multifunctional polymer brush surfaces*via* sequential and orthogonal thiol-click reactions. *J. Mater. Chem.* 22, 932–943 (2012).

CHAPTER 4

BLOCK COPOLYMERS: CONTROLLING NANOSTRUCTURE TO GENERATE FUNCTIONAL MATERIALS – SYNTHESIS, CHARACTERIZATION, AND ENGINEERING

Thomas H. Epps, III[a] and Rachel K. O›Reilly[b]

[a]Department of Chemical and Biomolecular Engineering and Department of Materials Science and Engineering, University of Delaware, Newark, Delaware 19716, USA

[b]Department of Chemistry, University of Warwick, Gibbet Hill, Coventry, CV4 7AL, UK

ABSTRACT

In this perspective, we survey recent advances in the synthesis and characterization of block copolymers, discuss several key materials opportunities enabled by block copolymers, and highlight some of the challenges that currently limit further realization of block copolymers in promising nanoscale applications. One significant challenge, especially as the complexity and functionality of designer macromolecules increases, is the requirement of multiple complementary techniques to fully characterize the resultant polymers and nanoscale materials. Thus, we highlight select characterization and theoretical methods and discuss how future advances can improve understanding of block copolymer systems. In particular, we consider the application of theoretical/simulation methods to the rationalization, and prediction, of observed experimental self-assembly phenomena. Finally, we explore several next steps for the field and emphasize some general areas of emerging research that could unlock additional opportunities for nanostructure-forming block copolymers in functional materials.

INTRODUCTION TO BLOCK COPOLYMERS

Nature uses molecular self-assembly to create precision nanostructures, craft unique compartmentalized environments, and build large constructs through hierarchical assembly. Indeed, recent developments in nanotechnology have mimicked natural approaches by utilizing nucleic acid sequence specificity to create higher order structures through the creation of DNA origami tiles, other three-dimensional structures, and nanomechanical molecular devices.[1] Inspired by these recent advances and motifs, considerable efforts have been initiated to recreate such concepts using synthetic building blocks such as small-molecule surfactants and block copolymers (BCPs).

Small-molecule surfactants and BCPs are both fashioned from two or more chemically dissimilar constructs that are covalently-bonded into a single molecule. Through a delicate mix of molecular interactions and materials processing, these molecules form a variety of nanoscale structures. The linking of constitutionally different units permits the combination of distinct properties within a macromolecule and enables interesting nanoscale assembly phenomena, and ultimately, unique macroscale behavior.[2] Importantly, many of the unique physical properties inherent to BCP materials are a result of the nanoscale hybridization of their components and cannot be accessed through simple blending of non-bonded blocks. Furthermore, the increased number of repeat units in BCPs, compared to small molecules, leads to dramatically improved morphological stability. This stability provides significant opportunities for BCP utilization in a broad range of environments but also necessitates new

Citation: Thomas H. Epps III and Rachel K. O'Reilly, Block copolymers: controlling nanostructure to generate functional materials – synthesis, characterization, and engineering, DOI: 10.1039/c5sc03505h.

methodologies to control the precise assembly of organized nanostructures. Indeed, through advances in polymer synthesis, functionalization, processing, and characterization, it is now possible to design, fabricate, and explore a vast array of BCPs with diverse and sophisticated self-assembly potential in bulk, thin film, and dilute solution environments. In Fig. 1 for example, the combination of multiple polymerization methods and targeted coupling chemistries enables the synthesis of complex macromolecules such as tapered BCPs (top left), the use of solvent processing recipes on motorized stages permits the generation of directed nanostructures for thin film templating (top right), the cryogenic transmission electron microscopy of cylindrical BCP micelles allows one to visualize nanostructure formation in solution environments (bottom right), and self-consistent field theory simulations of solvent removal in a cylinder-forming BCP thin film informs structure/processing relationships (bottom left). *(We also note that many hydrogel and concentrated solution systems contain BCPs, but discussion of those materials is not included in this work.)*

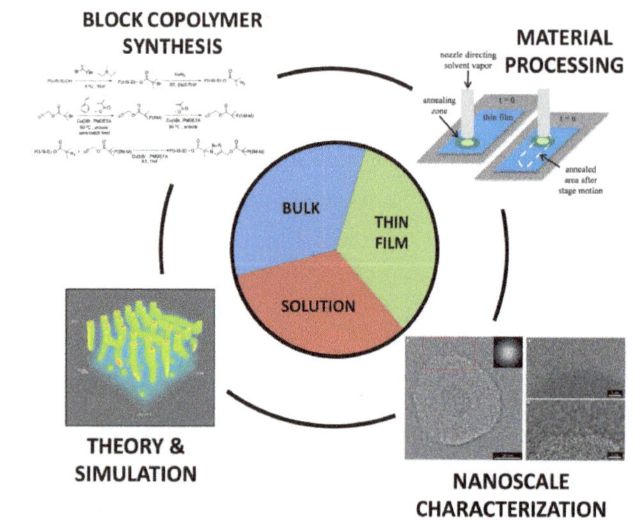

Figure 1: Key aspects of block copolymer materials design. The synthesis, material processing, nanoscale characterization, and theory/simulation of block copolymers are all crucial in the development of new hierarchically assembled structures. These factors are intimately linked to the informed design of materials for bulk, thin film, and solution applications. Top right image – reprinted with permission from ref. 89 © American Chemical Society; bottom right image – reprinted with permission from ref. 47 © Royal Society of Chemistry; bottom left image – reprinted with permission from ref. 88 © American Chemical Society; top left image – reprinted with permission from ref. 77 © American Chemical Society.

Bulk

Bulk assembly has been studied extensively for over 50 years, and the phase behavior of traditional A–B diblock copolymers is well-researched both theoretically and experimentally. In conventional bulk materials self-assembly processes are governed by an unfavorable mixing enthalpy coupled with entropic losses due to macromolecular junctions and chain stretching. Current commercial applications of bulk BCPs (diblock and multiblock) include thermoplastic elastomers for gaskets, cable insulation, footwear, blending, adhesives, automotive bumpers, snowmobile treads, *etc.* (*e.g.* Kraton™, Styroflex™, Solprene™, Hytrel™, Engage™, Sofprene™); thermoplastics for medical devices, protective headgear, and piping systems; and elastomers for car tires (*e.g.* poly(styrene-*b*-isoprene-*b*-butadiene) rubber [SIBR] from Goodyear) among others. For the case of thermoplastic elastomers, BCPs enable the facile and low-cost generation of a myriad of application-specific recyclable, flexible, thermoformed (or blow-moldable), creep-resistant, and durable materials as compared to conventional thermosets, primarily as a result of prescribed nanoscale phase separation.

Well-defined BCPs can phase separate into a variety of periodic and nanoscale morphologies according to the relative composition of the blocks, the overall degree of polymerization (*N*), the polymer–polymer interaction parameter ( ), and the ratio of statistical segment lengths.[3] The conceptual ability to tune morphology through adjustments in chemical composition allows one to generate materials tailored toward thermoplastic elastomer, membrane, and other applications. Though the mechanism underlying nanostructure formation for the simplest classes of BCPs with non-specific interactions is well understood, new macromolecular designs that incorporate multiblocks,[4] copolymer mixtures,[5] specific interactions (hydrogen-bonding,[6] π–π stacking,[7] *etc.*), engineered dispersity,[8] tapered segment profiles,[9] sequence-controlled monomer distributions,[10] small-molecule dopants (*e.g.* salts, plasticizers, and inorganic precursors),[11–13] and functional end-groups[14] significantly complicate the understanding of bulk macromolecular assembly, yet potentially yield exciting opportunities for novel material designs such as active membranes for batteries and fuel cells,[15] catalyst supports/scaffolds,[16,17] actuators, and self-healing or shape-memory systems.[18,19]

Thin Film

In addition to the factors that affect assembly in bulk, self-assembly processes in thin films (<several hundred nanometers in thickness) are strongly influenced by surface energetics and film thickness (commensurability) considerations.[20] The incorporation of these additional driving forces, as well as significant processing history effects due to kinetic limitations associated with thin film assembly, has a substantial impact on nanostructure formation and surface topology in thin films. Significant recent progress has been made in manipulating BCP thin film morphologies for media storage, photonics, nanolithography, nanotemplating, and ultrafiltration applications. A few examples for which thin film BCPs have received interest in high value industrial processing are: conventional chip manufacturing,[21] nanotemplating for dense bit-patterned media that could facilitate doubling of hard disk drive densities,[22] and nanolithography for patterning next-generation semiconductor devices.[23] Despite these industrial endeavors, each with its own challenges,[24] distinct hurdles that limit wide-spread usage of BCP thin films in emerging technologies remain.[25] These hurdles include precise control over the directed assembly of nanoscale domains through cost-effective and scalable approaches, understanding the influence of nanostructure formation dynamics and processing protocols (such as the influence of various annealing recipes,

film casting methods, surface energetics, and polymer molecular weights and architectures) on morphology and orientation, elimination (or significant reduction) of defects, and translation of nanopatterning techniques to non-traditional substrates (*e.g.* flexible substrates, porous scaffolds, graphene, metals).[24,26,27] A variety of surface fields, thermal and solvent annealing (uniform, gradient, and zone-annealing) protocols,[26] and other external fields (*e.g.* magnetic, electric, mechanical, *etc.*)[28–30] have been employed to manipulate nanoscale morphology, orientation, and ordering in BCP thin films. Further understanding of the many parameters associated with these tools will permit additional BCP applications in coatings,[31] nanoporous membranes,[32] anti-fouling materials,[33] and analytical and process-scale separation membranes,[34] electronics,[35,36] and optoelectronics,[7,37] including complex circuits,[38,39] stretchable/flexible electronics,[40] optic and acoustic wave guides,[41] sensors,[42] and holographic arrays:[43] all areas in which three-dimensional or hierarchical structures can transform materials design.[32]

Surface grafted (brush) block copolymers also are receiving significant interest as nanostructured surface coatings for drag reduction, surface energy modification, biosensing, and cellular manipulation applications.[44,45] These brush systems incorporate an additional variable in terms of block order (even in diblock copolymer systems) as a consequence of attachment to a substrate and typically do not possess the same kinetic limitations inherent in BCP films. Some key challenges include high-throughput synthesis and characterization of block copolymer brush systems with high reproducibility and accuracy, generating a detailed knowledge-base of factors that influence phase behavior and surface properties, and adapting block copolymer brush systems for specific applications.[44,45]

Dilute Solution

Dilute solution self-assembly is similarly complex because the introduction of a single solvent, or multiple solvents, dramatically affects the assembly process and the resultant nanostructures.[2,46] This added complexity is largely due to additional surface tension, interaction parameter, and entropy effects that can significantly alter the formation and stability of macromolecular assemblies in solution.[44,47] Thermodynamic and kinetic constraints can lead to the formation of various aggregates such as micelles and vesicles, related to the spherical, cylindrical, and lamellar morphologies, which are found in bulk systems;[2,48] although exotic structures such as helices,[49] toroids,[50] and networks[48,50] also have been reported. The majority of research has focused on aqueous self-assembly, for which the driving force is primarily hydrophobic interactions; however, there have been significant efforts to examine BCP assembly in organic solvents,[51] ionic liquids,[52] supercritical solvents,[53] mixed solvents, and during polymerization.[54] The free-energy landscape, as accessed through assembly protocols, solution processing (*e.g.* agitation and shear), hydrogen-bonding, reversible and irreversible stimuli-responses, salts, and cross-linking are critical in determining the final state of solution assemblies.[44,46,55] Several key advantages of BCP solution assemblies (as opposed to low molecular weight surfactant assemblies) include low critical aggregation concentrations (CAC)s and slow inter-aggregate chain exchange in highly selective solvents such as water.[2] Additionally, high loading capacities coupled with the ability to incorporate a myriad of functionalities and BCP compositions and architectures enables efficient bottom-up strategies to synthesize surfactants for interfacial stabilization such as commercially-relevant Pluronics™ and Tetronics™ and nanocontainers for biological (therapeutic agent delivery, imaging, diagnostics, theranostics) catalysis, separations, and self-healing applications.[44,46,56] Unfortunately, the slow dynamics that normally are advantageous in producing stable nanocarriers also lead to kinetically trapped structures,[46,57] thus such systems require careful optimization of preparation conditions to produce well-defined, uniform, and reproducible solution assemblies.

Overall, the substantial need for new materials with well-defined and predictable nanoscale and macroscale characteristics has stimulated further study of macromolecular assemblies in bulk, thin film, and solution environments, as all arenas are poised to engender ground-breaking technological and societal impacts. The on-going fabrication of more exotic, hierarchical, and nature-inspired BCPs provides tantalizing glimpses toward emerging applications enabled through complex morphology generation; however, further efforts linking synthesis, nanostructure fabrication, processing, characterization, and theory (see Fig. 2) are necessary to unlock the full potential of BCPs. In the following sections, we describe various methods for the synthesis of BCPs, highlight key tools that enable nanoscale characterization of self-assembling soft materials, discuss select contributions that link experiment with theory, simulation, and modelling, and emphasize several emerging directions for BCP activities. The focus is on recent literature and is not meant to ignore the wealth of seminal investigations that provide the inspiration for the work discussed herein. For the sake of brevity, much of the foundational work can be found in the references sections of the literature highlighted in each of the topics below. Additionally, the examples provided below are not meant to be all-inclusive, but instead are selected to provide a snapshot of the power and challenges associated with experimental and theoretical development of BCP nanostructures for wide-ranging materials.

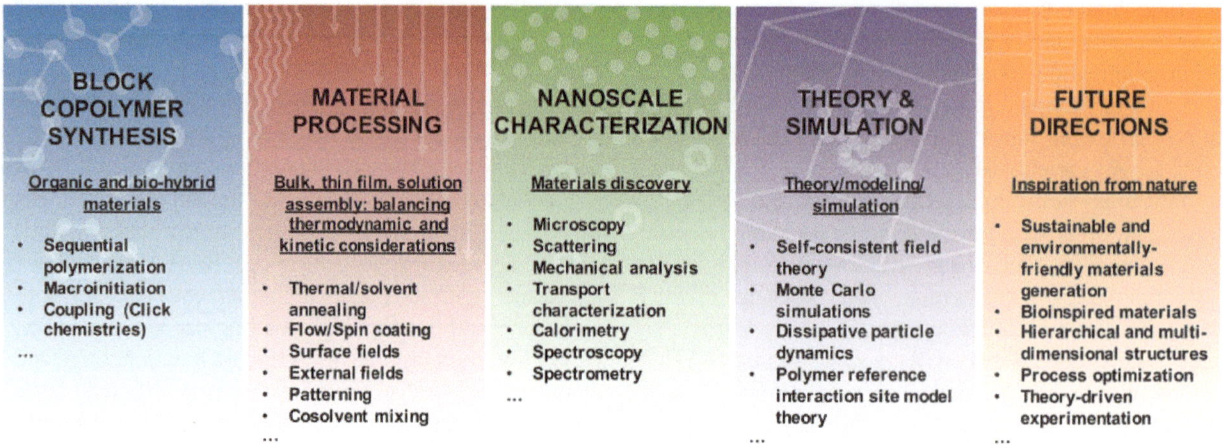

Figure 2: Generating functional materials from block copolymers. Highlighted features of block copolymers are categorized into synthetic approaches to produce novel materials, processing routes to manipulate nanoscale arrangement, characterization methods to obtain detailed nanoscale information, theoretical and simulation-based approaches to improve fundamental understanding, and directions for future research.

SYNTHESIS OF BLOCK COPOLYMERS – DESIGNER MACROMOLECULES WITH UNIQUE PROPERTIES

While, the synthesis of BCPs is an established field with many major advances over the last 50–60 years, more recent efforts in controlled polymer synthesis now enable the preparation of a wide range of BCP architectures including, linear, graft, dendritic, star-like, bottle-brush, hyperbranched, and cyclic BCPs.[4] All of these macromolecular structures have unique and interesting self-assembly behavior; however, given the diversity of available architectures, this review will focus only on the versatility of linear BCPs, which still possess a myriad of opportunities to advance functional materials design. Within the sub-class of linear BCPs, various block types have been prepared such as organic and bio-hybrid BCPs. The first type includes a range of organic (non-biological-based) BCPs that can assemble by minimizing the free energy landscape that primarily is influenced by mixing and chain stretching considerations. We also note that significant work has been carried out in BCP systems that have specific interactions such as hydrogen-bonding, metal-binding, π–π stacking, *etc.*, which leads to self-organizing structures with their own intricate features. Many of these materials have been reviewed extensively in the literature,[7,58] and while of interest, they will not be discussed in exhaustive detail herein. Instead, we will highlight a range of synthetic tools that can be used to prepare specific organic BCPs with unique properties.

The second type, nature-inspired or bio-hybrid BCPs, forms a burgeoning class of self-assembling materials, in which the potential for secondary structure formation and concerted specific interactions promises unparalleled opportunities in hierarchical and function-driven assembly.[59] These nature-inspired bio-hybrids often contain at least one constituent derived from a biomolecular building block such as a peptide, protein, nucleic acid, peptoid, or sugars.[60,61] The combination of synthetic and bio-blocks in a well-defined macromolecule potentially introduces distinct nanostructures, stimuli-responsive character, and specific functions that are difficult to generate in 'simpler' organic–organic BCP systems.[59] As a result, these bio-hybrid BCPs provide opportunities for the realization of diverse and highly targeted applications in self-assembled materials (Fig. 3).

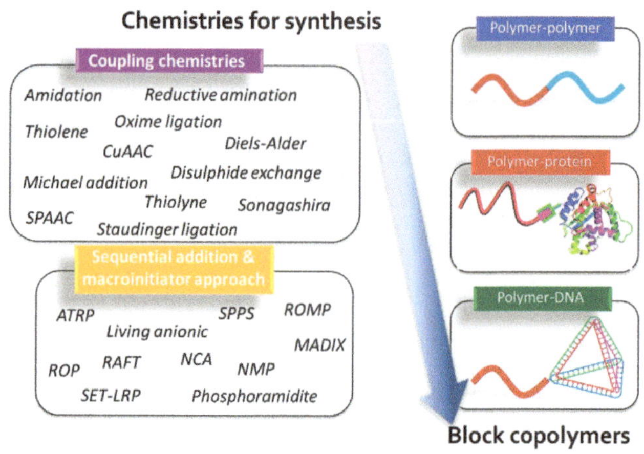

Figure 3: Chemistries for the synthesis of block copolymers. CuAAC, copper-catalyzed azide–alkyne cycloaddition; SPAAC, strain-promoted azide–alkyne cycloaddition; ATRP, atom-transfer radical polymerization; SPPS, solid phase peptide synthesis; ROMP, ring-opening metathesis polymerization; RAFT, reversible addition–fragmentation chain-transfer; ROP, ring-opening polymerization; SET-LRP, single-electron transfer living radical polymerization; NMP, nitroxide-mediated polymerization; NCA, *N*-carboxyanhydride; MADIX, macromolecular design by the interchange of xanthates.

In the following sections, we will highlight several noteworthy advances in the synthesis of linear BCPs that can lead to practical nanoscale assemblies for biomedicine, electronics, catalysis, nanotemplating, and responsive surface materials, among other applications. Furthermore, we also hint at several challenges, such as the need for the sustainable, efficient, and environmentally-friendly generation of functional macromolecules, which will be discussed in greater detail in the emerging directions section of this perspective.

Methods for Synthesis

There are numerous routes toward the synthesis of BCPs, which result in an extraordinarily broad range of macromolecules with tailorable and highly specific properties. These approaches can be broadly divided into three different classes: (continuous) sequential polymerization, macroinitiation, and coupling. Within these methods key considerations include the generation of well-defined polymers at high purity with controlled dispersity and high end-group fidelity, the incorporation of application-specific functional groups, and the ability to combine macromolecules synthesized from diverse polymerization methods, including the coupling of natural and synthetic building blocks. Each of these factors can have marked consequences on macromolecular self-assembly. For example, in many BCP generation approaches the final mixture can contain homopolymer or other "incomplete" contaminants due to loss of end-group fidelity, premature termination, incomplete end-group functionalization, or inefficient coupling. These contaminants can be difficult to remove or quantify; however, they can have a significant influence on macromolecular assembly.[62] Overcoming these limitations will facilitate copolymer design, as well as provide opportunities for the automated and high-throughput synthesis of complex architectures to possibly enable sustainable materials discovery.[63]

Continuous Sequential Polymerization

Perhaps the most straightforward method to prepare BCPs is through continuous sequential polymerization of two or more monomer sets using a single chain growth mechanism. That is to say, controlled polymerization of one monomer, followed by chain extension with a different monomer without intermediate termination/purification steps, can be used to prepare an AB polymer (or through further monomer addition, multiblock systems).[64] Sequential polymerization is especially applicable to methods such as living ionic, reversible-deactivation radical polymerization, and ring-opening polymerization. In particular, living anionic polymerization is somewhat restricted to this approach due to difficulties in reinitiating a terminated chain end; however, anionic techniques are still desirable for the ability to generate highly uniform polymers of extremely low dispersity and with excellent end-group fidelity,[3] as demonstrated by Goodyear (SIBR) for tires and by other companies in various applications. Additionally, continuous sequential methodologies can be tuned to produce tapered interfaces between polymer blocks.[9] These tapered block copolymers represent an emerging class of BCPs with unique and diverse self-assembly behavior.[65] Reversible-deactivation radical polymerization and ring-opening-based polymers also can be generated through continuous sequential polymerization, but these polymers are amenable to both macroinitiation approaches described below, due to the possibility of reinitiating a dormant chain end. Though sequential polymerization methods are useful for preparing a broad range of BCPs, they are somewhat limited in the polymerization of functional monomers, such as those containing nucleophilic or other reactive functionalities.

One approach to overcome this limitation in functional monomers involves the use of dual initiator (or protected initiator) species such as a hydroxyl-functionalized reversible addition–fragmentation chain-transfer (RAFT) chain transfer agent (CTA), which allows for the orthogonal polymerization of two or more distinct monomers. This route has been used most effectively for the combination of ring-opening and reversible-deactivation radical polymerization methods and unlocks access to a range of functionalizable BCPs.[66]

Macroinitiator Approaches

Although living anionic and anionic ring-opening polymerizations are useful for sequential approaches, the synthetically demanding nature of the reactions typically necessitates a macroinitiator approach to generate BCPs with the desired multiple block functionalities. This macroinitiation route can provide well-defined BCPs, but it normally involves a two-step polymerization process with the possible need for intermediate purification steps.[67] Fortunately, macroinitiation provides a ready means for generating a highly-defined library of macromolecules for systematic studies. A key consideration is that the polymerization mechanism utilized to synthesize the first block must result in a polymer with excellent end group fidelity to ensure that effective and efficient chain extension is possible.

Two major cases of macroinitiation can be defined; *case one*: the same polymerization mechanism used to generate the macroinitiator is used for chain extension, and *case two*: an orthogonal reaction scheme is used for chain extension. The first case is particularly amenable to ring-opening and reversible-deactivation radical polymerizations, in which the macroinitiator can be re-initiated following intermediate purification. However, challenges still remain in this case. As has been readily demonstrated in the literature, successful BCP formation may still not be possible due to blocking effects, which are based on the reactivity of the macroinitiator towards the chain-extending monomer. Macroinitiator reactivity is especially important when monomers with a more active dormant species (*e.g.* methacrylate or acrylonitrile) are utilized to extend a macroinitiator of lower activity (*e.g.* polystyrene or polyacrylate). This factor can be especially problematic when the specific functionality at the α and ω ends is important and hence reversing the order of polymerization is not possible. Additionally, macroinitiation schemes utilizing a single polymerization mechanism are not always amenable to the preparation of highly amphiphilic and hybrid BCPs, including those containing poly(ethylene oxide), polypeptides, or nucleic acids.

There have been several manipulations to controlled polymerizations to facilitate BCP synthesis from constituent monomers that cannot be polymerized using the same mechanism or initiating scheme (case two). Classic examples are macroinitiators synthesized through anionic or anionic ring-opening polymerization. Polystyrene-*b*-poly(ethylene oxide) [PS-*b*-PEO], polybutadiene-*b*-PEO [PB-*b*-PEO], PEO-*b*-poly(*N*-isopropylacrylamide) [PEO-*b*-PNIPAM], and PEO-*b*-ε-polycaprolactone [PEO-*b*-PCL] are several workhorse synthetic BCPs generated using this route.[48] Recent efforts have extended this approach to reversible-deactivation radical polymerizations, for which simple and effective chain end modification chemistries enable orthogonal polymerization mechanisms. Additionally, ring-opening metathesis polymerization (ROMP) provides an elegant tool for the synthesis of a diverse range of block copolymers.[68]ROMP is especially useful for the preparation of BCPs with interesting topologies including cycles,[69] grafts,[70] and bottle-brushes.[71]

Two areas that have received recent interest include degradable BCPs, containing polylactide and bio-hybrid systems such as polypeptide BCPs which have possible applications in biomedicine.[72,73] Additionally, the explosion of click chemistries as a model for polymer functionalization (*e.g.* polymer chain end modification) has facilitated the manipulation of macroinitiators to allow for a second polymerization mechanism. This functionalization avenue has significantly broadened the scope of BCPs accessible through a macroinitiator approach especially for such sustainable and degradable polymers.

Macroinitiator approaches also are very effective for the preparation of nature-derived BCPs, especially those from natural biopolymers such as cellulose, chitin, or proteins. Although these methods often lead to more complex architectures instead of simple and low dispersity linear BCPs, the manipulation of natural biopolymers to incorporate selective single site modifications has shown promise.[74] This method has been demonstrated for the modification of proteins through selective introduction of a functionality (*e.g.* a polymerization initiator) to enable polymerization of a second polymer in a 'grafting from' approach and to afford a bio-hybrid BCP (Fig. 4).[74] To extend this approach and ensure effective and high-yielding BCP synthesis, further efforts are required to selectively incorporate functionalities that enable BCP generation in nature-inspired or nature-derived systems.

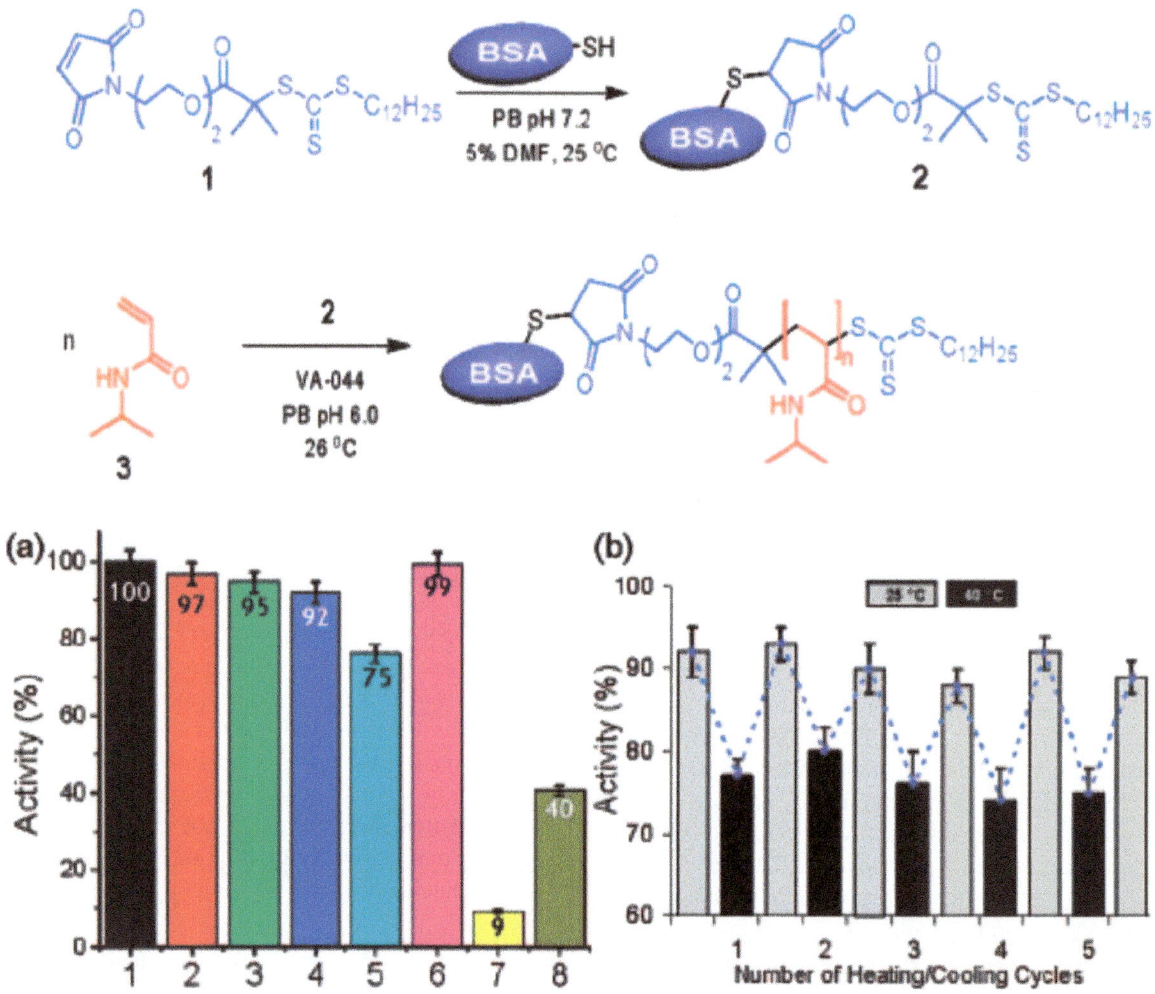

Figure 4: Bio-hybrid block copolymers generated through coupling approaches. (Top image) scheme for the synthesis of protein–polymer BCP. (Bottom image, left) activity of (1) BSA, (2) BSA–macro-CTA, (3) BSA–PNIPAM (free BSA present), (4) BSA–PNIPAM thermal precipitate, (5) BSA–PNIPAM thermal precipitate at 40 °C, (6) BSA physical mixture, (7) PNIPAM, (8) BSA after incubation at 75 °C for 3 h. (Right) activity of BSA–PNIPAM thermal precipitate during thermal cycling between 25 °C and 40 °C [adapted and reprinted with permission from ref. 74 © American Chemical Society].

Post-Polymerization Coupling

Perhaps the most versatile method for BCP synthesis is the post-polymerization coupling approach, which enables the conjugation of blocks with very distinct chemistries. This avenue for linking pre-synthesized macromolecules has been facilitated by the exploitation of click chemistries. The concept was introduced in 2001 by Sharpless,[75] and was later applied in BCP synthesis,[76] Click is exemplified by the copper(I)-catalyzed azide–alkyne cycloaddition reaction (CuAAC). This reaction, and other high-yielding and highly-efficient coupling approaches,[14]allow for the conjugation of a diverse array of end-functionalized polymers to afford a range of BCPs. One challenge in the translation of small molecule click approaches to polymer–polymer conjugation is the requirement for complete end group retention (and/or subsequent functionalization), which has been difficult to achieve for the vast majority of polymerization routes. However, methods such as living anionic polymerization and copper-mediated radical polymerization have demonstrated excellent end-group fidelity and have functionalities that can be readily modified to a click-like reactive group (*e.g.* the termination of the anionic polymerization with ethylene oxide to yield a terminal hydroxyl,[67] to then yield a terminal azide[77]).

Click concepts in macromolecular systems clearly must take into consideration the challenges in purification and detailed molecular characterization. For cases in which end-group fidelity is not maintained, the coupled BCPs mixture also will contain one or both of the un-coupled building blocks that can be difficult if not time-consuming to remove, unless the click-based reaction conditions and stoichiometry are defined to yield only the BCP and an easily separable building block. Furthermore, it is worth noting that not all small molecule click reactions are similarly effective in polymer systems. The radical-mediated thiol–ene reaction is an example of a reaction that has found application in small molecule and polymer modifications,[78] but it has not been as efficient in polymer–polymer couplings. This reaction was demonstrated clearly by DuPrez and Barner-Kowollik to have very limited effectiveness for the coupling of a range of chain-end functionalized polymers.[79] Studies such as these highlight the need for in-depth and careful characterization of such BCP reaction schemes to verify the effectiveness of the BCP formation (*i.e.* homopolymer contamination, *etc.*). Click-based approaches also have been employed in the post-polymerization modification of pre-formed BCPs. As demonstrated by Hammond and coworkers, alkyne side groups allow the creation of a versatile library of compounds from a single parent BCP through cycloadditions;[80] however, the efficiency of the coupling reactions remains a concern when dealing with multiple reaction sites on a long-chain macromolecule.

A further challenge to be overcome, which affects all of the synthesis methods described above, is the requirement for compatible solvents for the constituent blocks of the BCP. This constraint can be especially challenging for both macroinitiation and post-polymerization coupling, and if not properly considered can lead to low yields, incomplete BCP formation, and high dispersities, reducing the sustainability of BCP generation. Two classes of macromolecules in which these issues are particularly common are organic-biopolymers, such as nucleic acid–hydrophobic polymer conjugates, and conducting-organic BCPs, given that the conducting blocks often have limited solubilities in a wide range of solvents. An additional challenge in biomacromolecule coupling is that click-based groups must not undergo reactions with non-target sites, which is a concern in polymer/protein coupling (for example, through thiol–ene chemistry, when the protein construct may contain multiple cysteine residues). Thus, although small molecule click concepts are extremely useful in coupling macromolecular systems, care must be taken to generate well-defined and well-characterized BCPs.

CHARACTERIZATION OF COMPLEX NANOSCALE ASSEMBLIES

The self-assembly of complex macromolecules generated through various synthetic methods is of critical importance in the fabrication of materials targeted toward next-generation applications. Whereas conventional BCPs are described by a manageable set of parameters (typically N, block volume fractions, and statistical segment length ratios), many new systems incorporate multiple blocks with various architectures, dopants, and specific interactions.[4] These factors significantly complicate BCP self-assembly, as well as confound requisite nanoscale characterization.

Bulk Systems

For bulk systems, wide angle and small angle X-ray scattering (WAXS and SAXS), small angle neutron scattering (SANS) transmission electron microscopy (TEM) and TEM tomography, mechanical analysis, birefringence, calorimetry, and transport measurements are some of the common tools employed in materials characterization.[3] While these techniques are extremely useful for nanostructure determination, methods such as TEM, calorimetry, and mechanical analysis are not always amenable to rapid materials discovery. Thus, more detailed and customizable characterization tools are necessary to quickly elucidate the intricate structures in complex materials for which macromolecular constituents and processing steps are chosen, not to facilitate characterization, but to enable applications. Recent advances in that direction include the emergence of resonant soft X-ray scattering (RSoXS),[81] which can distinguish between nanoscale domains in BCPs that contain distinct chemical constituents without an over-reliance on strict electron density contrast. Another technique, energy-filtered TEM (EFTEM), has been used to probe chemical heterogeneity between domains caused by differences in block chemistries, as well as the locations of small molecule dopants.[82] When augmented with other newly pioneered methods in soft materials, such as X-ray Photoelectron Spectroscopy (XPS) with C_{60}^{+} ion sputtering, the elucidation of the spatial distributions of chemical species within multicomponent and polymer-based systems is facilitated in bulk and thin film (see below) materials.[83] Though RSoXS, EFTEM, and XPS-C_{60}^{+} sputtering provide substantial opportunities for the improved nanoscale characterization of soft materials, further progress is necessary to accurately probe three-dimensional and hierarchically-ordered nanomaterials in a rapid fashion.

Thin Films

Similar challenges and opportunities exist in the analysis of nanostructured thin film systems. Major characterization techniques include optical microscopy (OM, to analyze island/hole formation and wetting behavior), atomic force microscopy (AFM), TEM, scanning electron microscopy (SEM), grazing-incidence SAXS (GISAXS), reflectivity (neutron and X-ray), XPS, time-of-flight secondary-ion mass spectrometry (ToF SIMS), and contact angle (surface energy) measurements. While techniques such as OM enable rapid materials characterization when combined with high-throughput or gradient approaches,[84] the remaining nanostructure investigation methods typically are time-consuming and preparation-intensive. There have been substantive recent developments in several advanced tools for expedited thin film nanoscale characterization;[85,86] however, significant innovations are necessary to probe two-dimensional and three-dimensional structures over large areas, monitor defects, and provide real-time information during nanostructure processing. The incorporation of thermal and solvent annealing apparatuses into GISAXS and AFM experiments has begun to provide some insights into processing effects on nanostructure stability,[87] yet the role of thermal, solvent, and surface field gradients within thin films remains an underexplored area.[24,26,88,89] Neutron scattering techniques such as reflectometry, rotational SANS (RSANS), and multiple angle grazing incidence K-vector (MAGIK) off-specular reflectometry can provide significant insights into the thin film ordering processes,[90] in particular, by elucidating the spatial distribution of solvents in thin films during casting and processing. Insights such as these will be extremely useful in the design of appropriate conditions to maximize the desired nanostructure ordering. Furthermore, continued development of techniques such as ultra-fast AFM and *in situ* TEM provide key opportunities for monitoring the kinetic processes that are so influential in thin film behavior.[26,85,91]

Solution Assembly

The characterization of solution assemblies faces similar challenges, including the impact of kinetics and processing on the overall assembly process. However, solution assemblies face the additional difficulty of a dearth of nanoscale characterization tools that can perform *in situ* analysis of macromolecular aggregates and allow for determination of key nanostructure characteristics, such as the core radius, hydrodynamic radius and/or radius of gyration of a spherical micelle (Fig. 5). While methods such as dynamic and static light scattering (DLS and SLS), calorimetry, spectroscopy (*e.g.* nuclear magnetic resonance [NMR] and Raman), and rheology are capable of following gross aggregate evolution, techniques such as SAXS, SANS, TEM, and cryogenic-TEM (cryo-TEM) are capable of probing the fine details of aggregate structures, such as core and corona density profiles.[92] To fully understand the self-assembly process time-resolved measurements are required; however, methods such as microscopy (and scattering) are challenging to perform in this manner, largely due to the acquisition times and the larger background signals associated with probing a dilute solution environment. Additionally, approaches such as neutron scattering routinely rely on the ability to synthesize systems with the appropriate contrast (*i.e.* deuterated materials),[51] sometimes leading to fabrication for analysis and not application. Furthermore, methods such as TEM involve sample preparations that remove the assemblies from their native environment with sometimes unintended consequences;[92] thus, new approaches are necessary to capture the true behavior of amphipathic solution assemblies. Synchrotron radiation and pulsed neutron sources, along with stopped-flow techniques,[93] have the potential to provide detailed nanostructure information on the time scales of interest, but again, require well-designed systems. Recent studies have shown that real-time/solution state TEM (RT-TEM) is an intriguing alternative for monitoring assembly and stability, while maintaining access to detailed structural information;[94] however, it is worth noting that samples with sufficient contrast and suitable electron beam stability are necessary to fully take advantage of RT-TEM.

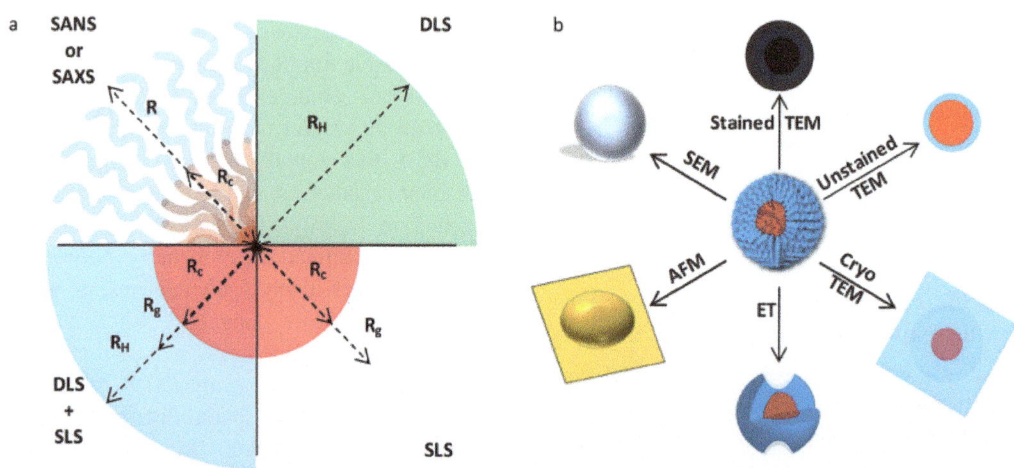

Figure 5: Characterization of block copolymer solution assemblies. Schematic showing (a) a subset of the structural information obtainable from various scattering techniques and (b) a subset of images extracted using different microscopy techniques for a spherical polymer micelle. Panel (a) highlights common structural dimensions that can be readily probed in a idealized spherical micelle, along with associated reciprocal-space scattering techniques; these dimensions include R_c (core radius), R_g (radius of gyration), R_h (hydrodynamic radius),R (micelle radius). Panel (b) illustrates complementary information that can be gleaned from real-space microscopy imaging [reprinted with permission from ref. 92 © Royal Society of Chemistry].

THEORY/SIMULATION/MODELLING

Numerous theoretical advances also can provide significant insights into BCP assembly and detail key information in the experimental development of new nanostructured systems. Self-consistent field theory (SCFT), density functional theory, molecular dynamics, and Monte Carlo (MC) simulations have been particularly useful in improving understanding of BCP nanostructure formation in bulk and thin films. Recent examples in bulk BCPs include the SCFT work of Hall and coworkers[65] that examined the influence of controlled heterogeneity between homogeneous blocks (*i.e.* tapering) on self-assembly of a diblock copolymer, and the lattice-based MC simulations of Matsen and coworkers[95] that explored the effects of controlled dispersity on copolymer phase behavior (Fig. 6). In each of these cases, the theoretical efforts complemented recent experimental work and provided tantalizing information useful in the fabrication of new nanoscale materials.[8] Additionally, reports by Wang and coworkers[96] (SCFT of salt-doped BCP melts) and Jayaraman and coworkers[97] (self-consistent Polymer Reference Interaction Site Model [PRISM] theory/MC simulations of copolymer coatings on nanoparticles in homopolymer matrices) again show direct experimental links that can aid materials discovery and provide practical trajectories for experimental investigations. In addition to the powerful trends highlighted in the above studies, further theoretical advances could eventually lead to specific predictive capabilities such as exact polymer constituents, molecular weights, architectures, dispersities, grafting densities, *etc.* that would produce the desired macromolecular behavior or specified nanostructure. These capabilities would dramatically streamline synthesis and characterization and lead to informed macromolecular design.

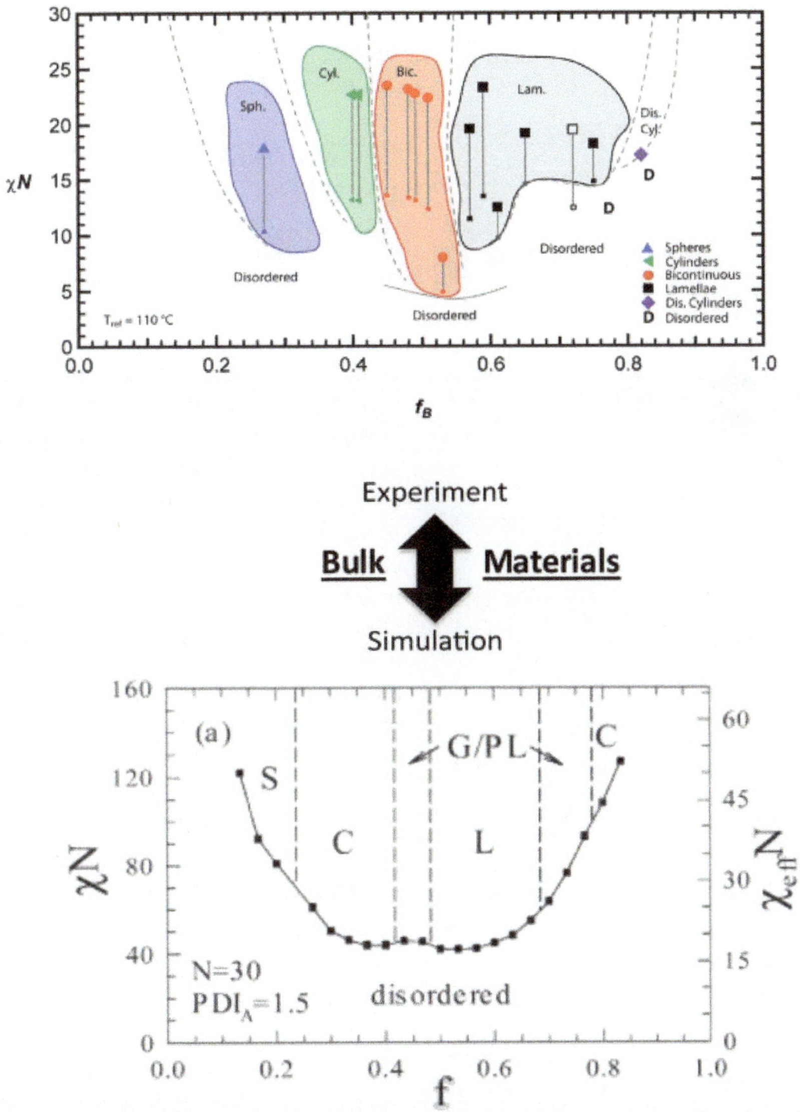

Figure 6: Linking theory and experiment in block copolymers. Bulk nanostructures – effects of polymer dispersity. (Top image) experimental phase diagram of a block copolymer system containing controlled dispersity [reprinted with permission from ref. 8 © American Chemical Society]. (Bottom image) Monte Carlo simulated phase diagram of block copolymer system with a similar dispersity concept [reprinted with permission from ref. 95 © American Chemical Society]. A comparison between the top (experimental) and bottom (simulated) images shows general agreement and illustrates how theory and simulations could potentially direct experimental efforts.

Similar insights have been made in BCP thin films, recent examples of which include the work of Alexander-Katz and coworkers (forward SCFT simulations in an inverse design algorithm to explore topographic templates for directed self-assembly)[98] and dePablo and coworkers[99] (MC simulations of BCP thin films on nanopatterned surfaces) that have produced detailed information to strengthen practical understanding of directed assembly on designer substrates. Furthermore, efforts by Frederickson and coworkers examined the influence of solvent removal rate on the stability of cylindrical orientations in ultrathin films using dynamical field theory simulations,[88] providing key knowledge that can be translated readily to experimental and application-oriented systems (Fig. 7).[100] The above examples demonstrate significant progress in linking theoretical studies to experimental investigations, and continued efforts in understanding the effects of thin film processing and fabrication methods, substrate interactions, and influences of macromolecular architectures on kinetic *vs.* thermodynamic assembly could drive significant advances in nanopatterning and sensing platforms using BCP thin films.

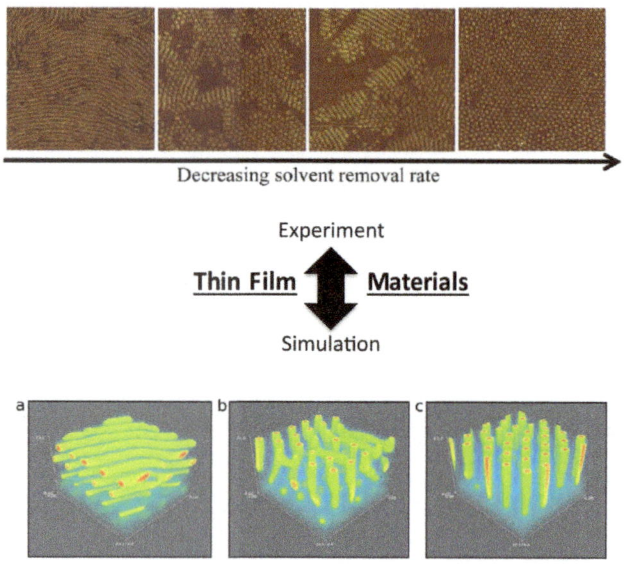

Figure 7: Linking theory and experiment in block copolymers. Thin film nanostructures – effects of processing on nanostructure orientation. (Top image) AFM images of parallel and perpendicular orientations of cylinders as a function of solvent evaporation rate in an ABA-triblock copolymer thin film [reprinted with permission from ref. 100 © American Chemical Society]. (Bottom image) parallel *vs.* perpendicular orientations generated from various solvent evaporation rates and solvent selectivities in BCP thin films using a dynamic extension of self-consistent field theory simulations [reprinted with permission from ref. 88© American Chemical Society]. In both the experimental and simulation efforts, faster solvent removal (evaporation) rates led to parallel orientations of cylinders, while slower solvent removal (evaporation) rates led to perpendicular cylinders. Refinement of general experimental and simulation trends, such as the correlation between solvent evaporation and nanostructure formation, will facilitate the continued development of nanostructure/processing relationships in thin film systems.

Solution assembly of BCPs presents great challenges for conventional theoretical/modelling/simulation methodologies. The intrinsic need to explicitly describe key interactions from angstroms to tens of nanometers, over relevant time-scales, necessitates multi-scale approaches; however, accurate descriptions of thermodynamic and kinetic processes are hampered by the inability to simultaneously access the necessary time-scales and length-scales. Several simulation/theory efforts have made significant in-roads into the understanding of BCP behavior in solution assemblies. Notable recent efforts include the exploration of equilibrium BCP chain exchange kinetics in dilute micellar solutions,[101] examination of the energetics of unimer insertion in concentrated micelle solutions,[102] and combined experimental and theoretical probing of the influence of BCP molecular weight and composition on critical aggregation concentration.[103] While these works indicate substantial progress, many challenges remain, especially related to the formation/processing of solution assemblies.[104] One path forward is to consider approaches currently applied in protein-engineering, especially in nature-inspired materials, to reconcile the influence of thermodynamics and kinetics (and processing) on protein folding that may be particularly applicable for many solution assembled BCP systems.[105]

EMERGING DIRECTIONS AND MAJOR CHALLENGES FOR BLOCK COPOLYMER ASSEMBLIES

Several challenges exist along the path toward accelerating the design of new nanostructured materials with positive societal and environmental impacts through the leveraging of continued advances in macromolecular synthesis, processing, and characterization. In particular, the desire to unlock exotic and hierarchically complex nanostructures for next-generation applications requires the multidimensional understanding of a myriad of chemistries, molecular architectures, fabrication protocols, and processing techniques.[106] This understanding can be facilitated by additional links between experiment and theory that provide true predictive capabilities. Furthermore, although not all-encompassing, three emerging areas that have been foreshadowed by the above discussions are: *(1) the sustainable and environmentally-friendly generation and processing of materials, (2) the optimization and detailed characterization of nature-inspired materials, and (3) the influence of processing and fabrication methods on nanoscale structures, in particular, solution assemblies.*

Sustainable and Environmentally-Friendly Generation and Processing of Materials

Major efforts in the sustainable and environmentally-friendly generation and processing of materials have focused on the synthesis of nanostructured polymers from bio-based or renewable feedstocks.[107] Many of these "green" systems have attractive functionalities (*e.g.* aldehydes, hydroxyls, and phenols) that permit the design of new monomers amenable to controlled polymerization techniques (or bio-based sources of "old" monomers) for applications such as thermoplastic elastomers, pressure-sensitive adhesives, nanocarriers for biological and catalysis applications, and blend compatibilizers; however, further efforts in cultivating sustainable and cost-effective feedstocks, achieving efficient syntheses (and purification), and obtaining suitable macroscale properties (glass transition temperature, degradation temperature, modulus, solubility, *etc.*) are necessary. As highlighted in the polymer synthesis sections, sustainable macromolecules generation also applies to the continued development of multi-component polymerization and coupling approaches that further reduce waste, solvent usage, and purification requirements.

Optimization and Detailed Characterization of Nature-Inspired Materials

As nature has become an inspiration for the exploration of self-assembled systems, bio-inspired materials embody many of the strengths and challenges in nanomaterials design. In contrast to biological polymers (such as proteins or DNA), synthetic polymers can be prepared from a much broader range of monomers, to afford polymers with a variety of structures and architectures and hence a vast range of properties and diversity of applications including diagnostics, therapeutic agent delivery, cell culture and tissue engineering, and biomaterials scaffolds and supports. However, a primary limitation of current synthetic polymers is the lack of general methods for producing precise chain structure (*i.e.* sequence control) and hence complex function (*i.e.* replication and evolution). Indeed, the development of new methods to enable the mimicking of biological function in macromolecules is an emerging area,[59] which holds great potential for the future.

While the basics of conventional BCP synthesis were established some time ago, it is only more recently that innovations including the fabrication of polymer–peptide, polymer–protein, and other stimuli-responsive materials that take advantage of the exquisite interactions facilitated by unique molecular organization and secondary structure formation are now possible.[59]Furthermore, forays into the mechanisms and energetics of peptide/protein folding and function have enabled the discovery of new methods to engineer macromolecules, in which the placement of individual repeat units is controlled to impart specific functions or directional interactions.[105] This approach is key as it allows for the preparation of biohybrid BCPs with specific polymer-biomolecular attachments. However, as the precise manipulation of functional groups can have substantial effects on material efficacy in complex systems, it is crucial that researchers continue to explore the necessary structure/property relationships by fully understanding the limitations inherent in many macromolecular and nanostructure characterization techniques.

Additionally, as stimuli-responsive and targeted assemblies become *en vogue* for therapeutics, diagnostics, and imaging applications, information to correlate the placement of designer functionalities on macromolecules to their spatial arrangement in solution assemblies is increasingly vital to truly design and optimize nanoscale materials for *in vivo* applications.[44,105] As one example, techniques such as anomalous SAXS (ASAXS) can probe this link in solution-assembled BCP systems,[108] but ASAXS comes with the added expense of incorporating the appropriate moieties (such as selenium labelling) to ensure adequate contrast, along with the necessity of a synchrotron source; thus, theoretical complements to predict exact constituent/architecture/nanostructure/property relationships are vital.

Influence of Processing and Fabrication Methods on Nanoscale Structures

As new materials are envisioned with intricate and precise self-organization potential, the role of nanostructure fabrication and processing becomes an even greater consideration in materials development. As demonstrated in peptide and protein-based systems, thermal history, mechanical processing, and exposure to external fields and environments have substantial and irreversible effects on macromolecular assembly leading to path dependent behavior.[109] In the case of solution assembly, information gleaned from the detailed literature on biomacromolecules on these processing effects can provide significant insights into the fabrication and stability of polymer solution constructs.[57] Studies examining the influence formulation and processing protocols are particularly relevant in light of work describing the impact of the above variables on solution assembled nanostructure size and shape, which ultimately will control nanocarrier delivery and function. We note that fabrication and processing effects are not limited to solution assemblies and biomaterials but also are inherent in bulk and thin film assemblies as highlighted in the previous sections. Similar key challenges also are present in nanocomposites,[110] organic electronic materials,[7] and resins,[111] among other arenas comprising block copolymers.

In summary, while efforts in high-throughput, automated, and gradient synthesis and characterization have accelerated materials development in conventional systems as well as the three areas highlighted above,[25,26] the sheer diversity of possibilities necessitates the intimate interfacing of experimental designs with theory/simulation/modelling. To facilitate this meshing of theory and experiment, it is vital that theoretical/modelling efforts continue to consider relevant processing protocols and molecular architectures in designing appropriate systems. However, it is also important that experimental studies take advantage of the complete suite of synthetic, molecular characterization, and nanostructure characterization tools to fully and accurately characterize macromolecular assemblies. The complexity of the chemical and biological units and the plethora of possible building blocks of the next-generation of block copolymers stretch the limits of nanomaterials characterization, which reinforces the urgent need for enhanced theoretical-experimental methods in *de novo* materials design. Thus, by harnessing the inherent strengths of soft materials chemistry, physics, processing, and theory to generate complex nanomaterials, new systems and tools will be developed to unlock the full potential of BCPs and continue to shrink the gap between concept and application.

AUTHOR CONTRIBUTIONS

T. H. E. and R. K. O. R. jointly discussed the content of the review and jointly wrote the manuscript.

ACKNOWLEDGEMENTS

The authors thank the University of Warwick, the University of Delaware (Thomas and Kipp Gutshall Professorship), and the Engineering and Physical Sciences Research Council for financial support. The authors thank Angela Holmberg and Ming Luo for helpful discussions.

REFERENCES

1. N. C. Seeman, C. Mao and H. Yan, Guest Editorial: Nucleic Acid Nanotechnology, *Acc. Chem. Res.*, 2014, 47, 1643–1644 .
2. Y. Y. Mai and A. Eisenberg, Self-assembly of block copolymers, *Chem. Soc. Rev.*, 2012, 41, 5969–5985 .
3. N. Hadjichristidis, S. Pispas and G. A. Floudas, *Block Copolymers: Synthetic Strategies, Physical Properties, and Applications*, Wiley-Inteience, 2003 .
4. F. S. Bates, M. A. Hillmyer, T. P. Lodge, C. M. Bates, K. T. Delaney and G. H. Fredrickson, Multiblock Polymers: Panacea or Pandora's Box?, *Science*, 2012, 336, 434–440 .
5. C.-T. Lo and P.-W. Chou, Effect of Molecular Properties of Random Copolymers on the Stability and Domain Dimension of Block Copolymer/Random Copolymer Blends, *J. Phys. Chem. B*, 2014, 118, 12763–12771 .
6. S. H. Han, J. K. Kim, V. Pryamitsyn and V. Ganesan, Phase Behavior of Binary Blends of Block Copolymers Having Hydrogen Bonding, *Macromolecules*, 2011, 44, 4970–4976 .
7. R. A. Segalman, B. McCulloch, S. Kirmayer and J. J. Urban, Block Copolymers for Organic Optoelectronics, *Macromolecules*, 2009, 42, 9205–9216 .
8. J. M. Widin, A. K. Schmitt, A. L. Schmitt, K. Im and M. K. Mahanthappa, Unexpected Consequences of Block Polydispersity on the Self-Assembly of ABA Triblock Copolymers, *J. Am. Chem. Soc.*, 2012, 134, 3834–3844 .
9. P. Hodrokoukes, G. Floudas, S. Pispas and N. Hadjichristidis, Microphase separation in normal and inverse tapered block copolymers of polystyrene and polyisoprene. 1. Phase state,*Macromolecules*, 2001, 34, 650–657 .
10. J.-F. Lutz, M. Ouchi, D. R. Liu and M. Sawamoto, Sequence-Controlled Polymers, *Science*, 2013, 341, 1238149 .
11. A.-V. G. Ruzette, P. P. Soo, D. R. Sadoway and A. M. Mayes, Melt-Formable Block Copolymer Electrolytes for Lithium Rechargeable Batteries, *J. Electrochem. Soc.*, 2001, 148, A537–A543 .
12. A. C. Finnefrock, R. Ulrich, G. E. S. Toombes, S. M. Gruner and U. Wiesner, The plumber's nightmare: a new morphology in block copolymer-ceramic nanocomposites and mesoporous aluminosilicates, *J. Am. Chem. Soc.*, 2003, 125, 13084–13093 .
13. C. E. Sing, J. W. Zwanikken and M. O. de la Cruz, Electrostatic control of block copolymer morphology, *Nat. Mater.*, 2014, 13, 694–698 .
14. P. L. Golas and K. Matyjaszewski, Marrying click chemistry with polymerization: expanding the scope of polymeric materials, *Chem. Soc. Rev.*, 2010, 39, 1338–1354 .
15. M. C. Orilall and U. Wiesner, Block copolymer based composition and morphology control in nanostructured hybrid materials for energy conversion and storage: solar cells, batteries, and fuel cells, *Chem. Soc. Rev.*, 2011, 40, 520–535 .
16. W. Q. Cai, J. G. Yu, C. Anand, A. Vinu and M. Jaroniec, Facile Synthesis of Ordered Mesoporous Alumina and Alumina-Supported Metal Oxides with Tailored Adsorption and Framework Properties, *Chem. Mater.*, 2011, 23, 1147–1157 .
17. P. D. Yang, D. Y. Zhao, D. I. Margolese, B. F. Chmelka and G. D. Stucky, Block copolymer templating syntheses of mesoporous metal oxides with large ordering lengths and semicrystalline framework, *Chem. Mater.*, 1999, 11, 2813–2826 .
18. J. Hentschel, A. M. Kushner, J. Ziller and Z. Guan, Self-Healing Supramolecular Block Copolymers, *Angew. Chem., Int. Ed.*, 2012, 51, 10561–10565 .
19. B. S. Lee, B. C. Chun, Y.-C. Chung, K. I. Sul and J. W. Cho, Structure and Thermomechanical Properties of Polyurethane Block Copolymers with Shape Memory Effect,*Macromolecules*, 2001, 34, 6431–6437 .
20. I. W. Hamley, Ordering in thin films of block copolymers: fundamentals to potential applications, *Prog. Polym. Sci.*, 2009, 34, 1161–1210 .
21. C. Hanisch and R. Bogue, Self-assembly: a review of recent developments, *Assemb. Autom.*, 2008, 28, 211–215 .
22. G. Liu, P. F. Nealey, R. Ruiz, E. Dobisz, K. C. Patel and T. R. Albrecht, Fabrication of chevron patterns for patterned media with block copolymer directed assembly, *J. Vac. Sci. Technol., B: Nanotechnol. Microelectron.: Mater., Process., Meas., Phenom.*, 2011, 29, 06F204 .
23. P. A. R. Delgadillo, R. Gronheid, C. J. Thode, H. Wu, Y. Cao, M. Neisser, M. Somervell, K. Nafus and P. F. Nealey, Implementation of a chemo-epitaxy flow for directed self-assembly on 300-mm wafer processing equipment, *J. Micro/Nanolithogr., MEMS, MOEMS*, 2012, 11, 031302 .
24. C. M. Bates, M. J. Maher, D. W. Janes, C. J. Ellison and C. G. Willson, Block Copolymer Lithography, *Macromolecules*, 2013, 47, 2–12 .
25. J. N. L. Albert and T. H. Epps, III, Self-assembly of Block Copolymer Thin Films, *Mater. Today*, 2010, 13, 24–33 .
26. M. Luo and T. H. Epps, III, Directed Block Copolymer Thin Film Self-Assembly: Emerging Trends in Nanopattern Fabrication, *Macromolecules*, 2013, 46, 7567–7579 .
27. J. W. Choi, M. Kim, N. S. Safron, M. S. Arnold and P. Gopalan, Transfer of Pre-Assembled Block Copolymer Thin Film to Nanopattern Unconventional Substrates, *ACS Appl. Mater. Interfaces*, 2014, 6, 9442–9448 CAS.
28. P. W. Majewski, M. Gopinadhan and C. O. Osuji, Magnetic field alignment of block copolymers and polymer nanocomposites: scalable microstructure control in functional soft materials, *J. Polym. Sci., Part B: Polym. Phys.*, 2012, 50, 2–8 .
29. S. Pujari, M. A. Keaton, P. M. Chaikin and R. A. Register, Alignment of perpendicular lamellae in block copolymer thin films by shearing, *Soft Matter*, 2012, 8, 5358–5363 .
30. M. Sano, S. Nakamura, M. Hara, S. Nagano, Y.

Shinohara, Y. Amemiya and T. Seki, Pathways toward Photoinduced Alignment Switching in Liquid Crystalline Block Copolymer Films, *Macromolecules*, 2014, 47, 7178–7186 .

31. S. H. Anastasiadis, Development of Functional Polymer Surfaces with Controlled Wettability, *Langmuir*, 2013, 29, 9277–9290 .
32. H. Sai, K. W. Tan, K. Hur, E. Asenath-Smith, R. Hovden, Y. Jiang, M. Riccio, D. A. Muller, V. Elser, L. A. Estroff, S. M. Gruner and U. Wiesner, Hierarchical Porous Polymer Scaffolds from Block Copolymers, *Science*, 2013, 341, 530–534 .
33. Z. L. Zhou, D. R. Calabrese, W. Taylor, J. A. Finlay, M. E. Callow, J. A. Callow, D. Fischer, E. J. Kramer and C. K. Ober, Amphiphilic triblock copolymers with PEGylated hydrocarbon structures as environmentally friendly marine antifouling and fouling-release coatings, *Biofouling*, 2014, 30, 589–604 .
34. A. Asatekin and C. Vannucci, Self-Assembled Polymer Nanostructures for Liquid Filtration Membranes: A Review, *Nanosci. Nanotechnol. Lett.*, 2015, 7, 21–32 .
35. C. T. Black, R. Ruiz, G. Breyta, J. Y. Cheng, M. E. Colburn, K. W. Guarini, H. C. Kim and Y. Zhang, Polymer self assembly in semiconductor microelectronics, *IBM J. Res. Dev.*, 2007, 51, 605–633 .
36. H. C. Kim, S. M. Park and W. D. Hinsberg, Block Copolymer Based Nanostructures: Materials, Processes, and Applications to Electronics, *Chem. Rev.*, 2010, 110, 146–177 .
37. S. B. Darling, Block copolymers for photovoltaics, *Energy Environ. Sci.*, 2009, 2, 1266–1273 .
38. A. F. Hannon, Y. Ding, W. Bai, C. A. Ross and A. Alexander-Katz, Optimizing Topographical Templates for Directed Self-Assembly of Block Copolymers via Inverse Design Simulations, *Nano Lett.*, 2014, 14, 318–325 .
39. M. P. Stoykovich, M. Müller, S. O. Kim, H. H. Solak, E. W. Edwards, J. J. de Pablo and P. F. Nealey, Directed Assembly of Block Copolymer Blends into Nonregular Device-Oriented Structures, *Science*, 2005, 308, 1442–1446 .
40. S. H. Kim, K. Hong, W. Xie, K. H. Lee, S. Zhang, T. P. Lodge and C. D. Frisbie, Electrolyte-Gated Transistors for Organic and Printed Electronics, *Adv. Mater.*, 2013, 25, 1822–1846 .
41. M. Ma, K. Titievsky, E. L. Thomas and G. C. Rutledge, Continuous Concentric Lamellar Block Copolymer Nanofibers with Long Range Order, *Nano Lett.*, 2009, 9, 1678–1683 .
42. E. Reimhult and F. Hook, Design of Surface Modifications for Nanoscale Sensor Applications, *Sensors*, 2015, 15, 1635–1675 .
43. F. H. Schacher, P. A. Rupar and I. Manners, Functional Block Copolymers: Nanostructured Materials with Emerging Applications, *Angew. Chem., Int. Ed.*, 2012, 51, 7898–7921 .
44. E. G. Kelley, J. N. L. Albert, M. O. Sullivan and T. H. Epps, III, Stimuli-responsive copolymer solution and surface assemblies for biomedical applications, *Chem. Soc. Rev.*, 2013, 42, 7057–7071 .
45. M. E. Welch and C. K. Ober, Responsive and patterned polymer brushes, *J. Polym. Sci., Part B: Polym. Phys.*, 2013, 51, 1457–1472 .
46. R. C. Hayward and D. J. Pochan, Tailored assemblies of block copolymers in solution: it is all about the process, *Macromolecules*, 2010, 43, 3577–3584 .
47. M. A. Dyson, A. M. Sanchez, J. P. Patterson, R. K. O'Reilly, J. Sloan and N. R. Wilson, A new approach to high resolution, high contrast electron microscopy of macromolecular block copolymer assemblies, *Soft Matter*, 2013, 9, 3741–3749 .
48. S. Jain and F. S. Bates, On the origins of morphological complexity in block copolymer surfactants, *Science*, 2003, 300, 460–464 .
49. S. Zhong, H. G. Cui, Z. Y. Chen, K. L. Wooley and D. J. Pochan, Helix self-assembly through the coiling of cylindrical micelles, *Soft Matter*, 2008, 4, 90–93 .
50. S. J. Holder and N. A. J. M. Sommerdijk, New micellar morphologies from amphiphilic block copolymers: disks, toroids and bicontinuous micelles, *Polym. Chem.*, 2011, 2, 1018–1028 .
51. J. Lu, F. S. Bates and T. P. Lodge, Chain exchange in binary copolymer micelles at equilibrium: confirmation of the independent chain hypothesis, *ACS Macro Lett.*, 2013, 2, 451–455 .
52. L. Meli, J. M. Santiago and T. P. Lodge, Path-dependent morphology and relaxation kinetics of highly amphiphilic diblock copolymer micelles in ionic liquids, *Macromolecules*, 2010, 43, 2018–2027 .
53. E. Buhler, A. V. Dobrynin, J. M. DeSimone and M. Rubinstein, Light-scattering study of diblock copolymers in supercritical carbon dioxide: CO_2 density-induced micellization transition, *Macromolecules*, 1998, 31, 7347–7355 .
54. S. Sugihara, A. Blanazs, S. P. Armes, A. J. Ryan and A. L. Lewis, Aqueous Dispersion Polymerization: A New Paradigm for In Situ Block Copolymer Self-Assembly in Concentrated Solution, *J. Am. Chem. Soc.*, 2011, 133, 15707–15713 .
55. V. A. Vasantha, S. Jana, S. S.-C. Lee, C.-S. Lim, S. L.-M. Teo, A. Parthiban and J. G. Vancso, Dual hydrophilic and salt responsive schizophrenic block copolymers - synthesis and study of self-assembly behavior, *Polym. Chem.*, 2015, 6, 599–606 .
56. A. Lu and R. K. O'Reilly, Advances in nanoreactor technology using polymeric nanostructures, *Curr. Opin. Biotechnol.*, 2013, 24, 639–645 .
57. E. G. Kelley, R. P. Murphy, J. E. Seppala, T. P. Smart, S. D. Hann, M. O. Sullivan and T. H. Epps, III, Size evolution of highly amphiphilic macromolecular solution assemblies via a distinct bimodal pathway, *Nat. Commun.*, 2014, 5, 3599 .
58. B. Rybtchinski, Adaptive Supramolecular Nanomaterials Based on Strong Noncovalent Interactions, *ACS Nano*, 2011, 5, 6791–6818 .

59. P. Theato, B. S. Sumerlin, R. K. O'Reilly and T. H. Epps, III, Stimuli responsive materials, *Chem. Soc. Rev.*, 2013, 42, 7055–7056 .

60. D. H. Zhang, S. H. Lahasky, L. Guo, C. U. Lee and M. Lavan, Polypeptoid Materials: Current Status and Future Perspectives, *Macromolecules*, 2012, 45, 5833–5841 .

61. D. Kedracki, I. Safir, N. Gour, K. Ngo and C. Vebert-Nardin, in *Bio-synthetic Polymer Conjugates Vol. 253 Advances in Polymer Science*, ed. Helmut Schlaad, Springer, Berlin Heidelberg, 2013, ch. 181, pp. 115–149 .

62. H. Benoit and G. Hadziioannou, Scattering theory and properties of block copolymers with various architectures in the homogeneous bulk state, *Macromolecules*, 1988, 21, 1449–1464 .

63. F. A. Leibfarth, J. A. Johnson and T. F. Jamison, Scalable synthesis of sequence-defined, unimolecular macromolecules by Flow-IEG, *Proc. Natl. Acad. Sci. U. S. A.*, 2015, 112, 10617–10622 .

64. A. H. Soeriyadi, C. Boyer, F. Nyström, P. B. Zetterlund and M. R. Whittaker, High-Order Multiblock Copolymers via Iterative Cu(0)-Mediated Radical Polymerizations (SET-LRP): Toward Biological Precision, *J. Am. Chem. Soc.*, 2011, 133, 11128–11131 .

65. J. R. Brown, S. W. Sides and L. M. Hall, Phase Behavior of Tapered Diblock Copolymers from Self-Consistent Field Theory, *ACS Macro Lett.*, 2013, 2, 1105–1109 .

66. K. J. Thurecht, A. M. Gregory, S. Villarroya, J. Zhou, A. Heise and S. M. Howdle, Simultaneous enzymatic ring opening polymerisation and RAFT-mediated polymerisation in supercritical CO_2, *Chem. Commun.*, 2006, 4383–4385 .

67. M. A. Hillmyer and F. S. Bates, Synthesis and characterization of model polyalkane-poly(ethylene oxide) block copolymers, *Macromolecules*, 1996, 29, 6994–7002 .

68. C. W. Bielawski and R. H. Grubbs, Living ring-opening metathesis polymerization, *Prog. Polym. Sci.*, 2007, 32, 1–29 .

69. A. J. Boydston, T. W. Holcombe, D. A. Unruh, J. M. J. Fréchet and R. H. A. Grubbs, Direct Route to Cyclic Organic Nanostructures via Ring-Expansion Metathesis Polymerization of a Dendronized Macromonomer, *J. Am. Chem. Soc.*, 2009, 131, 5388–5389 .

70. K. Nomura and M. M. Abdellatif, Precise synthesis of polymers containing functional end groups by living ring-opening metathesis polymerization (ROMP): efficient tools for synthesis of block/graft copolymers, *Polymer*, 2010, 51, 1861–1881 .

71. M. A. Sowers, J. R. McCombs, Y. Wang, J. T. Paletta, S. W. Morton, E. C. Dreaden, M. D. Boska, M. F. Ottaviani, P. T. Hammond, A. Rajca and J. A. Johnson, Redox-responsive branched-bottlebrush polymers for in vivo MRI and fluorescence imaging, *Nat. Commun.*, 2014, 5, 5460 .

72. N. Petzetakis, A. P. Dove and R. K. O'Reilly, Cylindrical micelles from the living crystallization-driven self-assembly of poly(lactide)-containing block copolymers, *Chem. Sci.*, 2011,2, 955–960 .

73. H. Robson Marsden and A. Kros, Polymer-Peptide Block Copolymers – An Overview and Assessment of Synthesis Methods, *Macromol. Biosci.*, 2009, 9, 939–951 .

74. P. De, M. Li, S. R. Gondi and B. S. Sumerlin, Temperature-Regulated Activity of Responsive Polymer–Protein Conjugates Prepared by Grafting-from via RAFT Polymerization, *J. Am. Chem. Soc.*, 2008, 130, 11288–11289 .

75. H. C. Kolb, M. G. Finn and K. B. Sharpless, Click Chemistry: Diverse Chemical Function from a Few Good Reactions, *Angew. Chem., Int. Ed.*, 2001, 40, 2004–2021 .

76. W. H. Binder and R. Sachsenhofer, 'Click' Chemistry in Polymer and Materials Science, *Macromol. Rapid Commun.*, 2007, 28, 15–54 .

77. W. F. Kuan, R. Roy, L. X. Rong, B. S. Hsiao and T. H. Epps, III, Design and Synthesis of Network-Forming Triblock Copolymers Using Tapered Block Interfaces, *ACS Macro Lett.*, 2012, 1, 519–523 .

78. C. E. Hoyle and C. N. Bowman, Thiol–Ene Click Chemistry, *Angew. Chem., Int. Ed.*, 2010, 49, 1540–1573 .

79. S. P. S. Koo, M. M. Stamenović, R. A. Prasath, A. J. Inglis, F. E. Du Prez, C. Barner-Kowollik, W. van Camp and T. Junkers, Limitations of radical thiol-ene reactions for polymer–polymer conjugation, *J. Polym. Sci., Part A: Polym. Chem.*, 2010, 48, 1699–1713 .

80. A. C. Engler, H.-I. Lee and P. T. Hammond, Highly Efficient "Grafting onto" a Polypeptide Backbone Using Click Chemistry, *Angew. Chem., Int. Ed.*, 2009, 48, 9334–9338 .

81. C. Wang, D. H. Lee, A. Hexemer, M. I. Kim, W. Zhao, H. Hasegawa, H. Ade and T. P. Russell, Defining the Nanostructured Morphology of Triblock Copolymers Using Resonant Soft X-ray Scattering, *Nano Lett.*, 2011, 11, 3906–3911 .

82. E. D. Gomez, A. Panday, E. H. Feng, V. Chen, G. M. Stone, A. M. Minor, C. Kisielowski, K. H. Downing, O. Borodin, G. D. Smith and N. P. Balsara, Effect of Ion Distribution on Conductivity of Block Copolymer Electrolytes, *Nano Lett.*, 2009, 9, 1212–1216 .

83. J. B. Gilbert, M. Luo, C. K. Shelton, M. F. Rubner, R. E. Cohen and T. H. Epps, III, Determination of Lithium-Ion Distributions in Nanostructured Block Polymer Electrolyte Thin Films by X-ray Photoelectron Spectroscopy Depth Profiling, *ACS Nano*, 2015, 9, 512–520 .

84. A. P. Smith, A. Sehgal, J. F. Douglas, A. Karim and E. J. Amis, Combinatorial mapping of surface energy effects on diblock copolymer thin film ordering, *Macromol. Rapid Commun.*, 2003, 24, 131–135 .

85. A. Toshio, High-speed atomic force microscopy coming of age, *Nanotechnology*, 2012, 23, 062001 .

86. Z. Jiang, X. Li, J. Strzalka, M. Sprung, T. Sun, A. R. Sandy, S. Narayanan, D. R. Lee and J. Wang, The dedicated high-resolution grazing-incidence X-ray scattering beamline 8-ID-E at the Advanced Photon

Source, *J. Synchrotron Radiat.*, 2012, 19, 627–636 .

87. X. Gu, I. Gunkel, A. Hexemer, W. Gu and T. P. Russell, An In Situ Grazing Incidence X-Ray Scattering Study of Block Copolymer Thin Films During Solvent Vapor Annealing, *Adv. Mater.*, 2014, 26, 273–281 .

88. S. P. Paradiso, K. T. Delaney, C. J. García-Cervera, H. D. Ceniceros and G. H. Fredrickson, Block Copolymer Self Assembly during Rapid Solvent Evaporation: Insights into Cylinder Growth and Stability, *ACS Macro Lett.*, 2014, 3, 16–20 .

89. J. E. Seppala, R. L. Lewis III and T. H. Epps, III, Spatial and Orientation Control of Cylindrical Nanostructures in ABA Triblock Copolymer Thin Films by Raster Solvent Vapor Annealing, *ACS Nano*, 2012, 6, 9855–9862 .

90. J. G. Son, H. M. Kang, K. Y. Kim, J. S. Lee, P. F. Nealey and K. Char, Orientation Change of Diblock Copolymer Thin Films by the Addition of Amphiphilic Surfactants: Effect of Film Thickness and Surfactant Concentration, *Macromolecules*, 2012, 45, 150–158 .

91. A. Jo, G. H. Gu, H. C. Moon, S. H. Han, S. H. Oh, C. G. Park and J. K. Kim, In situ TEM observation of phase transition of the nanoscopic patterns on baroplastic block copolymer films during nanoindentation, *Nanoscale*, 2013, 5, 4351–4354 .

92. J. P. Patterson, M. P. Robin, C. Chassenieux, O. Colombani and R. K. O'Reilly, The analysis of solution self-assembled polymeric nanomaterials, *Chem. Soc. Rev.*, 2014, 43, 2412–2425 .

93. Z. Y. Zhu, S. P. Armes and S. Y. Liu, pH-induced micellization kinetics of ABC triblock copolymers measured by stopped-flow light scattering, *Macromolecules*, 2005, 38, 9803–9812 .

94. M. T. Proetto, A. M. Rush, M.-P. Chien, P. Abellan Baeza, J. P. Patterson, M. P. Thompson, N. H. Olson, C. E. Moore, A. L. Rheingold, C. Andolina, J. Millstone, S. B. Howell, N. D. Browning, J. E. Evans and N. C. Gianneschi, Dynamics of Soft Nanomaterials Captured by Transmission Electron Microscopy in Liquid Water, *J. Am. Chem. Soc.*, 2014, 136, 1162–1165 .

95. T. M. Beardsley and M. W. Matsen, Monte Carlo Phase Diagram for a Polydisperse Diblock Copolymer Melt, *Macromolecules*, 2011, 44, 6209–6219 .

96. I. Nakamura and Z.-G. Wang, Salt-doped block copolymers: ion distribution, domain spacing and effective [small chi] parameter, *Soft Matter*, 2012, 8, 9356–9367 .

97. N. Nair and A. Jayaraman, Self-Consistent PRISM Theory – Monte Carlo Simulation Studies of Copolymer Grafted Nanoparticles in a Homopolymer Matrix, *Macromolecules*, 2010,43, 8251–8263 .

98. A. F. Hannon, Y. Ding, W. Bai, C. A. Ross and A. Alexander-Katz, Optimizing Topographical Templates for Directed Self-Assembly of Block Copolymers via Inverse Design Simulations, *Nano Lett.*, 2013, 14, 318–325 .

99. F. O. A. Detcheverry, G. Liu, P. F. Nealey and J. J. de Pablo, Interpolation in the Directed Assembly of Block Copolymers on Nanopatterned Substrates: Simulation and Experiments,*Macromolecules*, 2010, 43, 3446–3454 .

100. J. N. L. Albert, W. S. Young, R. L. Lewis, T. D. Bogart, J. R. Smith and T. H. Epps, III, Systematic Study on the Effect of Solvent Removal Rate on the Morphology of Solvent Vapor Annealed ABA Triblock Copolymer Thin Films, *ACS Nano*, 2012, 6, 459–466 .

101. Z. Li and E. E. Dormidontova, Equilibrium chain exchange kinetics in block copolymer micelle solutions by dissipative particle dynamics simulations, *Soft Matter*, 2011, 7, 4179–4188 .

102. A. Halperin, On Micellar Exchange: The Role of the Insertion Penalty, *Macromolecules*, 2011, 44, 5072–5074 .

103. M. M. Mok, R. Thiagarajan, M. Flores, D. C. Morse and T. P. Lodge, Apparent Critical Micelle Concentrations in Block Copolymer/Ionic Liquid Solutions: Remarkably Weak Dependence on Solvophobic Block Molecular Weight, *Macromolecules*, 2012, 45, 4818–4829 .

104. E. B. Zhulina and O. V. Borisov, Theory of Block Polymer Micelles: Recent Advances and Current Challenges, *Macromolecules*, 2012, 45, 4429–4440 .

105. E. H. Kellogg, A. Leaver-Fay and D. Baker, Role of conformational sampling in computing mutation-induced changes in protein structure and stability, *Proteins: Struct., Funct., Bioinf.*, 2011, 79, 830–838 .

106. P. Kujawa and F. M. Winnik, Innovation in Nanomedicine through Materials Nanoarchitectonics, *Langmuir*, 2013, 29, 7354–7361 .

107. K. Yao and C. Tang, Controlled Polymerization of Next-Generation Renewable Monomers and Beyond, *Macromolecules*, 2013, 46, 1689–1712 .

108. I. Akiba, A. Takechi, M. Sakou, M. Handa, Y. Shinohara, Y. Amemiya, N. Yagi and K. Sakurai, Anomalous Small-Angle X-ray Scattering Study of Structure of Polymer Micelles Having Bromines in Hydrophobic Core, *Macromolecules*, 2012, 45, 6150–6157 .

109. C. R. Thomas and D. Geer, Effects of shear on proteins in solution, *Biotechnol. Lett.*, 2011, 33, 443–456 .

110. S. K. Kumar and R. Krishnamoorti, Nanocomposites: Structure, Phase Behavior, and Properties, *Annu. Rev. Chem. Biomol. Eng.*, 2010, 1(1), 37–58 .

111. T. Li, M. J. Heinzer, E. M. Redline, F. Zuo, F. S. Bates and L. F. Francis, Microstructure and performance of block copolymer modified epoxy coatings, *Prog. Org. Coat.*, 2014, 77, 1145–1154

CHAPTER 5 ROLE OF HIGHLY BRANCHED, HIGH MOLECULAR WEIGHT POLYMER STRUCTURES IN DIRECTING UNIFORM POLYMER PARTICLE FORMATION DURING NANOPRECIPITATION

Fiona L. Hatton , Pierre Chambon , Alison C. Savage and Steve P. Rannard

Department of Chemistry, University of Liverpool, Crown Street, L69 7ZD, UK

ABSTRACT

The new macromolecular architecture, hyperbranched polydendrons, are composed of a broad distribution of molecular weights and architectural variation; however, nanoprecipitation of these materials yields highly uniform, dendron-functional nanoparticles. By isolating different fractions of the diverse samples, the key role of the most highly branched structures in directing nucleation and growth has been explored and determined.

INTRODUCTION

Polymeric nanostructures have been formed using various techniques including direct synthesis, such as micro/mini-emulsion polymerisation[1,2] and polymerisation induced self-assembly,[3–6]or the post-synthesis self-assembly of polymers of varying architectures, hydrophilicity or hydrophobicity.[7–9] Alternatively, physical fabrication methods such as the "Particle Replication In Non-wetting Templates" (PRINT™) approach[10] may be employed. In many self-assembly routes, a "solvent switch" is used to modify the media environment and drive inter-chain interactions. This can be achieved by changes in pH,[11] addition of salt,[12] removal of a co-solvent or the slow exchange of good solvent with an anti-solvent[13,14] (often water); solution and emulsion approaches have been reported.

A rapid and scale-able solvent switch technique that has gained considerable interest is nanoprecipitation,[15] especially in the formation of degradable nanoparticles for biomedical and nanomedicine applications. In general, polymers are dissolved in a good water-miscible solvent and rapidly added to an aqueous anti-solvent. Descriptions of both spontaneous, energy-barrier-free nucleation (spinodal decomposition)[16] and nucleation-and-growth mechanisms have been reported. In the latter case, providing the polymer is above its thermodynamic solubility limit within the resulting mixed solution, homogeneous nucleation of particles will occur.[16] The particle radii will initially be small, with a free energy of formation that is governed by parameters such as surface tension, particle radius and the free energy difference between media and particle. The free energy will pass through a maximum at a critical nucleus radius leading to rapid growth of larger particles. In either mechanism, growth will continue by depletion of polymer from the surrounding environment or through a process known as diffusion-limited cluster-cluster aggregation, assuming the balance of attractive and repulsive forces will allow interaction and steric/charge stabilisation does not prevent particles from achieving sufficient proximity

Citation: Fiona L. Hatton, Pierre Chambon, Alison C. Savage and Steve P. Rannard, Role of highly branched, high molecular weight polymer structures in directing uniform polymer particle formation during nanoprecipitation, DOI: 10.1039/c6cc00611f.

for effective collision.[15] To ensure rapid polymer nucleation, several groups utilise microfluidic or rapid-mixing techniques that allow distinct separation of the nucleation and growth/aggregation stages.[17,18]

Studies of the impact of polymer architecture and molecular weight on nanoprecipitation outcomes are very rare.[15] We have reported a new complex polymer architecture – "hyperbranched polydendrons"[19–21] – that overcomes the complexity of dendrimer synthesis by branching between linear–dendritic hybrid polymer chains during propagation (Scheme 1); weight average molecular weights $>10^6$ g mol^{-1}, representing >90 conjoined primary linear–dendritic chains, have been achieved. Hyperbranched polydendrons contain considerable variation in: (1) the number of branched chains (Fig. 1), (2) the position of branched points (Fig. 1) and, (3) molecular weight; a significant concentration of un-branched primary chains (linear–dendritic hybrid polymers) is also present. Despite this diversity, the complex mixtures nanoprecipitate to form stable, highly uniform nanoparticle dispersions. Each particle is surrounded by dendron functionality and outcomes are tunable to allow the multi-valency benefits of dendrimers to be achieved at sizes unobtainable by dendrimer synthesis alone (50–200 nm). Hyperbranched polydendrons mimic our studies of branched polymers without dendron chain-ends,[22,23] forming more uniform nanoprecipitates with increased stability and reduced size compared to particles generated from their chemically-identical linear primary chain building blocks. As such, they are ideal candidates for detailed examination of the role of the very highly branched, very high molecular weight fraction that is present within the molecular weight distribution.

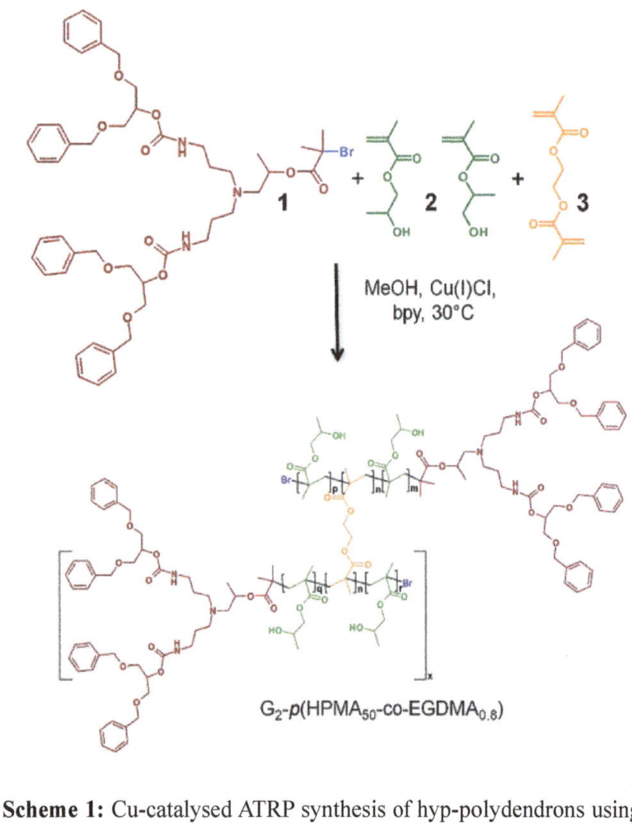

Scheme 1: Cu-catalysed ATRP synthesis of hyp-polydendrons using HPMA, **2** (both isomers shown), EGDMA, **3**, and a G_2 dendritic initiator, 1.

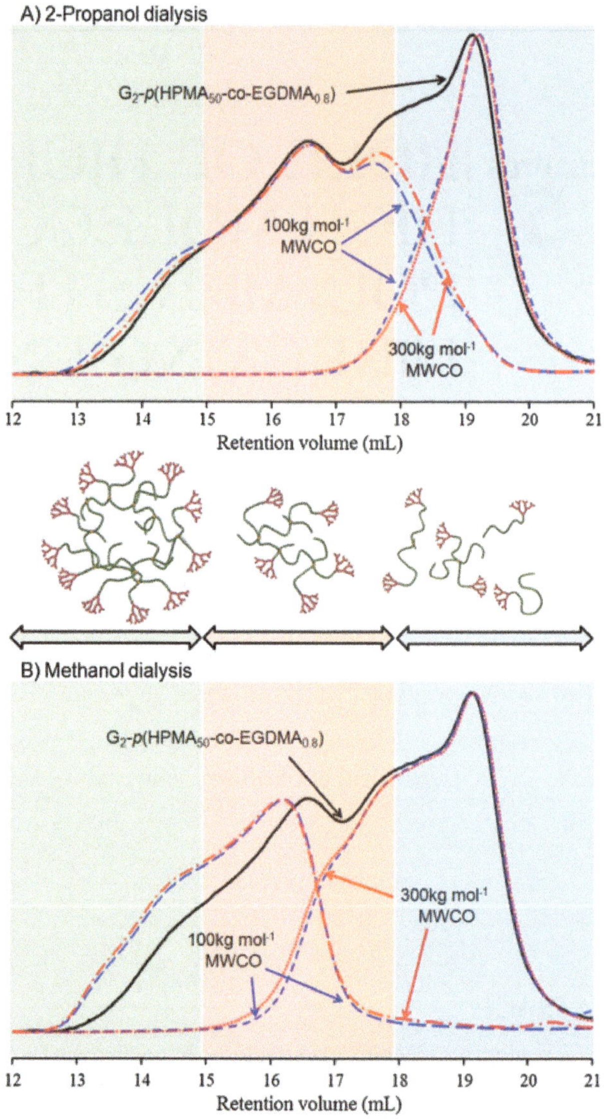

Figure 1: SEC (DMF/0.01 M LiBr eluent) refractive index chromatograms of hyperbranched polydendron samples before (solid black line) and after (red and blue dashed lines) dialysis using either (A) 2-propanol or (B) methanol as the dialysis solvent. Tubing with two different molecular weight cutoffs (MWCO) were utilized in each solvent condition.

Here we isolate the very high molecular weight fraction of hydrophobic hyperbranched polydendrons and study its role in directing successful nanoprecipitation to allow a greater mechanistic understanding and offer more general opportunities for enhanced nanoprecipitation processes.

The polymers were synthesised using previously reported approaches,[19–21] namely copper-catalysed methanolic atom transfer radical polymerisation at 30 °C, initiated using a new generation 2 (G_2) dendron initiator, 1, and a monomer mixture of 2-hydroxypropyl methacrylate, 2 (HPMA), and ethylene glycol dimethacrylate, 3 (EGDMA) in a 1 : 2 : 3 molar ratio of 1 : 50 : 0.8 (target number average degree of polymerisation for each primary chain of 50 monomer units). This ratio also ensures <1 brancher per primary chain and avoids gelation. The number average molecular weight (M_n) and weight average molecular weight (M_w) of the resulting branched polymer (G_2-*p*($HPMA_{50}$-*co*-$EGDMA_{0.8}$)) were determined by triple detection size exclusion chromatography

(DMF/0.01 M LiBr eluent; SEC) to be 115 700 g mol^{-1} and 1 538 000 g mol^{-1} respectively (dispersity ($Đ$) = 13.3; Fig. 1). The equivalent linear polymerisation in the absence of 3 yielded a polymer (G_2-p($HPMA_{50}$)) of M_n = 12 300 g mol^{-1} and M_w = 15 500 g mol^{-1} ($Đ$ = 1.26). Branching with 3 does not impact the formation of primary chains;[24] therefore, it is correct to consider the weight average structures within the G_2-p($HPMA_{50}$-co-$EGDMA_{0.8}$) as comprising approximately 100 conjoined G_2-p($HPMA_{50}$) chains.

This is highly important as M_w represents the molecular weight of macromolecules at the mean of the weight fraction and, therefore, approximately half of the physical mass of the sample comprises branched structures with a minimum of 100 conjoined chains; the M_n value indicates a large number of linear and lesser branched structures, as expected.

G_2-p($HPMA_{50}$-co-$EGDMA_{0.8}$) was fractionated using alcoholic dialysis (3 days) with either methanol or 2-propanol, and tubing with two nominal molecular weight cut-off values (MWCO = 100 kg mol^{-1} and 300 kg mol^{-1}). Two good solvents for p(HPMA) were studied due to the potential for variable wetting and pore swelling within the tubing; outer solutions were collected and fully replenished each day. Solutions within the dialysis tubing, and the combined solvent fractions collected from the reservoir, were dried and the recovered polymer studied by triple detection SEC (Fig. 1 and Table 1).

Table 1: SEC analysis of G_2-p($HPMA_{50}$-$EGDMA_{0.8}$) hyperbranched polydendrons fractions after alcoholic dialysis under varying conditions

Dialysis solvent (tubing MWCO)	Recovery site	G_2-p($HPMA_{50}$-co-$EGDMA_{0.8}$)		
		M_n (SEC, g mol^{-1})	M_w (SEC, g mol^{-1})	$Đ$
a Triple detection SEC using DMF/0.01 M LiBr eluent. *b* Polymer recovered from solution surrounding dialysis tubing. *c* Polymer recovered from inside dialysis tubing. See ESI for dn/dc values.				
—	—	115 700	1 538 000	13.3
2-Propanol				
100 kg mol^{-1}	External[b]	13 300	23 400	1.75
	Internal[c]	178 500	1 148 000	6.44
300 kg mol^{-1}	External[b]	14 000	21 700	1.55
	Internal[c]	136 400	955 800	7.01
Methanol				
100 kg mol^{-1}	External[b]	28 300	77 500	2.74
	Internal[c]	761 200	2 326 000	3.06
300 kg mol^{-1}	External[b]	25 400	99 500	3.92
	Internal[c]	615 800	1 990 000	3.23

The different MWCO dialysis tubing had a minimal impact on the separation, presumably due to non-aqueous solvent use; SEC chromatograms overlaid almost exactly (Fig. 1) and calculated M_n and M_w values were very similar (Table 1). Methanolic dialysis allowed enhanced isolation of the high molecular weight fraction with the lowest dispersity values.

Nanoprecipitation of the eight fractions (Table 1) and the linear–dendritic hybrid G_2-p($HPMA_{50}$) was conducted from THF solutions (5 mg mL^{-1}) using deionised water as the anti-solvent and targeting final nanoparticle concentrations of 1 mg mL^{-1} after THF removal. Dynamic light scattering analysis (DLS; Table 2) of the unfiltered samples yielded z-average diameter (D_z) values of 620–770 nm for nanoprecipitates of G_2-p($HPMA_{50}$) and the least branched materials fractionated using 2-propanol; similar methanol dialysis fractions generated significant macro-phase separated material. When the high molecular weight fractions isolated using 2-propanol or methanol were nanoprecipitated under identical conditions, considerably more uniform size distributions were observed with D_z values ranging from 135–205 nm and polydispersity (PdI) values ranging from 0.068–0.095 (Table 2).

Table 2: Dynamic light scattering

Polymer sample (dialysis solvent; tubing MWCO)	Recovery site	D_z (nm)	D_n (nm)	PdI
a DLS measured as unfiltered 1 mg mL^{-1} aqueous dispersions. *b* Polymer recovered from solution surrounding dialysis tubing. *c* Polymer recovered from inside dialysis tubing. *d* Significant macro-phase separation.				
G_2-p($HPMA_{50}$)	—	670	555	0.215
Dialysis fractions				
2-Propanol				
100 kg mol^{-1}	External[b]	620	525	0.214
	Internal[c]	190	165	0.070
300 kg mol^{-1}	External[b]	770	390	0.424
	Internal[c]	205	175	0.095

Methanol				
100 kg mol^{-1}	External^{b,d}	—	—	—
	Internal^{c}	145	115	0.075
300 kg mol^{-1}	External^{b,d}	—	—	—
	Internal^{c}	135	105	0.068

To investigate the role of the very high molecular weight species in directing the nanoprecipitation process of highly disperse samples, the high molecular weight fraction of G_2-*p*($HPMA_{50}$-*co*-$EGDMA_{0.8}$) collected using the 100 kg mol^{-1} MWCO tubing and methanolic dialysis (G_2-*p*($HPMA_{50}$-*co*-$EGDMA_{0.8}$)$_{MeOH}$) was systematically added at varying weight ratios to the G_2-*p*($HPMA_{50}$) sample prior to nanoprecipitation. This fraction was selected as it possessed the highest M_n and M_w values and the lowest $Đ$ value of the high molecular weight fractions, thereby allowing the most extreme case of molecular weight and architectural diversity after mixing with G_2-*p*($HPMA_{50}$). Nanoprecipitation of a G_2-*p*($HPMA_{50}$)/G_2-*p*($HPMA_{50}$-*co*-$EGDMA_{0.8}$)$_{MeOH}$ mixture containing just 1 wt% branched material generated significantly lower D_z and PdI values than those seen for G_2-*p*($HPMA_{50}$) alone; 198 nm (PdI = 0.034)*versus* 670 nm (PdI = 0.215) respectively (Fig. 2A).

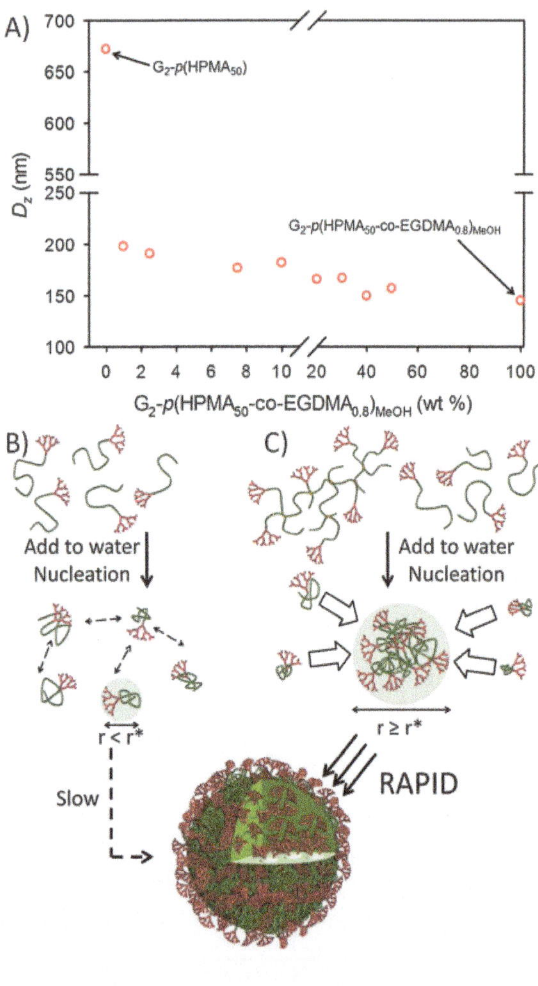

Figure 2: Effect of addition of highly branched, high molecular weight G_2-*p*($HPMA_{50}$-*co*-$EGDMA_{0.8}$)$_{MeOH}$ to G_2-*p*($HPMA_{50}$). (A) Observed modification of nanoprecipitate D_z (DLS) with increasing branched polymer; schematic representation of nucleation from (B) linear–dendritic hybrid polymers with slow growth stage and (C) hyperbranched polydendron generating large nucleus and rapid growth.

Further addition of G_2-*p*($HPMA_{50}$-*co*-$EGDMA_{0.8}$)$_{MeOH}$ led to a range of nanoparticles with steadily decreasing D_z values (PdI = 0.034–0.110), with the lowest value (145 nm) achieved from nanoprecipitation in the absence of linear G_2-*p*($HPMA_{50}$). The impact of highly branched, high molecular weight material within the nanoprecipitation of linear G_2-*p*($HPMA_{50}$) can be rationalised by consideration of the nucleation process in the mixed solvent/anti-solvent environment. In the absence of branched polymer, the diffusion of good solvent will lead to the media becoming increasingly less able to solvate the linear–dendritic hybrid polymer chains; these will collapse to form small particles comprising single or small numbers of collapsed macromolecules. It is unlikely that these small nuclei will exceed the critical radius (r^*), and slow assembly of chains will lead to poorly defined nucleation and growth stages (Fig. 2B); clearly distinct nucleation and growth are key to the formation of narrow size distributions.[15]

Conversely, the population of unimolecular structures comprising >100 conjoined chains will lead to the rapid formation of considerably larger nuclei, many of which would be expected to exceed r^* and allow a distinct nucleation stage; the linear–dendritic polymer would require a considerable number of single chains to coalesce in a rapid and concerted multi-body collision to achieve the same outcome. We recently reported a lack of discernible light scattering immediately after addition of linear polymer solutions to anti-solvents, in contrast to the observed strong scattering from branched polymers under identical conditions; this would support the absence of rapid nucleation in lower molecular weight linear materials.[23] Fractionation to generate G_2-*p*($HPMA_{50}$-*co*-$EGDMA_{0.8}$)$_{MeOH}$ also increased the number of conjoined chains within the weight average branched structure to >150 (number average structures contain >60 chains) (Table 1).

Decreasing D_z values with increasing content of G_2-*p*($HPMA_{50}$-*co*-$EGDMA_{0.8}$)$_{MeOH}$ (Fig. 2A) supports a nucleation-driven mechanism as increased nucleation density from additional highly branched polymer will lead to a larger number of smaller nanoparticles at constant polymer mass.[25]

The number of nanoparticles formed can be approximated by assuming the polymer density and utilising the number average diameter (D_n) values (Table S1, ESI†), and it is important to note that the number of chains and chain ends does not vary throughout this study. Previous studies have shown that increasing polymer molecular weight alone will lead to little difference in D_z values within nanoprecipitations of linear polymer systems, despite considerable variation in solution viscosity, with increased particle sizes resulting for some polymers.[26–28] The effects seen here are, therefore, clearly driven by the number of conjoined chains and the architecture of the high molecular weight fraction.

The ability to direct nanoprecipitation outcomes by introducing small fractions of highly branched polymers may have considerable value in ensuring consistency in medical products, decreasing complexity of mixing processes and in controlling nanoparticle sizes and polydispersities. To achieve such results with as little as 1 wt% of branched polymer opens significant opportunities, offers simplification within large scale processes and provides mechanistic insight.

The authors gratefully acknowledge financial support from the University of Liverpool and the Engineering and Physical Sciences Research Council for a PhD studentship (FL) and grant funding (EP/I038721/1) that underpinned the research.

NOTES AND REFERENCES

1. D. Qi, Z. Cao and U. Ziener, *Adv. Colloid Interface Sci.*, 2014, 211, 47 .
2. K. Landfester, A. Musyanovych and V. Mailaender, *J. Polym. Sci., Part A: Polym. Chem.*, 2010, 48, 493 .
3. L. P. D. Ratcliffe, B. E. McKenzie, G. M. D. Le Bouedec, C. N. Williams, S. L. Brown and S. P. Armes, *Macromolecules*, 2015, 48, 8594 .
4. R. J. J. Williams, C. E. Hoppe, I. A. Zucchi, H. E. Romeo, I. E. dell'Erba, M. L. Gomez, J. Puig and A. B. Leonardi, *J. Colloid Interface Sci.*, 2014, 431, 223 .
5. N. J. Warren and S. P. Armes, *J. Am. Chem. Soc.*, 2014, 136, 10174 .
6. B. Charleux, G. Delaittre, J. Rieger and F. D'Agosto, *Macromolecules*, 2012, 45, 6753 .
7. T. W. Nam, J. W. Jeong, M.-J. Choi, K. M. Baek, J. M. Kim, Y. H. Hur, Y. J. Kim and Y. S. Jung, *Macromolecules*, 2015, 48, 7938 .
8. M. S. Onses, L. Wan, X. Liu, N. B. Kiremitler, H. Yilmaz and P. F. Nealey, *ACS Macro Lett.*, 2015, 4, 1356 .
9. Z. L. Tyrrell, Y. Shen and M. Radosz, *Prog. Polym. Sci.*, 2010, 35, 1128 .
10. S. E. A. Gratton, P. D. Pohlhaus, J. Lee, J. Guo, M. J. Cho and J. M. DeSimone, *J. Controlled Release*, 2007, 121, 10 .
11. C. Maiti, R. Banerjee, S. Maiti and D. Dhara, *Langmuir*, 2015, 31, 32 .
12. C. Pinto Reis, R. J. Neufeld, A. J. Ribeiro and F. Veiga, *J. Nanomed. Nanotechnol.*, 2006, 2, 8 CrossRef .
13. J. P. Rao and K. E. Geckeler, *Prog. Polym. Sci.*, 2011, 36, 887 .
14. J. Jiang, Y. Liu, Y. Gong, Q. Shu, M. Yin, X. Liu and M. Chen, *Chem. Commun.*, 2012, 48, 10883 .
15. E. Lepeltier, C. Bourgaux and P. Couvreur, *Adv. Drug Delivery Rev.*, 2014, 71, 86 .
16. D. Horn and J. Rieger, *Angew. Chem., Int. Ed.*, 2001, 40, 4330 .
17. N. Anton, F. Bally, C. A. Serra, A. Ali, Y. Arntz, Y. Mely, M. Zhao, E. Marchioni, A. Jakhmola and T. F. Vandamme, *Soft Matter*, 2012, 8, 10628 .
18. K. M. Pustulka, A. R. Wohl, H. S. Lee, A. R. Michel, J. Han, T. R. Hoye, A. V. McCormick, J. Panyam and C. W. Macosko, *Mol. Pharmacol.*, 2013, 10, 4367 .
19. F. L. Hatton, P. Chambon, T. O. McDonald, A. Owen and S. P. Rannard, *Chem. Sci.*, 2014, 5, 1844 .
20. F. L. Hatton, L. M. Tatham, L. R. Tidbury, P. Chambon, T. He, A. Owen and S. P. Rannard, *Chem. Sci.*, 2015, 6, 326 .
21. H. E. Rogers, P. Chambon, S. E. R. Auty, F. Y. Hern, A. Owen and S. P. Rannard, *Soft Matter*, 2015, 11, 7005 .
22. R. A. Slater, T. O. McDonald, D. J. Adams, E. R. Draper, J. V. M. Weaver and S. P. Rannard, *Soft Matter*, 2012, 8, 9816 .
23. J. Ford, P. Chambon, J. North, F. L. Hatton, M. Giardiello, A. Owen and S. P. Rannard, *Macromolecules*, 2015, 48, 1883 .
24. J. Rosselgong, S. P. Armes, W. R. S. Barton and D. Price, *Macromolecules*, 2010, 43, 2145 .
25. Q. Wang, S. Samitsu, Y. Fujii, C. Yoshikawa, T. Miyazaki, H. Banno and I. Ichinose, *J. Polym. Sci., Part B: Polym. Phys.*, 2015, 53, 615 .
26. P. Legrand, S. Lesieur, A. Bochot, R. Gref, W. Raatjes, G. Barratt and C. Vauthier, *Int. J. Pharm.*, 2007, 344, 33 .
27. M. Beck-Broichsitter, E. Rytting, T. Lebhardt, X. Wang and T. Kissel, *Eur. J. Pharm. Sci.*, 2010, 41, 244 .
28. R. Stepanyan, J. G. J. L. Lebouille, J. J. M. Slot, R. Tuinier and M. A. Cohen Stuart, *Phys. Rev. Lett.*, 2012, 109, 138301

CHAPTER 6 PREPARATION OF THE INCLUSION COMPLEX-TYPE NONLINEAR OPTICAL POLYMER

Li-Fen Wang, Jyun-Wun Chen, and Jyun-Wei Chen

Department of Applied Chemistry and Material Science, Fooyin University, 151 Chin-Hsueh Road Ta-Liao Hsiang, Kaohsiung, 83102, Taiwan

ABSTRACT

This study uses the inclusion complex method to import nonlinear optical (NLO) chromophores, disperse red1 (DR1), and spiropyran (SP), into the γ-CD cavity of the γ-cyclodextrin polymer (γ-CDP) to prepare orderly aligned nonphotocontrollable and photocontrollable nonlinear optical polymers. Calculations support the ultraviolet/visible analyses and suggest the formation of the 1 : 2 DR1/γ-CDP and 1 : 2 SP/γ-CDP inclusion complexes. Upon complexation, the DR1 and SP molecules are free to align themselves along an applied electric field and show high order parameters of approximately 0.48 and 0.20, respectively. Reversible photochromic reactions exhibit that the SP/γ-CDP complex still retains the photochromic properties following corona poling.

INTRODUCTION

Nonlinear optical (NLO) material possessing optical modulation capabilities including changing optical phase, changing optical frequency, or coupling more than two light beams to increase or decrease frequency has recently become the most significant development goal for photonic devices [1, 2]. Through the development of NLO materials, photonics can be used as a future carrier of information and overcome transmission bottlenecks for current electronic transmission methods, including the resistive losses that occur with electrons moving through metal wires as the frequency of information increases [3].

Citation: Li-Fen Wang, Jyun-Wun Chen, and Jyun-Wei Chen, "Preparation of the Inclusion Complex-Type Nonlinear Optical Polymer," Journal of Spectroscopy, vol. 2013, Article ID 631946, 7 pages, 2013. doi:10.1155/2013/631946.

Although general NLO organic molecules have excellent NLO properties, their commercial applications are limited because of the difficulty of their processing. By combining NLO molecules with polymers, the high NLO susceptibility, fast response time, low dielectric constant, small dispersion in refractive index, structural flexibility, and ease of material processing have brought NLO polymers to the forefront of NLO research [4]. NLO polymers can be divided into guest/host polymer [5–7] and covalently linked polymers (side- and main-chain polymers) [8–10]. Guest/host polymers blend NLO chromophores (guest) to the polymer matrices (host). They can achieve a higher noncentral symmetry through poling, but are prone to phase separation [11]. Side- and main-chain polymers bond the NLO chromophores to the polymer chains covalently and exhibit good phase mixing. However, the mobility of the polymer chain hinders the alignment of NLO chromophores, influencing the efficiency of electric field poling [12, 13]. Crosslinking of the polymer which contains chromophore groups [14–16] or formation of interpenetrating polymer networks [17, 18] has been developed to reduce the mobility of the polymer segments; consequently, the materials require longer poling times and larger poling voltages [19, 20].

Disperse red 1 (DR1, Figure 1(a)) is very promising as NLO chromophore, because it is easy to create polar order of them by external electric field poling. To improve its processing and application characteristics, most studies have focused on the blending or covalent bonding methods to import it into the polymer [21–24]. Cui et al. [22] used a sol-gel process to copolymerize alkoxysilane dye and DR1 to prepare organic-inorganic hybrid materials. Aldea et al. [23] used the layer-by-layer electrostatic self-assembly structure to create multilayer thin film-based NLO materials. Tu et al. [24] explored the dynamic molecular behavior for mixing DR1 and poly(methyl methacrylate).

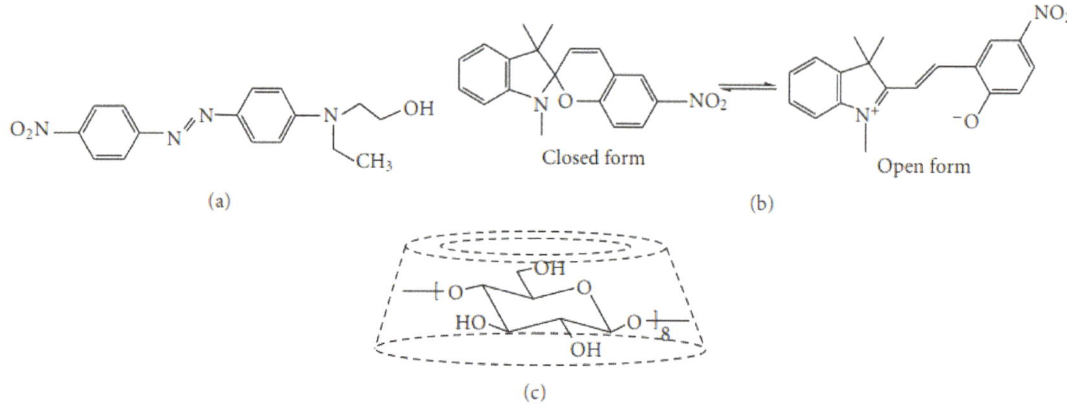

Figure 1: Schematic structures for (a) DR1, (b) SP, and (c) γ-CDP.

Spiropyran (SP, Figure 1(b)) plays a critical role in the photochromism field because the colorless SP form will convert to a red-colored photomerocyanine (PM) under ultraviolet (UV) irradiation. The PM form can revert to the SP form either thermally or photochemically [25–27]. Typically, SP is included in polymer matrices to obtain various forms of solid-state photochromic materials to extend its commercial application range [28–30]. Kim et al. [28] prepared a series of cationic polymers using spirooxazines as side-chain groups. All of these synthesized photochromic polymer materials, whether in solution phase or solid-state thin films, have good photochromic properties.

Photomerocyanine (PMs) are comprised of strong electron donor and acceptor groups connected by a -conjugated bridge. They possess delocalized -electron systems capable of exhibiting charge transfer resonances, which leads to a preferential orientation of the molecules along the applied electric field [31–33]. Experimental values of dipole moments and first-order hyperpolarizability β were reported for various PMs [34–37]. Replacing the benzopyryl moiety by 1,3-diethylthiobarbituric acid, the indoline-based PM exhibits a negative β value ($\beta = -46 \times 10^{-30}$ esu) and a dipole moment of 7.4×10^{-18} esu [34]. Theoretically, considerable efforts have been directed toward the molecular engineering of such chromophore structures [38–40]. We used the density functional theory (DFT) calculation [41–43], CAM-B3LYP/6-31+G(d,p), and found that PM has a high dipole moment of approximately 2.8×10^{-30} esu and shows high second-order NLO properties with first hyperpolarizability of approximately 38.4×10^{-30} esu (commonly used stilbene-series NLO molecules are approximately $7 \times 10^{-30} \sim 38 \times 10^{-30}$). Because colorless SP form does not have NLO properties, different optical wavelengths can convert it between the two forms and produce second-order NLO property changes that can be applied to photocontrollable components for data storage or optooptical switching [44].

Therefore, this study uses the inclusion complex method to separately import DR1 and SP into the γ-cyclodextrin (γ-CD) cavity of γ-cyclodextrin polymer (γ-CDP, Figure 1(c)) to prepare DR1/γ-CDP and SP/γ-CDP complexes for nonphotocontrollable and photocontrollable NLO polymers. Incorporation of chromophore in the cavity of γ-CDP increases the free volume surrounding the chromophore and decreases the spatial constraint from polymer chains. Additionally, the physical bonding between chromophore and γ-CDP enables the materials to simultaneously possess the guest/host NLO material's advantage of being simple to prepare and easy to polarize and the covalently linked NLO material's advantage of having a stable phase mixing structure.

EXPERIMENTAL

Preparation of Inclusion Complexes

N-Ethyl-N-(2-hydroxyethyl)-4-(4-nitrophenylazo)aniline (Disperse red1, DR1) and γ-cyclodextrin polymer (γ-CDP) were purchased from Sigma-Aldrich, and 1′-3′-3′-Trimethyl-6-nitrospiro [2H-1-benzopyran-2,2′-indoline] (Spiropyran, SP) was purchased from TCI, respectively, and used as received. Solutions (250 μL, 1×10^{-2} M) of DR1 and SP in ethanol were injected separately into 5 mL of aqueous γ-CDP (1×10^{-2} mol γ-CD/L) solution at 30°C with sonication for 10 min [45, 46]. After 24 h of stirring, the DR1/γ-CDP and SP/γ-CDP inclusion complexes were deposited separately by spin coating onto ITO (indium tin oxide) glass plates and dried at 40°C in vacuo overnight for Ultraviolet/Visible (UV/VIS, Hitachi U-3900) analysis. Because of the extreme light sensitivity of SP, all operations for SP/γ-CDP complex preparation must be conducted in a dark room.

To calculate the complex stoichiometry, solutions of DR1 and SP (1×10^{-5} M) in ethanol were prepared separately. Samples of 0~7 mL γ-CDP solution with concentrations of 1×10^{-3} M were added to each 1 mL sample of DR1 or SP solution. The samples were then diluted to 10 mL by adding water.

Poling Process

External electrical field (corona) poling was used to break the symmetry of the inclusion complex. The corona field between −3 and −10 kV was applied for 15 min after the film was heated to the desired temperature of approximately 150–200°C [14, 24, 47]. Then the sample was cooled to room temperature under a continuous electrical field. During the poling procedure, the corona current was maintained at less than 50 μA, and the distance between the ITO and the needle was set at 1 cm. For the SP/γ-CDP complex, the poling process had to be conducted under the irradiation of UV light because a high dipole moment value only appears after the formation of the PM form. The changes in the UV/VIS spectroscopic absorbance were used to examine the alignment of the γ-CDP-complexed NLO chromophore (DR1 and SP) before and after poling by order parameter Φ [48]:

$$\Phi = 1 - \left(\frac{A_f}{A_o}\right), \tag{1}$$

where A_0 and A_f are the UV/VIS absorbance for the polymer film before and after poling, respectively. After the oriented arrangement of molecules, the UV/VIS absorbance is reduced, causing the Φ value to increase.

Quantum Chemical Calculations

The geometries of complexed compounds were firstly optimized using Parameterized Model number 3 (PM3) [49], and then the two-layered hybrid ONIOM method [50], ONIOM(B3LYP/6-31G+(d,p):PM3), was adopted to find the ground-state geometry. The high-level layer, DR1 and SP, and the low-level layer, γ-CD, were calculated by B3LYP/6-31G+(d,p) and PM3, respectively. For comparison, B3LYP/6-31G+(d,p) and PM3 alone were also used to optimize the geometrical structures of isolated DR1 (or SP) and γ-CD, respectively. All calculations were made using the Gaussian09 program package [51].

RESULTS AND DISCUSSION

Stoichiometry of Inclusion Complexes

The stoichiometry of the inclusion complex was determined using the Benesi-Hildebrand method [52–54] from the UV/VIS data:

$$\frac{A_0}{\Delta A} = \frac{[1 + 1/(\mathrm{K} \cdot [\gamma - \mathrm{CD}]^n)]}{(\Delta\varepsilon \cdot [\mathrm{chromophore}])}, \tag{2}$$

where ΔA represents the difference between the absorptions of the chromophore (DR1 or SP) solution in the presence and absence (A_0) of γ-CD; ; $\Delta\varepsilon$ denotes the difference between the molar absorption coefficients of chromophore/γ-CD and chromophore; n is the molar ratio of γ-CD to SP; and [chromophore] and [γ-CD] are the concentrations of chromophore and γ-CD, respectively. The linear relationships of ΔA^{-1} against $[\gamma\text{-CD}]^{-2}$ for both DR1 and SP reveal the formation of 1:2 DR1/γ-CD and 1:2 SP/γ-CD complexes (Figure 2).

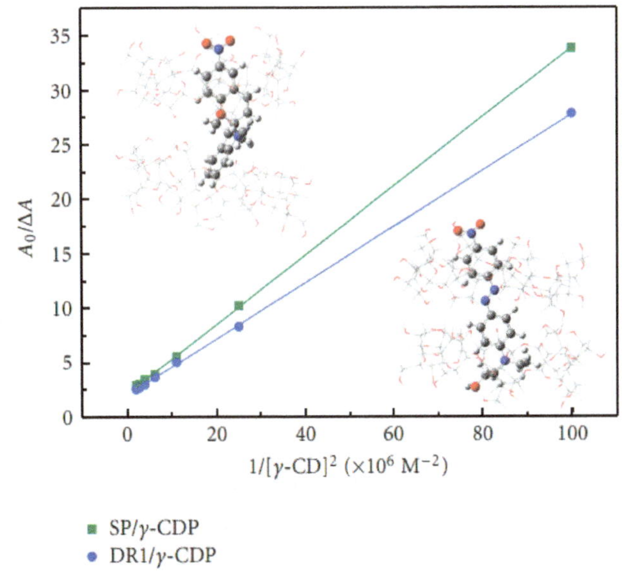

Figure 2: The linearity of the data when 1:2 complexation of DR1:γ-CD and SP:γ-CD is assumed. The insets show the optimized complexes.

Quantum chemical calculations by using the two-layered hybrid ONIOM method were performed for each of the complexes of interest to yield geometries (Figure 2 inset) and energies that were determined exclusively by the inclusion processes. The calculations of binding energies revealed that for both DR1 and SP, the 1:2 γ-CD inclusion complexes (22.6 and 15.4 kcal mol^{-1} for DR1/γ-CD and SP/γ-CD complexes, resp.) are more stable than the 1:1 γ-CD inclusion complexes (7.7 and 5.2 kcal mol^{-1} for DR1/γ-CD and SP/γ-CD complex, resp.). This result agrees with experimental UV/VIS analyses: the hydrophobic γ-CD cavity encourages the end moieties of chromophore studied each entering one γ-CD cavity from the polar solvent.

Nonphotocontrollable NLO Materials

Figure 3 shows the UV/VIS spectra of DR1 in ethanol and in γ-CDP. After γ-CDP complexation, the absorption maxima of DR1 moved from 481 nm to 514 nm, thereby confirming the formation of the DR1/γ-CDP complex. The increase in absorption wavelength upon complexation is attributable to the interaction of DR1 and γ-CDP. To achieve an overall noncentrosymmetry, the corona poling method was used to enable the orderly alignment of DR1 through an external electric field.

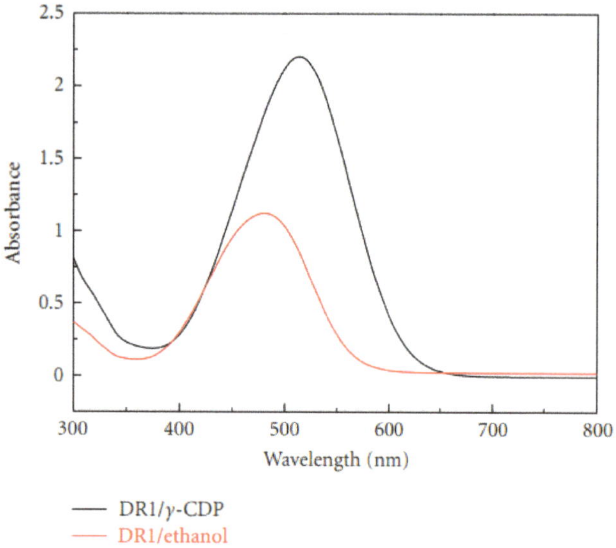

Figure 3: UV/VIS absorption spectra of DR1 in ethanol and in γ-CDP.

The poling voltage (−6.0, −7.0, −7.5 and −8.0 kV), temperature (170, 180, 190, and 200°C), and time (15, 25 and 30 min) were altered, respectively, to determine the optimal poling conditions. After the oriented arrangement of molecules, the UV/VIS spectrum exhibits a decrease in absorption through which the order parameter (Φ) of the poled film can be estimated. The experimental results indicate that the Φ value increases with the poling time until 15 min. However, the Φ value was no longer affected after 15 min.

Nevertheless, the poling voltage and temperature range used in this study have a significant impact on the value (Figure 4). As the poling temperature is fixed, the UV/VIS absorbance is reduced significantly (Φ value increased) with the increase of poling voltage (Figure 5), but the Φ value increases only slightly after the poling voltage is increased to −7.5 kV. At a fixed poling voltage (Figure 4), when the temperature increases from 170°C to 180°C, the Φ value changes more significantly, increasing approximately 12% (−8.0 kV) to 29% (−6.0 kV).

If the temperature continues to increase until 190°C, the Φ value would increase accordingly (only slightly), at approximately 4% (−8.0 kV) to 8% (−6.0 kV). At a poling temperature of 200°C, the largest Φ value is obtained at a low-voltage of −6.0 kV, showing approximately twice the Φ value compared to other temperatures. However, the Φ value increased only slightly when the voltage continued to increase. The optimal operating conditions for DR1/γ-CDP are −8.0 kV/190°C to obtain an Φ value of 0.48, which is higher than the Φ value between 0.16 and 0.30 for DR1 doped poly(methyl methacrylate) [55, 56].

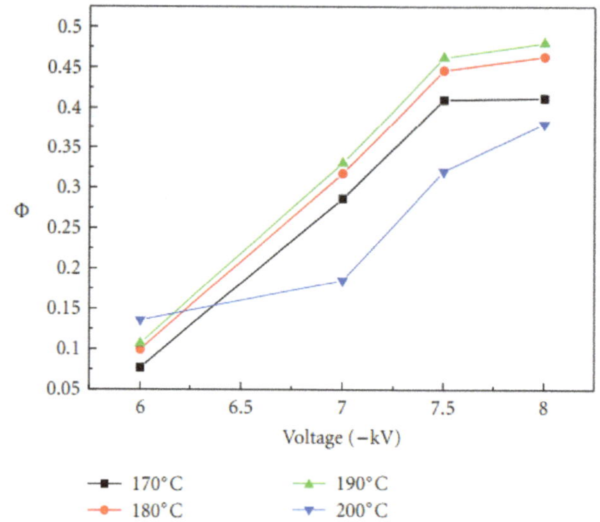

Figure 4: Effect of poling conditions on the order parameter (Φ) of DR1/γ-CDP complex.

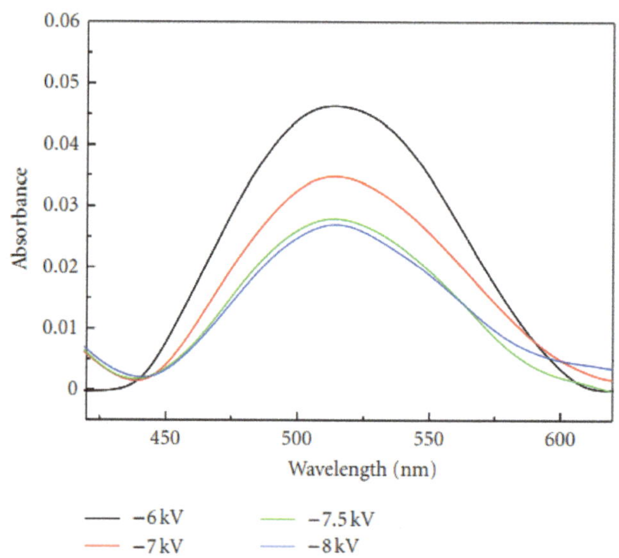

Figure 5: UV/VIS absorption spectra of DR1/γ-CDP complex after electric-field poling.

Photocontrollable NLO Materials

The UV/VIS absorption spectra for the prepared SP/γ-CDP inclusion complex indicate that, after SP was combined with γ-CDP, the absorption maximum of SP moved from 337 nm to 382 nm, confirming the formation of the SP/γ-CDP complex. The experimental results showed that the SP/γ-CDP complex had a higher Φ value at a poling temperature of 150°C. At this temperature, the Φ value increased with poling voltage increases from −1.0 kV to −3.0 kV, but decreased as the poling voltage increased further from −3.0 kV to −6 kV (Figure 6). Among them, the poling voltage of −3.0 kV demonstrated the highest Φ value, at approximately 0.20.

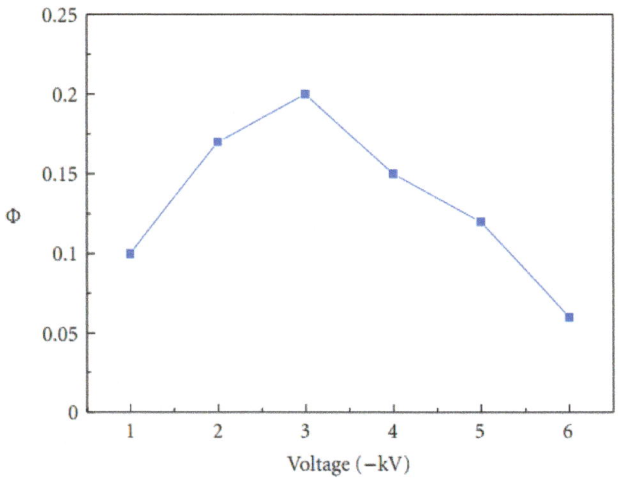

Figure 6: Order parameter (Φ) of SP/γ-CDP complex at 150°C.

To clarify whether SP/γ-CDP still retains photochromic properties after high voltage poling, the study applied repeated irradiation of 366 nm on the samples. The results indicated that after the materials absorbed UV light, a 568 absorption peak emerged, which was the absorption position of open-form PM. After the UV light was removed, the 568 absorption peak disappeared and the material returned to its original colorless state. Therefore, the SP/γ-CDP retains photochromic properties after high voltage poling.

The Φ values of all poled polymer placed in a drying oven for 120 days under ambient temperature and atmospheric pressure are maintained. This indicates that γ-CDP complexation produces excellent stability for the oriented arrangement of NLO chromophore. The hydrogen bonds between the hydroxyl groups laced on the rims of γ-CD increase the rigidity of the polymer chain preventing the relaxation of induced dipoles and further enhancing the stability of the poled complex (Figure 7). Additionally, the γ-CD cavity avoids the mobility of polymer segments interrupting the alignment of the NLO chromophore; therefore, the inclusion-complexed DR1 and SP molecules are free to align themselves along an applied electric field.

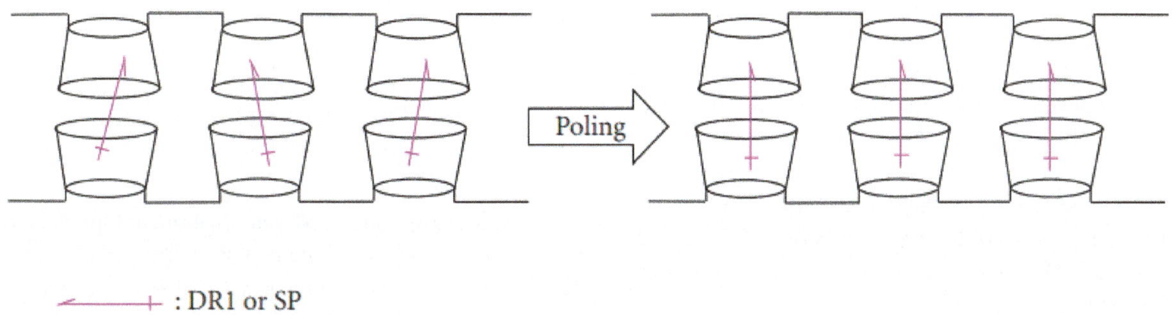

Figure 7: Schematics of aligned NLO chromophores in γ-CDP complexes.

CONCLUSION

Nonlinear optical polymers with oriented and orderly alignment have been prepared by forming inclusion complexes of disperse red 1 (DR1) and spiropyran (SP) in the γ-cyclodextrin cavity of γ-cyclodextrin polymer (γ-CDP). Among them, the SP/γ-CDP complex retains photochromic properties after poling and can be applied in photocontrollable NLO materials. Consistent with experimental ultraviolet/visible (UV/VIS) spectra, quantum chemical calculations provided valuable insight into the possible 1:2 inclusion geometries and the relative stabilities of the DR1/γ-CD and SP/γ-CD complexes. The UV/VIS spectral analyses demonstrate the correlation between poling voltage, poling temperature, poling time, and the NLO properties. The orderliness of the molecular alignment (order parameter Φ) within the material is significantly affected by the poling voltage and temperature. Voltage has a more significant effect than temperature.

Increasing voltage between −6.0 kV and −8.0 kV increases the Φ value of DR1/γ-CDP at a fixed temperature. At a fixed voltage, the Φ value increases with temperature at the ranges of 170°C and 190°C. The highest Φ value of approximately 0.48 is obtained under the condition where the voltage is −8.0 kV and temperature is 190°C. When the temperature is 200°C, the Φ value obtained under the voltage of −6.0 kV is approximately double the Φ value obtained from any other temperatures. However, continued increases of voltage increments may cause material degradation.

SP/γ-CDP still retains photochromic properties after high voltage poling, implying that γ-CDP provides adequate space to preserve the photochromic conversion of SP. The optimal operating conditions for SP/γ-CDP are −3.0 kV/150°C to obtain an Φ value of 0.20. Materials maintained the same Φ value after poling for 120 days. This may be attributed to the fact that the γ-CDP forms extensive hydrogen bonds in adjacent γ-CD and thereby increases rigidity, preventing the relaxation of induced dipoles and allowing the achievement of a high degree of stability.

ACKNOWLEDGMENTS

The authors would like to thank the National Science Council of Taiwan (Grant NSC 99-2221-E-242-001-) for financial support. Also, the computer time and facilities provided by the National Center for High-Performance Computing (NCHC) are highly appreciated.

REFERENCES

1. M. P. Balanay and D. H. Kim, "Optical properties of porphyrin analogues for solar cells: an NLO approach," Current Applied Physics, vol. 11, no. 1, pp. 109–116, 2011.
2. L. R. Dalton, P. A. Sullivan, and D. H. Bale, "Electric field poled organic electro-optic materials: state of the art and future prospects," Chemical Reviews, vol. 110, no. 1, pp. 25–55, 2010.
3. P. A. Sullivan and L. R. Dalton, "Theory-inspired development of organic electro-optic materials," Accounts of Chemical Research, vol. 43, no. 1, pp. 10–18, 2010.
4. M. J. Cho, D. H. Choi, P. A. Sullivan, A. J. P. Akelaitis, and L. R. Dalton, "Recent progress in second-order nonlinear optical polymers and dendrimers," Progress in Polymer Science, vol. 33, no. 11, pp. 1013–1058, 2008.
5. E. Jecs, J. Kreicberga, V. Kampars, A. Jurgis, and M. Rutkis, "Novel azobenzene precursors for NLO active polyuretanes: synthesis, quantum chemical and experimental characterization," Optical Materials, vol. 31, no. 11, pp. 1600–1607, 2009.
6. W. Shi, Z. Zhang, Q. Pan et al., "Nonlinear optical properties and chromophore electrostatic interactions for the poly(ether ketone) guest-host polymer films," Macromolecules, vol. 34, no. 6, pp. 2002–2007, 2001.·
7. W. J. Kuo, G. H. Hsiue, and R. J. Jeng, "Novel guest-host NLO poly(ether imide) based on a two-dimensional carbazole chromophore with sulfonyl acceptors," Macromolecules, vol. 34, no. 7, pp. 2373–2384, 2001. · ·
8. Z. Li, W. Wu, C. Ye, J. Qin, and Z. Li, "Two types of nonlinear optical polyurethanes containing the same isolation groups: syntheses, optical properties, and influence of binding mode," Journal of Physical Chemistry B, vol. 113, no. 45, pp. 14943–14949, 2009.
9. P. A. Ribeiro, D. T. Balogh, J. L. C. Fonseca, and J. A. Giacometti, "Chromophore relaxation in a side-chain methacrylate copolymer functionalized with 4-[N-Ethyl-N-(2-hydroxyethyl)]ammo-2′-chloro-4′-nitroazobenzene," Macromolecules, vol. 37, no. 7, pp. 2618–2624, 2004.
10. M. Ronchi, A. O. Biroli, D. Marinotto, M. Pizzotti, M. C. Ubaldi, and S. M. Pietralunga, "The role of the chromophore size and shape on the SHG stability of PMMA films with embebbed NLO active macrocyclic chromophores based on a cyclotetrasiloxane scaffold," Journal of Physical Chemistry C, vol. 115, no. 10, pp. 4240–4246, 2011.
11. K. H. Park, R. J. Twieg, R. Ravikiran et al., "Synthesis and nonlinear-optical properties of vinyl-addition poly(norbornene)s," Macromolecules, vol. 37, no. 14, pp. 5163–5178, 2004.
12. Y. Zhang, J. Martinez-Perdiguero, U. Baumeister et al., "Laterally azo-bridged H-shaped ferroelectric dimesogens for second-order nonlinear optics: ferroelectricity and second harmonic generation," Journal of the American Chemical Society, vol. 131, no. 51, pp. 18386–18392, 2009.
13. Z. Li, G. Yu, Y. Liu, C. Ye, J. Qin, and Z. Li, "Dendronized polyfluorenes with high azo-chromophore loading density: convenient synthesis and enhanced second-order nonlinear optical effects,"Macromolecules, vol. 42, no. 17, pp. 6463–6472, 2009.
14. C. Wang, C. Zhang, M. S. Lee, and L. R. Dalton, "Urethane-urea copolymers containing siloxane linkages: enhanced temporal stability and low optical loss for second-order nonlinear optical applications," Macromolecules, vol. 34, no. 7, pp. 2359–2363, 2001.
15. P. H. Chang, H. C. Tsai, Y. R. Chen, J. Y. Chen, and G. H. Hsiue, "Thermal stability and structural characterization of organic/inorganic hybrid nonlinear optical material containing a two-dimensional chromophore," Langmuir, vol. 24, no. 20, pp. 11921–11927, 2008.
16. J. A. F. Boogers, P. T. A. Klaase, J. J. de Vlieger, and A. H. A. Tinnemans, "Cross-linked polymer materials for nonlinear optics. 2. Polyurethanes bearing azobenzene dyes," Macromolecules, vol. 27, no. 1, pp. 205–209, 1994.
17. P. Prêtre, U. Meier, U. Stalder et al., "Relaxation processes in nonlinear optical polymers: a comparative study," Macromolecules, vol. 31, no. 6, pp. 1947–1957, 1998.
18. R. J. Jeng, C. C. Chang, C. P. Chen, C. T. Chen, and W. C. Su, "Thermally stable crosslinked NLO materials based on maleimides," Polymer, vol. 44, no. 1, pp. 143–155, 2003.
19. H. S. Nalwa and S. Miyata, Nonlinear Optics of Organic Molecules and Polymers, CRC press, 1997.
20. C. Zhang, C. Wang, L. R. Balten, H. Zhang, and W. H. Steier, "Progress toward device-quality second-order nonlinear optical materials. 4. a trilink high μβ NLO chromophore in thermoset polyurethane: a est-host approach to larger electrooptic coefficients," Macromolecules, vol. 34, no. 2, pp. 253–261, 2001.·
21. D. Möncke, G. Mountrichas, S. Pispas, and E. I. Kamitsos, "Orientation phenomena in chromophore DR1-containing polymer films and their non-linear optical response," Materials Science and Engineering B, vol. 176, no. 6, pp. 515–520, 2011.
22. Y. Cui, M. Wang, L. Chen, and G. Qian, "Synthesis and spectroscopic characterization of an alkoxysilane dye containing C. I. Disperse Red 1," Dyes and Pigments, vol. 62, no. 1, pp. 43–47, 2004.
23. G. Aldea, H. Gutiérrez, J. M. Nunzi, G. C. Chitanu, M. Sylla, and B. C. Simionescu, "Second harmonic generation diagnostic of layer-by-layer deposition from Disperse Red 1—functionalized maleic anhydride copolymer," Optical Materials, vol. 29, no. 12, pp. 1640–1646, 2007.
24. Y. Tu, Q. Zhang, and H. Ågren, "Electric field poled polymeric nonlinear optical systems: molecular dynamics simulations of poly(methyl methacrylate) doped with disperse red chromophores," Journal of Physical Chemistry B, vol. 111, no. 14, pp. 3591–3598, 2007.
25. K. Kinashi, Y. Ono, Y. Naitoh, A. Otomo, and Y. Ueda, "Time-resolved fluorescence study on the

photomerocyanine form of spiropyran and its derivative with azobenzene," Journal of Photochemistry and Photobiology A, vol. 217, no. 1, pp. 35–39, 2011.

26. S. V. Paramonov, V. Lokshin, and O. A. Fedorova, "Spiropyran, chromene or spirooxazine ligands: insights into mutual relations between complexing and photochromic properties," Journal of Photochemistry and Photobiology C, vol. 12, no. 3, pp. 209–236, 2011.
27. S. Kumar, K. Velasco, and A. McCurdy, "X-ray, kinetics and DFT studies of photochromic substituted benzothiazolinic spiropyrans," Journal of Molecular Structure, vol. 968, no. 1–3, pp. 13–18, 2010.
28. S. H. Kim, S. Y. Park, C. J. Shin, and N. S. Yoon, "Photochromic behaviour of poly[N,N-[(3-dimethylamino)propyl]methacrylamide] having spiroxazine pendant group," Dyes and Pigments, vol. 72, no. 3, pp. 299–302, 2007.
29. Z. S. Fu, B. B. Sun, J. Chen, and L. Yuan, "Preparation and photochromism of carboxymethyl chitin derivatives containing spirooxazine moiety," Dyes and Pigments, vol. 76, no. 2, pp. 515–518, 2008.
30. A. Yamano and H. Kozuka, "Perhydropolysilazane-derived silica-polymethylmethacrylate hybrid thin films highly doped with spiropyran: effects of polymethylmethacrylate on the hardness, chemical durability and photochromic properties," Thin Solid Films, vol. 519, no. 6, pp. 1772–1779, 2011.
31. L. Sanguinet, J. L. Pozzo, M. Guillaume et al., "Acidoswitchable NLO-phores; benzimidazolo[2,3-b] oxazolidines," Journal of Physical Chemistry B, vol. 110, no. 22, pp. 10672–10682, 2006.
32. L. Sanguinet, J. L. Pozzo, V. Rodriguez et al., "Acido- and phototriggered NLO properties enhancement,"Journal of Physical Chemistry B, vol. 109, no. 22, pp. 11139–11150, 2005.
33. Y. Atassi, J. A. Delaire, and K. Nakatani, "Coupling between photochromism and second-harmonic generation in spiropyran- and spirooxazine-doped polymer films," Journal of Physical Chemistry, vol. 99, no. 44, pp. 16320–16326, 1995.
34. R. Ortiz, S. R. Marder, L. T. Cheng, B. G. Tiemann, S. Cavagnero, and J. W. Ziller, "The dependence of the molecular first hyperpolarizabilities of merocyanines on ground-state polarization and length," Journal of the Chemical Society, Chemical Communications, no. 19, pp. 2263–2264, 1994.
35. G. J. Smith, A. P. Middleton, D. J. Clarke et al., "The effect of solvent on the excited vibronic states and first hyperpolarizability of "push-pull" merocyanines," Optical Materials, vol. 32, no. 9, pp. 1237–1243, 2010.
36. C. J. Wyss, G. J. Smith, A. D. Woolhouse et al., "The first-order hyperpolarizabilities of some charge transfer conjugated molecules with high transmission in the far red," Optical Materials, vol. 16, no. 3, pp. 341–347, 2001.
37. M. E. Reish, A. J. Kay, A. Teshome, I. Asselberghs, K. Clays, and K. C. Gordon, "Testing computational models of hyperpolarizability in a merocyanine dye using spectroscopic and DFT methods," Journal of Physical Chemistry A, vol. 116, no. 22, pp. 5453–5463, 2012.
38. C. Z. Zhang, H. Cao, C. Im, and G. Y. Lu, "Theoretical study of nonlinear optical properties of "parallel connection" chromophores containing parallel nonconjugated D-π-A units," Journal of Physical Chemistry A, vol. 113, no. 44, pp. 12295–12303, 2009.
39. F. Momicchioli, G. Ponterini, and D. Vanossi, "First- and second-order polarizabilities of simple merocyanines. An experimental and theoretical reassessment of the two-level model," Journal of Physical Chemistry A, vol. 112, no. 46, pp. 11861–11872, 2008.
40. Z. Sainudeen and P. C. Ray, "Nonlinear optical properties of zwitterionic merocyanine aggregates: role of intermolecular interaction and solvent polarity," Journal of Physical Chemistry A, vol. 109, no. 40, pp. 9095–9103, 2005.
41. A. D. Becke, "Density-functional thermochemistry. III. The role of exact exchange," The Journal of Chemical Physics, vol. 98, no. 7, pp. 5648–5652, 1993.
42. C. Lee, W. Yang, and R. G. Parr, "Development of the Colle-Salvetti correlation-energy formula into a functional of the electron density," Physical Review B, vol. 37, no. 2, pp. 785–789, 1988.
43. B. Miehlich, A. Savin, H. Stoll, and H. Preuss, "Results obtained with the correlation energy density functionals of becke and Lee, Yang and Parr," Chemical Physics Letters, vol. 157, no. 3, pp. 200–206, 1989.
44. J. A. Delaire and K. Nakatani, "Linear and nonlinear optical properties of photochromic molecules and materials," Chemical Reviews, vol. 100, no. 5, pp. 1817–1845, 2000.
45. S. X. Zhang, M. G. Fan, Y. Y. Liu, Y. Ma, G. J. Zhang, and J. N. Yao, "Inclusion complex of spironaphthoxazine with γ-cyclodextrin and its photochromism study," Langmuir, vol. 23, no. 18, pp. 9443–9446, 2007.
46. C. Elsässer, A. Vüllings, M. Karcher, and P. Fumagalli, "Photochromism of spiropyran-cyclodextrin inclusion complexes on Au(111)," Journal of Physical Chemistry C, vol. 113, no. 44, pp. 19193–19198, 2009.
47. M. He, T. Leslie, S. Garner, M. DeRosa, and J. Cites, "Synthesis of new electrooptic chromophores and their structure—property relationship," Journal of Physical Chemistry B, vol. 108, no. 25, pp. 8731–8736, 2004.
48. Z. Li, W. Wu, G. Yu et al., "Dendronlike main-chain nonlinear optical (NLO) polyurethanes constructed from "H"-type chromophores: synthesis and NLO properties," Applied Materials & Interfaces, vol. 1, no. 4, pp. 856–863, 2009.
49. P. J. Stephens, F. J. Devlin, C. F. Chabalowski, and M. J. Frisch, "Ab Initio calculation of vibrational absorption and circular dichroism spectra using density functional force fields," Journal of Physical Chemistry, vol. 98, no. 45, pp. 11623–11627, 1994.
50. V. Ruangpornvisuti, "Recognition of carboxylate and dicarboxylates by azophenol-thiourea derivatives:

a theoretical host-guest investigation," Journal of Molecular Structure, vol. 686, no. 1–3, pp. 47–55, 2004.· ·

51. M. J. Frisch, G. W. Trucks, H. B. Schlegel, et al., Gaussian 09, Gaussian, Wallingford, UK, 2009.

52. S. Prabhu, S. Saravanamoorthy, M. Ashok, and S. Velmathi, "Colorimetric and fluorescent sensing of multi metal ions and anions by salicylaldimine based receptors," Journal of Luminescence, vol. 132, no. 4, pp. 979–986, 2012.

53. Y. L. Shen, S. H. Yang, L. M. Wu, and X. Y. Ma, "Study on structure and characterization of inclusion complex of gossypol/beta cyclodextrin," Spectrochimica Acta A, vol. 61, no. 6, pp. 1025–1028, 2005. · ·

54. A. K. Ghosh, A. K. Mukherjee, and S. Bagchi, "Absorption spectrophotometric study of molecular complexation of asphaltene with bromanil," Energy & Fuels, vol. 23, no. 1, pp. 392–396, 2009.

55. K. Aljoumaa, Y. Qi, J. Ding, and J. A. Delaire, "Synthesis and characterizations of highly fluorinated poly(arylene ether)s for quadratic nonlinear optics," Macromolecules, vol. 42, no. 23, pp. 9275–9288, 2009.

56. M. Casalboni, F. Sarcinelli, R. Pizzoferrato, R. D›Amato, A. Furlani, and M. V. Russo, "Second-harmonic generation and absorption spectra of platinum organometallic complexes incorporated in PMMA films,"Chemical Physics Letters, vol. 319, no. 1-2, pp. 107–112, 2000

CHAPTER 7

POLY(3-ALKYLTHIOPHENE)S SHOW UNEXPECTED SECOND-ORDER NONLINEAR OPTICAL RESPONSE

S. Deckers[a], S. Vandendriessche[a], D. Cornelis[b], F. Monnaie[b], G. Koeckelberghs[b], I. Asselberghs[c], T. Verbiest[a] and M. A. van der Veen[d]

[a]Laboratory for Molecular Electronics and Photonics, University of Leuven, Celestijnenlaan 200D, 3001 Leuven, Belgium

[b]Laboratory for Polymer Synthesis, University of Leuven, Celestijnen 200F, 3001 Leuven, Belgium

[c]Imec, Kapeldreef 75, 3001 Leuven, Belgium

[d]Catalysis Engineering, Chemical Engineering Department, Delft University of Technology, Julianalaan 136, BL 2826 Delft, The Netherlands

Regioregular poly(3-hexylthiophene)s with chain lengths varying from 5 to 100 monomers are synthesized. Poly(3-hexylthiophene)s show in solution an unexpectedly significant second-order nonlinear optical response. The increase in transition dipole moment upon oligomerisation causes the significant second-order nonlinear optical response.

Second-order nonlinear optical materials are applied as electro-optical modulators, for frequency doubling, terahertz generation,[1] and are more intensively investigated as materials for optical computing.[2] So far, nearly all commercially used materials are inorganic materials. Organic materials however show faster response times,[3] and have more process flexibility, providing more potential for downscaling into functional devices. In this context, active polymers are of particular interest due to their increased stability over functionalised blends.

The versatile use of poly(3-alkylthiophene)s and conjugated polymers in general is widely explored in research fields such as organic field effect transistors (OFET), solar cells, organic light emitting diodes (OLED) and nonlinear optical devices.[4] The broad applicability of polythiophenes resulted in a large product diversity as well as in a wide range of synthetic routes to produce them in reasonable yields and obtain meso and nano scale architectures.[5] Although already investigated for their third order nonlinear optical response,[6] their second order nonlinear optical response has not been investigated thoroughly. This is largely due to the fact that the structure of polythiophenes does not follow the typical donor – conjugated structure – acceptor paradigm, and therefore no significant second-order nonlinear effect is expected. Typically, high second-order nonlinear optical responses of materials are expected in extended conjugated systems that are non-centrosymmetrically organized.[7] The dominant strategy is to utilize dipolar chromophores, either in the polymer backbone covalently bonded to the side chains or viadoping polymers with chromophores.[8]

Previously, a scaling study of the second-order nonlinear optical response has been conducted on non-conjugated oligonucleotides up to 50 monomers.[9] However, to the best of our knowledge, scaling studies of conjugated polymers are limited to 4–5 monomers.[10] Herein, we report for the first time the second-order nonlinear optical response of conjugated polymers upon increasing the chain length to 100 units. More specifically, regioregular poly(3-hexylthiophene) s (P3HT) are studied. Despite the limited donor strength of the alkyl substituents, an appreciably large second-order nonlinear optical response is observed. Upon increasing the

Citation: S. Deckers, S. Vandendriessche, D. Cornelis, F. Monnaie, G. Koeckelberghs, I. Asselberghs, T. Verbiest, and M. A. van der Veen, Poly(3-alkylthiophene)s show unexpected second-order nonlinear optical response, DOI: 10.1039/C3CC48099B.

polymerization degree the materials show a dramatic increase in the second-order response due to an increase in the transition dipole moment.

A series of varying chain lengths of regioregular head-to-tail coupled P3HT polymers (see Scheme 1) is synthesized. The small polydispersities (~1.2) were confirmed by gel permeation chromatography. The degree of polymerization was determined by ^{1}H NMR by end-group quantification.[11] We obtained a series ranging from 5 to 100 repeating monomer units (see ESI).

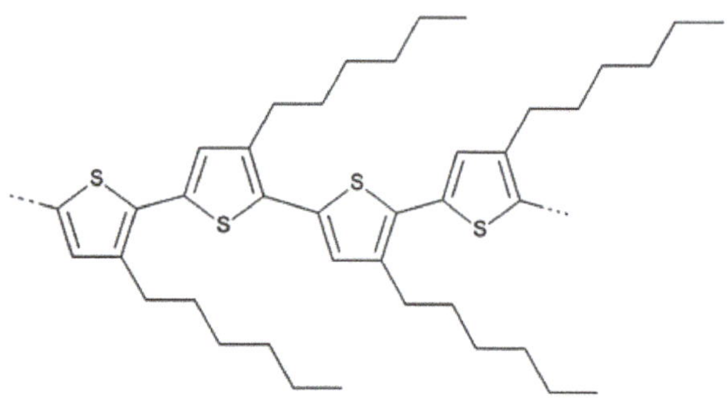

Scheme 1: Head-to-tail coupled regioregular poly(3-alkylthiophene).

The second-order nonlinear optical response of the polymers was determined in solution via a Hyper-Rayleigh Scattering (HRS) experiment at a fundamental input wavelength of 800 nm. The incoherent scattered light at the second harmonic wavelength of 400 nm is detected in a perpendicular geometry. To separate the frequency doubled light from two photon absorption fluorescence at the same frequency, we take advantage of the time-delay of fluorescence versus the quasi instant generation of frequency doubled light. This is done in the frequency domain according to the method described by Olbrechts et al.[12] Further details of the experimental set-up and conditions can be found in the ESI.†

We report the directly measured hyperpolarizability β, as well as the static hyperpolarizability β_0 in order to exclude contributions from resonance enhancement. The static hyperpolarizability β_0 is a frequency independent quantity that allows comparison of molecules that have different electronic resonances. It thus allows us to compare the performance of different organic molecules which is of relevance to optoelectronic applications. We thus adjust the red-shift in electronic resonance frequency upon increasing the chain length of P3HT by applying the homogeneously damped two-state model to derive the static hyperpolarizability β_0 from the measured hyperpolarizability β (see ESI†).[13]

The P3HT oligomers/polymers are dissolved in a good solvent, chloroform, at a typical concentration of 0.1 mM of monomer units to ensure complete solubility. In Fig. 1, we show the observed hyper-Rayleigh response per polymer chain β_{pol} as a function of the number of monomer units per polymer chain n. An apparent linear increase of the HRS-response upon chain lengthening is observed (the fitted line is guidance to the eye). However, when rescaling the data towards the number of monomer units in the polymer backbone (β_{pol}/n), a totally different trend is observed (Fig. 2). The depiction of β_{pol}/n is a means of comparing the amount of second-order nonlinear optical response generated by molecules of different molecular weight as it depicts the hyperpolarizability for the same density of material. Initially the HRS response β_{pol}/n increases upon chain lengthening, but saturation occurs for chain lengths longer than 15 repeating monomer units. The same trend is observed for the directly measured β_{pol}/n as well as for the static $\beta_{pol,0}/n$ (see Fig. 2). At this point, it is worthwhile mentioning that the monomer unit by itself has a very small hyperpolarizability that cannot be measured using our HRS set-up. This implies that the 2nd order NLO response does not originate from the 3-alkylthiophene monomer itself, but from a cooperative effect of the monomers within one oligomer–polymer chain such as the delocalized π-conjugated backbone.

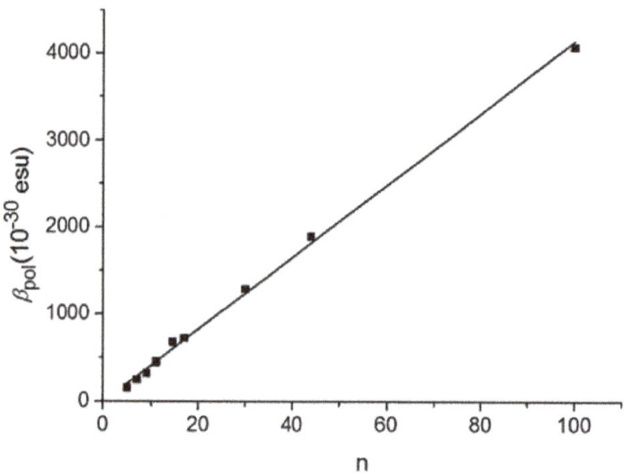

Figure 1: Hyperpolarizability β_{pol} versus the number of monomers n of P3HT.

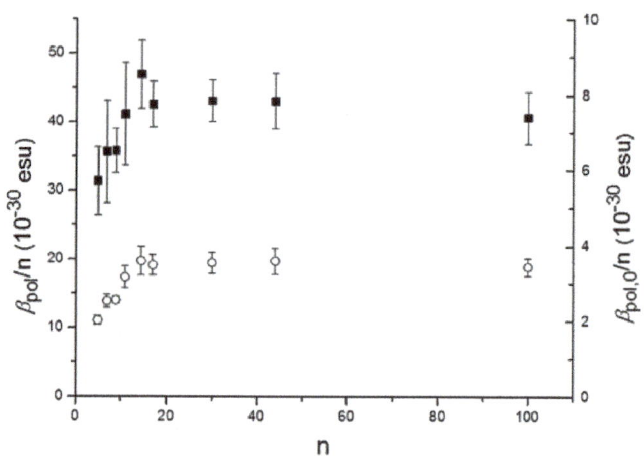

Figure 2: Hyperpolarizability per monomer unit β_{pol}/n (filled squares) and the static hyperpolarizability per monomer unit $\beta_{pol,0}/n$ (open circles) as a function of the number of monomer units n of P3HT.

To understand which properties of molecules contribute to the second-order nonlinear response, we provide the description of the static hyperpolarizability β_0 within the two-state model where two states provide the dominant contribution to hyper-Rayleigh scattering[14]

$$\beta_0 \sim (\mu_{ee} - \mu_{gg})\left(\frac{{\mu_{ge}}^2}{{E_{eg}}^2}\right)$$

with μ_{ee} and μ_{gg} being the dipole moment of the excited and ground states, respectively, μ_{ge} the transition dipole moment of the two states and E_{eg} the energy difference between the two states. The used model assumes one transition between states to be dominant, consistent with the HOMO–LUMO transition in polythiophenes. The transition dipole moment and the transition energy difference can be obtained from the linear UV-VIS spectra of the polymer solutions (see ESI). From Fig. 3, we can see that the hyperpolarizability and transition dipole moment per monomer unit both saturate around 15 (or 1/n = 0.7) monomer units, while the decrease in bandgap energy saturates only at much longer chain lengths. It is thus clear that the increase of the hyperpolarizability upon increasing the chain length follows the same trend as the increase of μ_{ge}^2/n. It is clear from the data that the cooperative enhancement upon incorporation of hexylthiophene into the polymer is mainly due to the increase of conjugation.

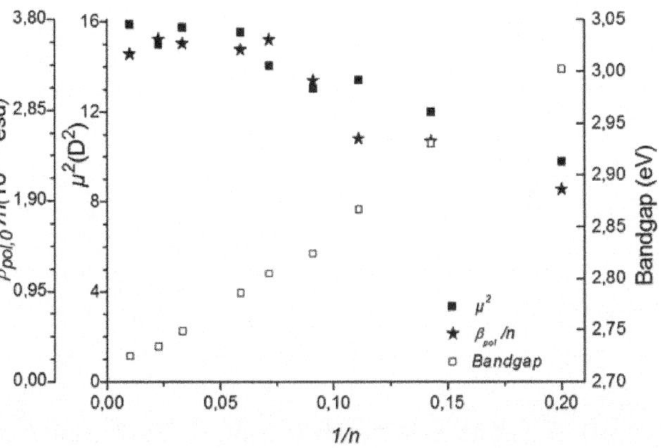

Figure 3: The transition dipole moment μ_{ge}^2 and the static hyperpolarizability per monomer unit$\beta_{pol,0}/n$ and the bandgap are given as a function of the inverse of the number of monomer units n of P3HT.

Several effects can cause the non-centrosymmetry required for the second order NLO effect. The alkyl substituents, all in the 3-position in the regioregular P3HT, cause breaking of non-centrosymmetry (see Scheme 1). The effect on the NLO response is expected to be relatively small due to the limited donor strength of alkyl substituents. A non-planar conformation of the backbone[15] with helical structure can also cause asymmetry. As non-centrosymmetry is a requirement for the second-order NLO effect, the conformation will have an effect on the second order NLO response. Experimental and theoretical studies have shown that the persistence length – a measure of the length of rodlike or straight conformation within a polymer – of P3HT in a good solvent is about 10 units.[16] Yet, the effect of conformation is not clear from the data. Indeed, as can be seen from Fig. 2, the trend in the hyperpolarizability is largely explained by electronic effects.

As mentioned, the second-order nonlinear optical response of the monomer could not be measured. This means that the extension of conjugation leads to a dramatic enhancement of the second-order nonlinear optical properties of the material. This is also in line with the theoretical results that predict a generally larger than 3-fold increase of the hyperpolarizability upon extending from the thiophene monomer to the trimer.[17] If we compare the static hyperpolarizability $\beta_{pol,0}$ for 5 units thus having 32‡ conjugated electrons, $17.7 \pm 1.2 \times 10^{-30}$ esu, with p-nitroaniline, a benchmark NLO molecule, extended with two additional benzene rings to 4-nitro-4″-amino-p-terphenyl, thus having 22 conjugated electrons, we find a very similar hyperpolarizability β, namely 16×10^{-30} esu.[18] The second-order nonlinear optical response of P3HT is thus very significant, while it was expected to be negligible. The best performing organic molecules typically have a hyperpolarizability one order of magnitude larger,[19] but polythiophenes outperform these complex organic molecules with respect to easy synthesis and processability into devices.[5,20] Moreover, in the solid-state polythiophenes display much more planar conformations with higher electron delocalization and less bending can be achieved.[21]Based on the results reported here, we conclude that such solid-state polythiophenes could have an even larger second-order nonlinear optical response.

In conclusion, we found for oligomers of P3HT in solution that they, despite the limited donor strength of the substituents, show an unexpectedly significant second-order nonlinear optical response. This response is in the order of a benchmark nonlinear optical molecule. The dramatic increase in hyperpolarizability with chain length is largely attributed to the increased conjugation length. This is the first systematic study of second-order nonlinear scattering response of a conjugated polymer as a function of chain length.

MAvdV and SV acknowledge the Scientific Research Fund Flanders (FWO) for their personal fellowships. We are also grateful to the Onderzoeksfonds KU Leuven/Research Fund KU Leuven for a GOA and CREA grant.

NOTES AND REFERENCES

1. (a) L. R. Dalton, P. A. Sullivan and D. H. Bale, Chem. Rev., 2010, 110, 25–55 ; (b) M. Lee, H. E. Katz, C. Erben, D. M. Gill, P. Gopalan, J. D. Heber and D. J. McGee, Science, 2002, 298, 1401–1403 ; (c) X. Zheng, C. V. McLaughlin, P. Cunningham and L. M. Hayden, J. Nanoelectron. Optoelectron., 2007, 2, 58–76 ; (d) L. R. Dalton, S. J. Benight, L. E. Johnson, D. B. Knorr, I. Kosilkin, B. E. Eichinger, B. H. Robinson, A. K.-Y. Jen and R. M. Overney, Chem. Mater., 2011, 23, 430–445 .
2. J. L. O'Brien, Science, 2007, 318, 1567–1570 .
3. J. M. Cole, Philos. Trans. R. Soc., A, 2003, 361, 2751–2770 .
4. (a) G. Wang, J. Swensen, D. Moses and A. J. Heeger, J. Appl. Phys., 2003, 93, 6137 ; (b) G. Li, V. Shrotriya, J. Huang, Y. Yao, T. Moriarty, K. Emery and Y. Yang, Nat. Mater., 2005, 4, 864 ; (c) N. R. Armstrong, W. Wang, D. M. Alloway, D. Placencia, E. Ratcliff and M. Brumbach, Macromol. Rapid Commun., 2009, 30, 717–731 .
5. (a) B. H. Cumpston, S. P. Ananthavel, S. Barlow, D. L. Dyer, J. E. Ehrlich, L. L. Erskine, A. A. Heikal, S. M. Kuebler, M. Rumi, X. Wu, S. R. Marder and J. W. Perry, Nature, 1999,398, 51–54 ; (b) G. Zotti, B. Vercelli and A. Berlin, Acc. Chem. Res., 2008, 41, 1098–1109 ; (c) I. Osaka and R. D. McCullough, Acc. Chem. Res., 2008, 41, 1202–1214 .
6. (a) S. Kishino, Y. Ueno, K. Ochiai, M. Rikukawa, K. Sanui, T. Kobayashi, H. Kunugita and K. Ema, Phys.

Rev. B, 1998, 58, 430–433 CrossRef; (b) M.-T. Zhao, B. P. Singh and P. N. Prasad, J. Chem. Phys., 1988, 89, 5535 ; (c) Y. Verbandt, H. Thienpont, I. Veretennicoff, P. Geerlings and G. L. J. a. Rikken, Chem. Phys. Lett., 1997, 270, 471–475 ; (d) T. Bjornholm, D. R. Greve, T. Geisler, J. C. Petersen, M. Jayaraman and R. D. McCullough, Adv. Mater., 1996, 8, 920 CrossRef; (e) T. Bjornholm, D. R. Greve, T. Geisler, J. C. Petersen, M. Jayaraman and R. D. McCullough, Synth. Met., 1997, 84, 531–532 .

7. T. Verbiest, K. Clays and V. Rodriguez, Second-Order Nonlinear Optical Characterization Techniques, CRC Press, Boca Raton, 2009 Search PubMed.

8. (a) P. N. Prasad and D. J. Williams, Introduction to nonlinear optical effects in molecules & polymers, Wiley-Interscience, New York, 1991 Search PubMed; (b) M. J. Cho, D. H. Choi, P. A. Sullivan, A. J. P. Akelaitis and L. R. Dalton, Prog. Polym. Sci., 2008, 33, 1013–1058 .

9. Thèse EPFL nr 2414 (2001). Rinuy, Juliette: Nonlinear optics of proteins at liquid interfaces and of DNA oligonucleotides in liquid phase.

10. M. G. Kuzyk, J. Pérez-Moreno and S. Shafei, Phys. Rep., 2013, 529, 297–398 .

11. M. Verswyvel, F. Monnaie and G. Koeckelberghs, Macromolecules, 2011, 44, 8489–8498 CrossRef.

12. G. Olbrechts, R. Strobbe, K. Clays and A. Persoons, Rev. Sci. Instrum., 1998, 69, 2233–2241 .

13. B. J. Orr and J. F. Ward, Mol. Phys., 1971, 20, 513 .

14. (a) C. B. Gorman and S. R. Marder, Proc. Natl. Acad. Sci. U. S. A., 1993, 90, 11297–11301 ; (b) J. L. Oudar and D. S. Chemla, J. Chem. Phys., 1977, 66, 2664 ; (c) J. Zyss and J. L. Oudar, Phys. Rev. A, 1982, 26, 2016 CrossRef; (d) J. L. Brdas, F. Meyers, B. M. Pierce and J. Zyss, J. Am. Chem. Soc., 1992, 114, 4928 CrossRef.

15. S. R. Bhatta, Y. Y. Yimer, M. Tsige and D. S. Perry, Comput. Theor. Chem., 2012, 995, 36–42 .

16. (a) G. M. Heffner and D. S. Person, Macromolecules, 1991, 24, 6295 ; (b) P. V. Shibaev, K. Schaumburg, T. Bjornholm and K. Norgaard, Synth. Met., 1998, 97, 97–104 ; (c) B. McCulloch, V. Ho, M. Hoarfrost, C. Stanley, C. Do, W. T. Heller and R. A. Segalman, Macromolecules, 2013, 46, 1899–1907 .

17. J. Waite and M. G. Papadopoulos, J. Phys. Chem., 1990, 94, 6244–6249 .

18. L.-T. Cheng, W. Tam, S. R. Marder, A. E. Stiegman, G. Rikken and C. W. Spangler, J. Phys. Chem., 1991, 95, 10643 .

19. (a) B. Coe, J. A. Harris, I. Asselberghs, K. Clays, G. Olbrechts, A. Persoons, J. T. Hupp, R. C. Johnson, S. J. Coles, M. B. Hursthouse and K. Nakatani, Adv. Funct. Mater., 2002, 12, 110–116 ; (b) J. Pérez-Moreno, Y. Zhao, K. Clays and M. Kuzyk, Opt. Lett., 2007, 32, 59–61 CrossRef.

20. (a) C. L. Gettinger, A. J. Heeger, J. M. Drake and D. J. Pine, J. Chem. Phys., 1994, 101, 1673–1678 ; (b) R. R. Tykwinski, U. Gubler, R. E. Martin, C. Bosshard and P. Gu, J. Phys. Chem. B, 1998, 102, 4451–4465 ; (c) J. Hou, M.-H. Park, S. Zhang, Y. Yao, L.-M. Chen, J.-H. Li and Y. Yang, Macromolecules, 2008, 41, 6012–6018 .

21. (a) M. Chang, D. Choi, B. Fu and E. Reichmanis, ACS Nano, 2013, 7, 5402–5413 ; (b) A. R. Aiyar, J.-I. Hong, R. Nambiar, D. M. Collard and E. Reichmanis,Adv. Funct. Mater., 2011, 21, 2652–2659

CHAPTER 8

FABRICATION AND TENSILE PROPERTY ANALYSIS OF POLYMER MATRIX COMPOSITES OF GRAPHITE AND SILICON CARBIDE AS FILLERS

Vikram G. Kamble[1, 2] and Naveen Kumar L.[3]

[1]Research Fellow, Proof and Experimental Establishment (DRDO), Balasore, Odisha-756025, India

[2]St Joseph Engineering College, Mangalore, Karnataka-575028, India

[3]MTech Scholar, Sri Jayachamarajendra College of Engineering, Mysore, Karnataka -570006, India

ABSTRACT

In a world there is ever-increasing demand for new materials, composite materials finds the topmost spot. Here in this work an attempt has been made to study the mechanical properties of Glass and Vinyl Ester composites with and without fillers. Laminates are prepared by Hand Layup technique with changing the volume percentage of Vinyl Ester and fillers in each. The laminates prepared are subjected to tensile test and systematically analyzed to study the behaviour and variation of properties with respect to their composition. The results obtained from the tests have been documented with great detail.

INTRODUCTION

Polymers and their composites are being used increasingly because of their good strength and low density. The advent of new technologies has increased the demand for superior materials for tribological applications like bearings, cams, gears, chutes, guides, etc (Pande et al., 1998). PMC's exhibit excellent frictional and wear characteristics even without external lubrication and can provide maintenance free operation, excellent corrosion resistance and operate with low noise emission due to good damping properties (Cha'o-Kuang and Tzu-Li, 1997). The knowledge of the relations between material parameters and tribological performance is important in the determination of strength and wear rate of the composite (Ismaila and Rozman, 1998). The characterization of engineering properties is a complex issue for fiberreinforced composites due to their inherent anisotropy and in homogeneity (Suresha, 2003). In terms of mechanical properties, advanced composite materials are evaluated by a number of specially designed test methods. These test methods are mechanically simple in concept but extremely sensitive to specimen preparation and test-execution procedures, often requiring complex data reduction analysis (Unal and Mimaroglu, 2004; Varada Rajulu et al., 2002). Auto clave based methods and to evaluate the Mechanical and Tribological properties of same using experimental methods as well as simulation methods through analytical and computational methods (Eldera, 1966). The properties of a composite material depend on the properties of the constituents, geometry and distribution of the phases. The contribution of a single constituent to the overall properties of the composite is determined by the concentration of the constituent, Volume fraction or weight fraction in that composite (Kalthoff, 2004). Experimental methods consists of tests methods involving in extraction of Mechanical property such as Tension test. Simulation methods consist of analytical

Citation: Vikram G. Kamble and Naveen Kumar L., Fabrication and Tensile Property Analysis of Polymer Matrix Composites of Graphite and Silicon Carbide as Fillers, DOI: 10.3844/ajeassp.2016.17.30.

as well as computational methods to validate experimental results of polymer composites by considering anisotropy and non homogeneity in their material properties (Swaminathan et al., 2007). Glass is generally a good impact resistant fibre but weighs more than carbon or mid. Glass fibres have excellent characteristics, equal to or better than steel in certain forms. The lower modulus requires special design treatment where stiffness is critical. Composites made from this material exhibit very good electrical and thermal insulation properties. Glass fibres are also transparent to radio frequency radiation and are used in antenna applications (Stevanovica et al., 2003). These composite materials are defined as material systems consisting of mixture of or combination of two or more micro constituents insoluble in each other and differing in form and or material composition (Kamble et al., 2015). The concept of composites such as fiber reinforced composites had its origin in nature for example in old trees, we can see the use of composites by the mankind in the weight history and even today in the continuing quest for improved performance such as less weight, more strength and lower cost: Composites are widely used in many applications as they satisfy the present day criteria for materials (Cybulskaa and Goswamib, 2001; Wu and Hahn, 1998). E-Glass has lower alkali content and stronger than AGlass (alkali), it has good tensile and compressive strength, stiffness, good electrical properties and relatively low cost but impact resistance is relatively poor. E-glass is the most common form of reinforcing fibre used in composites (Surendranathan et al., 2008).

Objective of the Work

The objective of the work is to fabricate polymer composite material using graphite and silicon carbide fillers and to investigate its mechanical tensile property. The manufacturing and fabrication of polymer composite material is done by Hand lay-up method.

SPECIMEN PREPARATION

Fabrication Processes

Reinforced plastics composites are a combination of a resin matrix, fibres and fillers. Which when cured produce a solid structure. The characterization of the desired product (such as size, shape and quantity) determines the method by which the basic holding blocks are combined and moulded. By varying the resin/glass ratio, the type and ratio of reinforcement, the type of resin and the type and amount of fillers, we can significantly alter (control) the physical properties. The variability of product composition and the latitude in size and design possibilities, enables a hand lay-up with vacuum bagging technique part be tailored to economically fit its desired function. The hand layup laminating procedure indicated in Fig. 1.

In the present work specimen were fabricated using hand lay-up and hydraulic press technique hence the procedure employed to prepare specimen is explained in detail.

Fabrication of Laminates Using Hand Lay-Up

- E-Glass Weave style: 2×2 twill Cross breaking strength: 6 to 7 kg/mn^2 Aerial density: 280 to 285 Gsm Thickness: 0.21 to 0.24 mm Finish type: FK 144 Areal Density: 8 Mil
- Matrix Vinyl ester Resin Amine Hardener The Resin and the Hardener are mixed in the ratio of 100:10. Catalyst: Methyl ethyl ketone peroxide Accelerator: Cobalt napthanate Promoter: N-N dimethyl aniline

Steps Involved in Fabrication of the Composite Laminate

Step 1: Preparation of matrix. The matrix contains a resin and hardener, which are mixed thoroughly by electric stirring process till they get mixed uniformly. The fiber weight fraction = 0.65 Weight of resin system = Weight of fabric × Matrix weight percentage. Both the resin and the hardener are mixed in the ratio of 100:10 at room temperature. The resin system is ready for hand lay-up technique,

Typical Calculation of Resin System

- Fibre weight fraction = 0.65
- Dimension of each layer of glass fabric = 400×400 mm
- Number of layers = 8 layers
- Weight of fabric (8 layers of 400×400 mm size) = 864 g
- Matrix weight percentage = 45±1 (for without filler)
- Weight of resin system = Weight of fabric × Matrix weight percentage = 864×0.45 = 388 g Resin: Hardener (by weight) = 100:10
- Weight of Hardener = 38 g

The interaction of components at the phase boundaries, which is also associated with the existence of a thick interface, known also as the inter phase; this is often considered as a separate phase, controlling adhesion between the components.

The method of fabrication both the resin and hardener are mixed at room temperature. The resin system is ready for use to prepare laminates by hand lay-up technique:

Step 2: Hand Lay-up Technique: The procedure involved in fabrication of laminate using hand lay-up technique is outlined below:

- The stone surface table was cleaned with acetone
- Wax as releasing agent was applied on the surface
- Glass fabric was cut to required size and number of layers
- Requirement of resin system was calculated based on the weight of fabric
- Quantity of resin and hardener required was calculated
- The surface table was first covered with a Teflon sheet
- Resin is applied onto the Teflon cloth with a brush
- One layer of glass fabric was kept on top of Teflon cloth and soaked with resin
- Subsequently remaining layers of glass fabric as per requirement were placed on top of the soaked fabric one by one and wetted with resin using a brush
- The wet layers were covered with a Teflon sheet, which acts as releasing layer
- The system was then ready for Hydraulic press

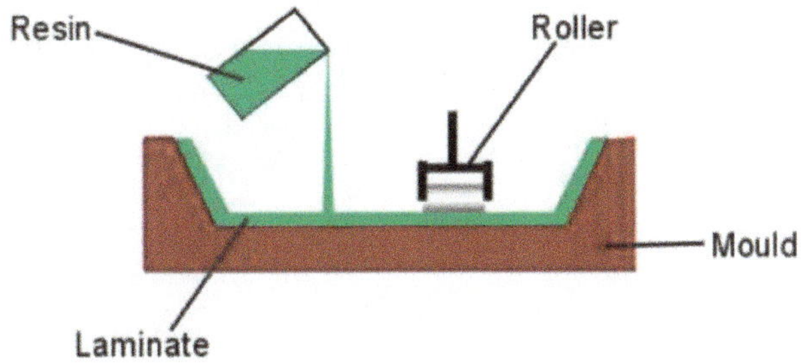

Figure 1: Hand lay-up technique.

Table 1: Testing compositions and calculations

Percentage of filler material	Matrix (%wt)	Resin	Filler (Graphite/SIC)	Glass
0%	45±1	388 gm	0 gm	55
1%	44±1	380 gm	4 gm	55
2%	43±1	371 gm	7 gm	55
3%	42±1	362 gm	10 gm	55
4%	41±1	354 gm	14 gm	55
5%	40±1	345 gm	17 gm	55
6%	39±1	336 gm	20 gm	55
7%	38±1	328 gm	22 gm	55
8%	37±1	319 gm	25 gm	55
9%	36±1	311 gm	27 gm	55
10%	35±1	302 gm	30 gm	55

Step 3: Hydraulic press In the hydraulic press the material obtained from Hand lay-up method is placed for the hardening purpose. Heat supplied is of around 600°C with a pressure of 1bar for more than half an hour. Then it is removed from hydraulic press and kept for cooling till it comes to room temperature.

Step 4: Post Curing The post-cure involves increasing the laminate temperature after the initial room temperature cure, which increases the amount of cross linking of the molecules that can take place. To some degree this post-cure will occur naturally at warm room temperatures, but higher properties and shorter post-cure times will be obtained if elevated temperatures are used:

- 50°C-30 Min
- 70°C-60 Min
- 85°C-120 Min

Burn-Out Test

This test is carried to find out the weight fraction of the Fibers and matrix in a composite. Steps:

- Small pieces of the composite are to be cut from different locations
- Weigh the empty ceramic boats
- Weigh the boats with composite
- Place the boats with composite in a furnace at a temperature of around 600°C for duration of 45 to 60 min
- Weight the boats with fibers

The laminate is prepared using the procedure described previously the laminate is trimmed at all the four edges. The specimens were cut from this laminate hence markings were made on the laminate using the parallel bars and a combination square as shown in the Fig. 2. Figure 3 shows the ready samples to be tested. The compression after impact test requires a perfect rectangular specimen with the opposite sides being perfectly parallel; at least the edges which are compressed should be parallel. Hence due care was taken in specimen preparation during the course of the work. The marked laminate was cut using a band saw of 24 tpi during the cutting process it was ensured that the cut is given exactly at a uniform rate and exactly on the lines which were marked. Around 15 specimen were obtained in each laminate subsequently the edges of each specimen were rubbed using a coarse sand paper (cloth emery) of J297 grade; to ensure that the edges do not contain any defects, as these defects may itself induce failure or edge crushing during the time of tensional testing. Once the specimens were ready inspection was carried on each specimen to check whether the specimen is perfectly rectangular and also to ensure that the opposite edges are perfectly parallel. The specimen of size 400×400 mm which is inspected is then used for further testing as shown in Fig. 2.

Figure 2: Fabricated laminates of standard size (without filler, Graphite filler and SIC filler).

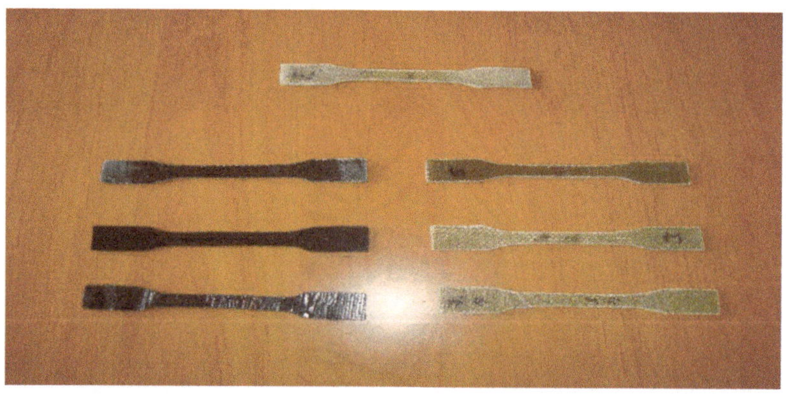

Figure 3: Tensile specimen for testing-Without filler (centre), Graphite series (left), silicon series (right).

TESTING AND CALCULATIONS

Tensile Test (ASTM D 638M)

Whenever some external forces acts on a body, it undergoes some deformations if a body is stressed within its elastic limit, the deformation completely disappears as soon as the forces are removed. And beyond the elastic limit, the deformation does not disappear completely, even after the removal of the forces and there remains some residual deformation. This test method is designed to produce tensile properties such as tensile strength, modulus of elasticity (young's modulus) and percentage of elongation for control and specification of composite materials. These data are useful for qualitative characterization, engineering design and R&D purposes. Tensile test, in a broad sense is measurement of the ability of a material to withstand forces that trend to pull it apart and to determine to what extent the material stretches before breaking. Tensile modulus, an indication of the relative stiffness can be determined from stress strain diagram. For every test three sample specimens were prepared. Table 1 shows testing compositions and calculations.

Procedure

- Measure the width and thickness of specimens with a suitable micrometer to the nearest 0.02 mm at several points along their narrow sections within the gauge boundaries. Record the minimum values of cross sectional area so determined
- Mark gauge length on the specimen
- Set the grip separation speed of the machine; place the specimen straight in the grips
- Tighten the grips evenly and firmly to the extent necessary to prevent slippage of the specimen during test and not the point where the specimen would be crushed
- Start the machine and record the load and extension curve
- Record the load and extension at yield and break point

RESULTS

Superimposed Graphs of Tensile-Graphite Series Superimposed Graphs of Tensile Silicon Series

The results obtained from test series involving an elongation history are shown in Fig. 4-24. The dynamic flow curve of the interrupted strain rate jump tests of dynamic testing. For every sample, three trails have been done and it's shown in each diagram. The lines on these diagrams represent the stress-elongation relationship obtained by interrupting the dynamic-tensile tests at different levels of strain rate change.

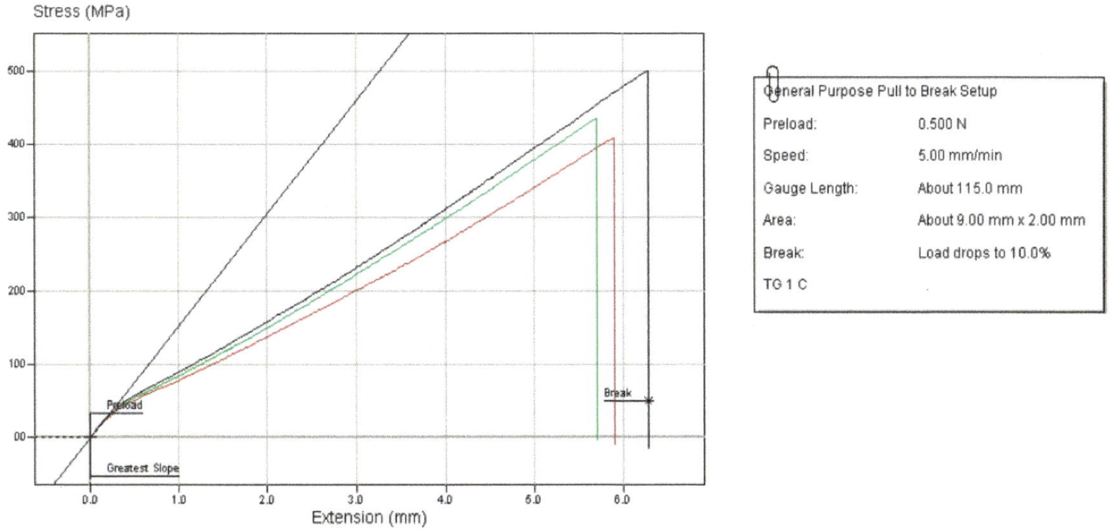

Figure 4: Superimposed graphs of tensile-graphite series TG1.

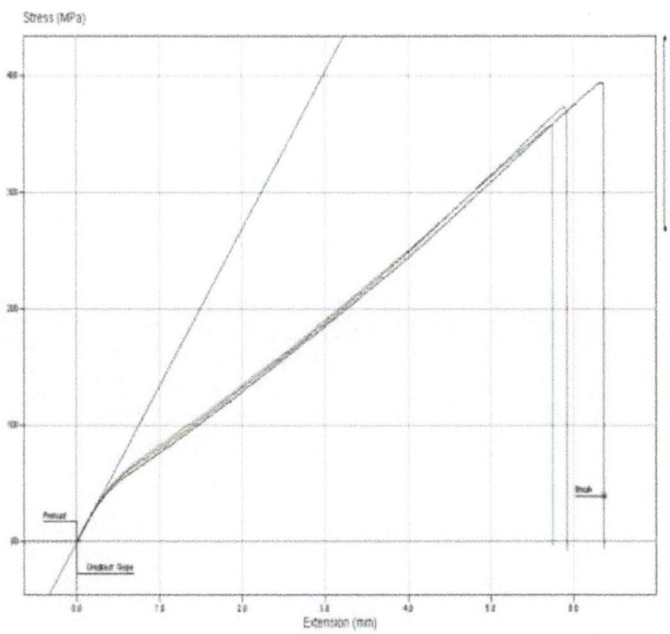

General Purpose Pull to Break Setup
Preload: 0.500 N
Speed: 5.00 mm/min
Gauge Length: About 115.0 mm
Area: About 10.0×2.25 mm
Break: Load drops to 10.0%

Figure 5: Superimposed graphs of tensile-graphite series TG2.

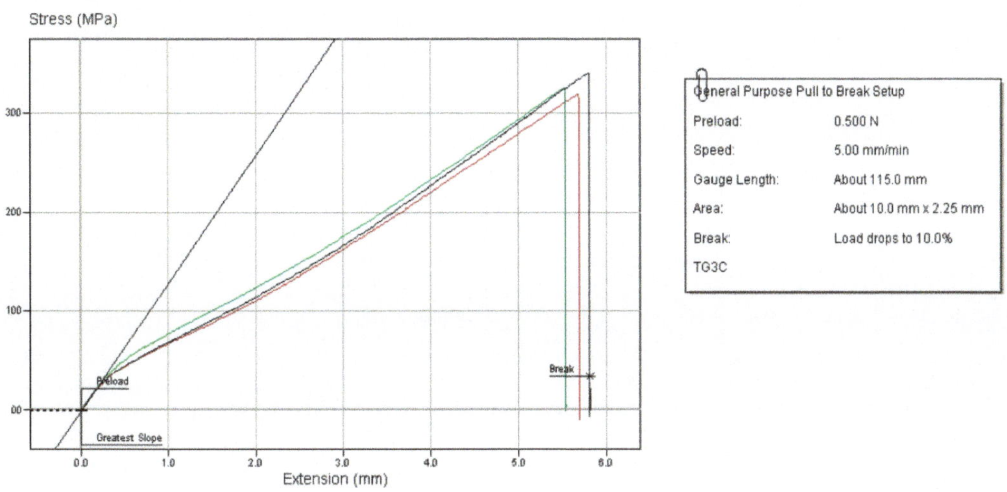

Figure 6: Superimposed graphs of tensile –graphite series TG3.

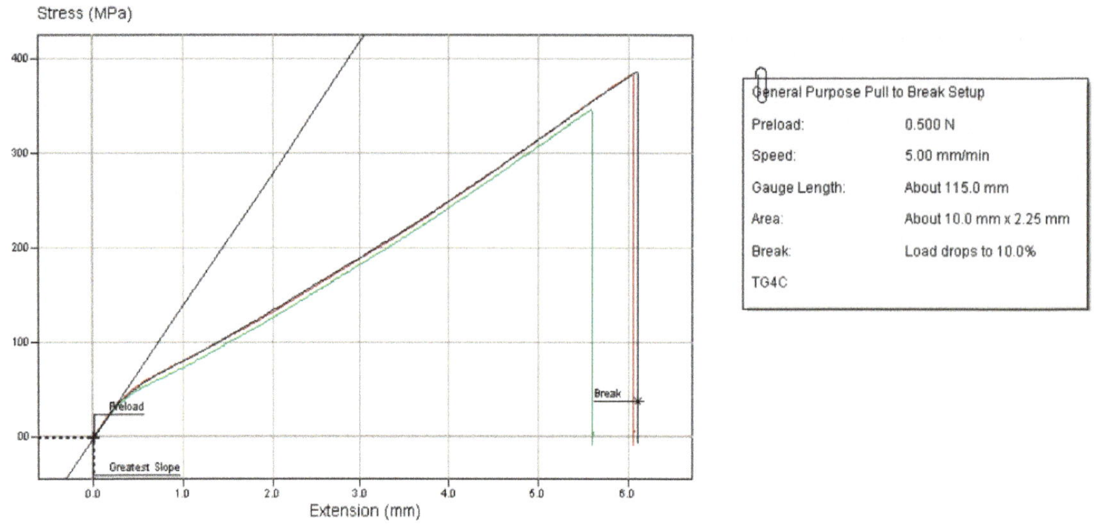

Figure 7: Superimposed graphs of tensile-graphite series TG4.

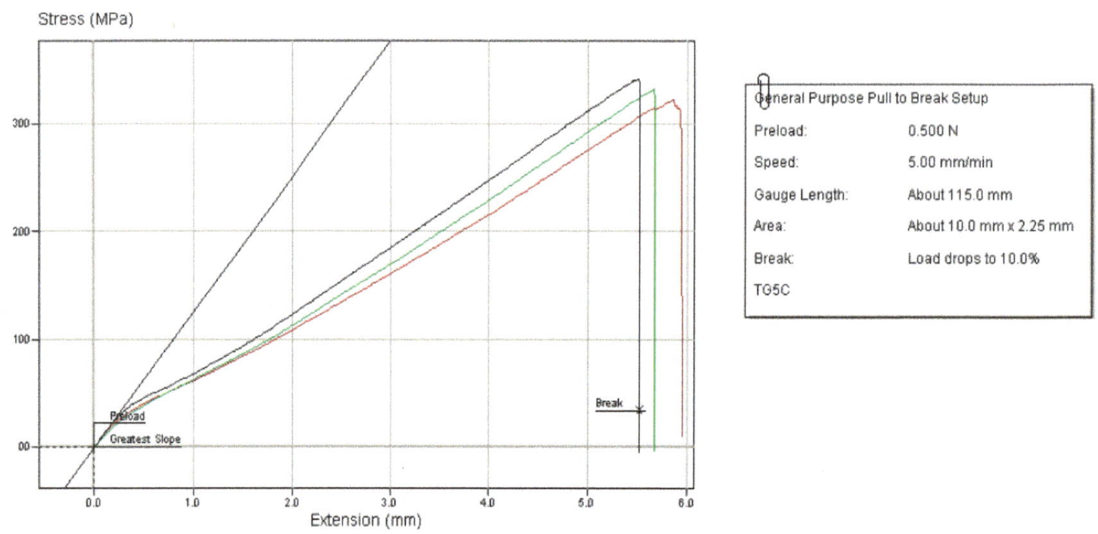

Figure 8: Superimposed graphs of tensile-graphite series TG5.

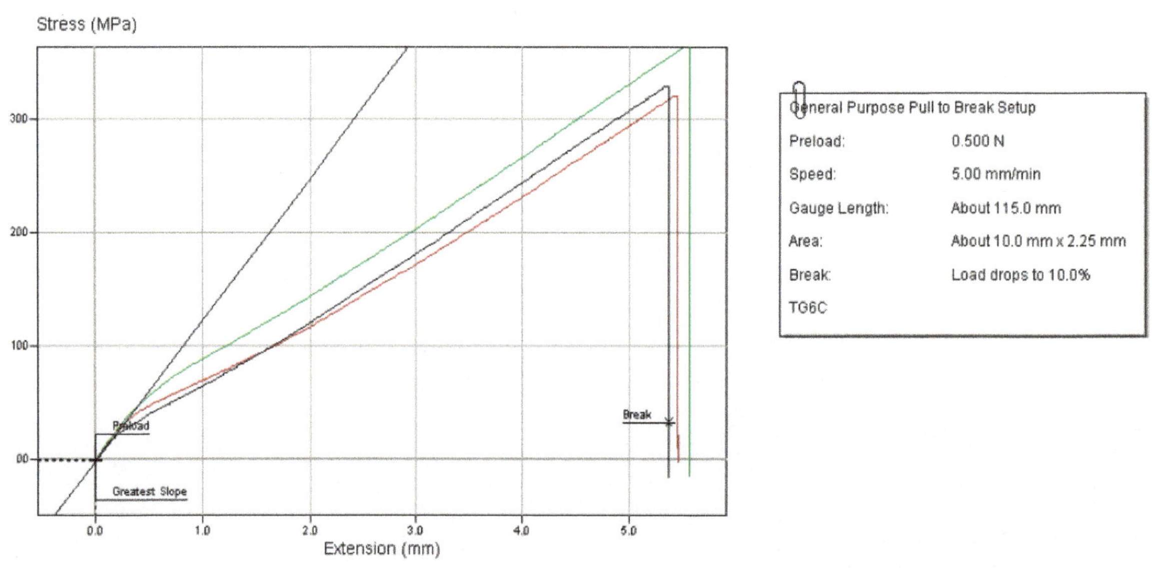

Figure 9: Superimposed graphs of tensile-graphite series TG6

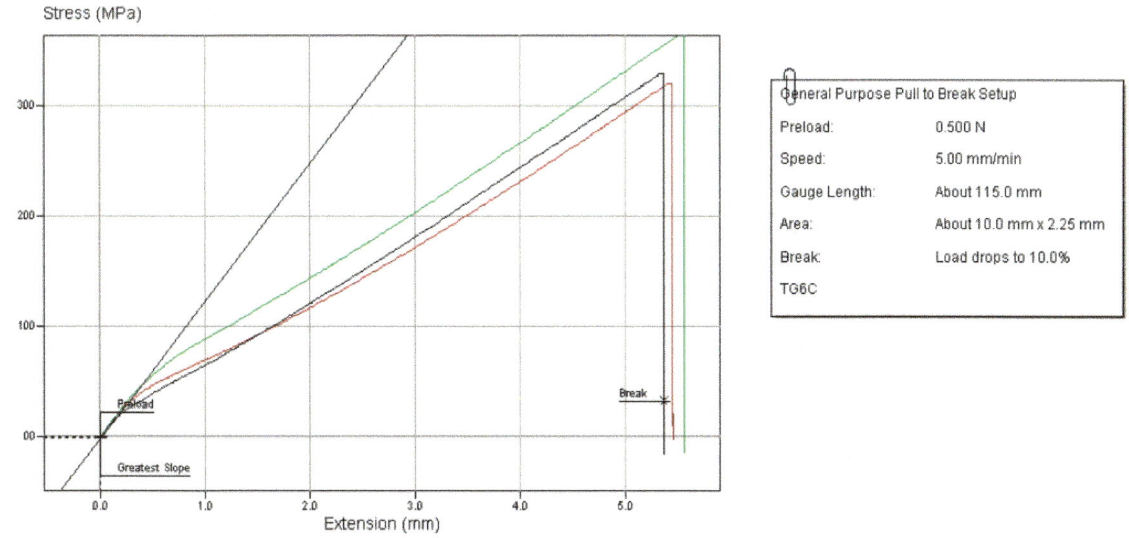

Figure 10: Superimposed graphs of tensile-graphite series TG7.

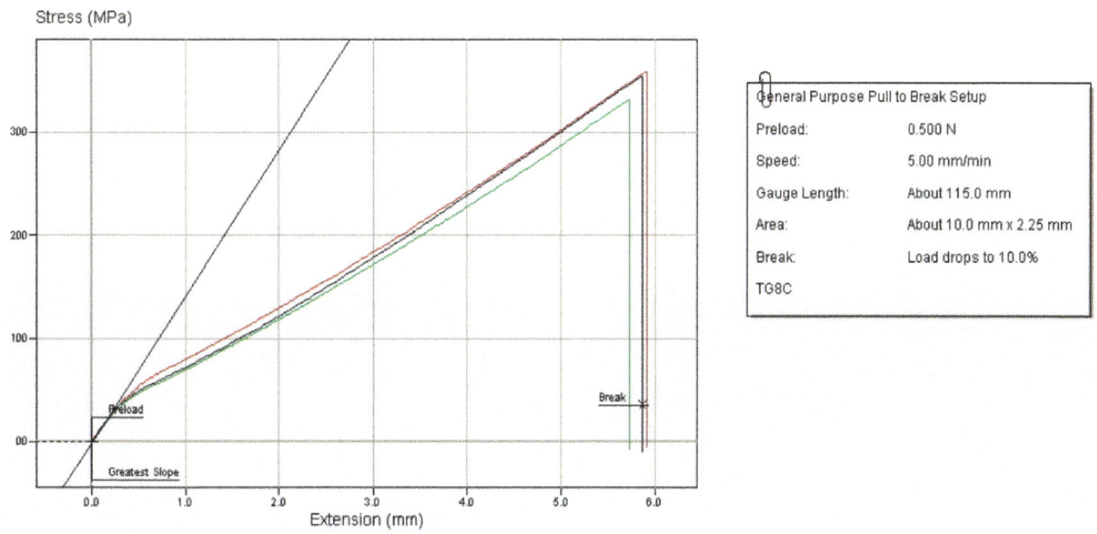

Figure 11: Superimposed graphs of tensile-graphite series TG8.

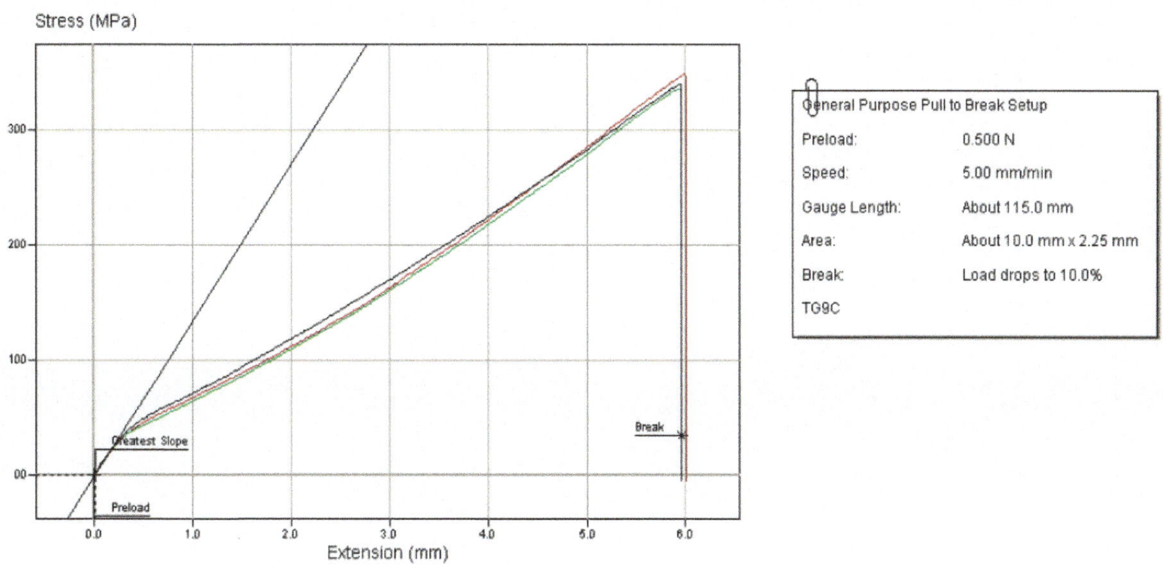

Figure 12: Superimposed graphs of tensile-graphite series TG9.

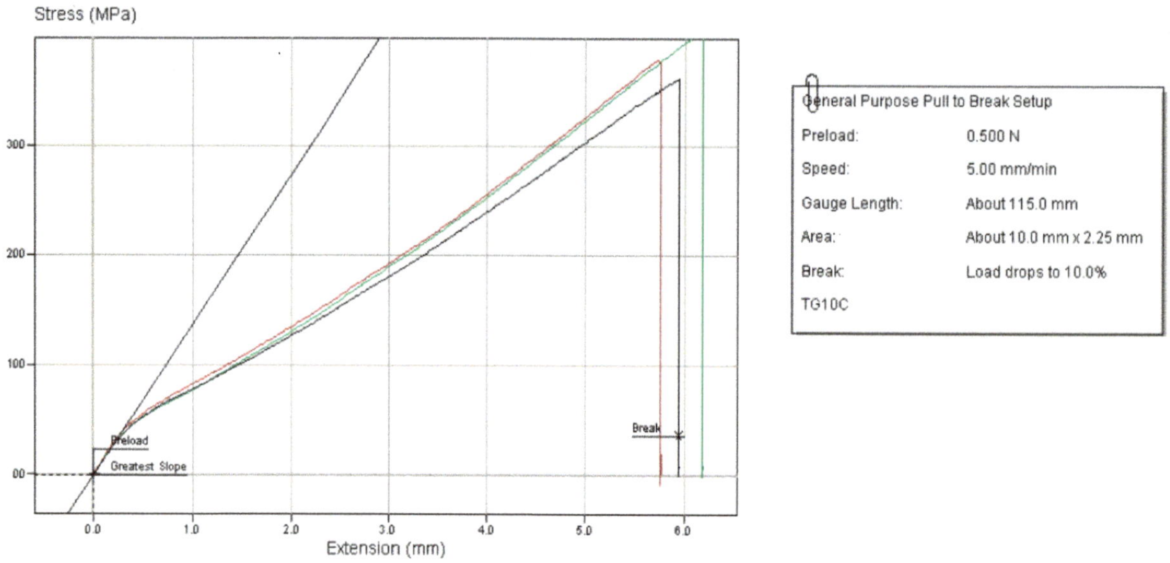

Figure 13: Superimposed graphs of tensile-graphite series TG10.

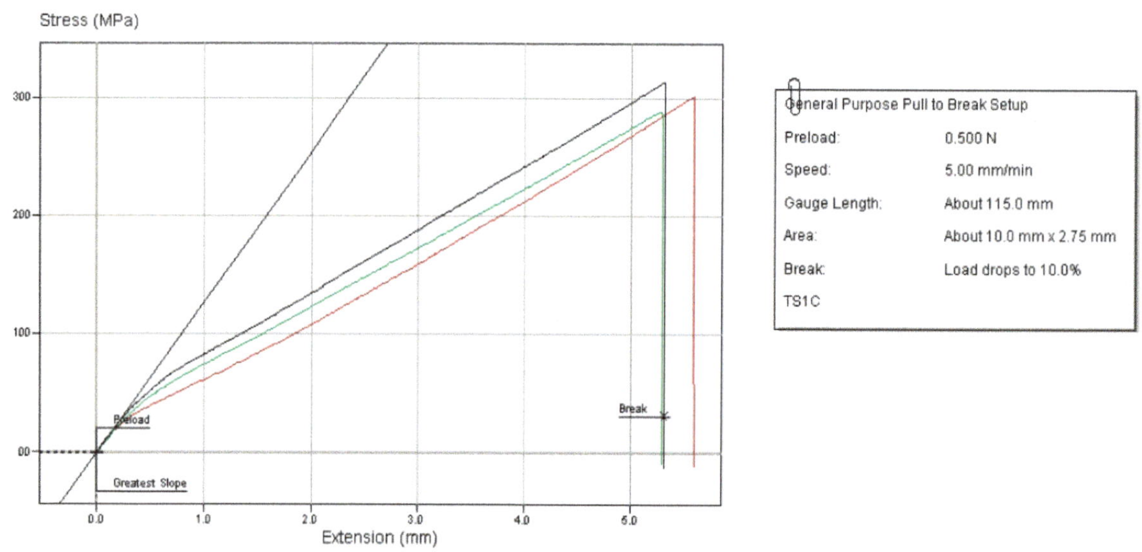

Figure 14: Superimposed graphs of tensile silicon series TS1.

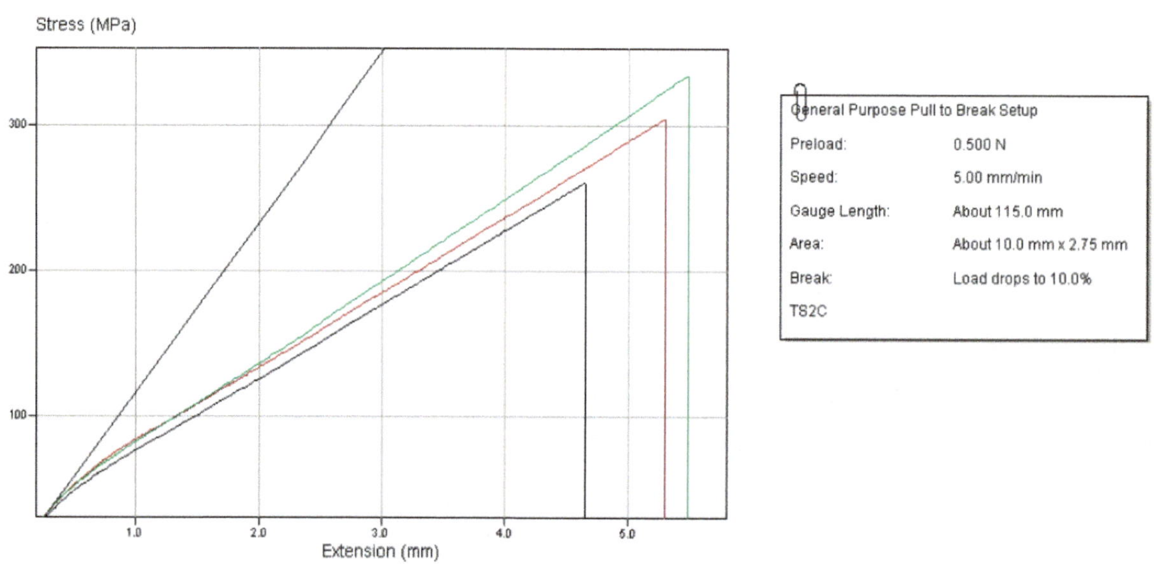

Figure 15: Superimposed graphs of tensile silicon series TS2.

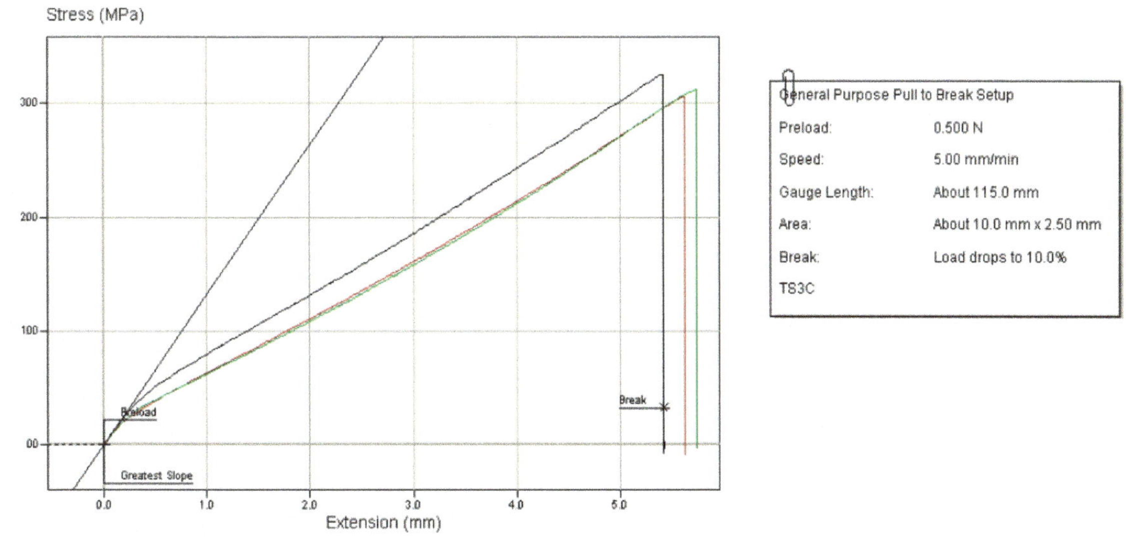

Figure 16: Superimposed graphs of tensile silicon series TS3

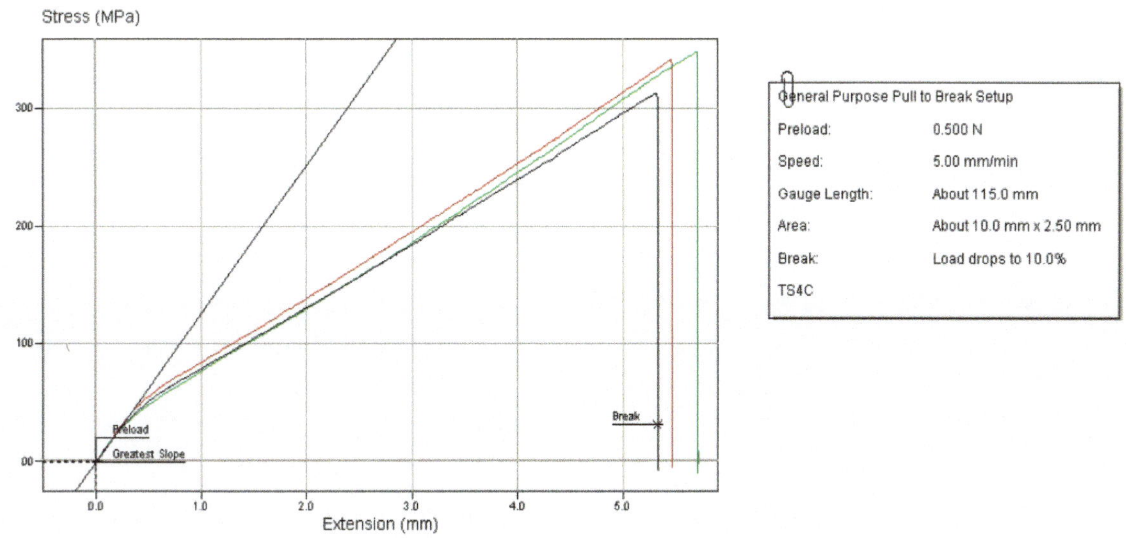

Figure 17: Superimposed graphs of tensile silicon series TS4.

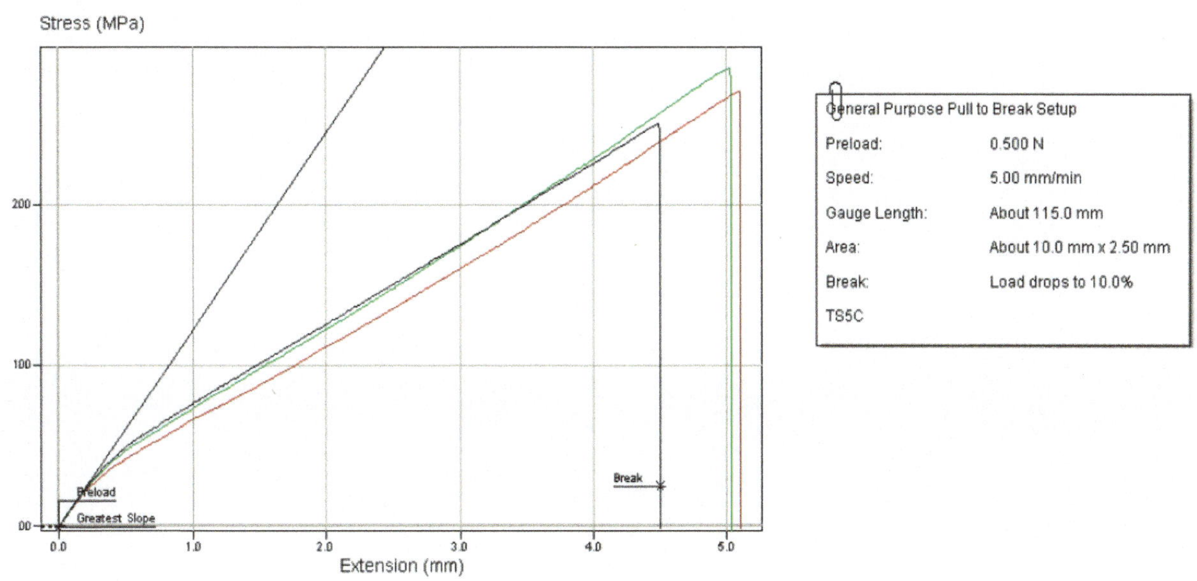

Figure 18: Superimposed graphs of tensile silicon series TS5.

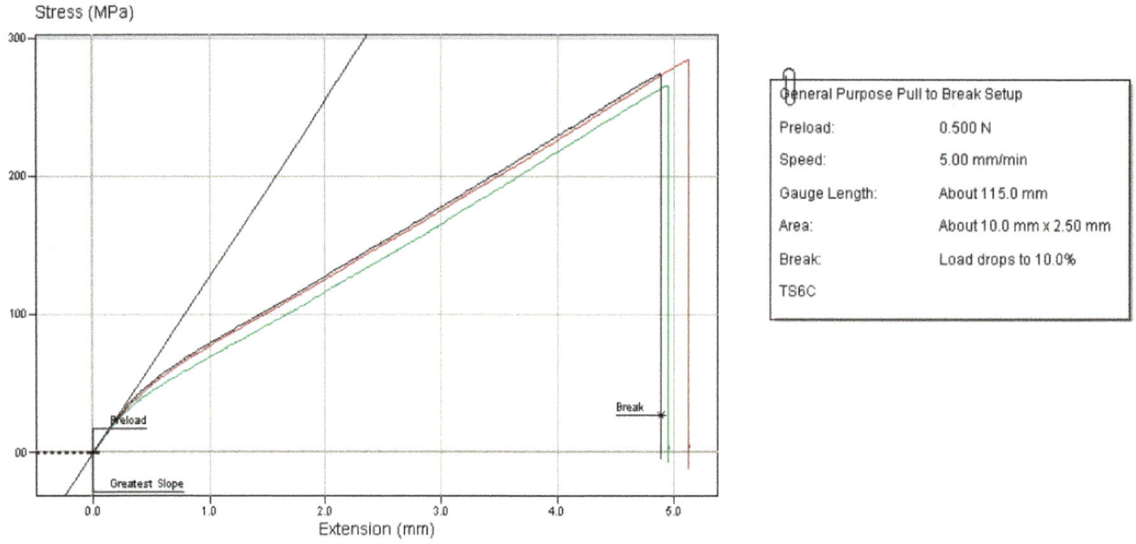

Figure 19: Superimposed graphs of tensile silicon series TS6.

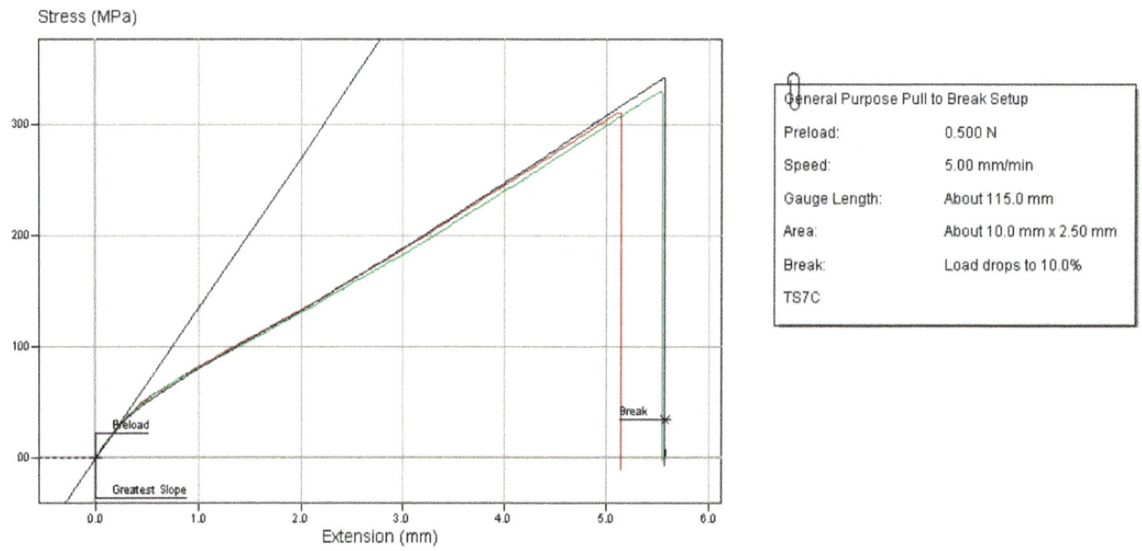

Figure 20: Superimposed graphs of tensile silicon series TS7.

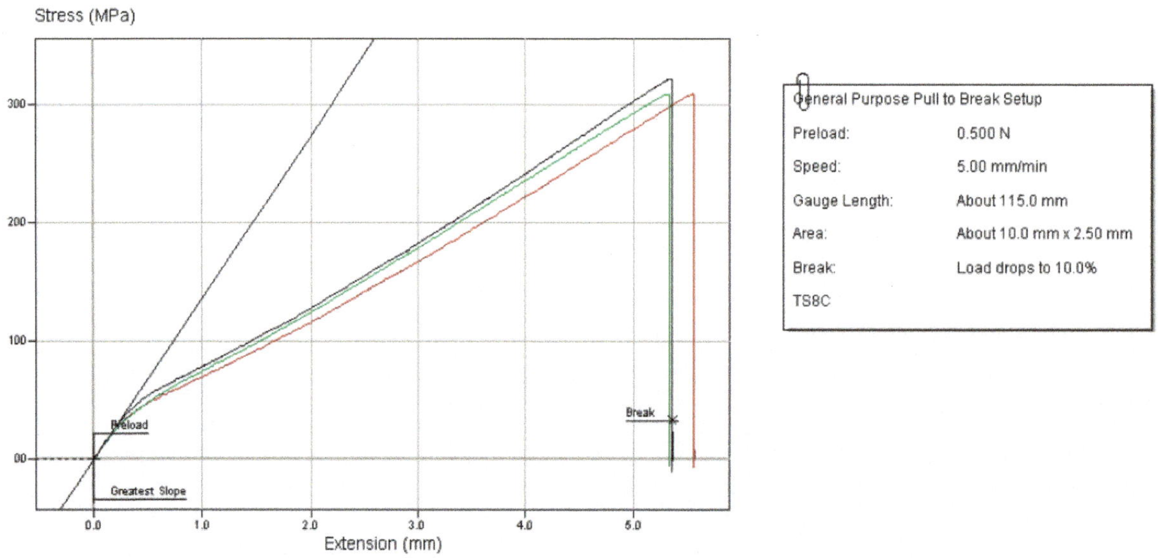

Figure 21: Superimposed graphs of tensile silicon series TS8.

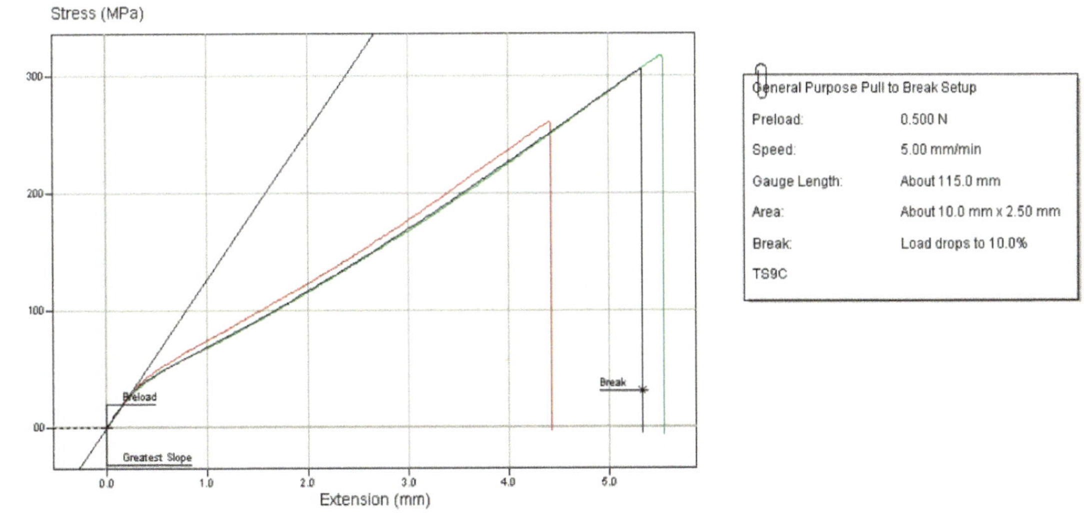

Figure 22: Superimposed graphs of tensile silicon series TS9.

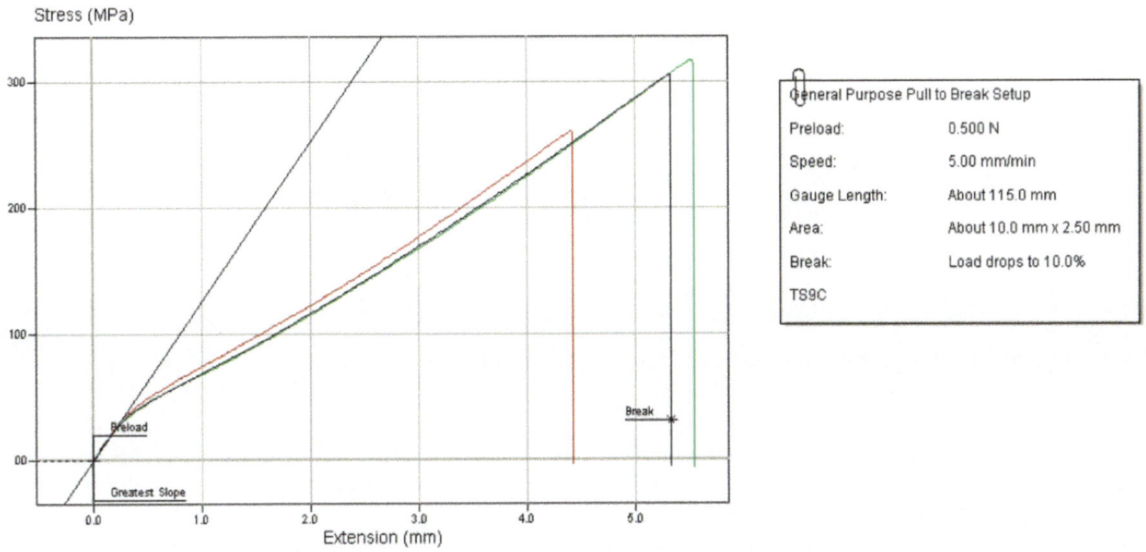

Figure 23: Superimposed graphs of tensile silicon series TS10.

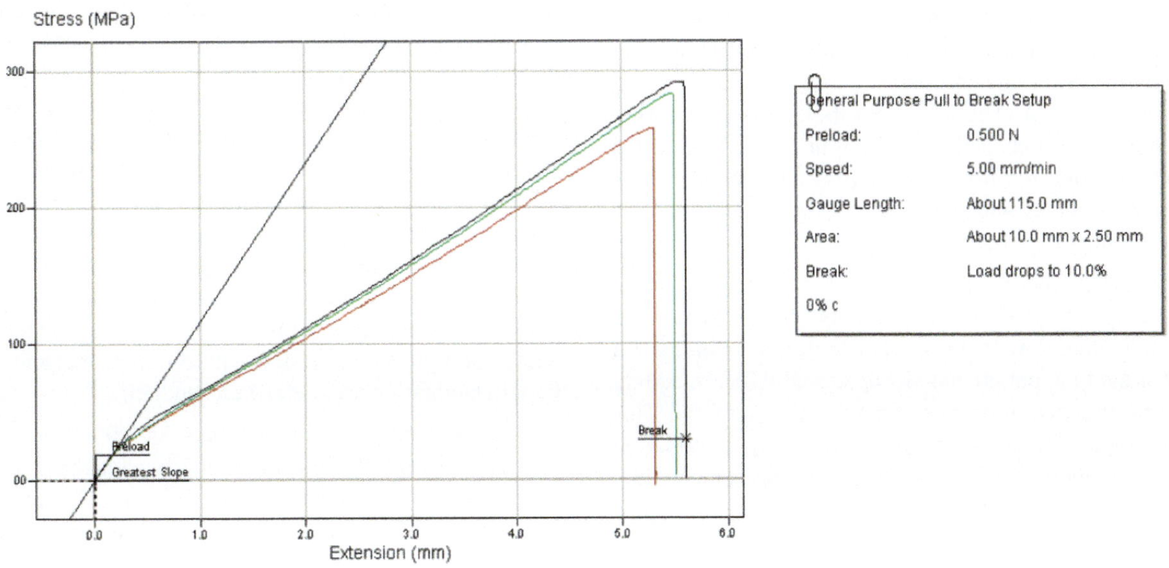

Figure 24: Without filler.

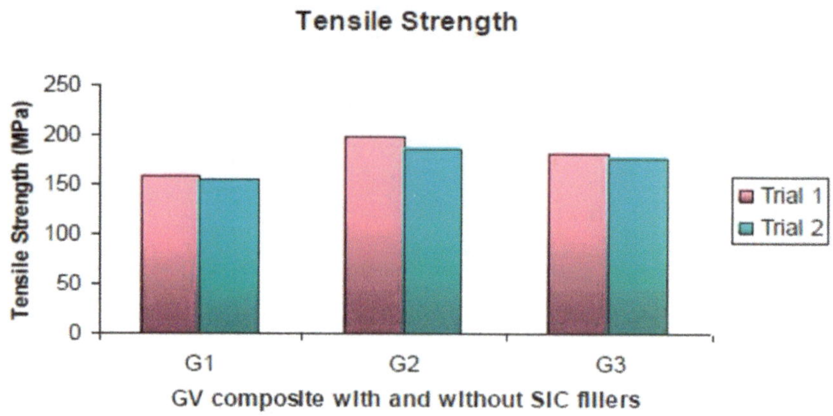

Figure 25: Standard graph showing tensile strength for different percentage of filler.

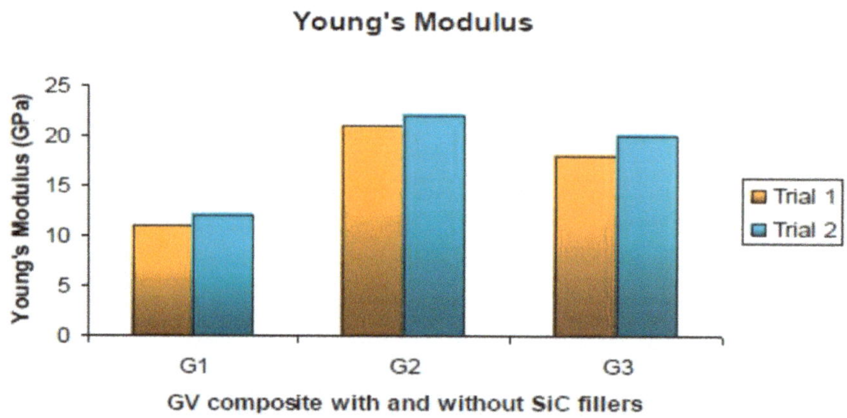

Figure 26: Standard graph showing young's modulus for different percentage of filler.

The quasi-static to dynamic strain rate shown in Fig. 4-24, because of strain extensional in material takes place. For prestrains less than about 1mm, the dynamic reloading curve is characterized by a well defined yield point whose level is higher than the maximum reached in quasi-static prestraining, but which is very close to the flow stress level reached at the same strain in these tests conducted exclusively at the dynamic rate. In general trend of a quasi-static reloading behaviour is complex and presumably governed, like with many other composites by strain ageing phenomenon, which depend on dynamic prestrains. And further more the quasi-static curves are also show phenomenon of dynamic strain ageing manifested by serrations along the flow curves. It is interesting to note that the same strain ageing phenomenon was also observed on static prestraining Fig. 4-24.

The present experimental results put in evidence that the static strain ageing effects are present in quasi-static and in the dynamic reloading flow curves obtained after prestraining at a different strain rate, but are practically absent in the reloading flow curves obtained after prestraining at the same rat. This fact would reflect the difference in the governing deformation modes and the microstructures developed during high and low rate prestraining. It is the passage from one deformation mode to the other which provokes the state and the dynamic strain ageing instabilities.

Figure 25 shows the Standard graph of tensile strength for different percentage of filler of G1, G2 and G3. And it is observed that the G2 exhibits the highest tensile strength. Similarly young's modulus is highest in the G2 which can be seen in Fig. 26.

CONCLUSION

Test specimens of constant fibre weight fraction (63 to 65%) and thickness 3 mm were prepared to study the effect of mechanical properties on the tensile strength of the Glass/epoxy composite. Tensile tests were carried out on these test specimens with different energy levels. The properties reduced from tension tests on composite materials are effective (averaged) properties. Tensile test, in a broad sense is measurement of the ability of a material to withstand forces that trend to pull it apart and to determine to what extent the material stretches before breaking. Tensile modulus, an indication of the relative stiffness can be determined from stress strain diagram. The residual tensile strengths of the damaged test specimens were obtained and their relationship with different incident energy levels, delamination areas etc. were studied:

- Introduction of graphite and SiC fillers in E-glass reinforced GV composites, improved in the mechanical properties, thus emphasizing the importance for the need to introduce fillers into glass reinforced GV composites
- Addition of fillers increases tensile strength, modulus, flexural strength, impact strength and hardness of Glass Vinyl ester composites

- Correlating the tensile results and observations of the fractured surface of the graphite filled GV composites, it is verified that the interfacial adhesion between fiber/matrix, which leads to, improved mechanical properties
- The tests showed linear elastic behaviour and brittle fracture for the test samples along with fibers in a composite fail at different stress levels as the applied tensile load increases
- The friction and slide wear behaviour of 7.5 wt.% graphite filled glass-epoxy composite is better compared to glass epoxy composites with lower graphite filler content
- The experimental results show that the self lubricating graphite filled G-E composite material investigated may be an adequate candidate for wear application in bearings

ACKNOWLEDGEMENT

We are very grateful to SJCE, Mysore (INDIA) for providing all facilities for testing our samples.

AUTHOR'S CONTRIBUTIONS

Vikram G Kamble: Planned the work and conducted the experiments. Naveen Kumar L: Did result analysing and all other mouse work.

REFERENCES

1. Cha'o-Kuang, C. and T. Tzu-Li, 1997. The indirect measurement of tensile strength by the deterministic grey dynamic model DGDM. Int. J. Syst. Sci., 28: 683-690. DOI: 10.1080/00207729708929428
2. Cybulskaa, M. and B.C. Goswamib, 2001. Tensile behaviour of staple yarns. J. Textile Inst., 92: 26-37. DOI: 10.1080/00405000108659612
3. Eldera, H.M., 1966. The tensile, compressive and bending moduli of some monofilament materials. J. Textile Inst. Transact., 57: T8-T14. DOI: 10.1080/19447026608662331
4. Ismaila, H. and H.D. Rozman, 1998. Studies on the tensile properties of rubberwood fibre-natural rubber composites. Int. J. Polymeric Mater., 41: 325-333. DOI: 10.1080/00914039808041053
5. Kalthoff, J.F., 2004. Characterization of the dynamic failure behaviour of a glass-fiber/vinyl-ester at different temperatures by means of instrumented Charpy impact testing. Composites, 35: 657-663. DOI: 10.1016/j.compositesb.2003.11.008
6. Kamble, V.G., S.G. Kamble and K.D. Ramesh, 2015. Prediction of frictional and wear behavior of aluminium matrix composites by artificial neural network. J. Molecular Eng. Mater., 2: 1450004- 1450014. DOI: 10.1142/S225123731450004X
7. Pande, G.N., J. Middleton and B. Kralj, 1998. Measurement of tensile strain softening and its influence on bond tensile test results. Proceedings of the 4th International Symposium Computer Methods in Structural Masonry, (MAM' 00).
8. Stevanovica, D., S. Kalyanasundaramb, A. Lowe and Jar, P.Y.B. 2003. Mode I and mode II delamination properties of glass/vinyl-ester composite toughened by particulate modified interlayers. Compos. Sci. Technol., 63: 1949-1964. DOI: 10.1016/S0266-3538(03)00162-3
9. Surendranathan, A.O., M. Savitha and K.V.
10. Sreenivasaprasad, 2008. Study on effect of incorporation of SiC fillers on Tensile and Flexural behavior of GV composites by experimental and simulation methods. MS&T.
11. Suresha, B., 2003. The mechanical and tribological properties.
12. Swaminathan, G., K.N. Shivakumar and M. Sharpe, 2007. RETRACTED: Evaluation of specific properties of glass and carbon/vinyl ester composites. Composites, 39: 27-35. DOI: 10.1016/j.compositesb.2007.02.012
13. Unal, H. and A. Mimaroglu, 2004. Influence of filler addition on the mechanical properties of nylon-6 polymer. J. Reinforced Plast. Compos., 23: 461-469. DOI: 10.1177/0731684404031977
14. Varada Rajulu, A., S.V. Sanjeev Kumar, G. Babu Rao, G.M. Shashidhara and J. He et al., 2002. Tensile properties of glass rovings/hydroxyl terminated polyester toughened epoxy composites. J. Reinforced Plast. Compos., 21: 1591-1596. DOI: 10.1177/0731684402021017494
15. Wu, T.J. and H.T. Hahn, 1998. The bearing strength of e-glass/vinyl-ester composites fabricated by vartm. Comp. Sci. Technol., 58: 1519-1529. DOI: 10.1016/S0266-3538(97)00180-2

CHAPTER 9

POLYPROPYLENE/HEMP WOODY CORE FIBER COMPOSITES: MORPHOLOGY, MECHANICAL, THERMAL PROPERTIES, AND WATER ABSORPTION BEHAVIORS

Chakaphan Ngaowthong[1,2], Vilai Rungsardthong[3] and Suchart Siengchin[1,4]

[1]Department of Materials and Production Engineering (MPE), The Sirindhorn International Thai – German Graduate School of Engineering (TGGS), King Mongkut's University of Technology North Bangkok, Bangkok, Thailand

[2]Department of Agricultural Engineering for Industry, King Mongkut's University of Technology North Bangkok, Prachinburi Campus, Prachinburi, Thailand

[3]Department of Agro-Industrial Technology, Faculty of Applied Science, King Mongkut's University of Technology North Bangkok, Bangkok, Thailand

[4]Center of Innovation in Design and Engineering for Manufacturing (CoI-DEM), King Mongkut's University of Technology North Bangkok, 1518 Pracharaj 1, Wongsawang Road, Bangsue, Bangkok 10800, Thailand

ABSTRACT

Natural fiber composites composed of polypropylene, maleic anhydride-graft-polypropylene, and hemp woody core fiber were produced by two-roll mill mixing. The hemp woody core fiber was treated by alkaline. The morphology of the polypropylene/hemp woody core fiber composites was studied by scanning electron microscopy technique. The mechanical and thermo-mechanical properties of the polypropylene/hemp woody core fiber composites were determined in tensile, flexural tests, and thermogravimetric analysis, respectively, and discussed. It was found that incorporation of hemp woody core fiber in polypropylene increased the tensile and flexural modulus. Note that the stiffness of short hemp woody core fiber/polypropylene composites was markedly improved. However, the flexural and impact strength of polypropylene/hemp woody core fiber composites was reduced compared to the neat polypropylene. Incorporation of hemp fiber increased the water uptake compared to the neat polypropylene. It was established that the Fick's law was fairly applicable to the water absorption results. The stiffness and thermogravimetric analysis behavior of the polypropylene/hemp woody core fiber composites were markedly affected by the surface treatments.

INTRODUCTION

Polypropylene (PP) is one of the most widely used polyolefin polymers which has several advantages, including strength, good corrosion, weather resistance, light weight, and low cost. Note that the low weight and good mechanical properties of PP have made it an ideal material in automotive construction. However, PP is not biodegradable and a significant part of municipal solid waste which has caused increasing environmental concerns. Recently, PP composed with natural fiber was widely used for making biocomposite materials to solve this issue.1⇓3 Due to the cost saving aspects and expected ecological benefits, the main motivation driving these researches and developments is related to the mechanical and thermal properties' profiles of natural fibers and wood, which indicate a substantial reinforcement potential.4 Hemp is a high-quality plant fiber which is flexible, strong, and durable. Currently, there are more than 30 countries around the world that produce industrial hemp including the hemp utilized extensively for furniture, building materials, textiles, food, cosmetics, pharmaceuticals, and paper.5·6 Thailand's textile and garment industry requires much more bark of the hemp long fibers for textiles producing shirts, pants, shoes, scarves, curtains, tablecloths, and so on. The reinforcement with woody core (xylem) for natural fiber composites7 has

Citation: Chakaphan Ngaowthong, Vilai Rungsardthong and Suchart Siengchin, Polypropylene/hemp woody core fiber composites: Morphology, mechanical, thermal properties, and water absorption behaviors, doi: 10.1177/1687814016638318.

attracted considerable attention in the recent years due to help the environment as well as to reduce the use of plastic in related composites.

Recently, one of the methods to reduce the amount of synthetic polymers is to add natural material into polymer. The use of natural fiber polymer composites was increasing substantially in engineering applications due to their outstanding properties such as tensile strength, modulus, and impact strength, compared to the pure polymers.[8,9] The major problems in natural fiber/polymer composites are the result of the difference in compatibility between a hydrophobic nature of polymer and a hydrophilic nature of natural fiber. Various methods have been adopted to enhance an adhesion between those natural fiber and polymer matrix such as the surface treatment of natural fiber and modification in chemical structure of polymer material.

Of the previous methods, alkalization and chemical grafting using maleic anhydride (MA) treatment are the most popular ways to enhance an interface. Alkalization is even considered a standard step to increase the efficiency of any further treatment.[5,7,10–13] MA grafting is used to modify not only fiber surface but also the polymer matrix to achieve better interfacial bonding and mechanical properties in polymer composites.[14] Several research activities have been conducted to improve fiber adhesion properties with the polymer matrix and found that the natural fiber modified with alkaline (ALK) treatment can be increased approximately 20%–40% in tensile strength and modulus of related composites[15–17] and approximately 30%–72% in flexural strength and modulus using maleate coupling agent.[18,19] Also, the effect of the natural fiber length and their surface modification on a poly(hydroxybutyrate-co-hydroxyvalerate) (PHBV) biocomposite was the topic of a recent contribution.[20] Moreover, Elkhaoulani et al.[21] investigated the mechanical and thermal properties of hemp fibers/PP. They found that MA clearly improved adhesion of the hemp fiber/PP matrix as a result of enhanced stiffness and thermal resistance.

The aim of this work was to produce PP and hemp woody core (HWC) fiber and to study their properties. A further aim of this work is to check the effect of fiber length, content, and modification to make artificial wood for various applications in the future market. Maleated PP (maleic anhydride-graft-PP (MA-g-PP)) was used to enhance the adhesion of HWC fiber in PP matrix. The mechanism promoting the improved interface induced by the MA-g-PP will be discussed through a scanning electron microscope and mechanical studies. The influence of the MA-g-PP addition on structural and mechanical properties of PP/HWC fiber composites will be discussed in conjunction with the water absorption studies.

EXPERIMENTAL

Materials

HWC chip was supplied by Highland Research and Development Institute (HRDI) (Chiangmai, Thailand). PP homopolymer (EL-Pro P702J; Thai Polyethylene Co., Ltd, Rayong, Thailand) was utilized as polymeric matrix for all composite systems. The density was 0.91 g/cm^3, the melting point of PP was 163°C, and the melt flow rate (230°C, 2.16 kg) was 12 g/10 min. Sodium hydroxide (analytical grade; Merck KGaA, Darmstadt, Germany) was served as treatment. Maleated PP (MA-g-PP), (Fusabond P353; DuPont™, Wilmington, DE, USA) was used as modifiers for PP systems. Its density and melting point were 0.904 g/cm^3 and 135°C, respectively (according to suppliers' information).

Analysis of the Chemical Composition of the HWC

Chemical analysis of the HWC age of 90 days and analysis of cellulose, hemicellulose, and lignin were tested according to TAPPI T203 om-88 and TAPPI T222 om-88 standard, respectively.

Preparation of the HWC Fiber

First, the HWC chips were washed using water for two to three times and then dried in an oven at 80°C for 48 h. The plastic cruncher machine and hammer mill machine used to prepare the size of chips into fiber form. The size of fibers was sorted by sieve shaker using 30 (large fiber (LF)), 60 (middle fiber (MF)), and 80 (small fiber (SF)) mesh, respectively. Fibers were then cleaned by soaking in hot water at a temperature of 50°C for 30 min and then dried in the oven at 80°C for 48 h.[22]

Examination of the Aspect Ratio of Fiber

An optical microscope (Microscope Leica DM2500 M, Wetzlar, Germany) was used to analyze the length and the diameter of the HWC fibers after size sorting. In order to determine the average length and diameter of fiber, at least 50 individual fibers were observed. Afterward, the aspect ratio for all sizes of hemp fiber was calculated.

Alkali Treatment

ALK was used to modify a surface of HWC fibers. HWCs were soaked in 5 wt% sodium hydroxide (NaOH) solution at ambient temperature and immersed for the fiber weight ratio of 20:1 for 30 min. After treatment, the fibers were washed with water to remove any traces of alkali on the fiber surface, until pH 7 of HWC was reached. The treated fibers (ALK) were then dried at 80°C for 48 h.[22]

Composite Processing and Samples' Preparation

PP and MA-g-PP additives in different ratios were prepared using two-roll mill machine (Labtech Scientific LRM-S-110/3E, Samutprakarn, Thailand) at a temperature of 190°C for 10 min. Afterward, the related compounds were grinded with the plastic cruncher machine. Then the resulting composites were made into dumbbell and rectangular shape by Cometech QC-601T (Taichung, Taiwan) manual compress molding machine at a temperature of 190°C for 10/10/5 min (preheat/heating/cooling) under a pressure of 50 MPa. The composition and designation of the compounds studied are given in Tables 1 and 2.

Table 1: Mixing ratio of polypropylene with different sizes of HWC fibers

Code	Sizes of fibers	Ratio of compounds (wt%)	
		PP	Fiber
PP/LF	LF	90	10
PP/MF	MF		
PP/SF	SF		

PP: polypropylene; LF: large fiber; MF: middle fiber; SF: small fiber.

Table 2: Recipes and designations of the PP-based composite systems

No.	Code	Ratio of compounds (wt%)			
		PP	Fiber treatment		Coupling agent (Fusabond 613)
			None	NaOH 5%	
1	PP	100			
2	PP9010	90	10		
3	PP7030	70	30		
4	PP6040	60	40		
5	PP9010ALK	90		10	
6	PP7030ALK	70		30	
7	PP6040ALK	60		40	
8	PP9010MAPP	87	10		3
9	PP7030MAPP	67	30		3
10	PP6040MAPP	57	40		3

PP: polypropylene; ALK: alkaline; MAPP: maleic anhydride-graft-polypropylene.

Mechanical Testing Methods

Tensile and flexural tests were performed on according to ASTM D638 and ASTM D790 standards, respectively, using the universal testing machine (Cometech). The impact strength test (Ceast 6957000, Torino, Italy-equipped) was performed on notched Izod specimens according to ASTM D256 standards and done by hammer mass of 2.75 J.

Scanning Electron Microscopy

A Hitachi Tabletop Microscope TM3000 (Tokyo, Japan) scanning electron microscope was used to investigate fiber/matrix adhesion and composite morphology. The scanning electron microscopy (SEM) fracture surfaces were used from dog-bone fracture sample for examination.

Water Absorption

Water absorption of PP/hemp fiber composites was carried out according to ASTM D570-98 standard test method. The PP/hemp fiber composites were cut from tensile specimens into rectangular shape ($19 \times 20 \times 3\,mm^3$) and then immersed in a distill water bath and then the specimens were removed periodically from water, and the increase in their weight was measured on a balance with a precision of 0.001 g over a period of 400 h. The percentage gain at any time (M_t) was calculated according toequation (1), where W_d and W_w denote the initial weight of material and weight of material after exposure to water absorption, respectively

$$M_t = \left(\frac{W_w - W_d}{W_w}\right) \times 100\% \quad (1)$$

Water Diffusion Coefficient

The average diffusion coefficient (D) was calculated by using equation (2) to measure the rate of moisture diffusion through all faces of the PP/hemp fiber composites

$$D = \pi\left(\frac{h}{4M_m}\right)^2\left(\frac{M_2 - M_1}{\sqrt{t_2} - \sqrt{t_1}}\right)^2 \quad (2)$$

where M_m is the maximum percent of moisture uptake, h is the sample thickness, $\sqrt{t}$ is the square root of time, and M_1 and M_2 are the initial slopes of the line of the moisture content.

Thermal Tests

Thermogravimetric analysis (TGA) was performed on a Mettler Toledo TGA/DSC1 (Greifensee, Switzerland) using nitrogen as purge gas with a flow rate of 40 mL/min at the temperature range of 40°C–700°C. The heating rate of TG measurements was 10°C/min and the weight of the samples was about 7 mg

RESULTS AND DISCUSSION

Analysis of the Chemical Composition of the HWC

The chemical composition data of hemp fibers data are listed in Table 3. The chemical composition data indicated that the cellulose of HWC fiber was 40.8% and the lignin achieved was only 21.7%. Note that the cellulose content of HWC was lower than that of the cellulose values of the related hemp bark and other natural fibers according to the recent reports.1·23 The finding maybe become further affected the below mechanical response.

Table 3: Results of chemical analysis by AOAC standard

Chemical composition	Wt%
Cellulose	40.8
Hemicellulose	14.4
Lignin	21.7

Analysis Size and Aspect Ratio of Fiber

The length and the diameter of the hemp fibers after sorting size are shown in Table 4. One can recognize that SF had a highest value of aspect ratio when compared to the LF and MF fibers.

Table 4: Results of dimension and aspect ratio of the hemp woody core after sorting size

Dimension	Fiber size (μm)		
	LF	MF	SF
Length	1507.91 ± 143.98	1314.76 ± 230.83	1016.91 ± 158.42
Diameter	962.77 ± 120.05	453.50 ± 91.79	208.53 ± 33.14
Aspect ratio	1.57	2.9	4.88

LF: large fiber; MF: middle fiber; SF: small fiber.

Effect on the Mechanical Properties of the Composite

The effects of hemp fiber length on the tensile and flexural mechanical properties of PP systems are displayed in Figure 1. One can see that the addition of hemp fiber to PP matrix improved the tensile and flexural stiffness of composites. Note that a natural fiber reinforcing effect is generally accompanied with reduced strength in most composite materials. This is the case also for the PP/HWC composite, the tensile and flexural strength value of which is below that of the neat PP. It can also be noticed that the stiffness in tensile and flexural tests of the related composites was strongly enhanced by adding high aspect ratio of SF. The tensile and flexural modulus of PP/SF composites was 20% higher than that of the neat PP. On the other hand, the effect of PP/MF composite on the mechanical properties was less sensitive than that of the PP/SF composite. An increase in tensile and flexural modulus of the PP/HWC composites can be well explained by the reinforcing effect of the SF. This is in accordance with the observation of PP/wood chip composites reported by Hietala et al.24 It was recognized earlier that the higher aspect ratio (SF) of the fiber composites has a strong impact on the mechanical properties.

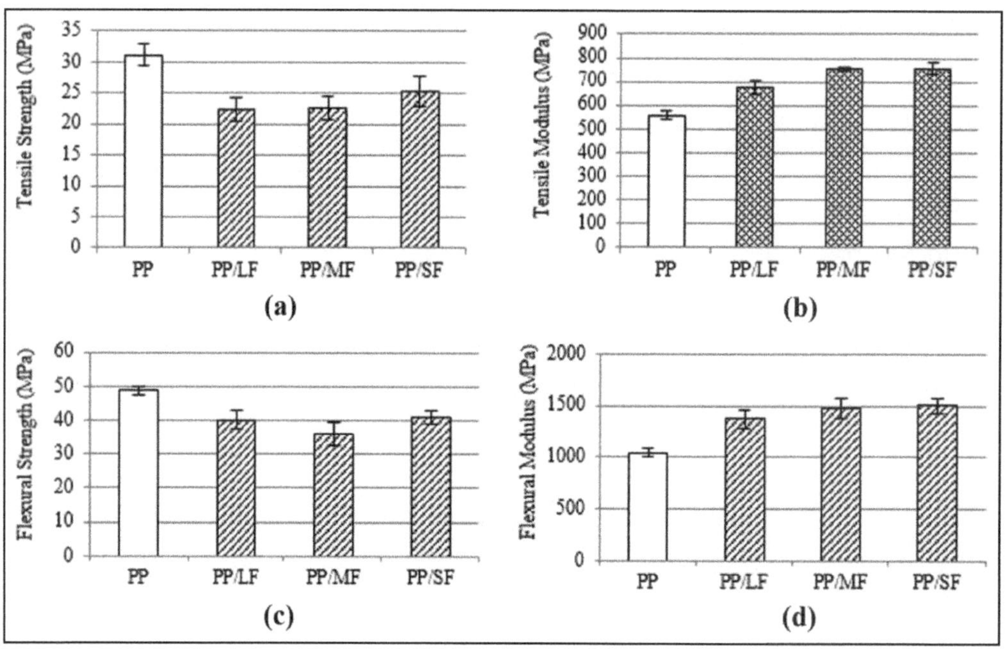

Figure 1: Mechanical properties of PP/hemp (90:10 wt%) fiber composite of long fiber (LF), middle fiber (MF), and short fiber (SF): (a) tensile strength, (b) tensile modulus, (c) flexural strength, and (d) flexural modulus.

The results of the tensile and flexural mechanical tests, caused by the SF fiber contents and surface treatments, are obvious in Figures 1 and 2. It is clearly seen that the incorporation of hemp fiber increased the tensile and flexural modulus of the related natural fiber composites and reduced the strength of the composites in the whole surface treatments. This result is in harmony with the effect of aspect ratio of the tensile mechanical data as shown above. It is noteworthy that the upgrade in tensile stiffness for SF untreated and flexural stiffness for SF ALK treated was not a linear function of the fiber content.

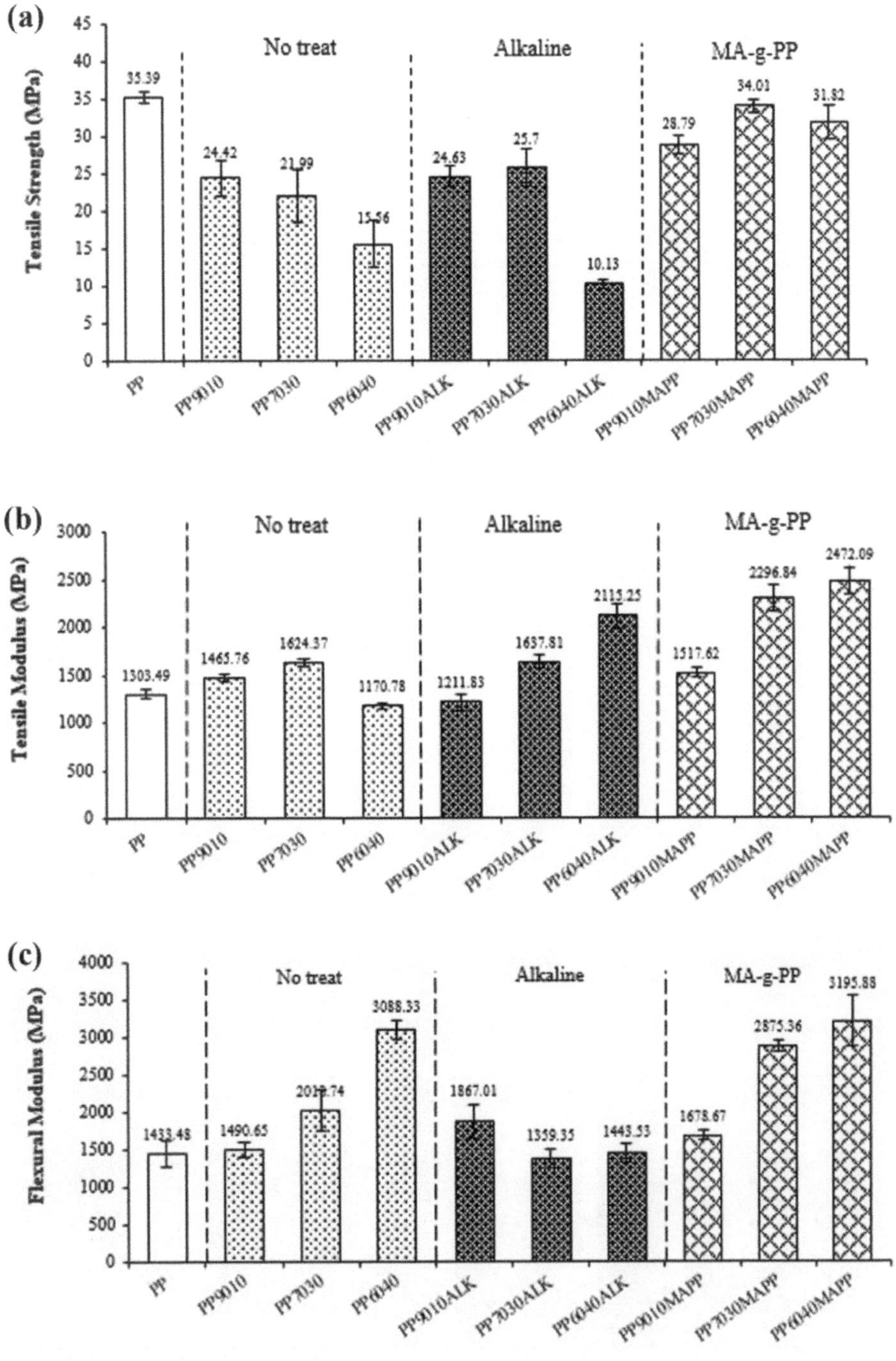

Figure 2: Mechanical properties of PP/SF composite different mixing ratios and surface treatments of HWC fiber: (a) tensile strength, (b) tensile modulus, and (c) flexural modulus.

Moreover, it was found that the modulus of the SF/MA-g-PP composites was markedly higher than that of the neat PP according to both tensile and flexural measurements. This finding hints for more restricted mobility of the PP segments with increasing amount of HWC fiber. This finding is in analogy with the results reported by Kabir et al.23 that when mixed with fiber composite material using MA-g-PP, it enhanced the adhesion between the fiber surface and the PP matrix. During a composite preparation, the temperature will cause MA to react with hydroxyl group (OH) on cellulose fibers and make covalent bonding between hydroxyl groups of fibers and MA which can act as a bridge between the hemp fiber and PP matrix, further reducing the hydrophilicity of the surface and the mechanical properties of the composites increase. The scheme of chemical reaction between the surface fibers and matrix is shown in Figure 3. On the other hand, there was no effect of ALK-treated SF on tensile strength compared with untreated SF. It has to be mentioned that the sodium hydroxide (NaOH) reacted with the hydroxyl group on fiber surface and thus decreased of hydroxyl group at surface.23Note that NaOH eliminates hemicellulose, lignin, pectin, and wax from the fiber surface; as a result, the strength of hemp fiber and tensile strength of composite decreased (cf. Figure 2(a)). For ALK treatment, the observed tensile and flexural modulus of PP/HWC composites was likely increased due to the increase in surface roughness resulting in better mechanical interlocking. Moreover, it is an open issue whether there is a correlation between the stiffness and an increase in the amount of cellulose exposed on the HWC fiber surface. To contribute to this topic, the first task is, however, to assess the amorphous cellulose content at the expense of crystalline cellulose description. Figure 4(a) and (c) displays the fracture morphology of the PP/HWC composites without treatment. The fracture surface became the typical feature of brittle fracture in composite prone to crazing and presented a gap between fiber and matrix with hints for pulled out HWC fiber. In the presence of ALK treatment, the fracture surface indicated an increase in HWC surface roughness and developed likely semi-brittle fracture crazes (cf. Figure 4(b) and (d)). Nevertheless, the impact strength of the PP composite systems was negative impact on these properties (Figure 5). However, the impact strength values suggest that the improvement in the stiffness and strength was achieved at the cost of the ductility of the PP/HWC fiber composites. A similar trend has been reported for oil palm–reinforced high-density polyethylene (HDPE) composites with the influence of coupling agent (maleated PP) that the impact strength value of HDPE/oil palm composites decreased markedly when compared to the neat HDPE matrix and slightly influenced with MAPP content.25

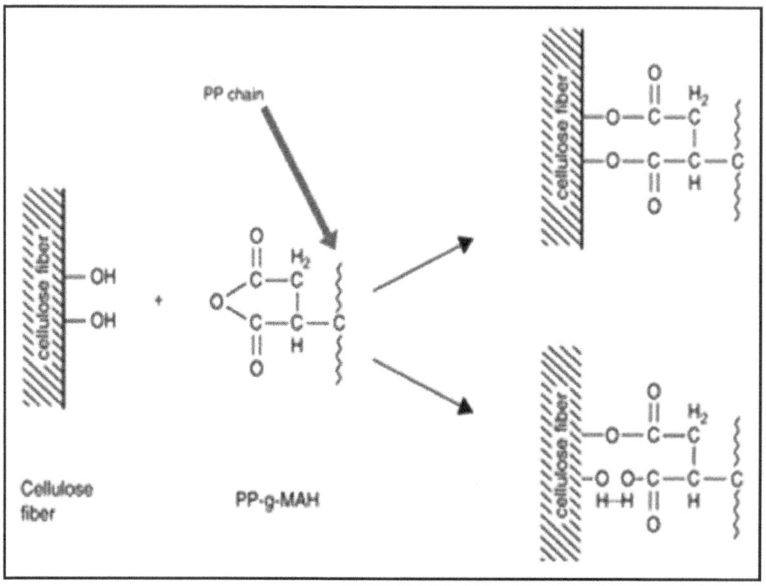

Figure 3: Chemical bonding between maleic anhydride-graft-polypropylene and fiber.22

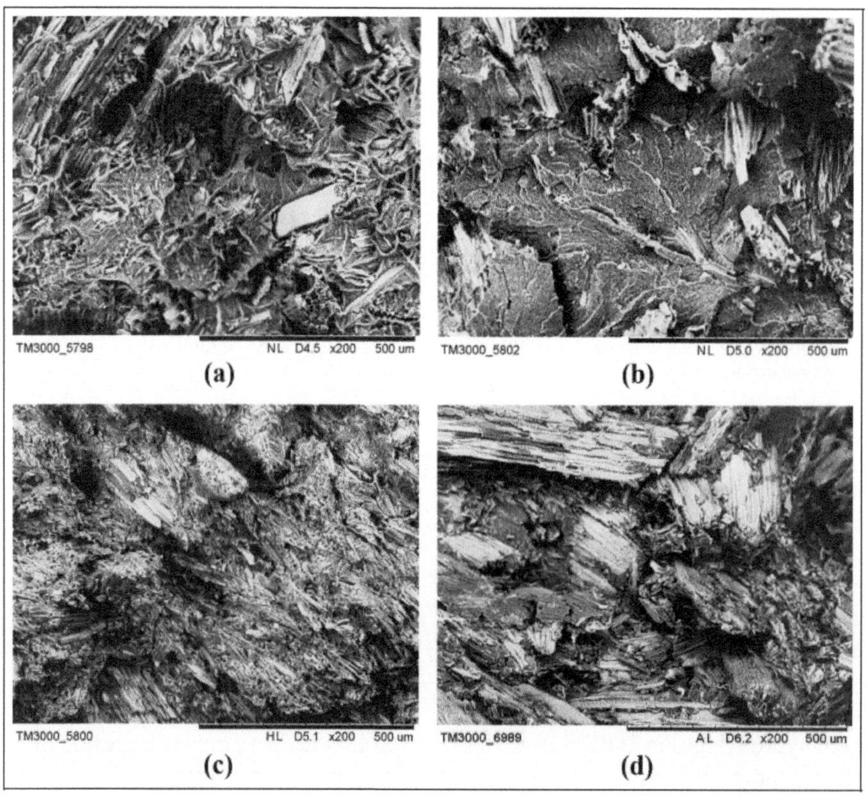

Figure 4: The morphology of fractured PP/HWC composites: (a) PP with untreated fiber (PP9010), (b) PP with alkalization fiber (PP9010ALK), (c) PP with untreated fiber (PP6040), and (d) PP with alkalization fiber (PP6040ALK).

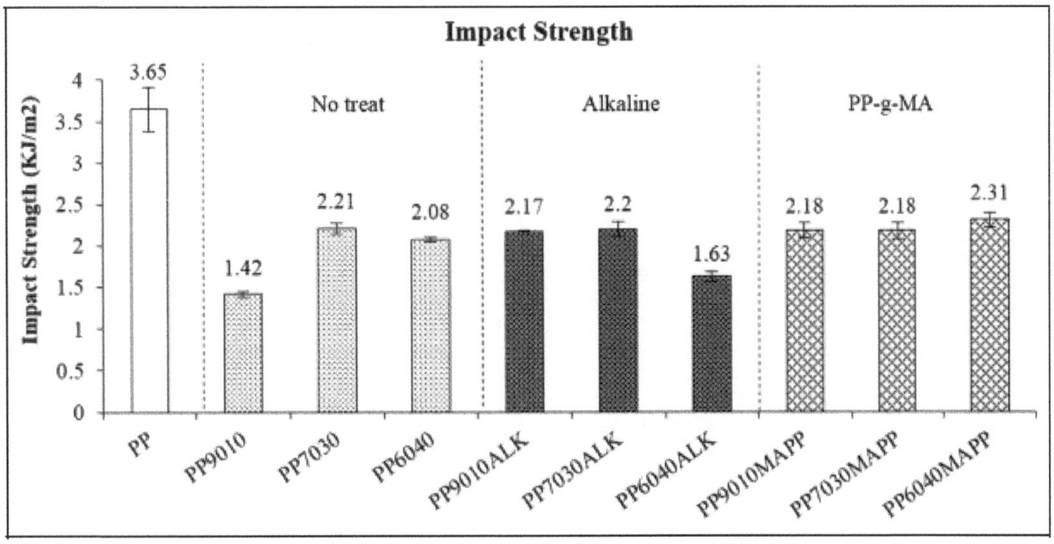

Figure 5: Impact strength of PP/SF composite different mixing ratios and surface treatments of hemp fiber.

Water absorption of PP/hemp fiber containing different amounts of hemp fiber composites as a function of time is demonstrated in Figure 6. One can recognize that the water sorption behavior was considered depending on the hemp contents. The neat PP recorded water uptake value at 0.002% after subjecting to water absorption at 417h. The PP/HWC composites exhibited remarkably large amount of water absorption. Especially, the water absorption of PP with 40 wt% of HWC fiber during the first 48 and 417h increased by 8.8% and 17.8%, respectively, compared to the neat PP. This can be explained by the chemical nature of cellulose and hemicellulose contents in HWC fiber according to the above result in Table 3. The water absorption of PP with treated hemp fiber and PP/hemp with MA-g-PP additives is also shown in Figure 6. As expected, the water absorption of PP/HWC fiber composites treated by alkali increased due to the weak interfacial bonding between HWC fiber and PP matrix and the roughness of HWC fiber, compared to the incorporation MA-g-PP in PP/HWC fiber composites. Accordingly, the observed water resistance was likely less due to the MA-g-PP themselves but more due to their adhesive covering polymer. Whether there is a correlation between the water resistance and the adhesion of HWC fiber and matrix. To contribute to this topic, the first task is to assess the dispersion by a suitable mathematical description. However, this aspect was not addressed in this work. The most striking and unexpected feature is that the water resistance increased in intensity after adding MA-g-PP; as in case of PP70/30 and PP60/40 composites, the opposite tendency should occur. The authors have no explanation for this behavior at present. Fick's law in equation (2) is adopted to calculate the diffusion coefficient (*D*) of water absorption value. It is well resolved that there is a fair agreement between the water uptake and those calculated by Fick's law. The calculation of diffusion coefficient is shown in Table 5. The *D* value of PP/hemp fiber composites increased with an increase in hemp fiber. This may be due to the hydrophilic character of the HWC fiber.

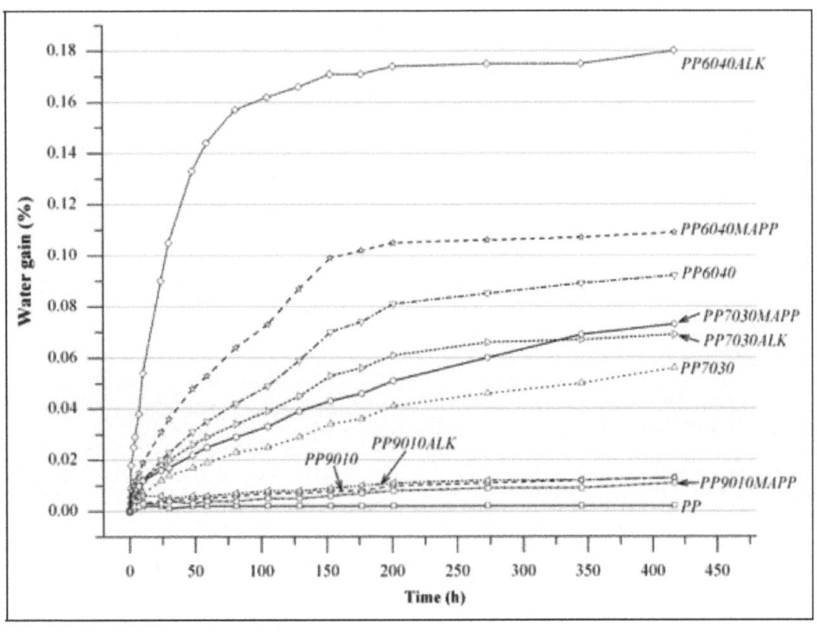

Figure 6: Effect of fiber contents and surface treatment of PP/hemp fiber composites on water absorption.

Table 5: Calculation of the diffusion coefficient (*D*)

No.	Code	*D* (mm^2/h)	R^2
1	PP	0.0000E + 00	N/A
2	PP9010	5.1350E−09	0.879
3	PP7030	7.3102E−08	0.995
4	PP6040	3.1477E−07	1.000
5	PP9010ALK	5.1719E−09	0.879
6	PP7030ALK	1.8979E−07	0.999
7	PP6040ALK	9.5513E−07	0.986
8	PP9010MAPP	2.5941E−09	0.593
9	PP7030MAPP	1.1845E−07	1.000
10	PP6040MAPP	6.1431E−07	0.998

N/A: not available; PP: polypropylene; ALK: alkaline; MAPP: maleic anhydride-graft-polypropylene.

In *thermal tests*, usually the decomposition temperature of natural fiber composite is reduced with increasing natural fiber content.26 Figure 7 presents the overall TG decomposition process for the PP composite system studied. Concerning the pattern of TGA processes, it is well known that the natural fiber composite exhibits three steps. The initial stage at low temperature is used to remove the moisture contained in the natural fiber composite; the second and third stages at high temperature are usually attributed to the decomposition of hemicellulose and non-cellulose, respectively.27 This is in accordance with our TGA observation. It is also clearly seen in Figure 7 that the thermal decomposition of the neat PP had only one step and started decomposition at the temperature of 443°C. Note that the thermal resistance of all PP/SF composites was lower than that of the neat PP at the whole temperature range. The resistance to thermal degradation at the high temperature of the SF/MA-g-PP composite was slightly higher than those of ALK-treated SF/PP composites. One possible explanation is due to the covalent bonding between hydroxyl groups of fibers and MA, but a definite explanation for this behavior can hardly be given.

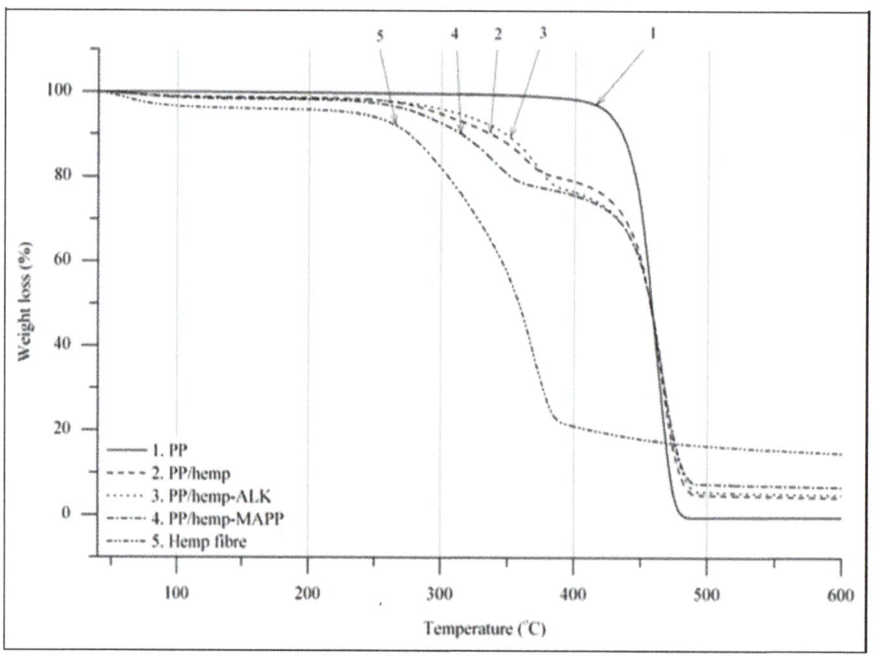

Figure 7: Weight loss percent of the PP/hemp fiber composite.

CONCLUSION

This research is devoted to study the thermal and mechanical properties and water absorption behavior of a PP and its aspect ratio/content/modification of HWC fiber in PP composites. The MA-g-PP was introduced in the HWC fiber/PP composites via two-roll mill mixing. It was found that incorporation SF into PP improved the modulus in tensile and flexural tests. The stiffness of PP/HWC fiber composites increased with an increase in the HWC fiber content and aspect ratio but at the cost of reduction in tensile and flexural strength. The tensile and flexural modulus of the SF/MA-g-PP composites was enhanced when compared to the neat PP. Incorporation of HWC fiber increased the water uptake of the PP composites. The thermal degradation of PP/HWC fiber composite systems studied decreased markedly compared with the neat PP.

ACKNOWLEDGMENTS

The authors are grateful to the Natural Composites Research Group (NCR) and King Mongkut's University of Technology

North Bangkok (KMUTNB) for laboratory support under the grant agreement KMUTNB-NRU-59-02.

REFERENCES

1. Faruk O, Bledzki AK, Fink H-P, et al. Biocomposites reinforced with natural fibers: 2000–2010. Prog Polym Sci 2012; 37: 1552–1596.
2. Bledzki AK, Letman M, Viksne A, et al. A comparison of compounding processes and wood type for wood fibre—PP composites. Compos Part A: Appl S 2005; 36: 789–797.
3. Beckermann GW and Pickering KL. Engineering and evaluation of hemp fibre reinforced polypropylene composites: fibre treatment and matrix modification. Compos Part A: Appl S 2008; 39: 979–988.
4. Sobczak L, Lang RW and Haider A. Polypropylene composites with natural fibers and wood—general mechanical property profiles. Compos Sci Technol 2012; 72: 550–557.
5. Pickering KL. Chapter 6. Assessing fibre surface treatment to improve the mechanical properties of natural fibre composites. In: Zafeiropoulos NE (ed.) Interface engineering of natural fibre composites for maximum performance. Cambridge: Woodhead Publishing, 2011, pp.186–202.
6. Salentijn EMJ, Zhang Q, Amaducci S, et al. New developments in fiber hemp (Cannabis sativa L.) breeding. Ind Crop Prod 2015; 68: 32–41.
7. Sgriccia N, Hawley MC and Misra M. Characterization of natural fiber surfaces and natural fiber composites. Compos Part A: Appl S 2008; 39: 1632–1637.
8. Teo PS and Chow WS. Water vapour permeability of poly (lactic acid)/chitosan binary and ternary blends. KMUTNB: Int J Appl Sci Technol 2014; 7: 23–27.
9. Dzˇ alto J, Medina L and Mitschang P. Volumetric interaction and material characterization of flax/furan bio-composites. KMUTNB: Int J Appl Sci Technol 2014; 7: 11–21.
10. Liu J, Jiang M, Wang Y, et al. Tensile behaviors of ECRglass and high strength glass fibers after NaOH treatment. Ceram Int 2013; 39: 9173–9178.
11. Gu H. Tensile behaviours of the coir fibre and related composites after NaOH treatment. Mater Des 2009; 30: 3931–3934.
12. Gandini A and Belgacem MN. Chapter 1. Modifying cellulose fiber surfaces in the manufacture of natural fiber composites. In: Zafeiropoulos NE (ed.) Interface engineering of natural fibre composites for maximum performance. Cambridge: Woodhead Publishing, 2011, pp.3–42.
13. Arbelaiz A and Mondragon I. Chapter 5. Testing the effect of processing and surface treatment on the interfacial adhesion of single fibres in natural fibre composites. In: Zafeiropoulos NE (ed.) Interface engineering of natural fibre composites for maximum performance. Cambridge: Woodhead Publishing, 2011, pp.146–185.
14. Kalia S, Kaith BS and Kaur I. Cellulose fibers: bio- and nano-polymer composites: green chemistry and technology. Berlin: Springer, 2011.
15. Prasad SV, Pavithran C and Rohatgi PK. Alkali treatment of coir fibres for coir-polyester composites. J Mater Sci 1983; 18: 1443–1454.
16. Mwaikambo LY, Tucker N and Clark AJ. Mechanical properties of hemp-fibre-reinforced euphorbia composites. Macromol Mater Eng 2007; 292: 993–1000.
17. Kabir MM, Wang H, Lau KT, et al. Chemical treatments on plant-based natural fibre reinforced polymer composites: an overview. Compos Part B: Eng 2012; 43: 2883–2892.
18. Mohanty S, Nayak SK, Verma SK, et al. Effect of MAPP as a coupling agent on the performance of jute–PP composites. J Reinf Plast Comp 2004; 23: 625–637.
19. Mohanty S, Verma SK, Nayak SK, et al. Influence of fiber treatment on the performance of sisal–polypropylene composites. J Appl Polym Sci 2004; 94: 1336–1345.
20. Dangtungee R, Tengsuthiwat J, Boonyasopon P, et al. Sisal natural fiber/clay-reinforced poly (hydroxybutyrateco-hydroxyvalerate) hybrid composites. J Thermoplast Compos. Epub ahead of print 10 December 2014. DOI: 10.1177/0892705714563128.
21. Elkhaoulani A, Arrakhiz FZ, Benmoussa K, et al. Mechanical and thermal properties of polymer composite based on natural fibers: Moroccan hemp fibers/polypropylene. Mater Des 2013; 49: 203–208.
22. Sawpan MA, Pickering KL and Fernyhough A. Effect of various chemical treatments on the fibre structure and tensile properties of industrial hemp fibres. Compos Part A: Appl S 2011; 42: 888–895.
23. Kabir M, Wang H, Lau K, et al. Tensile properties of chemically treated hemp fibres as reinforcement for composites. Compos Part B: Eng 2013; 53: 362–368.
24. Hietala M, Samuelsson E, Niinima¨ki J, et al. The effect of pre-softened wood chips on wood fibre aspect ratio and mechanical properties of wood–polymer composites. Compos Part A: Appl S 2011; 42: 2110–2116.
25. Kakou CA, Arrakhiz FZ, Trokourey A, et al. Influence of coupling agent content on the properties of high density polyethylene composites reinforced with oil palm fibers. Mater Des 2014; 63: 641–649.
26. Siengchin S and Dangtungee R. Effect of woven flax structures on morphology and properties of reinforced modified polylactide composites. J Thermoplast Compos 2013; 26: 1424–1440.
27. Luz S, Goncxalves A and Del'Arco A. Mechanical behavior and microstructural analysis of sugarcane bagasse fibers reinforced polypropylene composites. Compos Part A: Appl S 2007; 38: 1455–1461.

CHAPTER 10 SUPRAMOLECULAR POLYMER ADHESIVES: ADVANCED MATERIALS INSPIRED BY NATURE

Christian Heinzmann, Christoph Weder and Lu Montero de Espinosa

Adolphe Merkle Institute, University of Fribourg, Chemin des Verdiers 4, 1700 Fribourg, Switzerland

ABSTRACT

Due to their dynamic, stimuli-responsive nature, non-covalent interactions represent versatile design elements that can be found in nature in many molecular processes or materials, where adaptive behavior or reversible connectivity is required. Examples include molecular recognition processes, which trigger biological responses or cell-adhesion to surfaces, and a broad range of animal secreted adhesives with environment-dependent properties. Such advanced functionalities have inspired researchers to employ similar design approaches for the development of synthetic polymers with stimuli-responsive properties. The utilization of non-covalent interactions for the design of adhesives with advanced functionalities such as stimuli responsiveness, bonding and debonding on demand capability, surface selectivity or recyclability is a rapidly emerging subset of this field, which is summarized in this review.

INTRODUCTION

In principle, the phenomenon of adhesion is easily defined: if the work required to separate two dissimilar objects is higher than the work required to bring them into contact, then the objects will cling or stick to each other.[1] However, if the objects are more complex than a pair of well-defined molecules in a well-defined environment, in which attractive forces outweigh repulsion, adhesion is often a highly complex process that involves several mechanisms spanning different length scales.[1–3] Several theories have been developed over time to explain the chemical, physical and mechanical factors governing adhesion,[4,5] but so far no unified theory has been developed. Adhesion phenomena are central to many biological processes involving both chemical and physical mechanisms, the latter often being emergent. In nature, adhesion is, on a molecular level, often driven by reversible interactions that are controlled by a wide range of stimuli that are imparted by either the environment or released by living organisms to satisfy a specific need. Non-covalent interactions are found in many natural adhesive systems such as those displayed by the gecko footpads (van der Waals forces)[6] or the secretions of mussels and tubeworms (hydrogen bonds and/or metal–ligand complexes).[7] Additionally, non-covalent interactions are involved in various biological processes including cell adhesion to extracellular biomolecules[8] or bacterial adhesion.[9,10] These examples illustrate that nature uses various types of non-covalent interactions to control adhesive forces and access complex functions. The possibility to exploit similar, albeit significantly simplified, mechanisms to design

Citation: Christian Heinzmann, Christoph Weder and Lu Montero de Espinosa, Supramolecular polymer adhesives: advanced materials inspired by nature, DOI: 10.1039/C5CS00477B.

synthetic adhesives with advanced and stimuli-responsive properties has inspired a great deal of research activities in the last two decades. This contribution provides an overview of this exciting development, focusing on hydrogen bonding, host–guest interactions and metal–ligand complexation as the main non-covalent motifs that were utilized to design supramolecular adhesives. Mussel-inspired adhesives based on L-3,4-dihydroxyphenylalanine (DOPA) are intentionally excluded from this overview, since the related literature is vast and has been reviewed in several recent articles.[11–13] Apart from one specific example, materials in which self-adhesion (cohesion) effects are emphasized, *i.e.* supramolecular polymers developed to be (self)healable, are also not covered, and the reader is referred to several recent reviews dedicated to this topic.[14–17] Two general approaches to design reversible supramolecular adhesives can be distinguished, *i.e.* the modulation of the adhesive properties *via* the bulk material properties or, alternatively, the control of adhesion through a change of the surface interactions. While both approaches have been demonstrated to permit stimuli-responsive dynamic adhesion, the failure mechanisms are generally different; while a change of the bulk properties typically leads to cohesive failure, the modulation of surface interactions appears to cause adhesive failure.

The first report of the adhesion of a supramolecular polymer on a substrate dates back to 1997,[18–20] when Stupp and coworkers reported on a triblock oligomer that self-assembled into rod-like structures on account of long-range quadruple interactions (π–π stacking).[18] When solution-t onto a glass surface, the supramolecular material adhered tenaciously with a reported shear strength of 1.22 MPa, outperforming poly(vinyl phenol) (PVP) (0.53 MPa), whose side chains carry the same chemical functionality as the investigated supramolecular assemblies, and which is known for its hydrogen bonding ability[21] and coating properties.[22] The debonding on demand characteristics of the supramolecular polymer were not explored; however, its adhesion to the glass surface was shown to be resistant to washing with an aqueous HF solution, while PVP was washed from the glass surface within minutes under the same conditions.

Since the report of Stupp and coworkers, a broad range of supramolecular polymers have been explored as adhesives and numerous fundamental studies have been conducted, which are discussed in this review. The approach is also of interest in the context of technological exploitation, as evidenced by the number of patents filed in the field.[23–33] Despite the variety of non-covalent interactions that could be used to synthesize supramolecular polymers,[34,35] most of the examples found in the literature employ hydrogen bonding motifs to achieve reversible bonding and debonding. This may be explained by the relatively low synthetic efforts required to incorporate hydrogen bonding units into polymeric structures or polymerizable monomers.[34,36,37] Additionally, hydrogen bonds are compatible with many other functional groups and can be comparatively strong.[38] By contrast, the number of studies on supramolecular adhesives based on host–guest interactions, metal–ligand complexation, or other non-covalent interactions is much smaller. While this review is organized by the type of binding motif used, it appears that many of the concepts, functions and properties discussed could be realized by exploiting different types of non-covalent interactions.

SUPRAMOLECULAR ADHESIVES BASED ON HYDROGEN BONDING

Early work by Long and coworkers focused on the synthesis of a methacrylic monomer (2-isocyanatoethyl methacrylate) functionalized with the self-complementary quadruple-hydrogen bond forming ureido-4-pyrimidinone (UPy) motif developed by Meijer, Sijbesma and coworkers,[39] and its copolymerization with butyl acrylate to create polymers with different UPy content (Fig. 1a).[36]

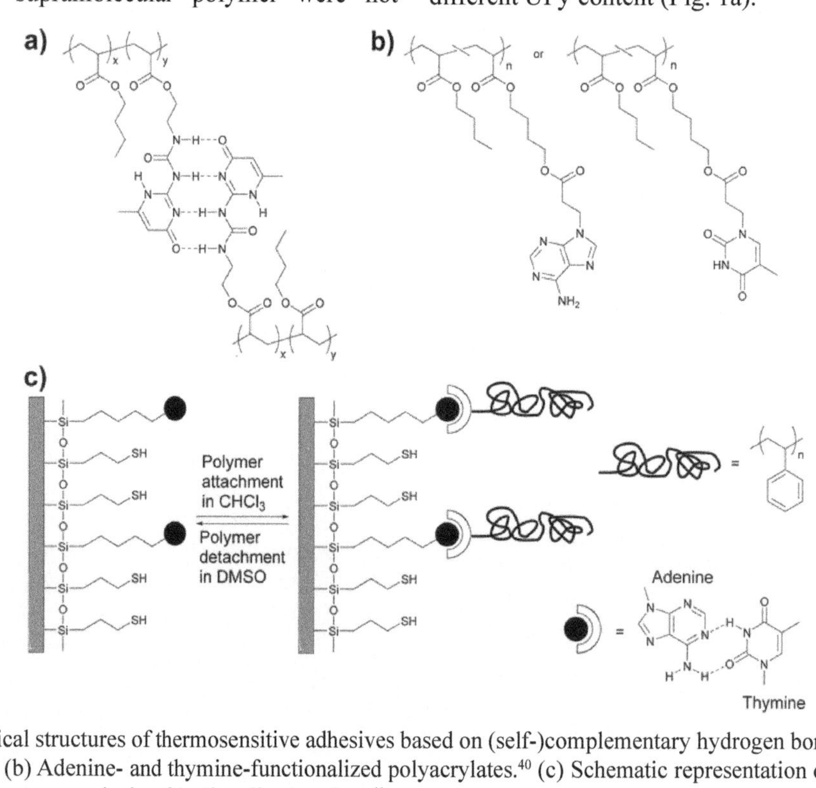

Figure 1: (a and b) Chemical structures of thermosensitive adhesives based on (self-)complementary hydrogen bonding motifs. (a) UPy-functionalized polyacrylates.[36] (b) Adenine- and thymine-functionalized polyacrylates.[40] (c) Schematic representation of the adhesion of thymine-functionalized polystyrene onto an adenine-functionalized surface.[41]

The incorporation of 2.5–10.4 mol% of the UPy-functionalized monomer caused an increase of the glass transition temperature (T_g) from −50 °C up to −23 °C, which is at least in part related to the formation of supramolecular cross-links *via* UPy dimerization. However, rheological measurements show clearly that the UPy dimers are completely dissociated at temperatures above 80 °C. The adhesion of these polymers to a glass surface was investigated by peel tests (Fig. 2),[36] which showed an increase of the adhesion strength with increasing UPy content on account of strong UPy–glass interactions. Further evidence of such interactions was found in thin layer chromatography (TLC) analyses, where the UPy-containing copolymers did not ascend the TLC plate if a non-coordinating solvent such as chloroform was used as mobile phase. On the other hand, polar solvents such as THF or ethyl acetate moved all (co)polymers to the top of the TLC plate due to disruption of the UPy–silica interactions. Remarkably, the addition of only 3.3 mol% of the UPy–monomer led to a three-fold increase of the peel strength. Rheological measurements data suggested that the higher viscosity of the UPy-containing polymers could contribute to increase their peel strength, as is known that cross-linked materials have an increased creep resistance.[42,43] The side-chain functionalization approach taken by Long and coworkers allows one to readily tune the supramolecular cross-link density as well as the glass transition temperature by simple variation of the polymer composition, in particular the content of the UPy–monomer and the choice of the comonomer.

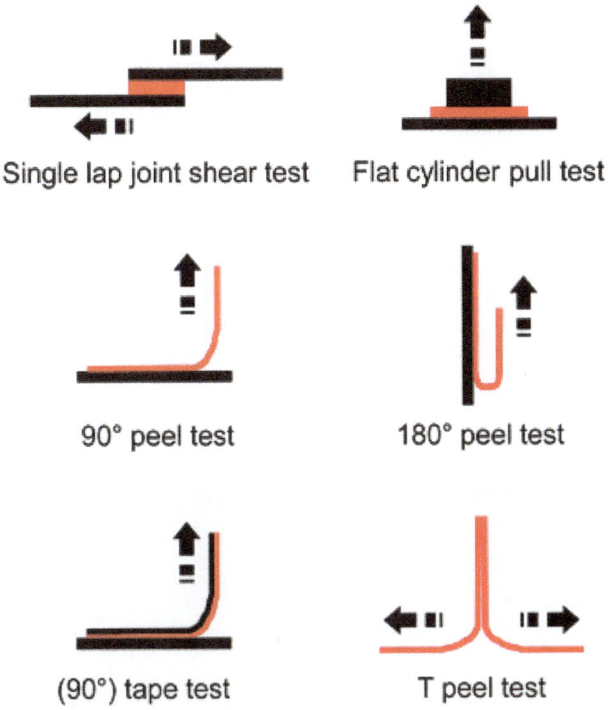

Figure 2: Overview of common test layouts to quantify the adhesive strength of an adhesive (red) to a substrate (black) or to itself.

Long, Ward and coworkers later extended this approach to nucleobase-containing polyacrylates (Fig. 1b)[40] and polystyrene (Fig. 1c).[41] While nucleobases can self-dimerize *via* hydrogen bonding, the complementary binding of nucleobases, which plays a key role in the functions of biopolymers such as DNA or RNA, is of course much stronger and presents an intriguing design approach for molecular recognition-based selective adhesion. In the first example,[41] thymine-terminated polystyrene was brought into contact with an adenine-functionalized silicon surface (Fig. 1c). The formation of adenine–thymine hydrogen bonds was confirmed with X-ray photoelectron spectroscopy (XPS) and through the increase of the water contact angle, which indicated the attachment of the hydrophobic polystyrene (PS) layer. When a hydroxyl-terminated polystyrene was used instead, only a very low adhesion to the adenine-functionalized surface was observed, due to the absence of specific interactions with the substrate. In a subsequent study, acrylate monomers carrying adenine (A) and thymine (T) motifs were synthesized and independently copolymerized with *n*-butyl acrylate (*n*BA) in different ratios.[40] While the adenine containing polyacrylate was found to assemble in needle-like nanostructures (*ca.* 100 nm × 10 nm) at an adenine content of as low as 7 mol%, the thymine-carrying polyacrylates did not show any self-assembly, even at a thymine-monomer content of over 30 mol%. The differences observed were explained with the higher tendency of adenine dimers to form π–π stacks that can aggregate into larger crystalline domains. Temperature-dependent small-angle X-ray scattering (SAXS) and wide-angle X-ray diffraction (WAXD) measurements showed a decrease of the Bragg peak intensity as the temperature was increased until no structure could be observed over 80 °C. These observations were correlated with dynamic mechanical analyses (DMA), which showed that the materials lost their mechanical integrity at *ca.* 80 °C. The macroscopic adhesive properties of poly(T-*co*-*n*BA) (10 mol% thymine) and poly(A-*co*-*n*BA) (4 mol% adenine) (Fig. 1b) were characterized by a combination of tape peel and shear tests (Fig. 2) and compared to poly(acrylic acid-*co*-*n*-butyl acrylate) (poly(AA-*co*-*n*BA), 4 mol% AA) and poly(4-vinylpyridine-*co*-*n*-butyl acrylate) (poly(VP-*co*-*n*BA), 10 mol% VP), which were used as hydrogen bonding analogues to study the effect of hydrogen bonding strength on the peeling strength. The adenine-based polyacrylate displayed a higher shear strength but a lower peel strength than the thymine-based polyacrylate. Interestingly, the adhesive strength of a blend of the adenine and thymine containing polyacrylates was up to two times higher than that of the individual polymers, on account of the much stronger binding of the complementary base pair. The individual nucleobase-containing polyacrylates displayed a 3–4 times higher peel strength than the poly(AA-*co*-*n*BA) and poly(VP-*co*-*n*BA) reference polymers, on account of their cross-linked structure, which led to reduced creep and cohesive failure. The blend of the adenine and thymine containing polyacrylates, in turn, showed a 3–4 times higher peel strength than a reference blend of the AA- and VP-based analogues, which are cross-linked through weaker self-complimentary motifs. Overall, these early studies by Long, Ward and coworkers demonstrated the suitability of nucleobases as reversible cross-links in adhesive systems and underlined the advantage of complementary over self-binding.

Selective adhesion was also investigated by Zimmerman and coworkers, who employed a synthetic DNA base pair analog formed by 2,7-diamidonaphthyridine (DAN) and ureido-7-deazaguanine (DeUG).[44] This quadruple hydrogen bonding pair was used to promote the adhesion of poly(styrene) (PS) bearing DAN as side chains (DAN–PS) onto a DeUG-functionalized glass surface (Fig. 3). The adhesion characteristics were investigated with a single lap joint layout (Fig. 2) and for comparison, three different control surfaces were tested, including the original primary amine-functionalized surface, an aminoalkyl-functionalized surface, and a surface displaying methylated DeUG units. It was shown that DAN–PS adhered better to the DeUG-functionalized surface (shear strength = 0.57 MPa) than to the control surfaces (shear strengths of below 0.30 MPa), reflecting a clear benefit of the complementary binding. When neat PS was used instead of DAN–PS, the shear strength decreased to 0.14 MPa for the DeUG-functionalized surface. Concerning the control surfaces, PS showed similar or lower adhesion than DAN–PS. Additional control surfaces were later prepared to rule out non-specific interactions by linking DeUG with a perfluorinated spacer.[45] To explore the healability of the adhesive, which requires disassembly and reassembly of the supramolecular assemblies, the failed lap joints were re-attached *via* addition of a small portion of dichloromethane (disassembly) and subsequent solvent evaporation (re-assembly). A recovery of 55% of the initial shear strength was observed for DAN–PS after the first re-bonding (decrease from 0.70 MPa to 0.39 MPa), and the system could withstand two further debonding/re-bonding cycles before the adhesion strength was as low as that of neat PS (0.17 MPa). The degrading adhesive properties were attributed to contamination or damage of the sample after each shear test cycle, which may reduce the extent of formation of the DAN–DeUG complexes. The concept shown by Zimmerman and coworkers should be broadly applicable to other substrates and/or binding motifs and represents a universal platform that permits the selective and specific coating of surfaces with functional supramolecular polymers.

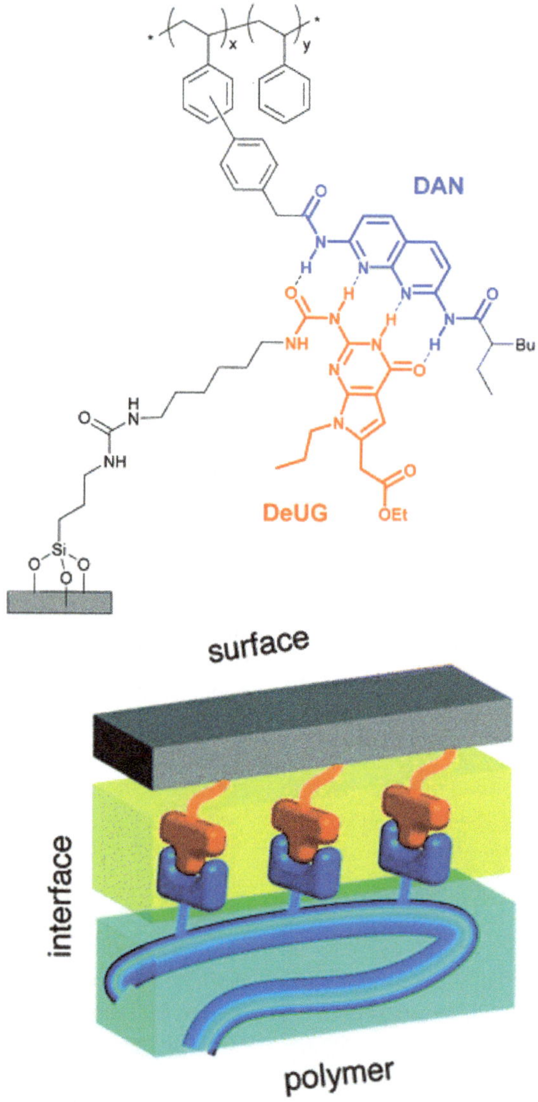

Figure 3: Schematic representation of polymer-surface selective adhesion based on DAN–DeUG recognition *via* hydrogen bonding. Adapted with permission from ref. 44 Copyright (2013) American Chemical Society.

A common approach to synthesize supramolecular polymers involves the functionalization of a low-molecular weight telechelic polymer with supramolecular units as end-groups. The ditopic macromonomers thus obtained are sufficiently long to minimize cyclization and supramolecular polymerization thus leads to high apparent molecular weights through linear chain extension (as opposed to low-molecular weight cyclic structures, which are observed in the e of low-molecular weight ditopic monomers).[34] In addition, if the polymer backbone and the supramolecular motifs are chemically dissimilar, the supramolecular polymer segregates into a biphasic morphology, much like a block-copolymer, and the mechanical properties of the material are enhanced if a physically cross-linking hard-phase is formed. Such reinforcement strategy is found in some of nature›s strongest materials such as spider silk,[46] where the segregation of protein's hydrophobic and hydrophilic domains contributes to their outstanding mechanical properties. Such well-defined structures are good models to study the effect that external factors such as the temperature have on the mechanical (and therefore adhesive) properties of supramolecular polymers. Hayes and coworkers highlighted the potential of urea- and urethane-terminated telechelic poly(isobutylene)s (PIBs) as thermosensitive adhesives.[47] These macromonomers (number-average molecular weight, $M_n \sim 5500$ g mol^{-1}) form extended chains through simple hydrogen bonding, resulting in materials displaying a glass transition at *ca.* −65 °C and a second transition between −25 and −10 °C. On the basis of rheological measurements, the authors assigned this transition to the temperature at which the hydrogen bonds dissociate, leading to a decreased flow resistance. This is supported by the fact that the low-temperature T_g is almost independent of the hydrogen bonding end-group of the functionalized PIB, whereas the second transition strongly depended on the latter.

Using a similar approach, Bosman and coworkers tested the adhesive properties of a telechelic poly(dimethylsiloxane) (PDMS) functionalized with UPy end-groups (Fig. 4a).[48] In this study, two glass slides were bonded with the supramolecular polymer by heating it to 120 °C, placing the melt between the two surfaces, and cooling the assembly to 20 °C. The joint thus formed (Fig. 2) was able to hold a 1 kg weight for 24 h, but failed after only 2 min when a 2 kg weight was mounted instead. The adhesive was shown to be fully reversible; this was shown in a qualitative experiment, in which the forcefully separated slides were brought again into contact and heated to 120 °C. After cooling, the original adhesive properties were restored. The pronounced temperature-response of these supramolecular polymers presents clear advantages with respect to their processability, and also promotes pronounced changes of the material›s properties in a narrow temperature regime.

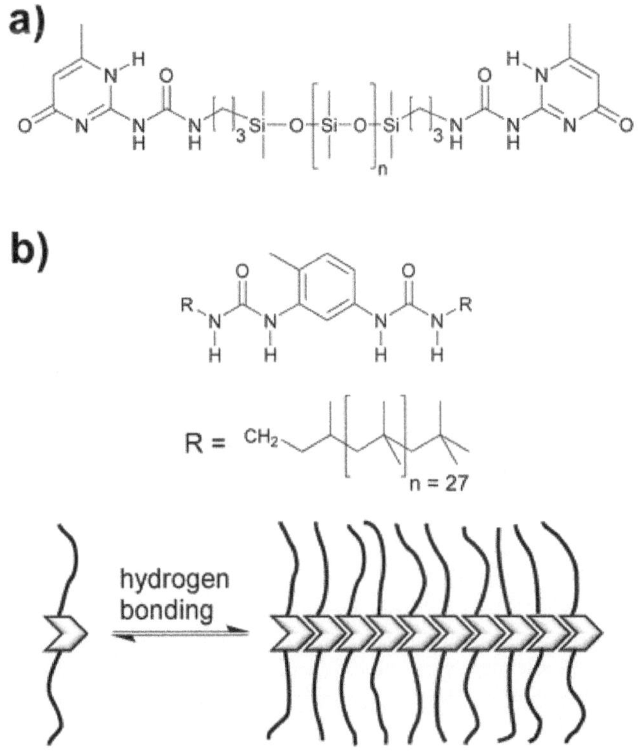

Figure 4: (a) Chemical structure of UPy-functionalized telechelic PDMS, which forms a supramolecular adhesive upon linear chain extension *via* hydrogen bonding.[48] (b) Schematic representation of the self-assembly of low-molecular weight PIB containing urea groups in the center of the chain molecules.[49]

The adhesive properties of assemblies of low-molecular weight polyisobutylene (PIB) containing two urea groups localized at the middle of the chain molecules (PIBUT) were investigated by Creton and coworkers (Fig. 4b).[49] The authors demonstrated that upon introduction of a small fraction of hydrogen bonding units, this otherwise amorphous polymer with low glass transition temperature (−73 °C) self-assembled into an ordered structure with a periodicity of 4 nm as measured with SAXS. Although no further structure analyses were performed, the authors suggested that this supramolecular polymer may assemble into a comb-like structure on the basis of previous solution studies with similar polymers.[50] Heating the polymer to temperatures above 100 °C caused dissociation of the hydrogen-bonded stacks (monitored by SAXS), which reformed after cooling to room temperature, restoring the material's properties. Interestingly, FTIR analysis suggests that the hydrogen bonds did not fully disassemble up to a temperature of 140 °C, which indicates that the observed order–disorder transition involves dissociation of van der Waals interactions rather than hydrogen bonds.[49] The adhesive properties of PIBUT were investigated with a contact mechanics method, also known as Johnson–Kendall–Roberts (JKR) method, using a flat-ended cylinder as contacting probe (Fig. 2). Stainless steel and PDMS, the latter being amongst the most difficult surfaces to bind, were used as test surfaces and the performance of the modified PIB was compared to reference materials (neat PIB and PBA are common pressure sensitive adhesives (PSA)). First, the adhesion of PIBUT on a stainless steel surface was compared to that of three different molecular weight PIBs (M_n = 2800, 85 000, and 400 000 g mol^{-1}). While PIB2800 behaves like an unentangled Newtonian liquid and the molecular weight of PIB400 000 is too high to form the fibrillary structures necessary for the tack test, the behavior of PIBUT is comparable to that of PIB85 000, which has the closest behavior to that of a standard PSA such as linear high-M_n PBA. In a second adhesive test, PIBUT was compared with PIB85 000 and PBA on stainless steel and PDMS. The modified PIB exhibited higher stress at debonding (0.85 MPa on steel and 0.6 MPa on PDMS) than the reference materials (neat PIB: 0.5 MPa and 0.1 MPa, respectively; PBA: 0.65 MPa and 0.2 MPa, respectively). According to the authors, the remarkably strong adhesion of PIBUT on PDMS, which is usually used as a release layer, make it potentially useful in applications involving difficult-to-bond surfaces.

Weder and coworkers recently reported the use of UV light as a remote stimulus to induce the reversible debonding of supramolecular adhesives.[51] A supramolecular system based on UPy-terminated telechelic poly(ethylene-*co*-butylene) (PEB) was synthesized and its adhesive properties were tested on steel and glass surfaces using a single lap joint shear test (Fig. 2 and 5a). As in the examples discussed above, UPy–PEB–UPy assembles into a supramolecular polymer *via* linear chain extension mediated by strong UPy dimerization.[53] The resulting material displays polymer-like mechanical properties (tensile storage modulus $E' = 7.1 \pm 1.8$ MPa, Young's modulus $E = 12 \pm 1$ MPa; measured at 25 °C, respectively) on account of its nanophase segregated lamellar morphology, in which the stacks formed by aggregating UPy dimers serve as the hard phase. In the adhesion tests, UPy–PEB–UPy displayed shear strengths of around 1 MPa on both steel and glass surfaces and, as in the previous examples, it underwent thermally induced reversible debonding; however, this material was also shown to debond on demand upon exposure to UV-light. Such functionality can be enabled through a light-heat conversion process whereby the UV-absorbing groups absorb UV light and produce a temperature increase on account of radiationless decay of the excited states. Since the UPy–PEB–UPy adhesive does not display a significant intrinsic absorption in the UV regime, the commercial UV absorber 2-(5-chloro-2*H*-benzotriazole-2-yl)-6-(1,1-dimethylethyl)-4-methyl-phenol was incorporated by simple blending to serve as light-heat converter and enable UV-light induced debonding, which took place after (optically) heating samples to *ca.* 70 °C. As a proof for reversibility/recyclability, the mechanically, thermally, or optically disassembled lap joints could be rebonded through exposure to light or heat, and the original adhesive properties were recovered. Further accounts of UPy–PEB–UPy as adhesive can be found in the patent literature. For example, an adhesive shear strength of 103.0 ± 23.7 MPa was reported for the adhesion to a zirconium ceramic substrate.[23]

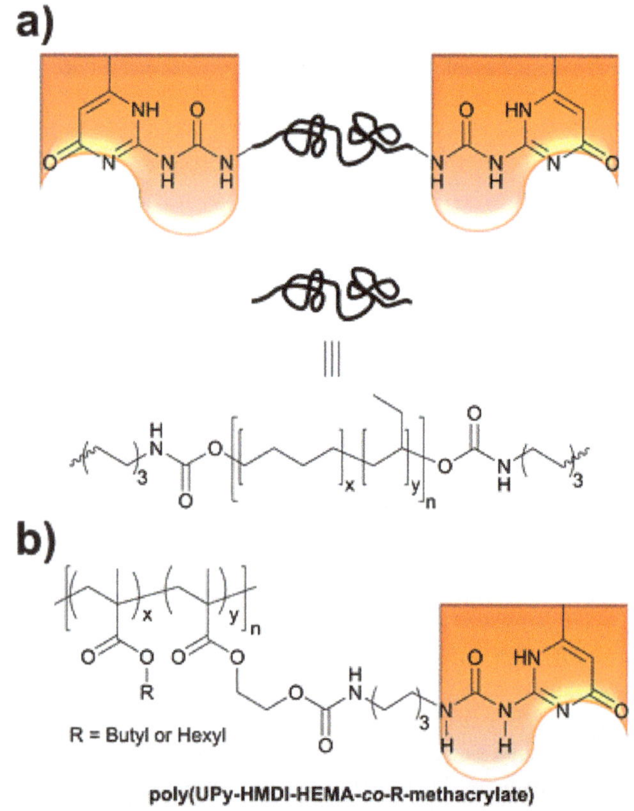

Figure 5: Chemical structures of the supramolecular polymers used by Weder and coworkers to achieve UV-light triggered debonding adhesives. (a) A telechelic poly(ethylene-*co*-butylene) (PEB) was functionalized with hydrogen bonding units (ureido-4-pyrimidinone; UPy), capable of intermolecular chain-extension *via* UPy–UPy-dimerization.[51] (b) A side-chain functionalized methacrylate copolymer with UPy as the supramolecular cross-linking unit.[52]

The Weder group recently extended the above concept of light- and temperature-responsive reversible adhesives to hydrogen-bonded, side-chain-functionalized supramolecular poly(alkyl methacrylate)s (Fig. 5b) that are reminiscent of the systems studied by Long (Fig. 1a),[36] and demonstrated their bonding and debonding on demand capabilities.[52] Thus, 2-hydroxyethyl methacrylate (HEMA) was functionalized with ureido-4-pyrimidinone (UPy) *via* a hexamethylenediisocyanate (HMDI) linker. The resulting (UPy–HMDI–HEMA) serves as a monomer that can, once polymerized, form supramolecular cross-links *via* hydrogen bonded dimers. Copolymers comprising 2.5, 5, or 10 mol% of the cross-linker were synthesized by copolymerizing UPy–HMDI–HEMA with hexyl methacrylate or butyl methacrylate. The adhesive properties of these copolymers were studied at temperatures between 20 and 60 °C by forming and testing single lap joints (Fig. 2) with stainless steel substrates. In a systematic study it was shown that increasing the concentration of UPy–HMDI–HEMA leads to improved mechanical and adhesive properties at elevated temperatures, on account of an increased concentration of supramolecular cross-links. The bond formation was reversible and debonded and subsequently rebonded lap joints displayed the same adhesive strength as pristine samples. The study also exemplarily confirmed the bonding and debonding on demand capabilities of one (co) polymer, which for light-induced debonding experiments was again blended with a UV-absorber. This work demonstrates the broad applicability of light-induced bonding and debonding on demand with supramolecular polymers and documents the ease with which the properties of such copolymers can be tailored.

Thermoplastic polyamides are frequently used as hot-melt adhesives due to their high thermal and mechanical resistance. On the other hand, their use is limited to heat-resistant surfaces as they require high processing temperatures. Leibler and coworkers tackled this problem by blending a commercial polyamide (synthesized by the reaction of a mixture of fatty acid dimers and trimers with ethylenediamine and piperazine, M_n = 20 000 g mol^{-1}, Fig. 6) with a hydrogen bonding semicrystalline supramolecular polymer.[54] The blend was designed to combine the good mechanical properties of polyamides with the highly temperature dependent properties of supramolecular polymers, yielding an easily processable material with low melt viscosity. The supramolecular polymer used consisted of a mixture of fatty acids dimers and trimers that had been modified by reaction with 1-(2-aminoethyl)-2-imidazolidinone (UDETA) in a 1 : 1 [COOH]/[NH_2] ratio (M_n ~ 800 g mol^{-1}, Fig. 6). The adhesive properties of the neat polyamide and a blend containing 20 wt% of the supramolecular polymer were compared in T-peel tests (Fig. 2) using a non-woven polypropylene web as surface. Notably, the blend withstood a maximum peeling force of 3 N for an adhesive thickness of 5 μm, surpassing the neat polyamide by up to a factor of 6. The adhesive failure for the blend was always cohesive, indicating a strong adhesion at the interfaces through extensive hydrogen bonding. The improved mechanical strength of the blend was reported to be the result of the crystallization of the supramolecular polymer and of the presence of a higher number of hydrogen bonding units that attach to the surface. Additionally, the blend could be processed at a lower temperature than the neat polyamide, due to the plasticizing effect of the supramolecular polymer, which disconnected into low-viscosity low-molecular weight species at temperatures over 120 °C. Apart from their direct use as adhesives, these materials have been also proposed as additives in adhesive formulations such as hot-melt adhesives, pressure sensitive adhesives (PSA), and solvent-based adhesives.[55,56]

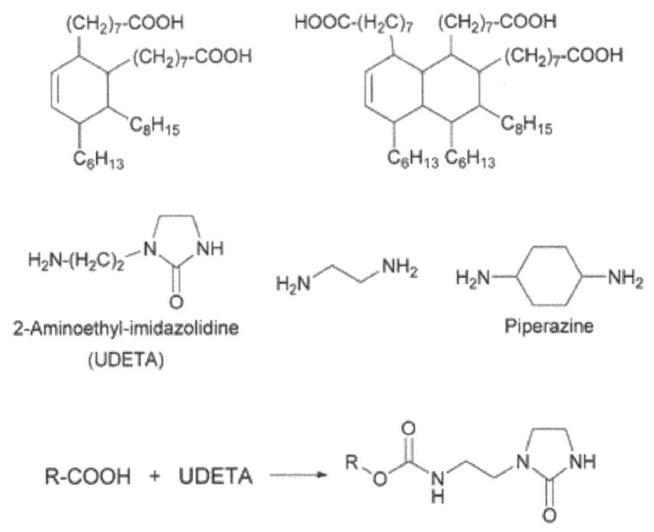

Figure 6: Chemical structures of fatty acid dimers and trimers and difunctional amines (top) that form the basis of the polyamides used by Leibler and coworkers in blends with supramolecular polymers formed by the reaction of fatty acids (and derivatives) with UDETA (bottom).[54]

The above discussed approach has also been used in various application-driven studies. Bostik, a company involved in the commercialization of adhesives, evaluated the adhesive properties of supramolecular polymers, based on carboxylic acids and amines as in Leibler's approach, using a 180° peel test layout (Fig. 2).[24] Depending on the degree of amine functionalization, the studied supramolecular polymers exhibited peel strengths between 0.7 (10 wt% functionalization) and 2 N cm^{-1} (30 wt% functionalization), which were superior to that of the unfunctionalized reference (0.4 N cm^{-1}). In another study, Bostik combined Leibler's supramolecular materials with styrene based-block copolymers such as poly(styrene-*b*-butadiene-*b*-styrene) (SBS), to obtain pressure-sensitive hot-melt adhesives.[25] In 180° peel tests (Fig. 2), the supramolecular formulations achieved higher adhesive strengths (between 5.7 and 6.4 N cm^{-1}) than a non-functionalized reference (4.0 N cm^{-1}). The debond-on-demand ability of these adhesives was demonstrated using heat as stimulus, which caused debonding of the supramolecular adhesive within seconds, whereas the reference material required several minutes. The polymer company Arkema also used Leibler's approach to prepare various materials for bituminous compositions, to increase processability and to enhance cohesion and adhesion.[26]

The non-covalent adhesion of two rigid surfaces by means of supramolecular interactions is kinetically disfavored as some (surface) flexibility is required for non-covalent bonds to be formed. An approach to enhance adhesion between such surfaces involves the introduction of long and flexible spacers between the surface and the hydrogen bonding units. Xie and coworkers addressed this problem by chemically modifying the surface of an epoxy resin with hydrogen bonding polymers.[57] Thus, the epoxy surface groups were functionalized with either low- or high-molecular-weight branched poly(ethyleneimine) (BPEI) (Fig. 7a). BPEI was chosen because of the high density of hydrogen bond donors and acceptors and on account of its branched architecture, which does not favor interfacial chain entanglements that would otherwise hinder reversible adhesion. In addition, a control sample was prepared by reacting the hydroxyl surface groups with acetic anhydride to investigate the role of hydroxyl hydrogen bonding in the adhesion, which was observed to be important. BPEI with M_n values of 1800, 10 000, and 60 000 g mol^{-1} were used, and the coverage of the epoxy polymer was dependent of the M_n, as demonstrated by AFM experiments. In the e of low M_n BPEI, AFM images showed isolated BPEI islands on the epoxy surface, whereas for the medium and high M_n BPEI, 2D network structures were observed. Additional XPS analysis confirmed the increase of the surface nitrogen content from 3% (non-functionalized epoxy) to 5, 10, and 12% (BPEI functionalized epoxies). In the adhesion experiments, all epoxy samples were bonded using the same procedure: heating to 90 °C (*i.e.* above T_g), adding a small amount of methanol (62 µL cm^{-2} per bonding area), pressing two surfaces together, and cooling under applied force. Methanol was shown to promote the dissociation of previously formed intra-surface hydrogen bonds, and this increases the formation of interfacial hydrogen bonds. As a result, the pull-off strength increased with increasing M_n from ~ 450 N cm^{-2} (non-functionalized epoxy and M_n = 1800 g mol^{-1}) to ~ 640 N cm^{-2} (M_n = 10 000 and 60 000 g mol^{-1}). Notably, these values are comparable to the adhesion achieved with a conventional superglue (500–900 N cm^{-2}). When the methanol treatment was omitted, a substantially lower pull-off strength (132 N cm^{-2}) was observed. In an additional experiment, the pH dependency of the pull-off strength was investigated with the BPEI of the highest M_n (60 000 g mol^{-1}) and was observed to slightly decrease with increasing pH from ~640 N cm^{-2} (pH = 2.5) to ~450 N cm^{-2} (pH = 11.5). This behavior was explained on the basis of the different BPEI ionization degrees at the different pH values. At low pH, the concentration of cationized nitrogen atoms (strong hydrogen bond acceptors) is higher than that of neutral nitrogen atoms (weaker acceptors), which leads to the observed stronger adhesion. Reversibility was investigated by heating the bonded samples to 90 °C, peeling them off, and reattaching them. The adhesive strength after two attach–detach cycles decreased to 67% of the original adhesive pull-off strength, which the authors related to minor changes in the surface morphology. This work documents a very useful approach to enhance adhesion to the surface of epoxy resins without changing their bulk properties.

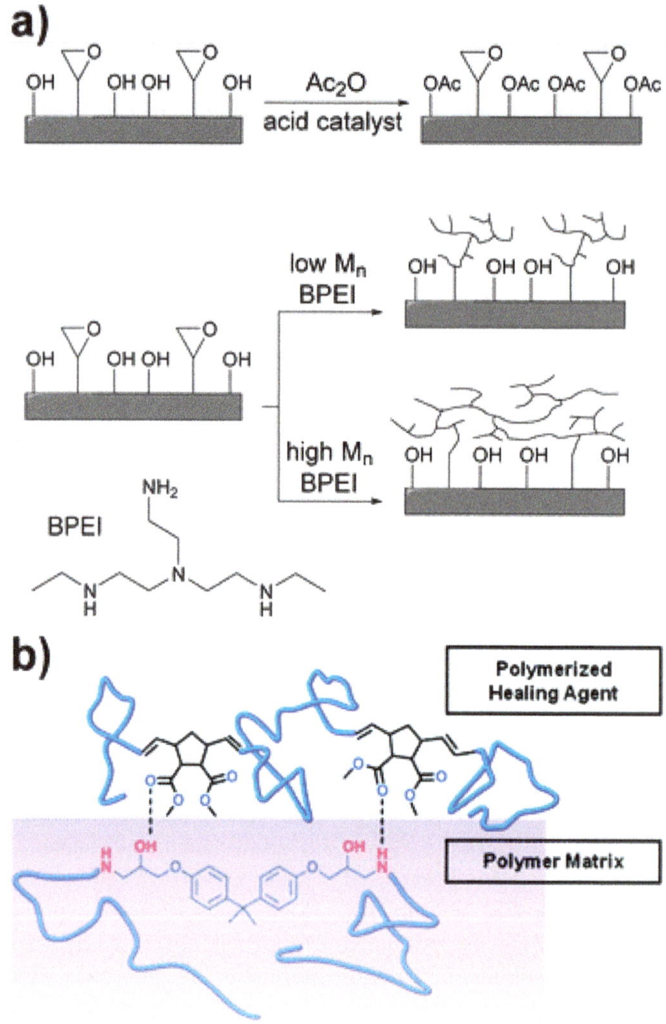

Figure 7: Approaches to achieve adhesion between rigid surfaces. (a) Surface functionalization of an epoxy resin with branched and hydrogen bonding BPEI to promote self-adhesion.[57] (b) Enhancement of the adhesive properties of an *in situ* epoxy healing agent *via* hydrogen bonding.[58] Part (b) was adapted with permission from ref. 58 Copyright (2011) American Chemical Society.

A complementary study was reported by Moore and coworkers,[58] who investigated the role of non-covalent interactions in the assisted surface adhesion of fractured epoxy resins. Taking the epoxy system EPON® 828/diethylenetriamine as model, the group optimized the composition of a healing agent mixture consisting of dimethylnorbornene ester (DNE), dicyclopentadiene (DCPD) and a ring-opening metathesis polymerization (ROMP) catalyst (Fig. 7b). DNE was selected to study the effect of hydrogen bonding between the ester groups and the amine functionalities of the epoxy resin, and DCPD was used to yield mechanical properties comparable of poly(DCPD) alone. A pre-crack was initiated in the epoxy resin with a razor blade and the specimens were pin loaded and shear tested. Once completely fractured, the healing agent mixture (DNE/DCPD/catalyst) was added inside capsules to the fractured surface at different concentrations and the adhesive strength after a curing/hardening step was measured. At 25 °C, the healing mixtures containing 5, 10 and 20 wt% of DNE displayed better adhesive properties than poly(DCPD) alone. In contrast, the mixture containing 2.5 wt% DNE showed limited healing, probably due the plasticizing effect of unreacted DNE counteracting any gain of adhesion promotion. At DNE contents of 10 and 20 wt%, the decrease of adhesive properties relative to 5 wt% can be explained by domination of non-covalent adhesion promotion over plasticization, yielding a slightly lower overall adhesion. The temperature at which curing/healing was carried out (25 and 50 °C) had also an effect of the adhesive properties, with higher adhesive properties at 50 °C (*ca.* 35 N *vs.* 10 N).

J. Zhang, X. Zhang and coworkers recently synthesized multi stimuli-responsive supramolecular hydrogels with reversible adhesive properties.[59] These networks were prepared by addition of poly(ethylene polyamine) (PPA) to an aqueous dispersion of oxidized multi-walled carbon nanotubes (ox-MWCNTs). The resulting materials displayed a combination of weak (N–H···N) and strong (N–H···O from PPA–CNT interaction) hydrogen bonding motifs, and their ratio could be tuned to obtain different cross-linking degrees, which provided a means to control the gel–sol transition temperature of the supramolecular hydrogels between 23 °C and 48 °C as the CNT content was varied from 0.015–0.5 wt%. In addition to being highly thermosensitive, these materials were shown to be sensitive to pH and to NIR irradiation on account of the presence of ox-MWCNTs, which display photothermal energy conversion. Thus, the supramolecular hydrogels underwent a gel–sol transition in 2 min upon exposure to NIR light and returned to the gel state after 3 min at room temperature. The presence of ox-MWCNTs further contributed to an increase of the hydrogels' viscosity and enhanced their adhesive properties, which were investigated using partially dried gels on Teflon® and plasma-treated glass surfaces. In one experiment, a weight was glued with the dried gel onto a Teflon® plate. Once inverted, the device could hold 320 g (2.8 N cm^{-2}) at 12 °C for 15 min. In a second experiment, the adhesion between two glass slides with a loading of 1.77 N cm^{-2} was tested and the temperature-dependent behavior explored. By varying the temperature between 5 and 40 °C, the authors observed adhesion failure times between *ca.* 40 s and *ca.* 20 min.

As mentioned before, non-covalent interactions play a key role in cell adhesion to biomaterials, which is a critical process for the formation of functional tissues. Cell adhesion is governed by a complex combination of adhesion mechanisms, which include the recognition of specific peptide sequences such as the well-studied RGD (Arg–Gly–Asp). Seeking to increase the versatility of current biomaterials, Meijer and coworkers reported a modular approach for the preparation of bioactive scaffolds using reversible supramolecular interactions.[60] As building blocks, the authors synthesized a UPy-functionalized telechelic oligo(ε-caprolactone) (UPy–PCL–UPy, M_n = 2700 g mol^{-1}) and two UPy-terminated peptides (GRGDS–UPy and PHSRN–UPy, each a sequence of 5 amino acids) (Fig. 8). PCL was chosen as it has been approved by the Food and Drug Administration (FDA) for specific applications in the human body due to its biocompatibility and biodegradability. The material properties of the neat PCL changed from a waxy, brittle solid (T_m = 43 and 49 °C) to a strong, elastic material (T_g = −59; T_m = 41 and 64 °C) upon functionalization with UPy units. Due to the reversibility of UPy dimers, UPy–PCL–UPy could be easily processed into several scaffolds (films, fibers, electrospun scaffold and meshes) and its biocompatibility was tested *in vitro* with two water-soluble model compounds, which indicated that the UPy-moiety is non-toxic (further verified by injection in rats). When PCLdiUPy was mixed with 4 mol% of GRGDS–UPy and/or PHSRN–UPy, a bioactive material was obtained on which mouse 3T3 fibroblast cells (5 × 10^4 cells cm^{-2}) could be cultured. It was shown that the UPy-functionalized peptides are necessary to promote cell adhesion and spreading *in vitro* due to the dynamic binding of the peptides to the supramolecular material. The highest cell adhesion (lowest cell spreading) was observed when both UPy-functionalized peptides were mixed together (pointing to a synergistic effect). Different GRGDS–UPy concentrations (1, 2, or 4 mol%) were also tested; however, no clear correlation between peptide concentration and cell adhesion and spreading could be found. Control materials with UPy-free peptides were not able to promote cell adhesion and rapidly diffused out of the film.

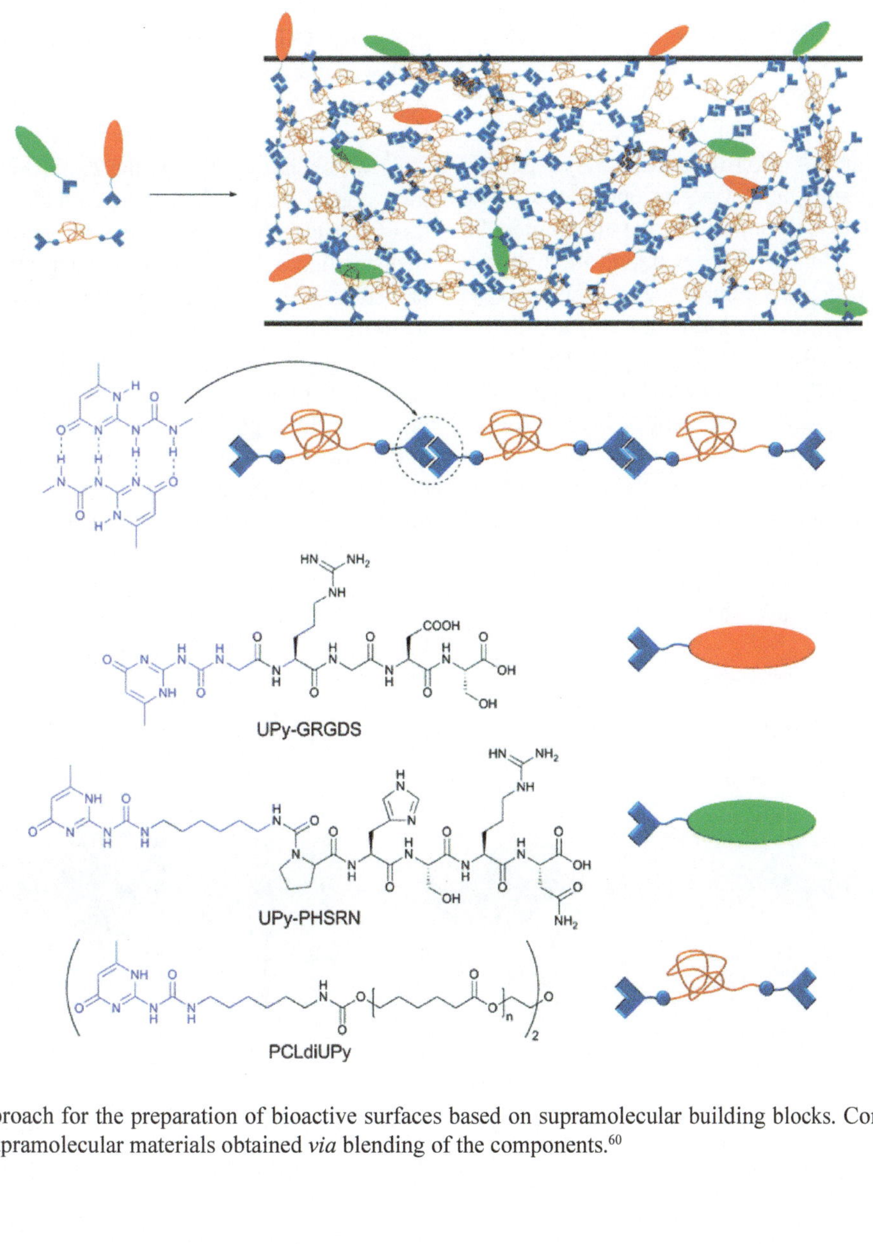

Figure 8: Modular approach for the preparation of bioactive surfaces based on supramolecular building blocks. Components and schematic representation of the supramolecular materials obtained *via* blending of the components.[60]

The strength of cell binding was determined by a wash-off test with or without prior incubation with fetal bovine serum (FBS), which acts as a detaching agent (competitive binding) of the cells to the bioactive material. While the cells attached to the blend containing both UPy-functionalized peptides could be washed off when incubated with FBS, the non-incubated cells were not washed off and started growing again after 1 day. These results reflect the significant potential of supramolecular polymers to reach new application fields if one fully exploits their dynamic character. The possibility to synthesize various bioactive materials in a modular approach is highly attractive, and the finding that the studied UPy-based building blocks are biocompatible, at least under a specific set of *in vitro* conditions, supports their usefulness in this context.

In anticipation of the possible technological exploitation of hydrogen-bonded polymers as adhesives, several patent applications have been filed in which the synthesis of such materials and the characterization of their adhesive properties is described. Apart from the examples mentioned throughout this section, of which some also formed the basis of pending patent applications or granted patents, the adhesive properties of several supramolecular polymers on steel by means of single lap joint tests (Fig. 2) were explored.[27,28] Among them are materials based on poly(propylene oxide) (M_w = 2000 g mol^{-1}) and 1,4-butanediol functionalized with various hydrogen bonding units (*e.g.* UPy, 2-aminopyrimidine, or isocytosine), which displayed adhesive strengths between 2.5 and 3.4 MPa.[27] The failure mode was in most es of adhesive nature as a result of limited interactions at the polymer–steel interface.

The patent literature also describes very general approaches, featuring C=O and/or C=S acceptor groups and N–H, O–H, and/or S–H donor groups, to utilize supramolecular polymers as adhesives.[29] Meijer and coworkers filed several patent applications in which the adhesive properties of UPy-based materials are at least mentioned, although only qualitative adhesive measurements were reported as these patents are mainly focussed on the synthesis and manufacturing/processing of the supramolecular polymers.[30] With a different application focus, Smith and Nephew investigated the adhesive properties of dihydroxybenzoic acid-functionalized telechelic aliphatic oligomers (weight-average molecular weight, M_w, between 500 and 2000 g mol^{-1}) in pull-tests (Fig. 2).[31] The adhesive strengths varied between 1.8 and 15.1 MPa, while a reference cyanoacrylate adhesive reached 7.1 MPa.

SUPRAMOLECULAR ADHESIVES BASED ON HOST–GUEST INTERACTIONS

While many supramolecular interactions display sufficiently high binding constants to serve as adhesives in a dry environment, most of them are disrupted in the presence of water, which acts as competitive binder. In nature, creatures such as mussels or sandtle worms secrete adhesive coacervates that allow them to adhere to surfaces in aqueous environments.[7] Kim and coworkers explored the potential of host–guest supramolecular interactions as a synthetic alternative to achieve underwater adhesion.[61] In contrast to most macrocyclic hosts such as cyclodextrins, which exhibit relatively modest association constants in water ($K_a \sim 10^6$ M^{-1}), cucurbit[*n*]uril (CB[*n*], *n* = 5–8, 10) macrocycles containing a hydrophobic cavity and two carbonyl-fringed portals were demonstrated to be useful candidates for underwater adhesion. Among them, cucurbit[7]uril (CB[7]) forms very stable complexes with aminomethylferrocene (Fc, $K_a \sim 10^{12}$ M^{-1}) and 1,1′-bis(trimethylammoniomethyl)ferrocene (BFc, $K_a \sim 10^{15}$ M^{-1}) in water (Fig. 9). The experimental set-up used to explore the adhesive properties of surfaces modified with polymer brushes carrying these motifs consisted of two polymer-grafted silicon surfaces [Si] in which the free ends of the polymer chains were functionalized with either CB[7] (thickness of the brush layer 6.3 nm) or Fc (thickness 3.0 nm) moieties. Both surfaces were smooth as per AFM analysis and wettable as confirmed by contact angle measurements. Adhesion was promoted by manually pressing the surfaces together; they had previously been immersed in water (pH 7.4), which is required as the expulsion of water drives the inclusion of Fc into CB[7]. The bonded interface was shown to outperform bonds formed with commercial adhesives by withstanding a 2 kg weight in water and a 4 kg weight in the dry state (Fig. 9). The adhesive characteristics were further quantified in a lap shear adhesion test (Fig. 2), which resulted in a maximum lap shear strength of 1.12 ± 0.06 MPa. The so-called “supramolecular VELCRO®” maintained its adhesive properties through 5 debonding–rebonding cycles. While the maximum lap shear strength was reduced by *ca.* 30% after the first debonding–rebonding cycle, it remained constant afterwards. The stimuli responsive behavior of the assembled CB[7]–[Si]/Fc–[Si] system was investigated by exposing it to a NaClO (10 mM) solution while supporting a 0.5 kg weight. Under such conditions, Fc is oxidized to Fc$^+$, which has a lower affinity to CB[7], and failure could be observed within minutes.

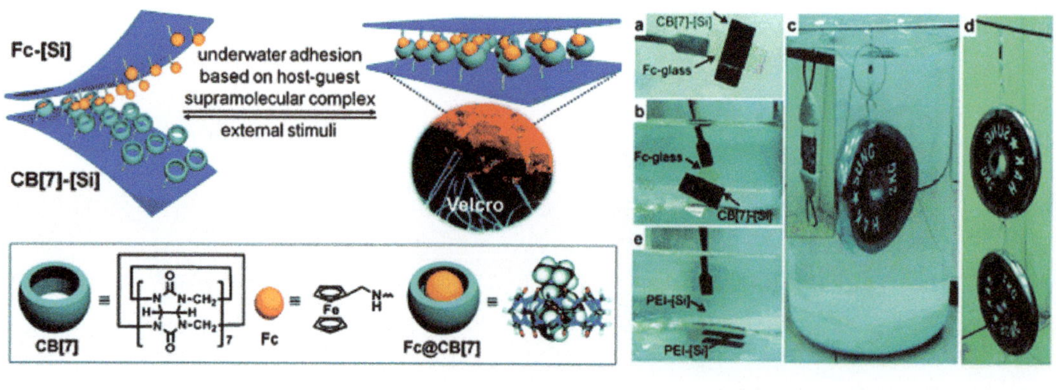

Figure 9: Redox-active adhesive based on CB[7]-Fc host–guest interactions and demonstration of underwater adhesion. Reprinted from ref. 61 with permission from John Wiley & Sons.

The surfaces were then immersed in ascorbic acid to reduce Fc^{+} to Fc and the reassembled system could support a load of 1 kg for at least 1 h. With this contribution, Kim and coworkers provided a robust and stimuli-responsive underwater adhesion mechanism that should be applicable to a diverse range of surfaces.

In another example of host–guest-based supramolecular adhesion, Harada and coworkers reported a supramolecular hydrogel displaying selective adhesive properties on the basis of a molecular recognition process.[62] The hydrogel studied was prepared by free-radical copolymerization of acrylamide with acrylamide derived monomers bearing β-cyclodextrin (βCD), bipyridine (bpy) and bisacrylamide (BAA) moieties. In water, the resulting copolymer (containing 2.2 mol% of βCD and 3.0 mol% of bpy moieties) forms a mixed network, where βCD–bpy inclusion complexes serve as reversible cross-linking points, whereas the bisacrylamide residues provide permanent cross-links. Upon addition of an appropriate metal salt, the bpy units bind to the metal ions to form $M(bpy)_n$ complexes and the degree of cross-linking decreases, although the mechanical integrity of the hydrogel is kept by the permanent cross-links. This allows modifying the properties of the hydrogel *via* the addition of metal salts. For instance, it was demonstrated that the adhesion of this hydrogel to another hydrogel containing *tert*-butyl units (which compete with bpy for the formation of inclusion complexes with βCD) could be switched on and off by addition of a metal salt (Fig. 10).

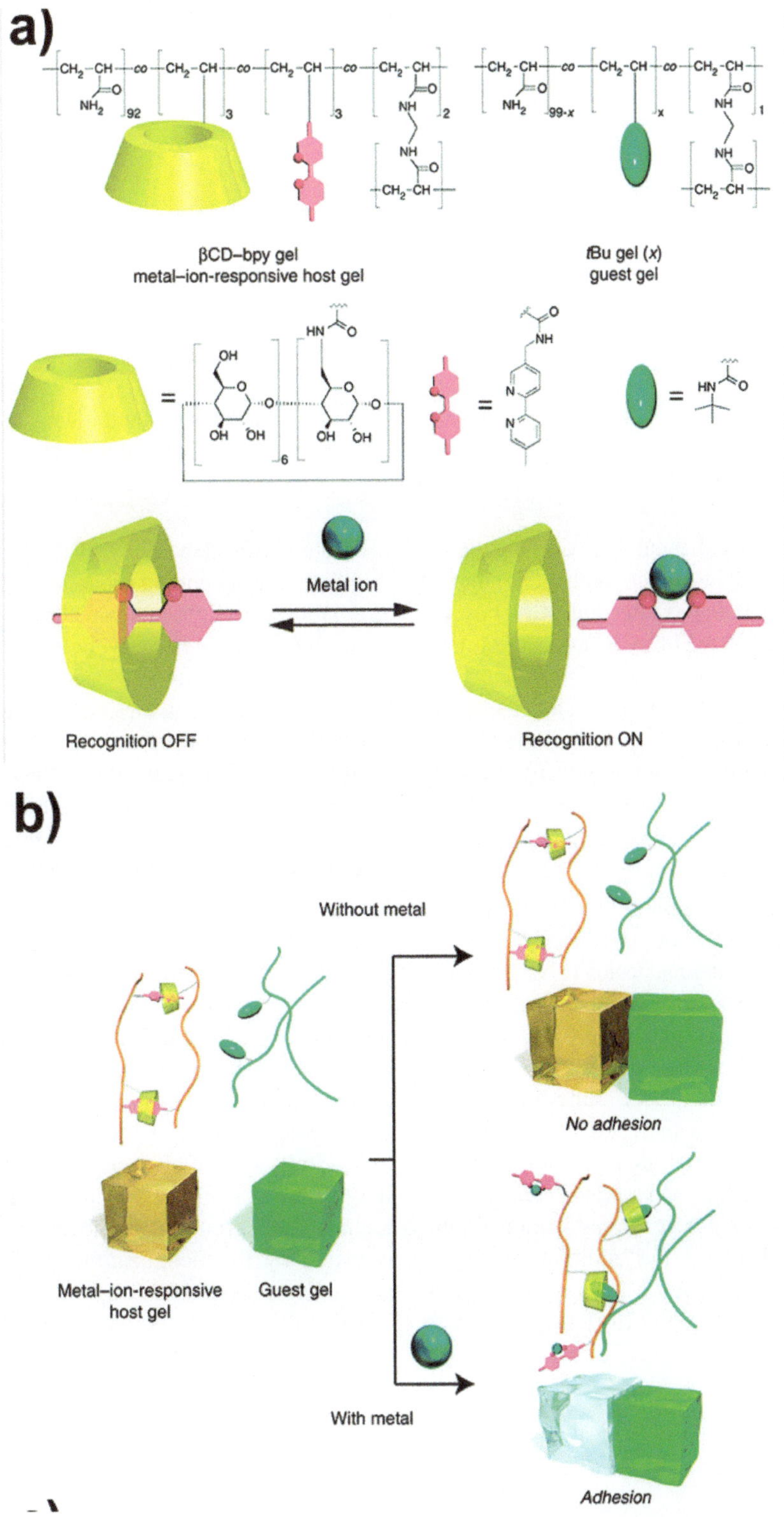

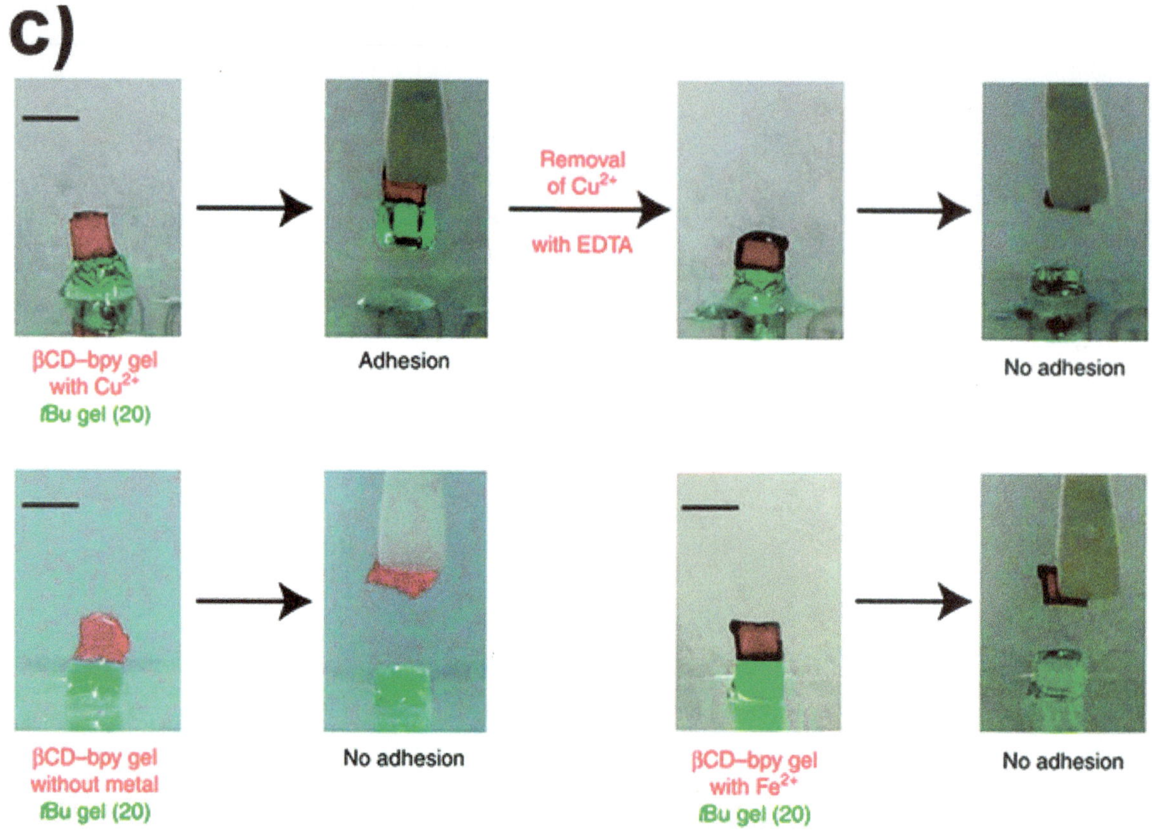

Figure 10: (a) Chemical structure of a βCD–bpy gel. (b) Schematic representation of its metal-triggered adhesion to a *tert*-butyl containing gel based on competing host–guest and metal–complex interactions. (c) Qualitative metal-triggered adhesion experiments. Figure reproduced from ref. 62 licensed under CC-BY 4.0.

The authors observed no adhesion between the βCD–bpy gel and the *tert*-butyl gel; however, both gels adhered (adhesion strength = 1000 ± 200 Pa) after the βCD–bpy gel was immersed for 1 min in a 100 mM $CuCl_2$ aqueous solution, on account of the formation of inclusion complexes of the now available βCD with the *tert*-butyl groups. The addition of other transition metals such as Co^{2+}, Ni^{2+}or Zn^{2+} in the form of MCl_2 was also shown to promote gel adhesion using the same βCD- and *tert*-butyl-based hydrogel systems.

The recent development of synthetic mimics of reptile's adhesive setae[63,64] motivated the same team to investigate the potential of host–guest interactions for hard surface adhesion.[65]They synthesized acrylamide-based αCD- and βCD-containing hydrogels and studied their adhesion to hard surfaces functionalized with adamantane, ferrocene or azobenzene units. As control, they synthesized an acrylamide hydrogel, which showed very low adhesion to all surfaces. By contrast, friction tests revealed selective hydrogel-surface adhesion for the pairs αCD-hydrogel/azobenzene surface and βCD-hydrogel/ferrocene surface. Furthermore, the αCD-azobenzene and βCD-ferrocene interactions could be suppressed *via* irradiation with UV light, which causes *trans*/*cis* isomerization of the azobenzene motif, (Fig. 11a) and the addition of $FeCl_3$, which causes oxidation of the ferrocene units to the weakly complexating ferrocenium species (Fig. 11b),[66] respectively to promote partial debonding of the hydrogel from the surface. Although the debonding process was in both es reversible upon irradiation with light of a wavelength of 430 nm and reduction with glutathione, respectively, some hysteresis was observed and the adhesion strength was somewhat lower than initially observed. While the adhesion of hydrogels to rigid surfaces is favored by their high flexibility, which facilitates the contact between complementary units, these results show the utility of host–guest interactions to achieve reversible adhesion in the presence of a hard surface and hints at the necessity of introducing certain mobility at the interface.

In the examples discussed above, host–guest interactions were used to promote selective (reversible) adhesion between two gels or between a gel and a rigid surface. Burdick and coworkers provided an additional application of such interactions by studying the host–guest-driven adhesion of two βCD-functionalized rigid surfaces mediated by an adamantane-based (Ad) adhesive.[67] The surfaces were created by functionalization of hyaluronic acid (HA) with βCD and methacrylate units, electrospinning of the functionalized HA into a fibrous mat-like structure and UV cross-linking of the methacrylate groups. On the other hand, the adhesive was prepared by functionalization of HA with Ad units. A first qualitative adhesion test was carried out by bonding two βCD-functionalized surfaces with the Ad-based adhesive, which was applied as a 0.5 wt% aqueous solution.

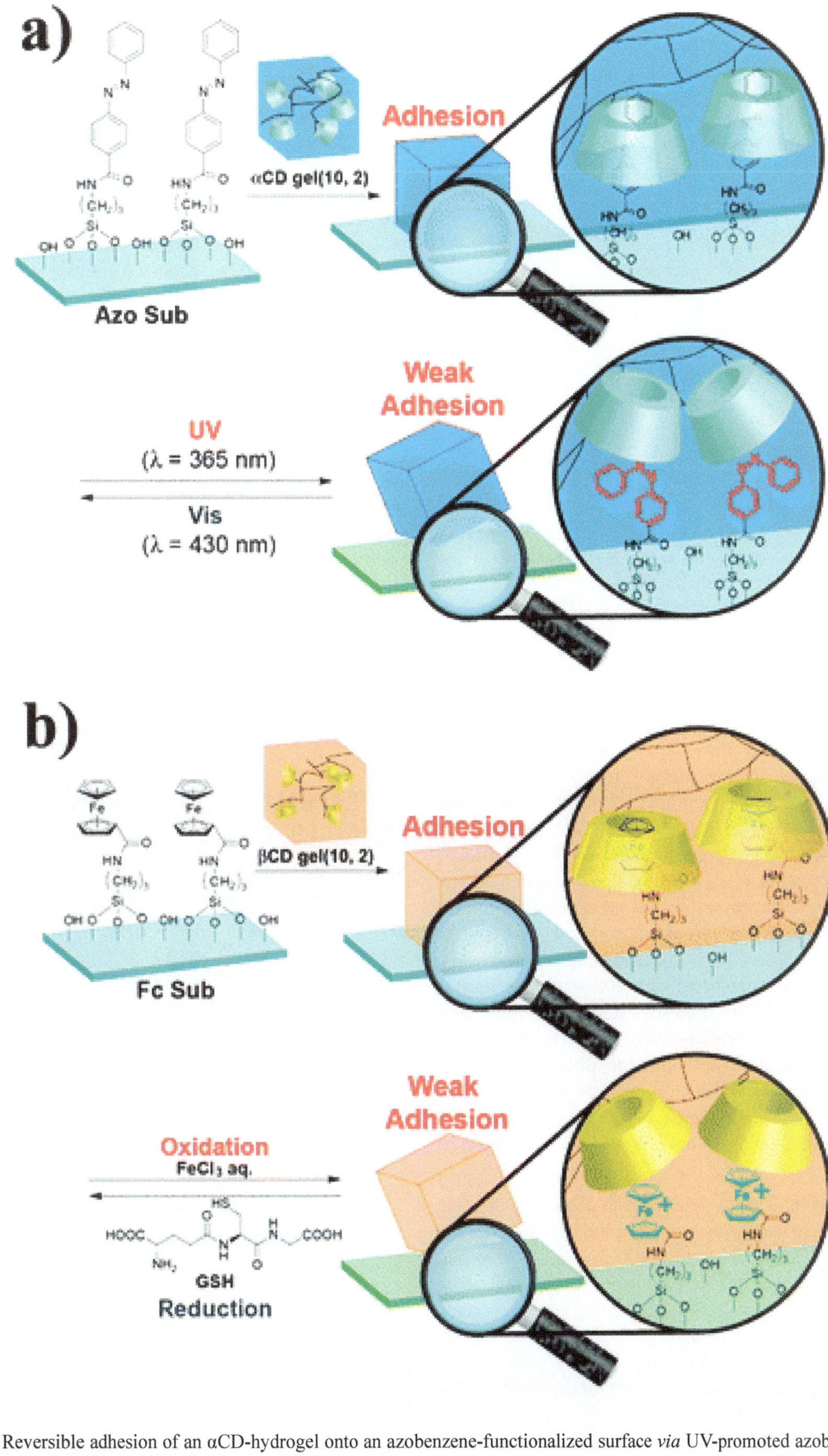

Figure 11: (a) Reversible adhesion of an αCD-hydrogel onto an azobenzene-functionalized surface *via* UV-promoted azobenzene isomerization. (b) Redox-responsive reversible adhesion of a βCD-hydrogel onto a ferrocene-functionalized surface. Reprinted from ref. 65with permission from John Wiley & Sons.

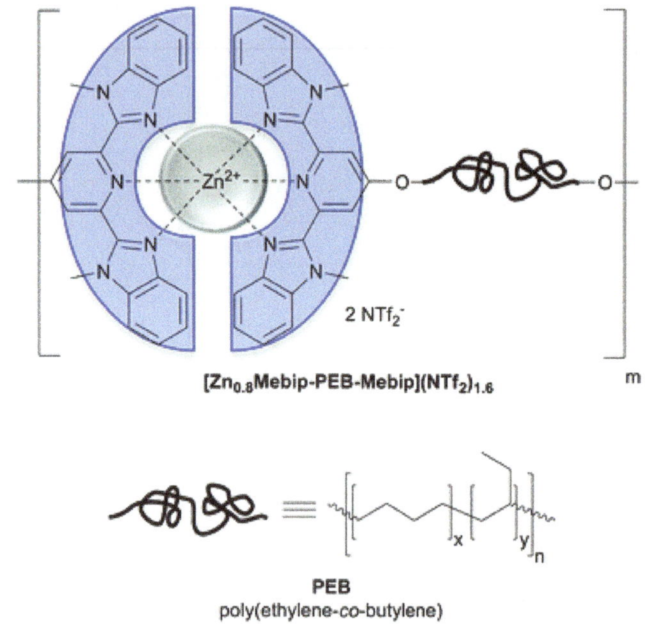

Figure 12: Chemical structure of the metallosupramolecular polymer used by Weder and coworkers to achieve UV-light triggered debonding adhesives. A telechelic poly(ethylene-*co*-butylene) (PEB) was functionalized with tridentate ligands (2,6-bis(1′-methylbenzimidazolyl)-pyridine; Mebip) capable of metal–ligand-coordination.[51]

After 15 min, the adhered surfaces were shown to remain bonded for up to seven weeks when kept in deionized water or phosphate-buffered saline (PBS) at 37 °C. A control experiment in which neat HA was used as adhesive resulted in immediate adhesive failure. The adhesive properties of this system were quantified by means of a single-lap joint test (Fig. 2), for which the βCD-methacrylate-functionalized HA was directly electrospun onto methacrylate-functionalized glass coverslips and then UV-cured. The coverslips were then bonded with the Ad-based adhesive (or neat HA as control) and a shear force was applied. The Ad-based adhesive showed a shear strength of 1 kPa; much higher than that of the control (*ca.* 20 Pa). According to the authors, this water-based adhesive system may be useful for *in vivo* applications such as the regeneration of structured tissues (cartilage, nervous tissue or cardiac muscle). The approach of Burdick and coworkers circumvents interface mobility issues with the aid of a supramolecular adhesive, offering an elegant alternative for reversible adhesion between hard surfaces.

The use of competitive interactions to achieve debonding on demand was also studied by Shi, Yang and coworkers.[68] The authors prepared PDMS cubes (3 mm × 3 mm × 3 mm) that were surface-functionalized with either αCD or azobenzene units with the aim of promoting the assembly of macroscopic objects through molecular recognition. To overcome surface rigidity-derived problems that can lead to insufficient molecular binding (*vide supra*), the surface of the PDMS cubes was decorated with a flexible coating made by a layer-by-layer approach. The composition of the multilayer coating consisted of a flexible poly(ethyleneimine)/poly(acrylic acid) (PEI/PAA) layer, a barrier layer of poly(diallyldimethylammonium chloride)/poly(sodium-*p*-styrenesulfonate) (PDDA/PSS) to lower the stickiness of PEI/PAA and prevent nonselective association of the building blocks, and a final layer of either PDDA/PAA–αCD (αCD-grafted PAA) or PDDA/PAA–Azo (azobenzene-functionalized PAA). PDMS cubes functionalized with αCD units were placed in water together with PDMS cubes functionalized with Azo groups and the container containing the system was oscillated in a shaker. After 1 min of oscillation, the αCD-cubes were found to adhere to the Azo-cubes, as a result of a host–guest molecular recognition process. This was confirmed with control tests where either αCD-cubes or Azo-cubes were shaken independently resulting in no assembly. The intended role of the flexible PEI/PAA layer was also confirmed in an additional control experiment in which PDMS cubes were directly functionalized with either αCD or Azo units without intermediate PEI/PAA layer. The PDMS cubes thus prepared did not assemble, confirming that a certain degree of surface mobility is necessary for the host–guest interactions to take place. The stimuli-triggered debonding ability of this system was investigated through addition of the competitive host 1-adamantane carboxylic acid, which displays a higher binding constant to αCD (10^5 M^{-1}) than the Azo units (10^3 M^{-1}). The adamantane monomer was added as a 0.5 mM aqueous solution to the pre-assembled PDMS cubes and debonding took place after shaking the cubes as in the assembly process. The bond strength between the macroscopic cubes was determined using a microbalance-based method specially designed for the system studied,[68]which showed that the adhesive force of the cubes having a flexible coating was 16 times higher than that between unmodified contact surfaces (3.05×10^{-3} N *vs.* 1.90×10^{-4} N). These results provide an interesting quantitative insight on the adhesion process that governs the oscillation-promoted self-assembly of complementary macroscopic objects.

SUPRAMOLECULAR ADHESIVES BASED ON METAL–LIGAND COMPLEXATION

Metallosupramolecular polymers are a well-studied class of stimuli-responsive materials,[69–71] but their adhesive properties have been rarely reported. One prominent example is the above-discussed work by Harada and coworkers, in which selective adhesion of hybrid gels was achieved using a combination of host–guest and metal–ligand (bpy-M^{2+}) interactions.[62] In another example, Weder and coworkers studied the stimuli-responsive adhesion of a metallosupramolecular polymer based on the linear chain extension of Mebip-terminated telechelic PEB *via*metal–Mebip complex formation (Fig. 12).[51] Chain extension of Mebip–PEB–Mebip takes place upon addition of a metal salt such as Zn^{2+}, resulting in a metallosupramolecular polymer that exhibits a nanophase segregated lamellar morphology with the metal complexes forming the crystalline hard phase. The adhesive properties of this material were tested on steel and glass surfaces using a single lap joint shear test (Fig. 2 and 12) that showed strong adhesion to both surfaces with shear strengths of up to 2.5 MPa. As for the hydrogen bonded polymers discussed above, reversible debonding could be thermally triggered, but most interestingly, UV-light could also be used as stimulus to promote debonding in the absence of a UV-absorber additive. In contrast to UPy dimers,[51] Mebip–metal complexes strongly absorb light in the UV regime and generate enough heat to cause complex dissociation upon radiationless decay of the excited states. A polymer prepared with a composition of [$Zn_{0.8}$Mebip–PEB–Mebip]$(NTf_2)_{1.6}$ was thus used as adhesive to bond two quartz glass slides and irradiated with light of a wavelength in the range of 320–390 nm and a power density of 900 mW cm^{-2}. This caused an increase of the temperature up to ~220 °C within 30 s and led to rapid debonding. The lap joints thus disassembled could be rebonded through exposure to light or heat to recover the original adhesive properties. This work thus expands the possibilities of UV-light-triggered debonding (*vide supra*) by using an intrinsically UV-active metal–complex as building block. The adhesion of $[Zn_{0.8}Mebip–PEB–Mebip](NTf_2)_{1.6}$ to a zirconium ceramic substrate was also studied (shear strength = 33.0 ± 8.7 N),[23] and the adhesive was shown to debond upon heating.

In another example, Xie and coworkers explored metal cation-π interactions, which are amongst the strongest non-covalent interactions (up to 40 kcal mol^{-1}), to bond two different rigid surfaces.[72] The authors brought into contact an epoxy polymer with cross-linked poly(styrene) containing sodium sulfonate surface groups (Fig. 13) at 80 °C and performed tensile tests after cooling the bonded parts to room temperature. The adhesive strength was 721 N cm^{-2}, which was almost double than that measured for a cross-linked poly(styrene) control (380 N cm^{-2}), and superior to that observed with sulfonated cross-linked poly(styrene). The failure mode was cohesive as per XPS analysis and the nature of the interactions responsible for the enhanced adhesion was confirmed by Raman spectroscopy using model sulfonate and aromatic compounds. The adhesion mechanism reported by Xie and coworkers relies on a very simple synthetic scheme, which makes it interesting from the application point of view.

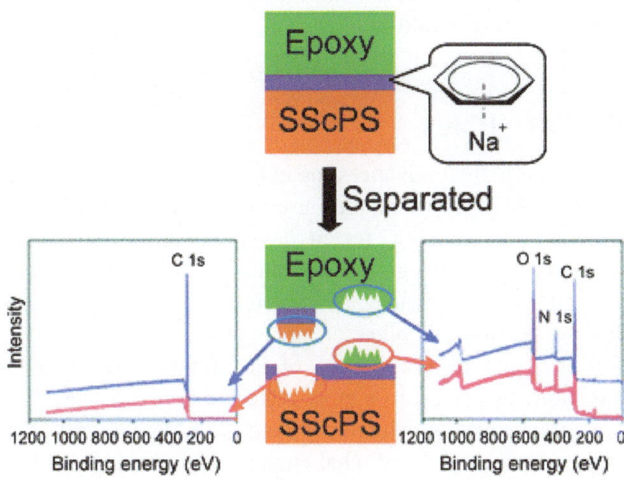

Figure 13: Approach to achieve adhesion between rigid surfaces with the use of metal cation-π interactions between epoxy resin and sulfonated polystyrene surfaces and determination of cohesive failure mechanism *via* XPS. Adapted from ref. 72 with permission from the Royal Society of Chemistry.

SUPRAMOLECULAR ADHESIVES BASED ON OTHER NON-COVALENT INTERACTIONS

This section discusses the use of non-covalent interactions such as hydrophobic/hydrophilic effects and ionic interactions to achieve adhesion, although it is noted that hydrogen bonds also play an important (complementary) role in many of these materials.

An illustrative example of selective and reversible adhesion was reported by Creton and coworkers on the basis of a switchable surface.[73] The authors grafted poly(styrene) (PS) and poly(2-vinyl pyridine) (P2VP) brushes and blends thereof onto a silicon surface and explored their adhesion to a blend of poly(vinyl pyrrolidone) and poly(ethylene glycol) (PVP–PEG) after different surface treatments. Upon exposing the surface to toluene, the PS chains swelled preferentially. After drying, the PS-enriched surface showed poor adhesion toward the PVP–PEG blend. However, when the surface was exposed to ethanol and subsequently to acidified water (pH 2) before drying, the protonated P2VP was brought to the surface and the PVP–PEG blend adhered well. This behavior was observed for surfaces with a P2VP brush content as low as *ca.* 2 mol% (relative to the content of PS) and the behavior was shown to be reversible. Thus, upon subsequent treatment of the protonated P2VP surface with aqueous ammonia, the adhesion of the PVP–PEG blend was significantly reduced.

Homogeneous hydrogels have been proposed as potential replacement for current two-phase dental adhesives.[74–76] Prud›homme and coworkers[77] reported hydrogels displaying tunable adhesion to both hydrophilic and hydrophobic surfaces relying on the coexistence of poly(methyl vinyl ether-*co*-maleic anhydride) (PMVE-MA) and poly(vinyl acetate) (PVAc) domains. Their synthetic approach involved the reaction of PVAc with hydrochloric acid to reach controlled hydrolysis degrees followed by cross-linking with PMVE-MA, which resulted in transparent gels when the reaction was carried out in a water/ethanol mixture. In order to elucidate the effect of molecular weight on the adhesive properties of these materials, 180° peel tests (Fig. 2) were performed with samples of different composition. The molecular weight of PMVE-MA and PVAc were shown to affect the adhesion in a non-linear fashion reaching a maximum peel strength of 15.4 N cm^{-2} with 500 kg mol^{-1} PVAc and 216 kg mol^{-1} PMVE-MA, which was ascribed to the optimum balance between entanglement and flow. The composition of the solvent mixture was shown to be critical for solubilizing the hydrophilic and hydrophobic polymeric components. While a 50 : 50 water/ethanol mixture provided homogeneous gels, dental applications require ethanol contents below 12% to avoid mucosal tissue irritation.[77] Propylene glycol, glycerol and poly(ethylene glycol) (PEG-400) were thus investigated as amphiphilic alternatives to reduce the ethanol content and, although they all provided good miscibility at certain compositions, a 22.0 : 11.5 : 16.5 mixture of water/ethanol/PEG-400 was selected as the best alternative as propylene glycol and glycerol display a high number of hydroxyl groups that may interfere in the cross-linking reaction.

Pressure sensitive adhesives (PSAs) are a special class of polymers that adhere strongly to surfaces upon application of a slight pressure force.[78] These materials have viscous liquid-like properties during adhesive bond formation but behave as slightly cross-linked rubbery polymers when a detaching force is applied. Thus, a complex combination of high fluidity in the bonding process and high cohesive strength to resist debonding stresses is required. At a molecular level, the strength of PSAs has been shown to depend on the balance between high cohesive strength and large free volume,[79] which motivated Feldstein and coworkers[80] to explore the adhesive properties of various mixtures of poly(*N*-vinyl pyrrolidone) (PVP) and PEG with an M_n of 400. In such mixtures, PEG-400 serves as a difunctional supramolecular cross-linker that bridges the carbonyl groups of PVP chains *via* hydrogen bonding (Fig. 14a). The resulting supramolecular network displays improved cohesive strength relative to PVP as a result of PVP/PEG-400 interactions, but also an increased free volume due to the introduction of spacers between PVP chains. The sub-nanoscale free-volume characteristics were studied with positron annihilation lifetime spectroscopy, which showed that an increase of the PEG-400 content from 0 to 43% led to an increase of *ca.* 40% in the free-volume fraction and of 15% in the mean hole radius. These observations indicate a reduction in number density of free-volume holes as the PEG-400 content is increased. Additionally, the effect of relative humidity (RH) on the free volume was found to be less pronounced. The adhesive properties of the blends were studied with peel and probe tack tests (Fig. 2). In order to study the effect that water has on the adhesive properties of these materials, the samples subjected to 180° peel tests were equilibrated at controlled vapor pressures and had a water content of 5–30%. On the other hand, the samples used in the probe tack tests were vacuum dried and contained *ca.* 3% water. In both tests, the maximum peel strength and probe tack adhesion were observed for a PVP–PEG-400 blend containing 36 wt% of PEG-400. Regarding the mechanical properties of these materials, tensile tests revealed an increasing linear correlation between the PEG-400 content and the elongation at break and, as expected, hydration caused a reduction of the tensile strength and an increase of the elongation at break on account of disrupting the hydrogen bonding between the PEG and the PVP. Overall, this study contributes to the understanding of the complex relationships between the nanoscopic free volume and the macroscopic adhesive and mechanical properties of PVP–PEG as PSAs models.

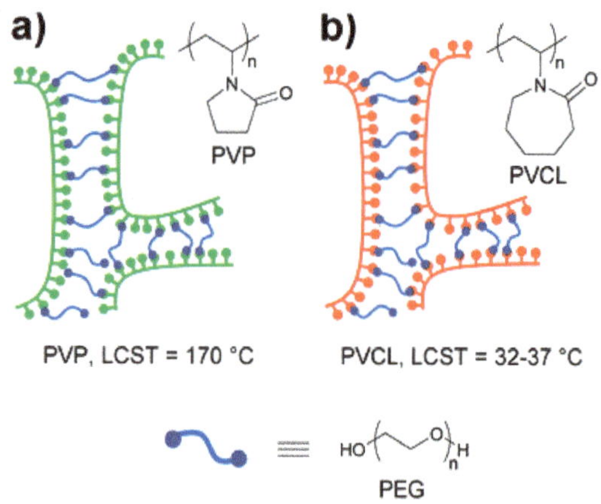

Figure 14: Temperature and water sensitive adhesives based on hydrogen bonded networks formed by blends of (a) PVP/PEG-400[80] and (b) PVCL/PEG-400.[81]

In the context of stimuli-responsive adhesives for medical purposes, ideal PSAs should display strong adhesion to skin but could ideally be removed when no longer needed without skin damage and pain. Feldstein and coworkers[81] continued their investigation of the blends discussed above by using poly(*N*-vinyl caprolactam) (PVCL) (Fig. 14b), a biocompatible polymer that shows a lower critical solution temperature (LCST) at 32–37 °C, *i.e.*, near physiological temperature, instead of PVP. As in their previous study, the content of PEG-400 had a clear effect on the properties of the blends at *ca.* 22 °C and 50% relative humidity, which varied from a non-tacky solid (at 30 wt% of PEG-400) to a tacky liquid (at 60 wt% of PEG-400). Within this compositional range, a PEG-400 content of 45–50% (with a water content of 3–5 wt%) resulted in a PSA with well-balanced solid- and liquid-like properties, displaying a debonding stress of ~0.67 MPa in a probe tack test at 22 °C. Interestingly, the same material shows a sharp decrease (~0.35 MPa) of debonding stress at 35 °C and behaves as a tacky liquid at temperatures above 45 °C. In this e, the low water content of the sample did not allow to clearly establish the role of the LCST of PVCL on the adhesive properties of the blend. The effect of temperature and water content on the adhesive properties of the blend containing 45% of PEG-400 was therefore studied by means of 180° peel tests (Fig. 2). The blends containing 10 or 20 wt% of water experienced a gradual decrease of peel strength with increasing temperature with the 20% water blend displaying the highest values. On the other hand, at a water content of *ca.* 30%, a fast and clear loss of adhesion was observed at *ca.* 55 °C, and a further increase of the water content led to a gradual decrease of the detaching temperature. This behavior could be correlated to the variations on the cloud point of a diluted PVCL/PEG-400 with the temperature. Additionally, these results indicate that the PVCL/PEG-400 blends can accommodate more water than the PVP–PEG-400 blends before losing their adhesive properties. In other words, PVCL/PEG-400 blends form a denser supramolecular network that provides stronger adhesion. Overall, the mechanical properties of these materials can be (reversibly) switched by exposure to water at temperatures close to that of the body, which has clear implications in the development of adhesives for medical applications that would readily detach by simple exposure to water.

CONCLUSIONS

Despite many great advances made in the burgeoning field of supramolecular polymers, their use in the context of adhesive materials is a relatively recent development, although also this subdomain seems now to be rapidly growing. The studies reviewed here evidence the potential of supramolecular adhesives in many application areas and in many es showe the role of nature as a source of inspiration. Non-covalent interactions are central to most adhesion processes used by living organisms, which allow them to adhere to different surfaces in a variety of challenging environments and to detach in response to external stimuli. The work reviewed in this contribution suggests that some of such advanced functionalities can now, if only in crude ways, be emulated by synthetic supramolecular polymers based on hydrogen bonding, host–guest interactions, metal–ligand complexation and others, and combinations of such interactions. Thus, supramolecular adhesives displaying (multi)stimuli-triggered debonding-on-demand, substrate selectivity, underwater adhesion or bioactivity have been reported. Among them, two main approaches to switchable adhesion can be distinguished, namely the engineering of the adhesive forces at the interface of two substrates, and the design of the materials with stimuli-responsive bulk properties. While the first approach allows for the synthesis of adhesives for high-tech applications where selective/orthogonal adhesion is required, the second approach appears to offer a simpler and perhaps more cost-effective alternative that still allows for accessing appealing features such as remotely triggered debonding. Non-covalent interactions can thus provide the basis for the development of adhesives with the potential to impact different application fields; however, this is to date only reflected in a few emerging products including the customized materials developed by Suprapolix,[82] Bostik,[24,25] and Arkema.[26] The main challenges in this field will therefore include the customer-driven design of supramolecular adhesives and the development of synthetic strategies leading to up-scalable productions. Nevertheless, considering the fast development of this research area, we expect an exciting future both in the fundamental and application arenas.

ACKNOWLEDGEMENTS

The authors acknowledge funding from the Commission for Technology and Innovation (CTI, Grant No. 13746.1), the U.S. Army Research Office (W911NF-12-1-0339) and the Adolphe Merkle Foundation for support of research on supramolecular adhesives. LME is grateful for support through the Ambizione program of the Swiss National Science Foundation.

REFERENCES

1. R. H. Lacombe, *Adhesion Measurement Methods: Theory and Practice*, CRC Press, Boca Raton, London, New York, 2005 .
2. K. W. Allen, *Int. J. Adhes. Adhes.*, 2003, 23, 87–93 .
3. A. V. Pocius, *Adhesion and Adhesives Technology*, Carl Hanser Verlag GmbH & Co. KG, Munich, 3rd edn, 2012 .
4. K. L. Mittal, *Adhesion Aspects of Thin Films*, VSP BV, Utrecht, 2001, vol. 1 .
5. K. L. Mittal, *Adhesion Aspects of Thin Films*, VSP BV, Utrecht, 2005, vol. 2 .
6. K. Autumn, Y. A. Liang, S. T. Hsieh, W. Zesch, W. P. Chan, T. W. Kenny, R. Fearing and R. J. Full, *Nature*, 2000, 405, 681–685 .
7. E. Hennebert, B. Maldonado, P. Ladurner, P. Flammang and R. Santos, *Interface focus*, 2015, 5, 20140064 .
8. E. A. Dubiel, Y. Martin and P. Vermette, *Chem. Rev.*, 2011, 111, 2900–2936 .
9. C. Rodriguez-Emmenegger, S. Janel, A. de los Santos Pereira, M. Bruns and F. Lafont, *Polym. Chem.*, 2015 10.1039/c1035py00197h.
10. H. J. Busscher and H. C. van der Mei, *PLoS Pathog.*, 2012, 8, e1002440 .
11. B. P. Lee, P. B. Messersmith, J. N. Israelachvili and J. H. Waite, *Annu. Rev. Mater. Res.*, 2011, 41, 99–132 .
12. S. C. Nicklisch and J. H. Waite, *Biofouling*, 2012, 28, 865–877 .
13. L. Li, W. Smitthipong and H. Zeng, *Polym. Chem.*, 2015, 6, 353–358 .
14. D. Habault, H. Zhang and Y. Zhao, *Chem. Soc. Rev.*, 2013, 42, 7244–7256 .
15. G. L. Fiore, S. J. Rowan and C. Weder, *Chem. Soc. Rev.*, 2013, 42, 7278–7288 .
16. Y. Yang and M. W. Urban, *Chem. Soc. Rev.*, 2013, 42, 7446–7467 .
17. L. Montero de Espinosa, G. L. Fiore, C. Weder, E. J. Foster and Y. C. Simon, *Prog. Polym. Sci.*, 2015 DOI:10.1016/j.progpolymsci.2015.1004.1003.
18. S. I. Stupp, V. LeBonheur, K. Walker, L. S. Li, K. E. Huggins, M. Keser and A. Amstutz, *Science*, 1997, 276, 384–389 .
19. A. E. Rowan, M. C. Feiters and R. J. M. Nolte, *Proc. Annu. Meet. Adhes. Soc.*, 1997, 20, 271–273 .
20. S. Abed, S. Boileau, L. Bouteiller, J. R. Caille, N. Lacoudre, D. Teyssie and J. M. Yu, *Polym. Mater.: Sci. Eng.*, 1997, 76, 45–46 .
21. Q. Chuan, A. T. N. Pires and L. A. Belfiore, *Polym. Commun.*, 1990, 31, 177–182 .
22. S. Tanaka, T. Aoki and M. Yoshida, *International Pat.*, WO9110756A1, 1991 .
23. N. Moszner, T. Hirt, K. Rist, U. Salz, C. Weder, G. Fiore, C. Heinzmann and V. Rheinberger, *International Pat.*, WO2013034778A2, 2013 .
24. N. Sajot, C. Brunet and M. Lachhab, *International Pat.*, WO2009150328A2, 2009 .
25. D. Goubard and N. Sajot, *Canada Pat.*, CA2740089A1, 2011 .
26. J. A. Gonzalez Leon, J.-P. Gillet, G. Barreto, M. Hidalgo and V. Luca, *International Pat.*, WO2011015773A2, 2011 .
27. C. I. Lindsay and S. A. Woutters, *US Pat.*, US20070149751A1, 2007 .
28. M. K. C. Moens, *EU Pat.*, EP1213309A1, 2002 .
29. J. Zhan and R. Du, *China Pat.*, CN201310168441A, 2013 .
30. G. M. L. Hoorne-van Gemert, S. Chodorowski-Kimmès, H. M. Janssen, E. W. Meijer and A. W. Bosman, *International Pat.*, WO2010002262A1, 2010 .
31. B. Greener, *International Pat.*, WO0107396A1, 2001 .
32. G. Fichman, L. Adler-Abramovich, E. Gazit and P. B. Messersmith, *International Pat.*, WO2014191997A1, 2014 .
33. N. Boden, A. Agelli, E. Ingham and J. Kirkham, *International Pat.*, WO2004007532A2, 2004 .
34. L. Brunsveld, B. J. B. Folmer, E. W. Meijer and R. P. Sijbesma, *Chem. Rev.*, 2001, 101, 4071–4097 .
35. T. Aida, E. W. Meijer and S. I. Stupp, *Science*, 2012, 335, 813–817 .
36. K. Yamauchi, J. R. Lizotte and T. E. Long, *Macromolecules*, 2003, 36, 1083–1088 .
37. P. Cordier, F. Tournilhac, C. Soulié-Ziakovic and L. Leibler, *Nature*, 2008, 451, 977–980 .
38. F. H. Beijer, R. P. Sijbesma, H. Kooijman, A. L. Spek and E. W. Meijer, *J. Am. Chem. Soc.*, 1998, 120, 6761–6769 .
39. R. P. Sijbesma, F. H. Beijer, L. Brunsveld, B. J. B. Folmer, J. H. K. Ky Hihberg, R. F. M. Lange, J. K. L. Lowe and E. W. Meijer, *Science*, 1997, 278, 1601–1604 .
40. S. Cheng, M. Zhang, N. Dixit, R. B. Moore and T. E. Long, *Macromolecules*, 2012, 45, 805–812 .
41. K. Viswanathan, H. Ozhalici, C. L. Elkins, C. Heisey, T. C. Ward and T. E. Long, *Langmuir*, 2006, 22, 1099–1105 .

42. C. Creton, *MRS Bull.*, 2003, 28, 434–439 .

43. S. D. Tobing and A. Klein, *J. Appl. Polym. Sci.*, 2001, 79, 2230–2244 .

44. C. A. Anderson, A. R. Jones, E. M. Briggs, E. J. Novitsky, D. W. Kuykendall, N. R. Sottos and S. C. Zimmerman, *J. Am. Chem. Soc.*, 2013, 135, 7288–7295 .

45. Y. Zhang, C. A. Anderson and S. C. Zimmerman, *Org. Lett.*, 2013, 15, 3506–3509 .

46. L. Römer and T. Scheibel, *Prion*, 2014, 2, 154–161 .

47. P. Woodward, D. Hermida Merino, I. W. Hamley, A. T. Slark and W. Hayes, *Aust. J. Chem.*, 2009, 62, 790–793 .

48. G. M. L. van Gemert, J. W. Peeters, S. H. M. Söntjens, H. M. Janssen and A. W. Bosman, *Macromol. Chem. Phys.*, 2012, 213, 234–242 .

49. J. Courtois, I. Baroudi, N. Nouvel, E. Degrandi, S. Pensec, G. Ducouret, C. Chanéac, L. Bouteiller and C. Creton, *Adv. Funct. Mater.*, 2010, 20, 1803–1811 .

50. S. Pensec, N. Nouvel, A. Guilleman, C. Creton, F. Boue and L. Bouteiller, *Macromolecules*, 2010, 43, 2529–2534 .

51. C. Heinzmann, S. Coulibaly, A. Roulin, G. L. Fiore and C. Weder, *ACS Appl. Mater. Interfaces*, 2014, 6, 4713–4719 .

52. C. Heinzmann, U. Salz, N. Moszner, G. L. Fiore and C. Weder, *ACS Appl. Mater. Interfaces*, 2015, 7, 13395–13404 .

53. R. P. Sijbesma, F. H. Beijer, L. Brunsveld, B. J. B. Folmer, J. H. K. Ky Hihberg, R. F. M. Lange, J. K. L. Lowe and E. W. Meijer, *Science*, 1997, 278, 1601–1604 .

54. F. Chabert, F. Tournilhac, N. Sajot, S. Tencé-Girault and L. Leibler, *Int. J. Adhes. Adhes.*, 2010, 30, 696–705 .

55. S. Grimaldi, J.-P. Gillet, M. Hidalgo, F.-G. Tournilhac, P. Cordier and L. Leibler, *International Pat.*, WO2008029065A2, 2008 .

56. D. Montarnal, L. Leibler, F.-G. Tournilhac and M. Hidalgo, *International Pat.*, WO2009141558A2, 2009 .

57. R. Wang and T. Xie, *Langmuir*, 2010, 26, 2999–3002 .

58. G. O. Wilson, M. M. Caruso, S. R. Schelkopf, N. R. Sottos, S. R. White and J. S. Moore, *ACS Appl. Mater. Interfaces*, 2011, 3, 3072–3077 .

59. R. Du, J. Wu, L. Chen, H. Huang, X. Zhang and J. Zhang, *Small*, 2014, 10, 1387–1393 .

60. P. Y. Dankers, M. C. Harmsen, L. A. Brouwer, M. J. van Luyn and E. W. Meijer, *Nat. Mater.*, 2005, 4, 568–574 .

61. Y. Ahn, Y. Jang, N. Selvapalam, G. Yun and K. Kim, *Angew. Chem., Int. Ed.*, 2013, 52, 3140–3144 .

62. T. Nakamura, Y. Takashima, A. Hashidzume, H. Yamaguchi and A. Harada, *Nat. Commun.*, 2014, 5, 4622 .

63. A. K. Geim, S. V. Dubonos, I. V. Grigorieva, K. S. Novoselov, A. A. Zhukov and S. Y. Shapoval, *Nat. Mater.*, 2003, 2, 461–463 .

64. B. Yurdumakan, N. R. Raravikar, P. M. Ajayan and A. Dhinojwala, *Chem. Commun.*, 2005, 3799–3801 .

65. Y. Takashima, T. Sahara, T. Sekine, T. Kakuta, M. Nakahata, M. Otsubo, Y. Kobayashi and A. Harada, *Macromol. Rapid Commun.*, 2014, 35, 1646–1652 .

66. T. Matsue, D. H. Evans, T. Osa and N. Kobayashi, *J. Am. Chem. Soc.*, 1985, 107, 3411–3417 .

67. C. B. Highley, C. B. Rodell, I. L. Kim, R. J. Wade and J. A. Burdick, *J. Mater. Chem. B*, 2014, 2, 8110–8115 .

68. M. Cheng, F. Shi, J. Li, Z. Lin, C. Jiang, M. Xiao, L. Zhang, W. Yang and T. Nishi, *Adv. Mater.*, 2014, 26, 3009–3013 .

69. G. R. Whittell, M. D. Hager, U. S. Schubert and I. Manners, *Nat. Mater.*, 2011, 10, 176–188 .

70. X. Yan, F. Wang, B. Zheng and F. Huang, *Chem. Soc. Rev.*, 2012, 41, 6042–6065 .

71. M. Burnworth, L. Tang, J. R. Kumpfer, A. J. Duncan, F. L. Beyer, G. L. Fiore, S. J. Rowan and C. Weder, *Nature*, 2011, 472, 334–337 .

72. R. Wang and T. Xie, *Chem. Commun.*, 2010, 46, 1341–1343 .

73. H. Retsos, A. Kiriy, V. Senkovskyy, M. Stamm, M. M. Feldstein and C. Creton, *Adv. Mater.*, 2006, 18, 2624–2628 .

74. K. Zhao, X. R. Cheng, Y. L. Chao, Z. A. Li and G. L. Han, *Dent. Mater.*, 2004, 20, 419–424 .

75. M. Ozcan, Y. Kulak, C. de Baat, A. Arikan and M. Ucankale, *J. Prosthodontics*, 2005, 14, 122–126 .

76. R. Koppang, E. Berg, S. Dahm, C. Real and F. Floystrand, *J. Prosthet. Dent.*, 1995, 73, 486–491 .

77. X. Guo, F. Deng, L. Li and R. K. Prud'homme, *Biomacromolecules*, 2008, 9, 1637–1642 .

78. M. M. Feldstein, *Polym. Sci., Ser. A*, 2009, 51, 1341–1354 .

79. M. M. Feldstein, *Handbook of Pressure-Sensitive Adhesives and Products*, CRC Press, Boca Raton, London, New York, 2009 .

80. M. M. Feldstein, E. V. Bermesheva, Y. C. Jean, G. P. Misra and R. A. Siegel, *J. Appl. Polym. Sci.*, 2011, 119, 2408–2421 .

81. M. M. Feldstein, K. A. Bovaldinova, E. V. Bermesheva, A. P. Moscalets, E. E. Dormidontova, V. Y. Grinberg and A. R. Khokhlov, *Macromolecules*, 2014, 47, 5759–5767 .

82. http://www.suprapolix.com/ .

CHAPTER 11

CORE CROSS-LINKED STAR POLYMERS FOR TEMPERATURE/PH CONTROLLED DELIVERY OF 5-FLUOROURACIL

Elizabeth Sánchez-Bustos[1,2], José M. Cornejo-Bravo[2], and Angel Licea-Claverie[1]

[1]Instituto Tecnológico de Tijuana, Centro de Graduados e Investigación en Química, Apartado Postal 1166, 22000 Tijuana, BC, Mexico

[2]Facultad de Ciencias Químicas e Ingeniería, Universidad Autónoma de Baja California, Calzada Universidad 14418, 22390 Tijuana, BC, Mexico

ABSTRACT

RAFT polymerization with cross-linking was used to prepare core cross-linked star polymers bearing temperature sensitive arms. The arms consisted of a diblock copolymer containing N-isopropylacrylamide (NIPAAm) and 4-methacryloyloxy benzoic acid (4MBA) in the temperature sensitive block and poly(hexyl acrylate) forming the second hydrophobic block, while ethyleneglycol dimethacrylate was used to form the core. The acid comonomer provides pH sensitivity to the arms and also increases the transition temperature of polyNIPAAm to values in the range of 40 to 46°C. Light scattering and atomic force microscopy studies suggest that loose core star polymers were obtained. The star polymers were loaded with 5-fluorouracil (5-FU), an anticancer agent, in values of up to 30 w/w%. In vitro release experiments were performed at different temperatures and pH values, as well as with heating and cooling temperature cycles. Faster drug release was obtained at 42°C or pH 6, compared to normal physiological conditions (37°C, pH 7.4). The drug carriers prepared acted as nanopumps changing the release kinetics of 5-FU when temperatures cycles were applied, in contrast with release rates at a constant temperature. The prepared core cross-linked star polymers represent advanced drug delivery vehicles optimized for 5-FU with potential application in cancer treatment.

INTRODUCTION

Recent research reports on stimuli-responsive polymers have been focused in the field of polymer topology, particularly in the synthesis of well-defined macromolecules with precisely controlled architecture by incorporating site-specific branching points and functionalities [1], in addition to their ability to respond to various stimuli like temperature, pH, or light [2]. The features mentioned above are desirable for the preparation of tailor-made polymeric systems for a broad range of applications covering the fields of drug delivery, diagnostics, biotechnology, sensors: micromechanical and optical systems, coatings, and textiles [3].

Star polymers represent a special case of branched macromolecules and are very attractive for polymer chemists due to their multiarm topological architecture, smaller hydrodynamic size, ideal rheological behavior, particular bulkiness, and solution properties compared to their linear counterparts [4]. Well-defined star polymers can be prepared by growing the polymer arms from a functional core (core-first) or by the "arm-first" approach, where end-functionalized

Citation: Elizabeth Sánchez-Bustos, José M. Cornejo-Bravo, and Angel Licea-Claverie, "Core Cross-linked Star Polymers for Temperature/pH Controlled Delivery of 5-Fluorouracil," Journal of Chemistry, vol. 2016, Article ID 4543191, 12 pages, 2016. doi:10.1155/2016/4543191.

polymeric chains are attached to a functional core by specific reactions or by copolymerization with a cross-linker. The "arm-first" approach exhibits several advantages over the "core-first" methodology, such as an easier procedure to achieve higher molecular weight star polymers without the formation of coupling star structures [5] and the possibility of having well-defined arms and combining different types of arms in a single star [6]. Star polymers were first synthesized by the "arm-first" approach in 1966 using anionic living polymerization to obtain arms followed by a reaction with a suitable linking agent to create the star polymer architecture [7]. Controlled radical polymerization (CRP) techniques, such as atom transfer radical polymerization (ATRP) [8], nitroxide mediated radical polymerization (NMP) [9], and reversible addition-fragmentation chain transfer polymerization (RAFT), have been versatile techniques for the synthesis of star polymers using both "arm-first" and "core-first" methods [10–13]. Star polymers can be divided into two structural categories, homogeneous (homoarm) or mixed (miktoarm) star polymers [6]. Arms may be built of homo-, co-, or even terpolymers, in such a way that the final properties of the resulting star-shaped polymers can be adjusted by choosing the respective chemical structure of arms and core, depending on the required application [14].

Amphiphilic polymers formed by hydrophilic and hydrophobic segments self-aggregate in aqueous solutions to form micelles and other types of aggregates [15–18]. Amphiphilic star polymers may also aggregate in a complex fashion depending on their chemical composition, relative size of core to arms, number of arms, and so forth [19–21].

Due to their small size (<200 nm), star polymers may accumulate in tumors or inflamed tissues through the enhanced permeability and retention effect (EPR) [22]. Several pathologies, such as inflammation, tumor, and infarcted tissue, show a local decrease in pH (1–2.5 pH units) [23–25]. The particular pH conditions at pathological sites, and also local heating by 2–5°C in solid tumors, can be used to enhance drug release from accumulated or locally administered stimuli-responsive polymeric materials [26–28].

An important issue for the biomedical application of temperature sensitive stars polymers is to adjust the transition temperature (LCST) above normal body temperature; thus, when the nanomaterials are injected into the body, they are in the swollen state (37°C < LCST) and circulate in blood vessels but collapse inside cancer tissue (T_{tumor} > LCST) due to the hydrophobic change at temperatures above their LCST [29]. The process of collapse (and possible aggregation) leads to drug release inside tumors [30, 31]. The LCST of polyNIPAAm (32°C) can be increased by copolymerization of NIPAAm with hydrophilic or ionized monomers [32, 33]. It was demonstrated that copolymerization of NIPAAm with amphiphilic weak acid monomers allows tuning the LCST of polyNIPAAm above body temperature [13]; this leads, in the case of nanohydrogels, to maintaining adequate swelling/deswelling properties [34, 35].

In a previous study, our group has reported the synthesis of core cross-linked (CCS) temperature sensitive star polymers with a random number of poly(N-isopropylacrylamide)-b-poly(hexyl acrylate) (PNIPAAm-b-PHA) or PNIPAAm arms using the RAFT polymerization technique. Furthermore, the release behavior of different drugs from those star polymers was also investigated observing that a star polymer with PNIPAAm-b-PHA arms is more effective in entrapping drugs than a star polymer formed by pure PNIPAAm arms since the PHA block forms a hydrophobic intermediate shell over the core where hydrophobic drugs can be entrapped [12]. Other examples of PNIPAAm-based star polymers have been developed for drug delivery applications [21,36–42]. However, most of them have been designed so as to be able to entrap and release the drug due to the ability of PNIPAAm star polymers to self-aggregate [21, 36–41], which may have the drawback of early aggregate disruption due to the dilution factor in the bloodstream, resulting in premature drug release. CCS star polymers represent intermediate architecture between branched polymer chains and polymer nanoparticles [43]. They are larger than conventional star polymers, having a single molecule or atom as a core, but smaller than nano/microgels. They have the advantage of presenting a relatively large core (cargo compartment) depending on the synthetic conditions and they are surrounded by a large number of polymer arms that stabilize them, making them dispersible in a suitable solvent for the arms and preventing self-aggregation.

In the present study, the RAFT technique was employed to synthesize poly(NIPAAm-co-amphiphilic weak acid) polymers to tailor the LCST in the range of 40 to 46°C at pH of 7.4. These polymeric chains were further activated to form block copolymers with HA. The block copolymers were used to prepare dual sensitive, temperature and pH, star polymers, cross-linked with ethyleneglycol dimethacrylate (EGDMA). These CCS star polymers were tested to load 5-fluorouracil (5-FU), an anticancer drug, and to study their drug delivery behavior under different in vitro conditions.

EXPERIMENTAL SECTION

Materials

The acid monomer, 4-methacryloyloxybenzoic acid (4MBA), the RAFT agent, 2-hydroxyethyl 2-phenylacetate dithiobenzoate (DFH), and the free-radical initiator, 4,4′-azobis(4-cyanopentanol) (ACP), were prepared as described in the literature [13]. N-Isopropylacrylamide (NIPAAm, 97%, Aldrich) was purified by recrystallization from n-hexane. 4,4′-Azobis(4-cyanopentanoic acid) (ACPA, 98%, Fluka) was recrystallized from methanol. n-Hexyl acrylate (HA, Aldrich) and ethyleneglycol dimethacrylate (EGDMA, Aldrich) were purified by passing the reagents through an inhibitor remover column for benzoquinones (Aldrich). p-Dioxane (ACS grade, Fermont), diethyl ether (ACS grade, Fermont), tetrahydrofuran (THF, HPLC grade, Aldrich), and 5-fluorouracil (5-FU, Sigma) were used without further purification.

Synthesis of Linear Poly(NIPAAm) Arms (Macro-CTAs of PolyNIPAAm) and Statistical Copolymers of NIPAAm with 4MBA

Linear poly(NIPAAm), used to form the arms of star polymers, was synthesized with prescribed molecular weights via RAFT polymerization (Figure 1(a)). Poly(NIPAAm) arms were prepared using molar ratios of monomer:CTA:initiator of

180 : 1.0 : 0.2, of 132 : 1.0 : 0.2, and of 88 : 1.0 : 0.2. Statistical copolymers were synthesized with various molar percentages of the acid comonomer in order to adjust the LCST above 37°C. Poly(NIPAAm-co-4MBA), aiming at 9% of 4MBA, was synthesized using the same molar ratios that were used in the synthesis of linear poly(NIPAAm) considering the total amount of monomers NIPAAm + 4MBA in the molar ratio.

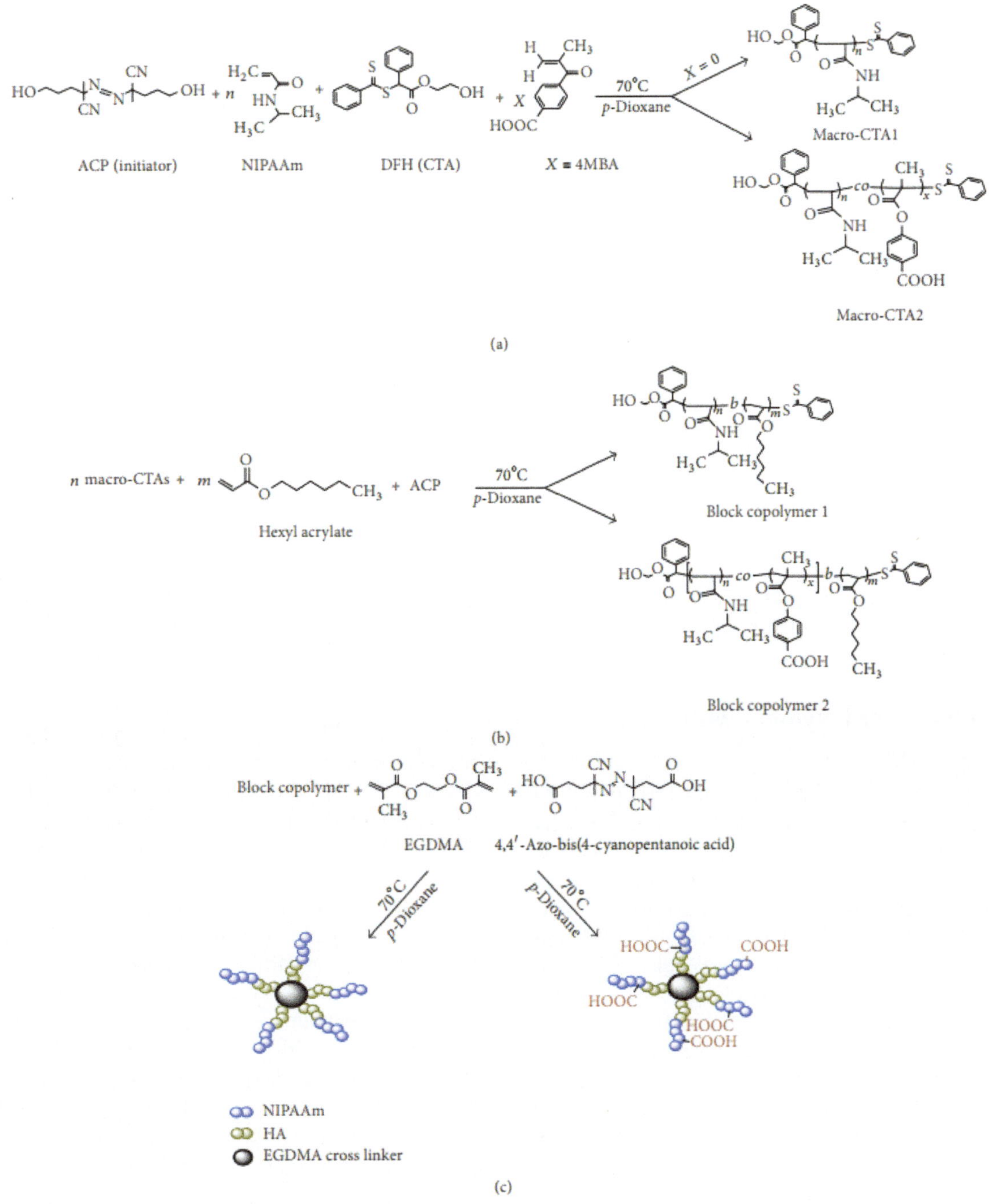

Figure 1: Synthetic routes to obtain pH and temperature sensitive star polymers.

The general procedure was performed as follows: 0.049 g (0.147 mmol) of DFH was stirred until it was completely dissolved in 3.0 mL of p-dioxane and poured into an ampoule containing a magnetic stir bar. 2.85 g (25.18 mmol) of NIPAAm, 0.2867 g (1.30 mmol) of 4MBA, 0.0074 g (0.0295 mmol) of ACP, and 3.0 mL of p-dioxane were added to that ampoule under continuous stirring. The ampoule was degassed by three freeze-thaw cycles alternating between vacuum and argon gas. The reaction mixture was heated to 70°C under argon atmosphere, and the polymerization was performed over 16 h. The reaction mixture was cooled down to 25°C and the polymerization product was precipitated in diethyl ether, then redissolved in acetone, and finally precipitated in cold diethyl ether. The purification process was repeated three times; then, the solvent was removed under vacuum and the product was allowed to dry overnight in a vacuum oven at 25°C.

Preparation of Linear Blocks of Poly(NIPAAm)-b-poly(hexyl acrylate, HA) and Poly(NIPAAm-co-4MBA)-b-poly(HA) Arms (Macro-CTAs of Block Copolymers)

The same procedure described above for the preparation of linear poly(NIPAAm) macro-CTAs was followed (Figure 1(b)). Briefly, 1 g (0.3435 mmol) of PNIPAAm macro-CTA (used instead of DFH) was stirred until it was fully dissolved in 2.5 mL of p-dioxane and the resulting mixture was added to an ampoule containing a magnetic stir bar; 0.1799 g (1.15 mmol) of HA monomer, 0.00173 (0.00687 mmol) of ACP, and 2.5 mL of p-dioxane were added to the ampoule under continuous stirring. The ampoule was degassed by three freeze-thaw cycles alternating between vacuum and argon gas. The reaction mixture was allowed to warm up to 70°C under argon atmosphere, and the polymerization was performed with stirring for 16 h. The reaction mixture was cooled down to 25°C and the polymerization product was precipitated in diethyl ether, dissolved in acetone, and precipitated again in cold diethyl ether. The purification process was repeated three consecutive times. The solvent was removed under vacuum and the product was allowed to dry overnight in a vacuum oven at 25°C. The same procedure was followed for the preparation of block copolymers starting with macro-CTAs of poly(NIPAAm-co-4MBA). The calculations were always done considering the same macro-CTA to initiator molar ratio and the same volume of p-dioxane as the solvent of choice.

Preparation of Star Polymers with a Cross-Linked Core

The synthesis of star polymers using the "arm-first" RAFT method with cross-linking was adapted from procedures reported in the literature [12, 13, 44, 45] (Figure 1(c)). Different molar ratios of cross-linker to arms were used to prepare stars polymers with PNIPAAm-b-PHA arms. An example is described as follows: 0.6 g (0.0187 mmol) of PNIPAAm-b-PHA macro-CTA was stirred until it was fully dissolved in 2.4 mL of p-dioxane and poured into an ampoule containing a magnetic stir bar; 0.037 g (0.187 mmol) of EGDMA, 0.00094 g (0.00374 mmol) of ACP, and 1.5 mL of p-dioxane were added to the ampoule under continuous stirring. The ampoule was degassed by three freeze-thaw cycles alternating between vacuum and argon gas. The reaction mixture was allowed to warm up to 70°C under argon atmosphere, and the polymerization was performed under stirring for 24 h. The reaction mixture was cooled down to 25°C and the polymerization product was precipitated in diethyl ether, redissolved in acetone, and precipitated again in cold diethyl ether. The purification process was repeated three times. The solvent was removed under vacuum and the product was allowed to dry overnight in a vacuum oven at 25°C.

For the synthesis of star polymers with poly(NIPAAm-co-4MBA)-b-PHA arms, a molar ratio of EGDMA : macro-CTA = 10 : 1 was used with macro-CTA poly(NIPAAm-co-$4MBA_{5\%}$)-b-PHA (M_n = 28,570 g mol^{-1}); and EGDMA : macro-CTA ratio = 20 : 1 was used with poly(NIPAAm-co-$4MBA_{9\%}$)-b-PHA (M_n = 21,730 g mol^{-1}). ACPA was used as the initiator and a 5 : 1 molar ratio of macro-CTA : initiator was used in all the cases. The polymerization procedure is the same as the one described above.

Characterization and Measurements

The polymerization yields were determined gravimetrically. The molecular weights and dispersity (Đ = M_w/M_n) were determined by gel permeation chromatography (GPC). The GPC system consists of a Varian 9002 HPLC pump equipped with a refractive index detector (Varian RI-4), a triangle light scattering detector (Mini Dawn, λ = 690 nm, Wyatt), and Phenomenex separation columns: Phenogel 10 μm 500 Å (00H-0643-K0); Phenogel 10 μm 104 Å (00H-0645-K0); Phenogel 10 μm 105 Å (00H-0646-K0) in series. The mobile phase used in the GPC was tetrahydrofuran (THF, HPLC grade) for NIPAAm polymers (homopolymers, block copolymers, and star polymers) and THF/CH_3COOH (50 : 1) v/v for copolymers, block copolymers, and star polymers containing acid comonomer units, while the flow rate was 0.5 mL min^{-1} at 25°C. Monodisperse polystyrene was used as the calibration standard to confirm the accuracy of the light scattering detector. dn/dc = 0.093 mL/g [46], reported for polyNIPAAm in THF, was used for molecular weight determinations of all polymers without HA. For polymers containing HA, an average dn/dc value between that of polyNIPAAm and that reported for polybutylacrylate in THF (0.065 mL/g [47]) was used, namely, 0.079 mL/g.

Dynamic light scattering (DLS) was used to determine the hydrodynamic diameters (D_h). Measurements were carried out at 20°C using a Zeta-sizer "Nano-ZS" from Malvern Instruments (ZEN3500) equipped with a green laser operating at λ = 532 nm. The angle of measurement was 173° (backscattering) and the size analysis was performed by CONTIN. The reported hydrodynamic diameters were calculated using the Stokes-Einstein equation for spheres [48]. Reported D_h values were the maxima in size distribution by volume from CONTIN analysis. The scattering intensity as a function of temperature was also used to determine the phase transition temperature of the prepared NIPAAm polymers at a given concentration, called, for the sake of simplicity, lower critical solution temperature (LCST). The size analysis was performed at different temperatures using 3 min of equilibrating time at each temperature in a heating cycle. The polymer concentration was 1 mg mL^{-1}and the value reported for the LCST was the temperature at which the scattering intensity increased sharply. The solutions were prepared, shaken for 12 h, and stored overnight in the refrigerator at 4°C. Before measurements, the solutions were filtered off using a 0.45-micron syringe filter for eliminating dust.

^{1}H-NMR spectra were recorded on a Varian Mercury 200 MHz nuclear magnetic resonance instrument with $CDCl_3$ or CD_3OD as the solvents and tetramethylsilane (TMS) as the internal reference. Atomic force microscopy (AFM) imaging experiments were performed using an Agilent Technologies SPM 5100 microscope equipped with N9520A high resolution scanner. The samples were prepared by dissolving the star polymer in a small volume of THF and then dispersing it in deionized water. The volumes and amounts were adjusted to obtain a concentration of 0.1 mg mL^{-1} in water. Then, the polymer solutions were filtered through a 0.45 μm syringe filter, followed by drop-casting on a freshly cleaved mica wafer, and dried in a controlled temperature room at 20°C for 24 h. The AFM experiments were carried out using the intermittent contact mode with silicon cantilevers at ambient conditions.

For each sample, three to four images were recorded from the same slide and the diameter of the assemblies was measured. The average value of the assemblies is reported as the average diameter of the star polymers.

Drug Loading and In Vitro Drug Release

5-FU (8 mg) was dissolved in 1.5 mL of dimethylacetamide (DMAc), while the star polymer (4 mg) was dissolved in 1.5 mL of DMAc in a separate vial. The drug solution was added dropwise to the star polymer solution with stirring at 20°C. The resulting solution was poured into a dialysis bag (molecular weight cutoff: 12,000–14,000 daltons) and it was subjected to dialysis against distilled water for 4 h to remove the nonencapsulated drug. The distilled water was replaced every 30 min for the first 2 h. The dialyzed solution was then lyophilized. The freeze-dried samples of drug-loaded star polymers were dispersed in a phosphate buffer (PBS) at pH of 7.4, followed by stirring them vigorously for 2 h and then sonicating them for 15 min. The loading amount of 5-FU in the star polymers was determined by measuring the UV absorbance at 267 nm using a Beckman Coulter spectrometer DU520. The drug loading (DL%) and entrapment efficiency (EE%) of the star polymer were calculated as follows:

$$DL\% = \left(\frac{\text{MDstar}}{\text{mass of stars} + \text{MDstar}}\right) \times 100,$$

$$EE\% = \left(\frac{\text{initial MD} - \text{MDsupernatant}}{\text{initial MD}}\right) \times 100, \quad (1)$$

where MD is the mass of drug (5-FU) and MDstar is the mass of drug in the star polymer. For the release experiments, the drug-loaded star polymers were placed into dialysis bags (10 mL) and dialyzed against 250 mL (sink conditions) of PBS at pH of 7.4 or at pH of 6, in a shaking bath (VWR International, model number 1217) with mild shaking motion (20 rpm) at 37 and 42°C, respectively. At predetermined time intervals, a 1 mL aliquot of the released medium was taken out and assayed for 5-FU content, and 1 mL of fresh buffer solution was replaced into the release media every predetermined time interval. The concentration of 5-FU released from the drug delivery system was monitored by measuring the UV absorbance at 267 nm using a Beckman Coulter spectrometer DU520. The cumulative drug release was calculated from the relationship based on the standard calibration curve. All the analyses were performed in triplicate and the results were reported with an average of three runs.

RESULTS AND DISCUSSION

Synthesis of Macro-CTAs

The chemical, molecular, and size analysis of polyNIPAAm and the statistical copolymer poly(NIPAAm-co-4MBA) as macro-CTAs were studied using FT-IR, ^{1}H-NMR, GPC, and DLS as analytical tools. The number average molecular weights (M_n) for polyNIPAAm-based macro-CTAs are defined by the NIPAAm to CTA molar ratio, as expected for a controlled radical polymerization method like RAFT. The dispersity values (Đ) obtained for polyNIPAAm were very low (Table 1, entries 1 to 4) demonstrating well behaving RAFT polymerization.

Table 1: Preparation of linear macro-CTAs at 70°C in p-dioxane for 12 h

Entries	Polymer	M : CTA : I[a]	M_n (target) (g/mol)	M_n (GPC) (g/mol)	Đ (M_w/M_n)	Yield (wt%)
1	PolyNIPAAm	180 : 1 : 0.2	20,673[b]	26,100	1.008	75
2	PolyNIPAAm	132 : 1 : 0.2	15,249[b]	14,400	1.022	70
3	PolyNIPAAm	88 : 1 : 0.2	10,277[b]	10,800	1.030	73
4	PolyNIPAAm	88 : 1 : 0.2	10,277[b]	8,200	1.040	70
5	Poly(NIPAAm-*co*-4MBA$_{5\%}$)	180 : 1 : 0.2	21,510[c]	22,000	1.157	78
6	Poly(NIPAAm-*co*-4MBA$_{9\%}$)	132 : 1 : 0.2	16,354[c]	15,200	1.149	70

[a]M: monomer; CTA: chain transfer agent; I: initiator.
[b]$M_n(\text{target}) = ([M]_0/[CTA]_0)M_{Mon} + M_{CTA}$.
[c]$M_n(\text{target}) = ([NIPAAm]_0/[CTA]_0)M_{NIPAAm} + ([4MBA]_0/[CTA]_0)M_{4MBA} + M_{CTA}$.

A similar behavior can be observed in the case of the poly(NIPAAm-co-4MBA) statistical copolymers, showing Đ values below 1.16 (entries 5-6 in Table 1). These values suggest that the polymerization proceeded in all cases with a high level of control. The polymerization yields ranged from 70 to 78%.

Synthesis of Block Copolymers

The main interest of introducing a second hydrophobic block constituted of n-hexyl acrylate (HA) units into polyNIPAAm or poly(NIPAAm-co-4MBA) linear polymers was to synthesize star polymers with a hydrophobic shell over the cross-linked core, which are able to retain drugs for transport in physiological conditions. The hydrophobic PHA block may also lead to star aggregate formation depending on the relative block size and number of arms of the star. Previously, the monomer HA had been successfully grown under RAFT polymerization conditions to form block copolymers with polyNIPAAm to generate polyNIPAAm-b-polyHA [12] and poly(NIPAAm-co-4MBA)-b-polyHA [13]. The molecular weights of the obtained products in this work are shown in Table 2.

Table 2: Preparation of linear block copolymer arms

Entries	Polymer	M : CTA : I[a]	M_n (target)[i] (g/mol)	M_n (GPC) (g/mol)	Đ (M_w/M_n)	Yield (wt%)
1	PolyNIPAAm-*b*-polyHA	22 : 1 : 0.2[b]	11,645	12,780	1.065	70
2	PolyNIPAAm-*b*-polyHA	32 : 1 : 0.2[c]	15,820	17,890	1.175	70
3	PolyNIPAAm-*b*-polyHA	50 : 1 : 0.2[d]	19,930	20,200	1.137	72
4	PolyNIPAAm-*b*-polyHA	32 : 1 : 0.2[e]	31,111	32,070	1.177	75
5	Poly(NIPAAm-*co*-4MBA$_{9\%}$)-*b*-polyHA	32 : 1 : 0.2[f]	21,160	21,730	1.030	71
6	Poly(NIPAAm-*co*-4MBA$_{9\%}$)-*b*-polyHA	64 : 1 : 0.2[g]	25,160	25,340	1.011	65
7	Poly(NIPAAm-*co*-4MBA$_{5\%}$)-*b*-polyHA	32 : 1 : 0.2[h]	27,200	28,570	1.038	60

[a]Reactions in *p*-dioxane at 70°C for 16 h, M = HA; [b]macro-CTA is PNIPAAm (M_n = 8,208 g/mol); [c]macro-CTA is PNIPAAm (M_n = 10,820 g/mol); [d]macro-CTA is PNIPAAm (M_n = 14,430 g/mol); [e]macro-CTA is PNIPAAm (M_n = 26,110 g/mol); [f]macro-CTA is P(NIPAAm-*co*-4MBA$_{9\%}$) (M_n = 15,160 g/mol); [g]macro-CTA is P(NIPAAm-*co*-4MBA$_{9\%}$) (M_n = 15,160 g/mol); [h]macro-CTA is P(NIPAAm-*co*-4MBA$_{5\%}$) (M_n = 22,000 g/mol); [i]M_n(target) = $([M]_0/[\text{macro-CTA}]_0)M_{\text{Mon}} + M_{\text{macro-CTA}}$.

Results in Table 2 showed that polyNIPAAm was reacted with different molar ratios of HA to form the second block. In each case, a comparison between observed and target molecular weights showed that they were similar, demonstrating good RAFT copolymerization control resulting in Đ values below 1.2 (entries 1 to 4). Indeed, when using the same molar ratio of HA to macro-CTA but polyNIPAAm macro-CTAs with different values, Đ values were not affected (entries 2 and 4). The same results regarding the high level of control can be observed for the formation of poly(NIPAAm-co-4MBA)-b-(polyHA) using as macro-CTAs statistical copolymers which have different molar ratios of 4MBA and even different lengths of the macro-CTA chain. In all the cases, good match was observed between the calculated and the observed molecular weights with low Đ values (entries 5 to 7). The obtained yields were in a range of 60 to 75%.

Synthesis of Star Polymers

The star polymers were prepared from the block copolymers presented in Section 3.2. The block copolymers represent the arms of the star. The molar ratios of cross-linker to arms EGDMA : polyNIPAAm-b-polyHA or EGDMA : poly(NIPAAm-co-4MBA)-b-polyHA for the preparation of the corresponding star polymers are shown in Table 3. The molecular weights that were obtained from GPC for the products prepared at 10 : 1 and 20 : 1 ratios are, in most of the cases, three to four times the molecular weight of the arms. These molecular weights are lower than the expected values according to literature reports for this type of star polymer, where more arms per star were expected [17, 43], nevertheless enough to suggest that star polymers are formed. Figure 2 shows a comparison between GPC traces for a linear copolymer, a block copolymer prepared from it (arm), and a star polymer as a result of the final reaction of the block copolymer with the cross-linker. Results reveal narrow size distributions for the linear and block copolymers and a wider distribution for the star polymer, while the retention time decreases with increasing size, as expected.

Table 3: Reactions for preparation of star polymers with a random number of arms and a cross-linked poly(EGDMA) core

Entries	Star polymer [cross-linker : macro-CTA : I][a]	Crosslinker : macro-CTA : I[a]	M_n (GPC) (g/mol)	Đ (M_w/M_n)	Yield (wt%)
1	[PolyNIPAAm-*b*-polyHA][b]-EGDMA	20 : 1 : 0.2	41,740	1.333	70
2	[PolyNIPAAm-*b*-polyHA][c]-EGDMA	10 : 1 : 0.2	136,500	1.200	77
3	[(Poly(NIPAAm-*co*-4MBA$_{9\%}$)-*b*-polyHA)][d]-EGDMA	20 : 1 : 0.2	76,600	1.175	65
4	[(Poly(NIPAAm-*co*-4MBA$_{5\%}$)-*b*-polyHA)][e]-EGDMA	10 : 1 : 0.2	100,700	1.290	70

[a]Cross-linker: EGDMA; macro-CTA: block copolymer as macro chain transfer agent; I: initiator. [b]Macro-CTA of M_n = 12,780 g/mol. [c]Macro-CTA of M_n = 32,070 g/mol. [d]Macro-CTA of M_n = 21,730 g/mol. [e]Macro-CTA of M_n = 28,570 g/mol.

The star polymers were characterized by spectroscopy analysis. FT-IR and ^{1}H-NMR, spectra of a 10 : 1 star product, are shown in Figure 3. The spectroscopic analysis confirmed the incorporation of both HA and EGDMA units into the star polymer; however, the EGDMA units were difficult to be identified separately from HA or 4MBA units (Figures 3(a) and 3(b)). The FT-IR spectrum of the star polymer with poly(NIPAAm-co-4MBA$_{5\%}$)-b-polyHA linear arms showed strong carbonyl stretching vibration of the polyNIPAAm amide groups at 1641 cm^{-1} and a strong carbonyl stretching band at 1731 cm^{-1} corresponding to the ester groups of both HA and EGDMA units. Furthermore, stretching vibration of aromatic C-H at 3074 cm^{-1} and at 3295 cm^{-1} (O-H) demonstrates the presence of the 4MBA units (Figure 3(a)). In the case of the ^{1}H-NMR spectrum, the characteristic signal at 4 ppm corresponds to the methine proton on the isopropyl groups present in polyNIPAAm (signal d) and to the methylene protons attached to the HA and EGDMA ester groups (signals l and q, q'), respectively, whereas in the case of the 4MBA units in the linear arms, the chemical shifts of aromatic protons (signals h and i) are observed at 7.2 and 8 ppm, respectively;

other signals are superimposed and were assigned as shown in that spectrum (Figure 3(b)).

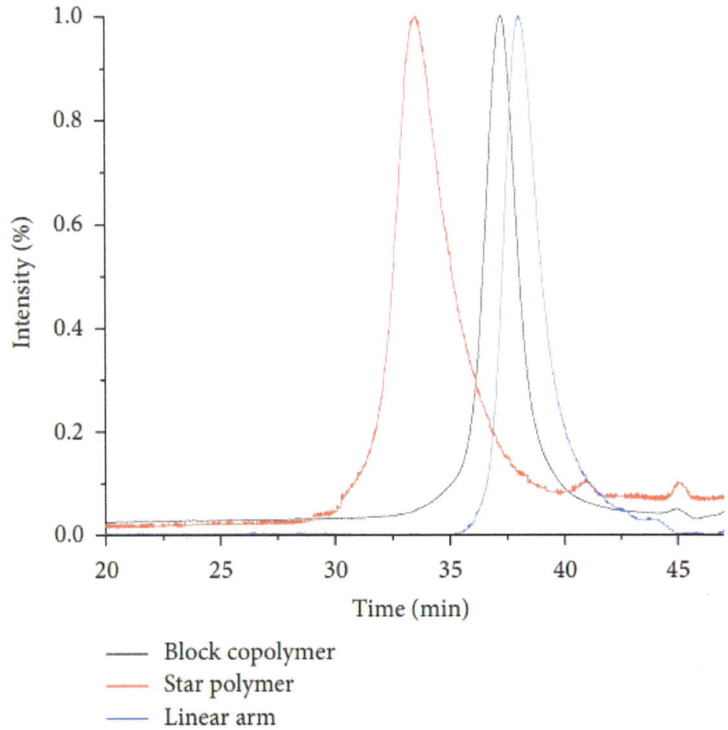

Figure 2: GPC traces of poly(NIPAAm-co-4MBA$_{5\%}$) (blue), poly(NIPAAm-co-4MBA$_{5\%}$)-b-polyHA (black), and [poly(NIPAAm-co-4MBA$_{5\%}$)-b-polyHA]$_1$-EGDMA$_{10}$ (red).

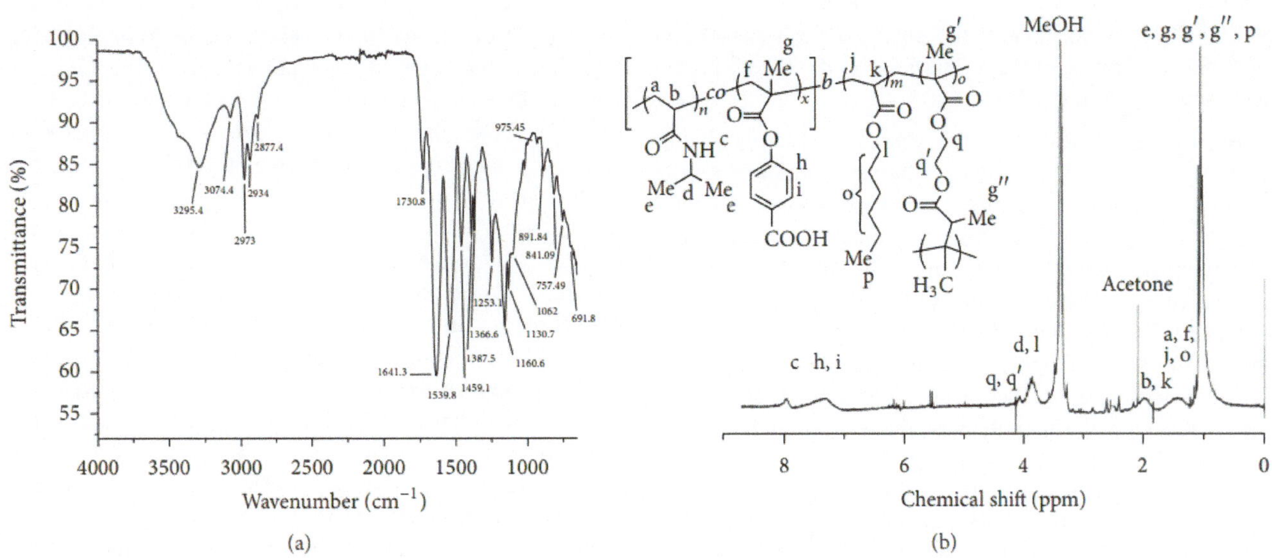

Figure 3: Spectra of star polymer [poly(NIPAAm-co-4MBA)-b-polyHA]$_1$EGDMA10: (a) FT-IR spectrum, (b) 1H-NMR spectrum (MeOD).

The DLS technique was also used to characterize the materials. Figure 4(a) shows the hydrodynamic diameters (D_h) of poly(NIPAAm-co-4MBA$_{5\%}$), poly(NIPAAm-co-4MBA$_{5\%}$)-b-polyHA, linear arms, and star polymer [poly(NIPAAm-co-4MBA$_{5\%}$)-b-polyHA]$_1$-EGDMA$_{10}$, all at 20°C. As can be seen, the size of the materials increases in each synthetic step. Furthermore, the size increases from polymer arm (~10 nm) to star polymer (~100 nm) suggesting that the star polymer is not a simple star polymer with a congested core.

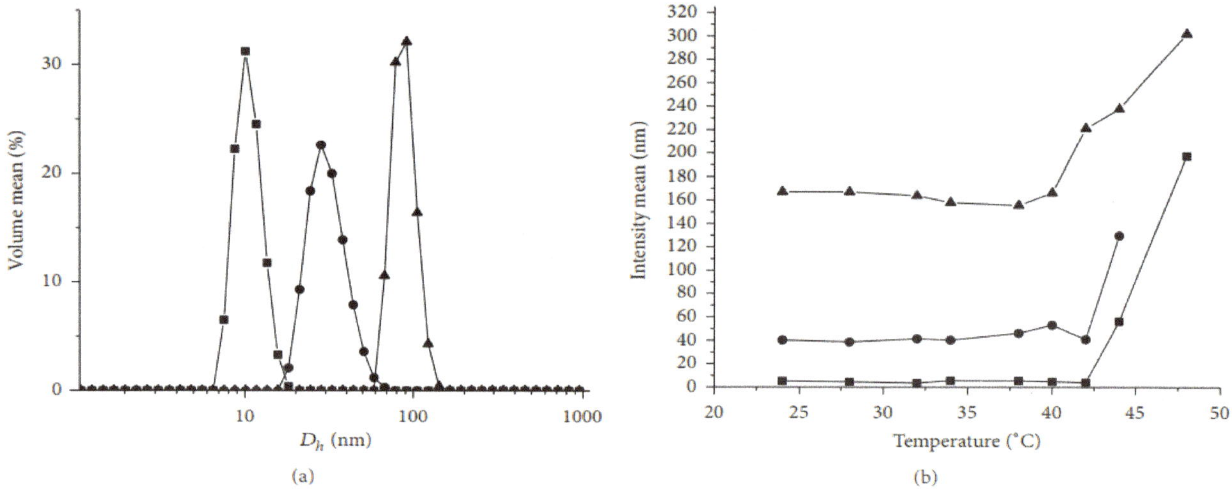

Figure 4: DLS results in PBS at pH 7.4: (a) hydrodynamic diameters at 20°C, (b) light scattering intensity as a function of temperature. Symbols disclosure: poly(NIPAAm-co-$4MBA_{5\%}$), M_n = 22,000 g/mol (■); poly(NIPAAm-co-$4MBA_{5\%}$)-b-polyHA, M_n = 28,570 g/mol (●); and star polymer [poly(NIPAAm-co-$4MBA_{5\%}$)-b-polyHA]$_1$-$EGDMA_{10}$, = 100,700 g/mol (▲).

The core cross-linked star architecture may contain a loose core as described in [12]. Figure 4(b) shows the light scattering intensity at different temperatures for the materials at pH of 7.4. At a given temperature, an increase in the intensity is observed due to aggregation of the particles. We consider this point as the LCST of the materials. As can be observed, the LCST is the same for the copolymer and the block copolymer, whereas in the case of the star polymer it is shifted to lower temperatures. The LCST is originated from the poly(NIPAAm-co-4MBA) copolymer, first block of the star arms, and modified slightly by cross-linking with EGDMA.

To further investigate the obtained star polymer structures, D_h was determined in buffer solutions; results are shown in Table 4. The observed hydrodynamic diameter depends on the size of the arms and on the arms-to-cross-linker ratio. The larger the arms, the longer the star diameter, in agreement with the molecular weight of the stars; but also the smaller the cross-linker-to-arms ratio, the longer the star diameter. Lower cross-linker content in the star synthesis may result in a less congested core which may lead to star/star coupling reactions during polymerization, while higher cross-linker content in the synthesis may result in a more congested core hindering star-to-star coupling during the synthesis. In the case of the star polymers containing acid units, that is, the star polymer prepared using poly(NIPAAm-co-$4MBA_{5\%}$)-b-polyHA (M_n = 28,570 g/mol) arms, the diameter (D_h) depends also on the ionization of those acid units. In PBS, at pH 7.4, the tendency of contraction of hydrophobic polyHA units is counterbalanced by the expansion of arms due to ionization of 4MBA units. It was observed that D_h value of star polymer in PBS at 20°C is smaller at pH 6 than at pH 7.4. This might be directly related to the fact that at this pH value the 4MBA units are not completely ionized [49], contributing, together with polyHA units, to the increase in the hydrophobic balance resulting in shrinkage of the star polymer arms. The star polymer reported in entry 3 (Table 4) also shows a smaller diameter at pH 6. Therefore, the size of star polymers under certain solvent conditions, either organic or aqueous at certain pH value, results from balance of different contributions from the units present in the star polymer. The effect of pH on the LCST values of the prepared star polymers is also shown in Table 4. In the case of the star polymer containing polyNIPAAm-b-polyHA arms in PBS at pH 7.4, in the absence of 4MBA units (entries 1 and 2), the LCST values are in the range from 30 to 32°C, which are typical values for the LCST of polyNIPAAm, whereas in the case of star polymers containing [poly(NIPAAm-co-4MBA)-b-polyHA] arms (entries 3 and 4), the observed LCST values were in the range from 40 to 46°C at pH of 7.4, depending on the 4MBA content in the arms of the star. In Figure 5, the effect of pH on the LCST values of star polymers is shown.

Table 4: Hydrodynamic diameters and LCST values of star polymers, in PBS at pH 6 and in PBS at pH 7.4

Entries	Star polymer	PBS, pH 6		LCST, pH 6	PBS, pH 7.4		LCST, pH 7.4
		D_h (nm)	*pdi*	Temp. (°C)	D_h (nm)	*pdi*	Temp. (°C)
1	[PolyNIPAAm-*b*-polyHA][a]$_1$-$EGDMA_{20}$	—	—	—	65.38 ± 4.4	0.397 ± $7e^{-3}$	32
2	[PolyNIPAAm-*b*-polyHA][b]$_1$-$EGDMA_{10}$	—	—	30	108.8 ± 3.6	0.256 ± 0.02	30
3	[Poly(NIPAAm-*co*-$4MBA_{9\%}$)-*b*-polyHA][c]$_1$-$EGDMA_{20}$	62.74 ± 0.62	0.257 ± $1e^{-3}$	36	94.68 ± 1.5	0.187 ± $8e^{-3}$	46
4	[Poly(NIPAAm-*co*-$4MBA_{5\%}$)-*b*-polyHA][d]$_1$-$EGDMA_{10}$	112.4 ± 4.8	0.326 ± 0.03	36	171.03 ± 0.6	0.302 ± 0.02	40

[a]Macro-CTA of M_n = 12,780 g/mol. [b]Macro-CTA of M_n = 32,070 g/mol. [c]Macro-CTA of M_n = 21,730 g/mol. [d]Macro-CTA of M_n = 28,570 g/mol.

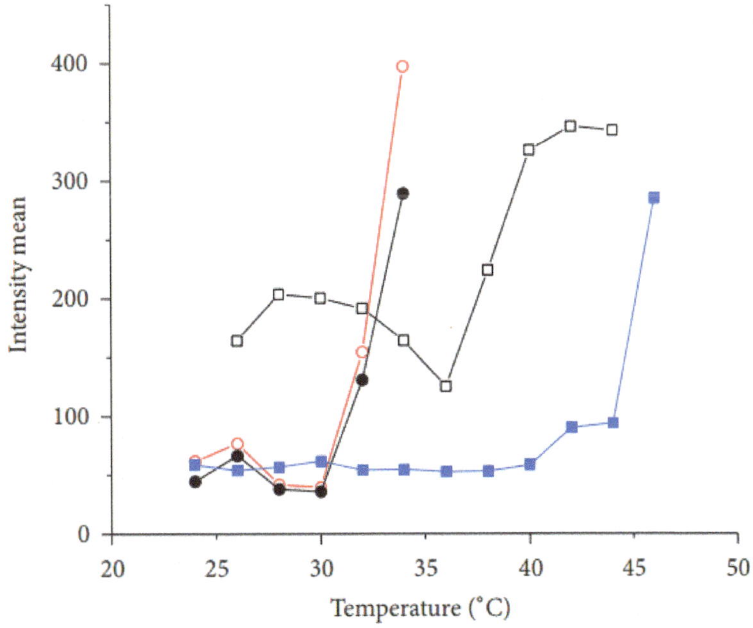

Figure 5: LCST of star polymers: star polymer [(PNIPAAm-b-polyHA)$_1$-EGDMA$_{10}$] in PBS at pH 7.4 (●) and in PBS at pH 6 (○); star polymer [poly(NIPAAm-co-4MBA)-b-polyHA]$_1$-EGDMA$_{10}$ in PBS at pH 6 (□) and in PBS at pH 7.4 (■).

The increase in the LCST at pH 7.4 can be attributed to the presence of the acid comonomer 4MBA, taking into consideration the fact that its pK_a value is 5.3 [49]. As a result, at pH 7.4, the majority of monomer units are ionized increasing the interaction with water and hence requiring a higher temperature for chain collapse to occur. The results of the LCST analysis in PBS at pH 6 are also shown in Table 4. In the case of star polymers containing [poly(NIPAAm-co-4-MBA)-b-polyHA] arms (entries 3 and 4), the observed LCST values are 36°C which are lower than the LCST at pH 7.4 (see also Figure 5).

This is attributed to the decrease in the ionization degree of 4MBA at pH 6 reducing the interaction with water. The results imply that this star polymer responds to changes in pH and temperature and that its LCST values at different pH values are optimized for drug delivery applications.

Further evidence of the star polymers morphology is obtained from the AFM analysis. In Figure 6, we can observe the morphology of [poly(NIPAAm-co-4MBA$_{5\%}$)-b-polyHA]$_1$-EGDMA$_{10}$ and [polyNIPAAm-b-polyHA]$_1$-EGDMA$_{10}$ star polymers. The average value was 100.1 ±12.4 nm of diameter with morphology like a well-defined oval for the star containing the acid units (4MBA) and 96.27 ± 17.2 nm for the star without acid units. The sizes obtained by AFM are in good agreement with the DLS measurements (112.4 ± 4.8, at pH ~ 6 of aqueous dispersion, and 108.8 ± 3.6 nm, resp.); the slightly smaller sizes by AFM can be attributed to the drying process of the star polymers on the mica surface before measurement.

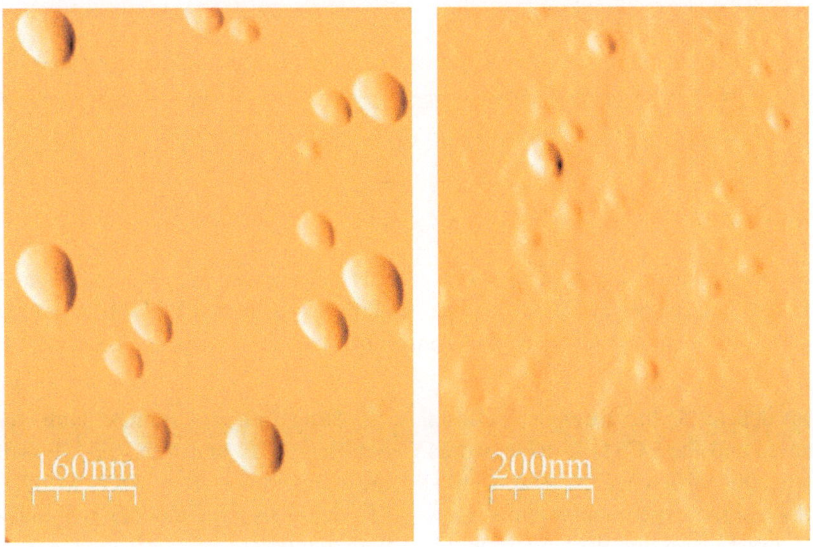

(a) (b)

Figure 6: AFM topography images of star polymers: (a) [poly(NIPAAm-co-4MBA$_{5\%}$)]$_1$-EGDMA$_{10}$, (b) [polyNIPAAm-b-polyHA]$_1$-EGDMA$_{10}$.

Drug Loading and Release Experiments

For the drug loading and release experiments, two star polymers were chosen with similar arm size and star architecture: the stars prepared with a 10:1 ratio of EGDMA:arms were selected since their arms are comparable in size: M_n is 32070 g/mol for polyNIPAAm-b-polyHA arms and 28570 g/mol for poly(NIPAAm-co-4MBA$_{5\%}$)-b-polyHA arms. As previously discussed, the LCST of the corresponding star polymers with polyNIPAAm-b-polyHA arms was 32°C, whilst the LCST of the [(polyNIPAAm-co-4MBA$_{5\%}$)-b-polyHA]$_1$-EGDMA$_{10}$ star was close to 40°C at pH of 7.4 and 36°C at pH of 6, which are suitable for a dual sensitivity drug delivery system. The selected star polymers were loaded with 5-FU by the dialysis method. The dialysis time plays an important role in the yield of loaded star polymers avoiding excess of nonbound drug. As the optimized dialysis time, a 4 h period was used to yield 30.5 w/w% loading with 25% efficiency for the acid containing star polymer and 25.3 w/w% loading and 16% efficiency for the star polymer without acid groups. Figure 7 shows a comparison of release profiles of 5-FU under different temperature and pH conditions. The free drug diffuses out of the dialysis bag in approximately 30 min (68% cumulative release) at 37°C in medium at pH of 7.4. In the case of the star that contains the acid monomer 4MBA, at pH 7.4 and 37°C (T < LCST), a burst of 20% release at the beginning of the process (20 min) was observed, which may be related to the fact that 5-FU molecules are entrapped and distributed in the arms of the star and also at the interface between core and shell. However, the drug release is halted to 25% even after 26 h, while at 42°C (T = LCST), the cumulative release increases to approximately 70% after 26 h. At pH 6 and 37°C (T > LCST at these conditions), cumulative release of about 84% in 26 h was observed. On the other hand, the star without acid units (4MBA) shows cumulative release at 26 h of about 90% at 37°C and 87% at 42°C (T > LCST), respectively. Both temperatures are much higher than the LCST of that star. The effect that the pH has on the release rate is clearly visible in the case of the star polymer containing 4MBA units.

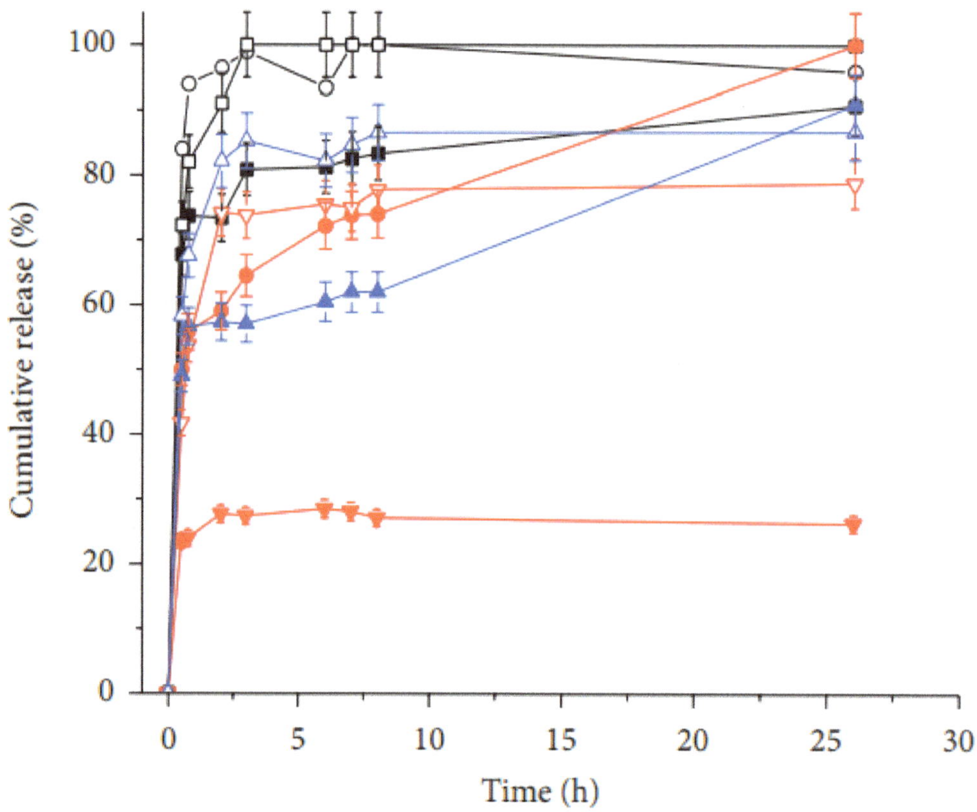

Figure 7: Release of 5-fluorouracil from different star polymers at constant temperatures: free 5-fluorouracil in PBS at pH 6 at 37°C (○) and at pH 7.4 at 37°C (■) and at 42°C (□); star polymer [poly(NIPAAm-co-4MBA)-b-polyHA]$_1$-EGDMA$_{10}$ in PBS at pH 6 at 37°C (●) and at pH 7.4 at 37°C (▼) and at 42°C (▽); star polymer [polyNIPAAm-b-polyHA]$_1$-EGDMA$_{10}$ in PBS at pH 7.4 at 37°C (▲) and 42°C (Δ).

The kinetics of release depend on the difference between the experimental temperature and the LCST; in other words, they depend on how large this difference is. The slowest release rate was obtained when the experimental temperature was lower than the LCST. In this case, the PHA hydrophobic shell over the core of the star polymer containing 4MBA units retained 75% of the 5-FU loaded. In the case of the star without 4MBA units, the release temperature was 5°C higher than the LCST; the star polymer tends to precipitate generating release channels for 5-FU.

Drug release studies with temperature jump cycles between 37°C and 42°C are shown in Figure 8. Increasing the temperature from 37°C to 42°C results in an increase of release rate for the star without acid units (T > LCST at both temperatures) indicating a temperature effect on drug diffusivity. However, the temperature release control is more efficient for the polymeric star containing the acid monomers when it goes from T < LCST at 37°C and pH 7.4 to T = LCST at 42°C and the same pH.

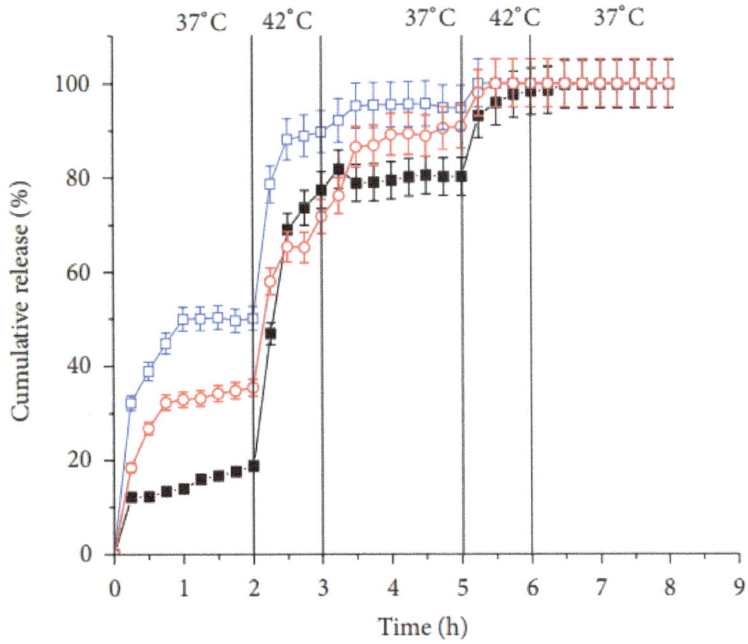

Figure 8: Release of 5-fluorouracil from star polymer {[poly(NIPAAm-*co*-4MBA)-*b*-polyHA]$_1$-EGDMA$_{10}$} in PBS at pH 7.4 (■) and PBS at pH 6 (□) and from star polymer [polyNIPAAm-b-polyHA]$_1$-EGDMA$_{10}$ in PBS at pH 7.4 (○) with temperature jump cycles from 37 to 42°C.

In this experiment, it was also observed that the star polymer with arms containing acid units (4MBA) acts as a nanopump since the temperature cycles increased the rate of release. Altogether, 100% drug release was attained after 6.5 h of applying temperature cycles, whilst at a constant temperature, the cumulative release was 70% after 26 h at 42°C (pH 7.4).

At pH of 6, faster release than at pH of 7.4 was observed, both at 37°C (Figure 8). A jump in temperature to 42°C produces even faster release at both pH values, delivering the payload in the first temperature cycle almost completely at pH 6 but not at pH 7.4. These results are very encouraging in regard to the potential application as release carriers of the temperature and pH sensitive star polymers that were prepared, considering that release occurs at conditions found in some pathologies such as in tumors. Based on these results, the star polymer containing acid units, namely, [poly(NIPAAm-co-4MBA$_{5\%}$)-b-polyHA]$_1$-EGDMA$_{10}$ (M_n= 100,700 g/mol), represents an advanced drug carrier exhibiting tunable release kinetics for 5-FU that can be modified by increasing the temperature above the normal body temperature at physiological pH, either by a pathology or by external induction, and/or at lower pH than physiological pH in the site of delivery at normal body temperature, as found in tumors or in endosomes.

CONCLUSIONS

It was shown that it is possible to synthetize star polymers with well-defined arms in specific composition by combining the RAFT technique with the use of cross-linkers. The ratio of cross-linker to macro-CTA (arms) is critical in order to obtain the desired product. The prepared star polymers containing (polyNIPAAm-b-polyHA) arms in different sizes and composition with a cross-linked core showed the same LCST as linear polyNIPAAm (30–32°C). The RAFT process allows for the incorporation of amphiphilic weak acids into the polyNIPAAm block producing pH and temperature sensitive materials. The LCST of the star polymers containing [poly(NIPAAm-co-4MBA)-b-polyHA] arms was tailored to values in the range of 40 to 46°C, at pH of 7.4, depending on 4MBA content, and about 36°C at pH of 6. The polymeric stars can be loaded with 5-FU up to 30 w/w%. In vitro release experiments showed that the star polymer materials release 5-FU at higher rates when the temperature is above their LCST. The loaded star polymers behave as nanopumps with off/on release if subjected to temperature cycles around the LCST. From these findings, we conclude that the temperature sensitive star polymers with an adjusted LCST, containing 4MBA, are promising intelligent drug carriers with the advantages of temperature and pH adjustable release kinetics.

ACKNOWLEDGMENTS

This work was supported by the National Council of Science and Technology of Mexico (CONACYT) Grants CB2010-C01-157173, CB2012-C01-178709, and INFR2014-C01-224284. Elizabeth Sánchez-Bustos thanks CONACYT for PHD fellowship (no. 208593). Furthermore, the technical support in NMR analysis by I. A. Rivero and A. Ochoa-Terán and in AFM analysis by P. Navarro is gratefully acknowledged. The authors are indebted to L. A. Licea-Higgs for English corrections in the paper.

REFERENCES

1. D. Kuckling and A. Wycisk, "Stimuli-responsive star polymers," Journal of Polymer Science, Part A: Polymer Chemistry, vol. 51, no. 14, pp. 2980–2994, 2013.
2. M. A. Cohen-Stuart, W. T. S. Huck, J. Genzer et al., "Emerging applications of stimuli-responsive polymer

materials," Nature Materials, vol. 9, no. 2, pp. 101–113, 2010.

3. H. Gao and K. Matyjaszewski, "Synthesis of functional polymers with controlled architecture by CRP of monomers in the presence of cross-linkers: from stars to gels," Progress in Polymer Science, vol. 34, no. 4, pp. 317–350, 2009.

4. N. Hadjichristidis, M. Pitsikalis, and H. Iatrou, "Polymers with star-related structures," inMacromolecular Engineering: Precise Synthesis, Materials Properties, Applications, K. Matyjaszewski, Y. Gnanou, and L. Leibler, Eds., chapter 6, pp. 909–971, Wiley-VCH, Weinheim, Germany, 2007.

5. A. Blencowe, J. F. Tan, T. K. Goh, and G. G. Qiao, "Core cross-linked star polymers via controlled radical polymerisation," Polymer, vol. 50, no. 1, pp. 5–32, 2009.

6. A. Vora, K. Singh, and D. C. Webster, "A new approach to 3-miktoarm star polymers using a combination of reversible addition-fragmentation chain transfer (RAFT) and ring opening polymerization (ROP) via 'Click' chemistry," Polymer, vol. 50, no. 13, pp. 2768–2774, 2009.

7. J.-G. Zilliox, D. Decker, and P. Rempp, "Herstellung von 'Sternpolymeren' auf anionischem Wege,"Angewandte Chemie, vol. 78, no. 11, p. 614, 1966.

8. K. Matyjaszewski and J. Xia, "Atom transfer radical polymerization," Chemical Reviews, vol. 101, no. 9, pp. 2921–2990, 2001.

9. C. J. Hawker, A. W. Bosman, and E. Harth, "New polymer synthesis by nitroxide mediated living radical polymerizations," Chemical Reviews, vol. 101, no. 12, pp. 3661–3688, 2001.

10. R. T. A. Mayadunne, J. Jeffery, G. Moad, and E. Rizzardo, "Living free radical polymerization with reversible addition-fragmentation chain transfer (RAFT polymerization): approaches to star polymers,"Macromolecules, vol. 36, no. 5, pp. 1505–1513, 2003.

11. A. B. Lowe and C. L. McCormick, "Reversible addition-fragmentation chain transfer (RAFT) radical polymerization and the synthesis of water-soluble (co) polymers under homogeneous conditions in organic and aqueous media," Progress in Polymer Science, vol. 32, no. 3, pp. 283–351, 2007.

12. J. Alvarez-Sánchez, A. Licea-Claveríe, J. M. Cornejo-Bravo, and C. W. Frank, "Star polymers with random number of temperature sensitive arms and crosslinked poly(EGDMA)-core and their application to drug delivery," Reactive and Functional Polymers, vol. 71, no. 11, pp. 1077–1088, 2011.

13. L. A. Picos-Corrales, A. Licea-Claverie, J. M. Cornejo-Bravo, K.-F. Arndt, and S. Schwarz, "Well-definedN-isopropylacrylamide dual-sensitive copolymers with LCST ≈38°C in different architectures: linear, block and star polymers," Macromolecular Chemistry and Physics, vol. 213, no. 3, pp. 301–314, 2012.

14. G. Lapienis, "Star-shaped polymers having PEO arms," Progress in Polymer Science, vol. 34, no. 9, pp. 852–892, 2009.

15. J. Li, J. Ren, Y. Cao, and W. Yuan, "Synthesis of biodegradable pentaarmed star-block copolymers via an asymmetric BIS-TRIS core by combination of ROP and RAFT: from star architectures to double responsive micelles," Polymer, vol. 51, no. 6, pp. 1301–1310, 2010.

16. X. Li, Y. Qian, T. Liu et al., "Amphiphilic multiarm star block copolymer-based multifunctional unimolecular micelles for cancer targeted drug delivery and MR imaging," Biomaterials, vol. 32, no. 27, pp. 6595–6605, 2011.

17. S. Aryal, M. Prabaharan, S. Pilla, and S. Gong, "Biodegradable and biocompatible multi-arm star amphiphilic block copolymer as a carrier for hydrophobic drug delivery," International Journal of Biological Macromolecules, vol. 44, no. 4, pp. 346–352, 2009.

18. M. L. Adams, A. Lavasanifar, and G. S. Kwon, "Amphiphilic block copolymers for drug delivery," Journal of Pharmaceutical Sciences, vol. 92, no. 7, pp. 1343–1355, 2003.

19. E. He, C. Y. Yue, and K. C. Tam, "Association behavior of star-shaped pH-responsive block copolymer: four-arm poly(ethylene oxide)-b-poly(methacrylic acid) in aqueous medium," Langmuir, vol. 25, no. 9, pp. 4892–4899, 2009.

20. A. Kowalczuk, B. Trzebicka, S. Rangelov, M. Smet, and A. Dworak, "Star macromolecules with hyperbranched poly(arylene oxindole) cores and polyacid arms: synthesis and solution behavior," Journal of Polymer Science Part A: Polymer Chemistry, vol. 49, no. 23, pp. 5074–5086, 2011.

21. Q. Bian, Y. Xiao, and M. Lang, "Thermoresponsive biotinylated star amphiphilic block copolymer: synthesis, self-assembly, and specific target recognition," Polymer, vol. 53, no. 8, pp. 1684–1693, 2012.

22. H. Kobayashi, R. Watanabe, and P. L. Choyke, "Improving conventional enhanced permeability and retention (EPR) effects; what is the appropriate target?" Theranostics, vol. 4, no. 1, pp. 81–89, 2014.

23. K. Engin, D. B. Leeper, J. R. Cater, A. J. Thistlethwaite, L. Tupchong, and J. D. Mcfarlane, "Extracellular ph distribution in human tumours," International Journal of Hyperthermia, vol. 11, no. 2, pp. 211–216, 1995.

24. I. F. Tannock and D. Rotin, "Acid pH in tumors and its potential for therapeutic exploitation," Cancer Research, vol. 49, no. 16, pp. 4373–4384, 1989.

25. J. C. Garbern, E. Minami, P. S. Stayton, and C. E. Murry, "Delivery of basic fibroblast growth factor with a pH-responsive, injectable hydrogel to improve angiogenesis in infarcted myocardium," Biomaterials, vol. 32, no. 9, pp. 2407–2416, 2011.

26. A. M. Ponce, Z. Vujaskovic, F. Yuan, D. Needham, and M. W. Dewhirst, "Hyperthermia mediated liposomal drug delivery," International Journal of Hyperthermia, vol. 22, no. 3, pp. 205–213, 2006.

27. W. Xiong, W. Wang, Y. Wang et al., "Dual temperature/pH-sensitive drug delivery of poly(N-isopropylacrylamide-co-acrylic acid) nanogels conjugated with doxorubicin for potential application in tumor hyperthermia

therapy," Colloids and Surfaces B: Biointerfaces, vol. 84, no. 2, pp. 447–453, 2011.

28. A. Chilkotia, M. R. Drehera, D. E. Meyera, and D. Raucherb, "Targeted drug delivery by thermally responsive polymers," Advanced Drug Delivery Reviews, vol. 54, no. 5, pp. 613–630, 2002.

29. J. E. Chung, M. Yokoyama, M. Yamato, T. Aoyagi, Y. Sakurai, and T. Okano, "Thermo-responsive drug delivery from polymeric micelles constructed using block copolymers of poly(N-isopropylacrylamide) and poly(butylmethacrylate)," Journal of Controlled Release, vol. 62, no. 1-2, pp. 115–127, 1999.

30. L. Zhang, G. Rui, Y. Mi, J. Xiqun, and L. Baorui, "Thermo and pH dual-responsive nanoparticles for anti-cancer drug delivery," Advanced Materials, vol. 19, no. 19, pp. 2988–2992, 2007.

31. F. M. Winnik, A. R. Davidson, G. K. Hamer, and H. Kitano, "Amphiphilic poly(N-isopropylacrylamides) prepared by using a lipophilic radical initiator: synthesis and solution properties in water,"Macromolecules, vol. 25, no. 7, pp. 1876–1880, 1992.

32. M. A. Ward and T. K. Georgiou, "Thermoresponsive polymers for biomedical applications," Polymers, vol. 3, no. 3, pp. 1215–1242, 2011.

33. L. Klouda and A. G. Mikos, "Thermoresponsive hydrogels in biomedical applications," European Journal of Pharmaceutics and Biopharmaceutics, vol. 68, no. 1, pp. 34–45, 2008.

34. A. Serrano-Medina, J. M. Cornejo-Bravo, and A. Licea-Claverie, "Synthesis of pH and temperature sensitive, core-shell nano/microgels, by one pot, soap-free emulsion polymerization," Journal of Colloid and Interface Science, vol. 369, no. 1, pp. 82–90, 2012.

35. C. Obeso-Vera, J. M. Cornejo-Bravo, A. Serrano-Medina, and A. Licea-Claverie, "Effect of crosslinkers on size and temperature sensitivity of poly(N-isopropylacrylamide) microgels," Polymer Bulletin, vol. 70, no. 2, pp. 653–664, 2013.

36. W.-Q. Chen, H. Wei, S.-L. Li et al., "Fabrication of star-shaped, thermo-sensitive poly(N-isopropylacrylamide)-cholic acid-poly(ε-caprolactone) copolymers and their self-assembled micelles as drug carriers," Polymer, vol. 49, no. 18, pp. 3965–3972, 2008.

37. H. Wei, W.-Q. Chen, C. Chang et al., "Synthesis of star block, thermosensitive poly(l-lactide)-star block-poly(N-isopropylacrylamide-co-N-hydroxymethylacrylamide) copolymers and their self-assembled micelles for controlled release," The Journal of Physical Chemistry C, vol. 112, no. 8, pp. 2888–2894, 2008.

38. S. J. T. Rezaei, M. R. Nabid, H. Niknejad, and A. A. Entezami, "Folate-decorated thermoresponsive micelles based on star-shaped amphiphilic block copolymers for efficient intracellular release of anticancer drugs," International Journal of Pharmaceutics, vol. 437, no. 1-2, pp. 70–79, 2012.

39. K. Miao, H. Liu, and Y. Zhao, "Thermo, pH and reduction responsive coaggregates comprising AB_2C_2star terpolymers for multi-triggered release of doxorubicin," Polymer Chemistry, vol. 5, no. 10, pp. 3335–3345, 2014.

40. L. Saleh-Ghadimi, M. Fathi, and A. A. Entezami, "Heteroarm star-shaped poly (N-isopropylacryamide-co-itaconic acid) copolymer prepared by glucose core as ATRP initiator," International Journal of Polymeric Materials and Polymeric Biomaterials, vol. 63, no. 5, pp. 246–255, 2014.

41. F. Xu, J.-W. Xu, B.-X. Zhang, and Y.-L. Luo, "Self-assembly micelles from novel tri-armed star C3-(PS-b-PNIPAM) block copolymers for anticancer drug release," AIChE Journal, vol. 61, no. 1, pp. 35–45, 2015.

42. S.-P. Rwei, Y.-Y. Chuang, T.-F. Way, W.-Y. Chiang, and S.-P. Hsu, "Preparation of thermo- and pH-responsive star copolymers via ATRP and its use in drug release application," Colloid and Polymer Science, vol. 293, no. 2, pp. 493–503, 2015.

43. Q. Chen, X. Cao, Y. Xu, and Z. An, "Emerging synthetic strategies for core cross-linked star (CCS) polymers and applications as interfacial stabilizers: bridging linear polymers and nanoparticles,"Macromolecular Rapid Communications, vol. 34, no. 19, pp. 1507–1517, 2013.

44. J. Ferreira, J. Syrett, M. Whittaker, D. Haddleton, T. P. Davis, and C. Boyer, "Optimizing the generation of narrow polydispersity 'arm-first' star polymers made using RAFT polymerization," Polymer Chemistry, vol. 2, no. 8, pp. 1671–1677, 2011.

45. J. Liu, H. Duong, M. R. Whittaker, T. P. Davis, and C. Boyer, "Synthesis of functional core, star polymers via RAFT polymerization for drug delivery applications," Macromolecular Rapid Communications, vol. 33, no. 9, pp. 760–766, 2012.

46. R. Salgado-Rodríguez, A. Licea-Claverie, and K.-F. Arndt, "Random copolymers of N-isopropylacrylamide and methacrylic acid monomers with hydrophobic spacers: pH-tunable temperature sensitive materials," European Polymer Journal, vol. 40, no. 8, pp. 1931–1946, 2004.

47. J. Brandrup, E. H. Immergut, and E. A. Grulke, Eds., Polymer Handbook, John Wiley & Sons, New York, NY, USA, 4th edition, 1999.

48. K.-F. Arndt and G. Müller, Polymer Charakterisierung, Carl Hanser, München, Germany, 1996.

49. A. Licea-Claverie, E. Rogel-Hernández, J. A. Lopéz-Sanchez, L. A. Castillo-Arambula, J. M. Cornejo-Bravo, and K. F. Arndt, "A facile synthesis route for carboxyaryl-methacrylates; a way to obtain aromatic polyelectrolytes," Designed Monomers and Polymers, vol. 6, no. 1, pp. 67–80, 2003.

CHAPTER 12

PROTEIN-POLYMER NANO-MACHINES. TOWARDS SYNTHETIC CONTROL OF BIOLOGICAL PROCESSES

Sivanand S Pennadam[2], Keith Firman[1], Cameron Alexander[2] and Dariusz C Górecki[2]

[1]School of Biological Sciences, Institute of Biomedical and Biomolecular Sciences, University of Portsmouth, Portsmouth PO1 2DT, UK

[2]School of Pharmacy and Biomedical Sciences, Institute of Biomedical and Biomolecular Sciences, University of Portsmouth, Portsmouth, PO1 2DT UK

ABSTRACT

The exploitation of nature's machinery at length scales below the dimensions of a cell is an exciting challenge for biologists, chemists and physicists, while advances in our understanding of these biological motifs are now providing an opportunity to develop real single molecule devices for technological applications. Single molecule studies are already well advanced and biological molecular motors are being used to guide the design of nano-scale machines. However, controlling the specific functions of these devices in biological systems under changing conditions is difficult. In this review we describe the principles underlying the development of a molecular motor with numerous potential applications in nanotechnology and the use of specific synthetic polymers as prototypic molecular switches for control of the motor function. The molecular motor is a derivative of a TypeI Restriction-Modification (R-M) enzyme and the synthetic polymer is drawn from the class of materials that exhibit a temperature-dependent phase transition.

The potential exploitation of single molecules as functional devices has been heralded as the dawn of new era in biotechnology and medicine. It is not surprising, therefore, that the efforts of numerous multidisciplinary teams [1, 2]. have been focused in attempts to develop these systems. as machines capable of functioning at the low sub-micron and nanometre length-scales [3]. However, one of the obstacles for the practical application of single molecule devices is the lack of functional control methods in biological media, under changing conditions. In this review we describe the conceptual basis for a molecular motor (a derivative of a TypeI Restriction-Modification enzyme) with numerous potential applications in nanotechnology and the use of specific synthetic polymers as prototypic molecular switches for controlling the motor function [4].

TYPE I RESTRICTION-MODIFICATION ENZYMES

Type I R-M enzymes are multifunctional, multisubunit enzymes that provide bacteria with protection against infection by DNA-based bacteriophage [5] They accomplish this through a complex restriction activity that cuts the DNA at random locations, which can be extremely distal (>20 kbp) from the enzyme's recognition sequence. In fact, the enzyme is capable of two opposing functions (restriction and modification), which are controlled enzymatically through an allosteric effector (ATP) and temporally through the assembly of the holoenzyme. In addition, the R-M enzyme has a powerful ATPase activity, which is associated with DNA translocation prior to cleavage; it is this translocation process that leads to random cleavage sites. Therefore, these enzymes are unusual

Citation: Sivanand S Pennadam, Keith Firman, Cameron Alexander and Dariusz C Górecki, Protein-polymer nano-machines. Towards synthetic control of biological processes, DOI: 10.1186/1477-3155-2-8.

molecular motors that bind specifically to DNA and then move the rest of the DNA through this bound complex (Fig 1).

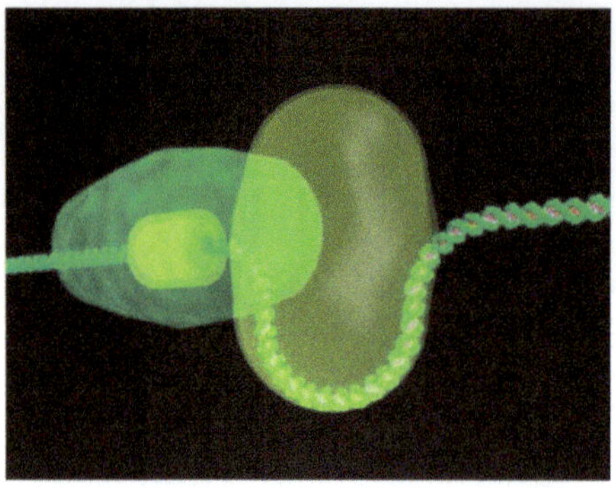

Figure 1: DNA Translocation by TypeI Restriction-Modification enzyme. The yellow block represents the recognition sequence for the enzyme. The enzyme binds at this site and upon addition of ATP, DNA translocation begins. During translocation, an expanding loop is produced.

Type I R-M enzymes fall into families based on complementation grouping, protein sequence similarities, gene order and related biochemical characteristics [6, 7, 8]. Within one sub-type (the IC family) there are three well-described members, including EcoR124I, which is the focus of our interest. This enzyme recognises the DNA sequence GAAnnnnnnRTCG [9] and is comprised of three subunits (HsdR,M,S) in a stoichiometric ratio of R_2M_2S [10, 11], (Fig 2). However, Janscák *et al.* also showed that the *Eco* R124I R-M holoenzyme exists in equilibrium with a sub-assembly complex of stoichiometry R_1M_2S [11] which is unable to cleave DNA, but retains the ATPase and motor activity [12]. The HsdS subunit is responsible for DNA specificity; HsdM is required for DNA methylation (modification activity) and together they can produce an independent DNA methyltransferase (M_2S) [13, 14]. HsdR, along with the core MTase is absolutely required for DNA cleavage (restriction activity) and is also responsible for ATP-binding and subsequent DNA translocation. Therefore, the HsdR subunit is the motor subunit of the enzyme and this subunit is associated with helicase activity [15, 16,17, 18]. However, the precise mechanism of DNA translocation is uncertain and the true nature of the motor function has yet to be fully determined but a number of important functional units – nuclease, helicase and assembly domains have been identified within the HsdR subunit [19].

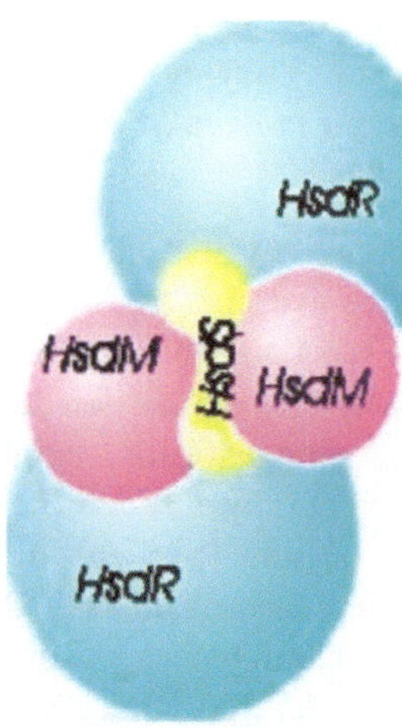

Figure 2: Schematic of the motor subunits. HsdS denotes the DNA binding subunit; HsdM – is the subunit responsible for DNA methylation and HsdR subunit, together with the core enzyme acts to restrict DNA.

A VERSATILE MOLECULAR MOTOR

The motor activity of Type I R-M enzymes is the mechanism through which random DNA cleavage is accomplished. Szczelkun *et al.* [20] showed that cleavage only occurs in a *cis* fashion indicating that the motor component of the HsdR subunit is able to ‹grasp› adjacent DNA and pull this DNA through the enzyme-DNA-bound complex. According to the Studier model [21] cleavage occurs when two translocating enzymes collide (Fig 3). However, highly efficient cleavage of circular DNA carrying only a single recognition sites for the enzyme suggests collision-based cleavage is not the whole story [20, 22].

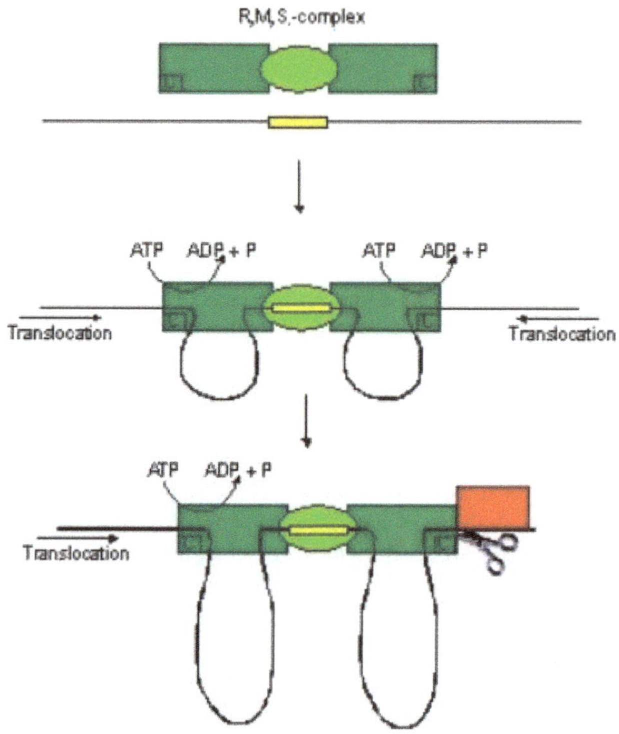

Figure 3: Mechanism of DNA cleavage. The enzyme subunits are represented by: green ellipse – M2S complex, green box – HsdR subunit (with ATPase and restrictase activities; C denoting cleavage site). The black line represents DNA with the yellow box denoting the recognition sequence. Arrow shows direction of DNA translocation. For more details see text.

DNA translocation has been assayed in bulk solution using protein-directed displacement of a DNA triplex and the kinetics of one-dimensional motion determined. The data shows processive DNA translocation followed by collision with the triplex and oligonucleotide displacement. A linear relationship between lag duration and inter-site distance gives a translocation velocity of 400 ± 32 bp/s at 20°C. Furthermore, this can only be explained by bi-directional translocation. An endonuclease with only one of the two HsdR subunits responsible for motion could still catalyse translocation. The reaction is less processive, but can 'reset' in either direction whenever the DNA is released (Fig 4).

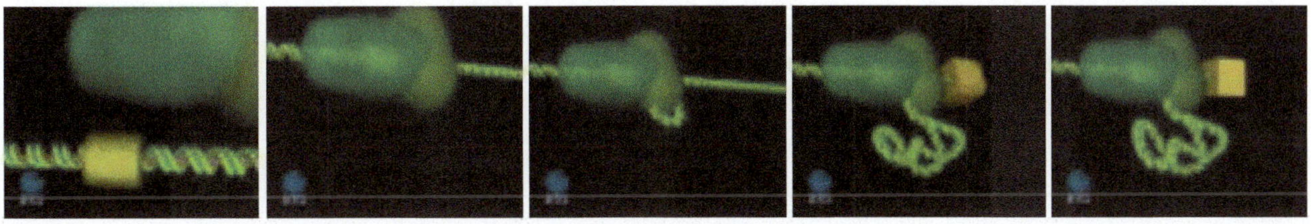

(a) (b) (c) (d) (e)

Figure 4: Motor activity of type I R-M Enzyme. (a) The yellow block represents the DNA-binding (recognition) site of the enzyme, which is represented by the green object approaching from the top of the diagram and about to dock onto the recognition sequence. (b) The motor is bound to the DNA at the recognition site and begins to attach to adjacent DNA sequences. (c) The motor begins to translocate the adjacent DNA sequences through the motor/DNA complex, which remains tightly bound to the recognition sequence. (d) Translocation produces an expanding loop of positively supercoiled DNA. The motor follows the helical thread of the DNA resulting in spinning of the DNA end (illustrated by the rotation of the yellow cube). (e) When translocation reaches the end of the linear DNA it stops, resets and then the process begins again.

As previously mentioned, the final step of the subunit assembly pathway of the Type I Restriction-Modification enzyme EcoR124I produces a weak endonuclease complex of stoichiometry $R_2M_2S_1$. We have produced a hybrid HsdR subunit combining elements of the HsdR subunits of the EcoR124I and EcoprrI [23, 24, 25] Type I Restriction-Modification enzymes. This subunit has been shown to assemble with the EcoR124I DNA methyltransferase (MTase) to produce an active complex with low-level restriction activity. We have also assembled a hybrid REase and the data obtained show that the hybrid endonuclease (REase) containing only HsdR(prrI) is an extremely weak complex, producing primarily R_1-complex. The availability of the hybrid REase produced from core MTase(R124I) and HsdR(prrI), which provides a stable R_1-complex, also gives a useful molecular motor that will not cleave the DNA that it translocates.

SUB-CELLULAR LOCALISATION OF R-M ENZYMES

As can be seen from the above, DNA cleavage by Type I restriction enzymes occurs by means of a very unusual, and highly energy-dependent, mechanism. Therefore, these enzymes are believed to be involved not only as a defence mechanism for the bacterial cell, but also in some types of specialised recombination system controlling the flow of genes between bacterial strains [26, 27]. A periplasmic location would be well adapted for the restriction activity of R-M enzymes, but recombination requires a cytoplasmic location. Restriction enzymes protect the cells by cutting foreign DNA and could be assumed to be located at the cell periphery. Using immunoblotting to analyse subcellular fractions, Holubova *et al.* [28] detected that the subunits of the R-M enzyme were predominantly in the spheroplast extract. The HsdR and HsdM subunits were found in the membrane fraction only when co-produced with HsdS and, therefore, part of a complex enzyme, either methylase or endonuclease. Further studies have shown that the R-M enzyme is bound to the membrane *via*the HsdS subunit and that for some enzymes this may involve DNA [29].

USES OF THE ECOR124I MOLECULAR MOTOR: POLYMER-PROTEIN CONJUGATES IN NANOBIOTECHNOLOGY

One of the major obstacles for the practical application of single molecule devices is the absence of control methods in biological media, where substrates or energy sources (such as ATP) are ubiquitous. Synthetic polymers offer a robust and highly flexible means by which devices based on single biological molecules can be controlled. They can also be used to link individual biomacromolecules to surfaces, package them or to control their specific functions, thus expanding the applicability of the natural molecules outside conventional biological environments.

Moreover, a number of synthetic polymers have been recently developed that can potentially perform nanoscale operations in a manner identical to natural and engineered biopolymers. A key property of these materials is 'smart' behaviour, especially the ability to undergo conformational or phase changes in response to variations in temperature and/or pH. Synthetic polymers with these properties are being developed for applications ranging from microfluidic device formation, [30] through to pulsatile drug release [31, 32, 33, 34], control of cell-surface interactions [35, 36, 37, 38, 39], as actuators [40] and, increasingly, as nanotechnology devices [41].

In the context of bio-nanotechnology we focus here on the uses of one particular subclass of smart materials, i.e. substituted polyacrylamides, but it should be noted that there are many more examples of synthetic polymers and engineered/modified biopolymers that exhibit responsive behaviour and new types and applications of smart materials are constantly being reported.

Poly(N-isopropylacrylamide) (PNIPAm) is the prototypical smart polymer and is both readily available and of well-understood properties [42]. PNIPAm undergoes a sharp coil-globule transition in water at 32 °C, being hydrophilic below this temperature and hydrophobic above it. This temperature (the Lower Critical Solution Temperature or LCST) corresponds to the region in the phase diagram at which the enthalpic contribution of water hydrogen-bonded to the polymer chain becomes less than the entropic gain of the system as a whole and thus is largely dependent on the hydrogen-bonding capabilities of the constituent monomer units (Fig 5). Accordingly, the LCST of a given polymer can in principle be "tuned" as desired by variation in hydrophilic or hydrophobic co-monomer content.

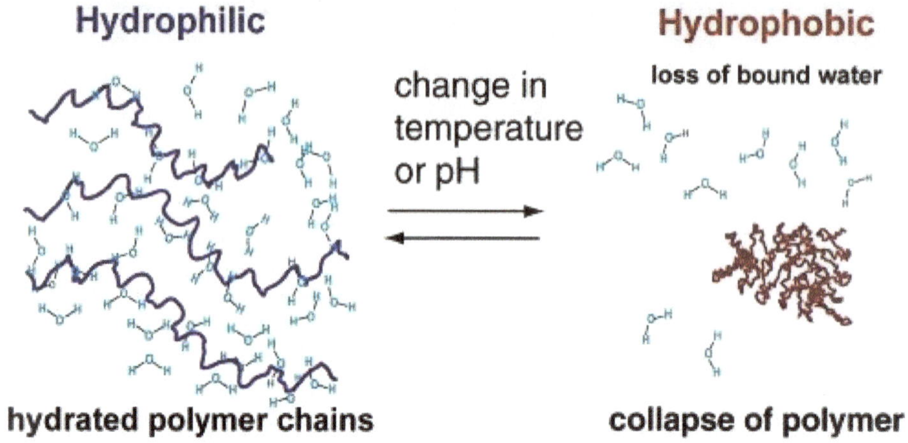

Figure 5: Inverse temperature solubility behavior of responsive polymers at the Lower Critical Solution Temperature (LCST). Left hand side shows hydrated polymer below LCST with entropic loss of water and chain collapse above LCST (right hand side).

Soluble PNIPAm-Biopolymer Conjugates

Covalent attachment of single or multiple responsive polymer chains to biopolymers offers the possibility of exerting control over their biological activity as, in theory at least, the properties of the resultant polymer-biopolymer conjugate should be a simple additive function of those of the individual components. This principle is now being widely exploited in pharmaceutical development, as covalent attachment of, for example, PEG chains to therapeutic proteins has been shown to stabilize the proteins without losing their biological function [43, 44, 45, 46, 47, 48]. Polymer-biopolymer conjugates can be prepared as monodisperse single units, or as self-assembling ensembles depending on the chemistries used for attaching the synthetic component and on the associative properties of the polymer and/or biopolymer. Furthermore, by altering the response stimulus of the synthetic polymer, and how and where it is attached to the biopolymer, the activity of the overall conjugate can be very closely regulated. These chimeric systems can thus be considered as true molecular-scale devices.

Pioneering work in this area has been carried out by Hoffman, Stayton and co-workers, who engineered a mutant of cytochrome b5 such that a single cysteine introduced *via* site-directed mutagenesis was accessible for reaction with maleimide end-functionalised PNIPAm [49]. Since the native cytochrome b5 does not contain any cysteine residues this substitution provided a unique attachment point for the polymer. The resultant polymer-protein conjugate displayed LCST behaviour and could be reversibly precipitated from solution by variation in temperature. This approach has proved to be very versatile and a large number of polymer-biopolymer conjugates have now been prepared, incorporating biological components as diverse as antibodies, protein A, streptavidin, proteases and hydrolases [50, 51, 50, 51]. The biological functions or activities of these conjugate systems were all similar to their native counterparts, but were switched on or off as a result of thermally induced polymer phase transitions. Of especial note have been the recent reports of a temperature and photochemically switchable endoglucanase, which displayed varying and opposite activities depending on whether temperature or UV/Vis illumination was used as the switch [52].

Controllable DNA Packaging and Compartmentalization Devices

We are currently developing responsive polymers as a switch to control the EcoR124I motor function and are investigating this polymer-motor conjugate as part of an active drug delivery system. We aim for the practical demonstration of a nano-scale DNA packaging/separation and delivery system uniting the optimal features of both natural and synthetic molecules. In essence, we assemble a supramolecular device containing the molecular motor capable of binding and directionally translocating DNA through an impermeable barrier. To control the process of translocation in biological systems, where a constant supply of ATP is present, we have added to the motor subunit of EcoR124I the thermoresponsive poly(N-isopropylacrylamide) (PNIPAm), which, through its coil-globule transition, acts as a temperature-dependent switch controlling motor activity.

PNIPAm copolymers with reactive end-groups are being attached to a preformed R subunit of the motor *via*coupling of a maleimide-tipped linker on the synthetic polymer terminus to a cysteine residue. This residue has been selected, as it is both accessible and located close to the active centre on the R subunit of the motor. The protein-polymer conjugates are stable to extensive purification and, when combined with M2S complex, the activity of this conjugate motor system is similar to the native counterpart, but can be switched on or off as a result of thermally induced polymer phase transitions [53, 54].

Thus the conjugation of the responsive polymer to the molecular motor generates a nano-scale, switchable device

(Fig 6), which can translocate DNA under one set of conditions (i.e. into a protective capsule or into a compartment). Conversely, in another environment (e.g. inside cells), in response to changed conditions (e.g. changed temperature, pH) the polymer switch will change its conformation, allowing ATP to power the motor, releasing DNA from capsules or compartments.

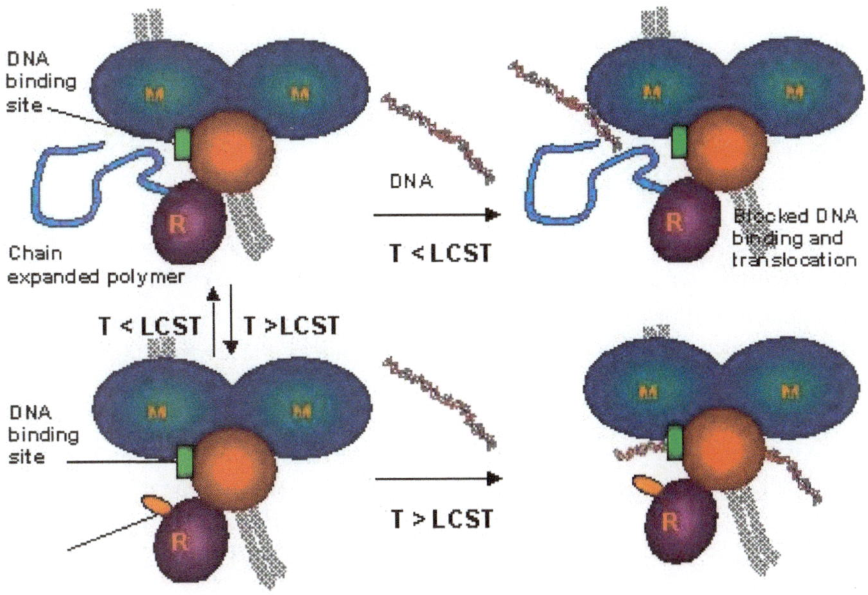

Figure 6: Schematic representation of the molecular motor function controlled by a thermoresponsive polymer switch. R, M and S denote the specific motor subunits. Chain-extension of the polymer below LCST provides a steric shield blocking the active site. Chain collapse (above LCST) enables access to the active site and restoration of enzyme function. For more details see text.

The conjugation of the motor with synthetic polymers brings additional advantages. One such benefit arises from the ability to functionalise the polymer side chains or terminus in a way that allows attachment of the entire complex to surfaces for sensing and device applications.

Therefore, although our hybrid polymer-protein conjugate was originally aimed at gene targeting (as it has the potential to increase the delivery of intact DNA to cell nuclei and thereby increase gene expression) this system may also be used in building automated nano-chip sensors, therapeutic and diagnostic devices, where DNA itself would be a target, or where DNA might be used as a 'conveyor-belt' for attached molecules. The strength of the molecular motor has proven sufficient to disrupt most protein-DNA interactions and thus numerous processes and applications where highly localised force is required can also be envisaged.

CONCLUSIONS

The use of synthetic polymers offers a number of possibilities, which otherwise could not be exploited or would be difficult to take advantage of, if purely biological systems were used. Moreover, the combination of the properties of molecular motors with "smart" polymers has hitherto been unexplored and represents a novel concept in nanotechnology, which could ultimately lead to a wholly new class of molecular devices. Nanoscale control of molecular transport *in vitro* and especially *in vivo* opens up a whole host of possibilities in medicine, including drug or DNA delivery (e.g. gene therapy), but also where protection of a therapeutic is required under one biological regime and release in another (e.g. prodrugs conjugated to DNA which can be released by nuclease-mediated degradation at the site of action). In addition, this system may allow the generation of switchable nanodevices and actuators, controllable by changes in the synthetic copolymer structure as well as ATP-mediated DNA motion and may pave the way for biofeedback-responsive nanosystems. It can be used for nano-scale isolation of various biochemical processes in separate compartments connected *via* a tightly controlled shuttle device.

In essence, this concept bridges the disciplines of chemistry and biology by using a biological motor to control chemistry and a synthetic polymer to regulate biological processes.

AUTHOR'S CONTRIBUTIONS

KF conceived the idea of using the modified R-M enzyme as a molecular motor and carried out, with co-workers, the molecular studies of the motor components, SSP carried out the polymer synthesis, polymer-motor conjugations and functional studies, CA designed and participated in the synthesis of smart polymers and DCG conceived of the study. All authors participated in study design and coordination as well as the reading and approval of the final manuscript

ACKNOWLEDGEMENTS

This work is supported by a Wellcome Trust Showcase grant (Grant Reference 067484) and Institute of Biomedical and Biomolecular Sciences, University of Portsmouth.

REFERENCES

1. Ishii Y, Ishijima A, Yanagida T: Single molecule nanomanipulation of biomolecules. Trends in

Biotechnology. 2001, 19: 211-216. 10.1016/S0167-7799(01)01635-3.

2. Wang MD: Manipulation of single molecules in biology. Current Opinions in Biotechnology. 1999, 10: 81-86. 10.1016/S0958-1669(99)80015-9.
3. Badjic JD, Balzani V, Credi A, Silvi S, Stoddart JF: A Molecular Elevator. Science. 2004, 303: 1845-1849. 10.1126/science.1094791.
4. Firman K, Dutta C, Weiserova M, Janscak P: The role of subunit assembly in the functional control of type I restriction-modification enzymes. Molecular Biology Today. 2000, 1: 1-8.
5. Murray NE: Type I restriction systems: sophisticated molecular machines (a legacy of Bertani and Weigle). Microbiology and Molecular Biology Reviews. 2000, 64: 412-434. 10.1128/MMBR.64.2.412-434.2000.
6. Barcus VA, Titheradge AJB, Murray NE: The diversity of alleles at the *hsd* locus in natural populations of *Escherichia coli*. Genetics. 1995, 140: 1187-1197.
7. Lee NS, Rutebuka O, Arakawa T, Bickle TA, Ryu J: KpnAI, a new type I restriction-modification system in Klebsiella pneumoniae. J Mol Biol. 1997, 271: 342-8. 10.1006/jmbi.1997.1202.
8. Titheradge AJB, Ternent D, Murray NE: A third family of allelic *hsd* genes in *Salmonella enterica*: sequence comparisons with related proteins identify conserved regions implicated in restriction of DNA. Mol Microbiol. 1996, 22: 437-447. 10.1046/j.1365-2958.1996.00126.x.
9. Price C, Shepherd JCW, Bickle TA: DNA recognition by a new family of type I restriction enzymes: a unique relationship between two different DNA specificities. European Molecular Biology Organisation Journal. 1987, 6: 1493-1498.
10. Dryden DTF, Davies GD, Martin I, Powell LM, Murray NE, Ellis DJ, Berge T, Edwardson JM, Henderson RM: The assembly of the *Eco* KI type I DNA restriction/modification enzyme and its interaction with DNA. Biochem Soc Trans. 1999, 27: 691-696.
11. Janscák P, Dryden D, Firman K: Analysis of the subunit assembly of the type IC restriction-modification enzyme *Eco*R124I. Nucleic Acids Res. 1998, 26: 4439-4445. 10.1093/nar/26.19.4439.
12. Firman K, Szczelkun M: Measuring motion on DNA by the type I restriction endonuclease *Eco* R124I using triplex dissociation. European Molecular Biology Organisation Journal. 2000, 19: 2094-2102. 10.1093/emboj/19.9.2094.
13. Dryden DTF, Cooper LP, Murray NE: Purification and characterisation of the methyltransferase from the type I restriction and modification system of *Escherichia coli* K12. J Biol Chem. 1993, 268: 13228-13236.
14. Taylor I, Patel J, Firman K, Kneale GG: Purification and biochemical characterisation of the *Eco* R124 modification methylase. Nucleic Acids Res. 1992, 20: 179-186.
15. Davies GP, Powell LM, Webb JL, Cooper LP, Murray NE: *Eco* KI with an amino acid substitution in any one of seven DEAD-box motifs has impaired ATPase and endonuclease activities. Nucleic Acids Res. 1998, 26: 4828-36. 10.1093/nar/26.21.4828.
16. Davies GP, Kemp P, Molineux IJ, Murray NE: The DNA translocation and ATPase activities of restriction-deficient mutants of *Eco* KI. J Mol Biol. 1999, 292: 787-796. 10.1006/jmbi.1999.3081.
17. Gorbalenya AE, Koonin EV: Endonuclease (R) subunits of type I and type III restriction-modification enzymes contain a helicase-like domain. FEBS Letts. 1991, 291: 277-281. 10.1016/0014-5793(91)81301-N.
18. Murray NE, Daniel AS, Cowan GM, Sharp PM: Conservation of motifs within the unusually variable polypeptide sequences of type I restriction and modification enzymes. Mol Microbiol. 1993, 9: 133-143.
19. Cooper LP, Dryden DTF: The domains of a type I DNA methyltransferase. Interactions and role in recognition of DNA methylation. J Mol Biol. 1994, 236: 1011-1021. 10.1016/0022-2836(94)90008-6.
20. Szczelkun MD, Dillingham MS, Janscák P, Firman K, Halford SE: Repercussions of DNA tracking by the type IC restriction endonuclease *Eco* R124I on linear, circular and catenated substrates. European Molecular Biology Organisation Journal. 1996, 15: 6335-6347.
21. Studier FW, Bandyopadhyay PK: Model for how type I restriction enzymes select cleavage sites in DNA. Proceedings of the National Academy of Sciences USA. 1988, 85: 4677-4681.
22. Szczelkun MD, Janscák P, Firman K, Halford SE: Selection of non-specific DNA cleavage sites by the type IC restriction endonuclease *Eco* R124I. J Mol Biol. 1997, 271: 112-123. 10.1006/jmbi.1997.1172.
23. Linder P, Doelz R, Gubler M, Bickle TA: An anticodon nuclease gene inserted into a *hsd* region encoding a type I DNA restriction system. Nucleic Acids Res. 1990, 18: 7170-
24. Tyndall C, Lehnherr H, Sandmeier U, Kulik E, Bickle TA: The type IC *hsd* loci of the enterobacteria are flanked by DNA with high homology to the phage P1 genome: implications for the evolution and spread of DNA restriction systems. Mol Microbiol. 1997, 23: 729-736. 10.1046/j.1365-2958.1997.2531622.x.
25. Tyndall C, Meister J, Bickle TA: The *Escherichia coli prr* region encodes a functional type IC DNA restriction system closely integrated with an anticodon nuclease gene. J Mol Biol. 1994, 237: 266-274. 10.1006/jmbi.1994.1230.
26. Price C, Bickle TA: A possible role for restriction in bacterial evolution. Micro Sci. 1986, 3: 296-299.
27. Bickle TA, Krüger DH: Biology of DNA restriction. Microbiological Reviews. 1993, 57: 434-450.
28. Holubová I, Janscák P, Weiserová M, Benada O, Hubácek J, Firman K: Cellular localisation of the type I restriction and modification system of *Escherichia coli* K-12. 1994, A76-
29. Holubová I, Vejsadová S, Weiserová M, Firman K: Localisation of the type I restriction-modification enzyme *Eco* KI in the bacterial cell. Biochem Biophys Res Commun. 2000, 270: 46-51. 10.1006/bbrc.2000.2375.

30. Barker SLR, Ross D, Tarlov MJ, Gaitan M, Locascio LE: Control of Flow Direction in Microfluidic Devices With Polyelectrolyte Multilayers. Anal Chem. 2000, 72: 5925-5929. 10.1021/ac0008690.

31. Kikuchi A, Okano T: Pulsatile drug release control using hydrogels. Advanced Drug Delivery Reviews. 2002, 54: 53-77. 10.1016/S0169-409X(01)00243-5.

32. Li SK, D›emanuele A: On-Off Transport Through a Thermoresponsive Hydrogel Composite Membrane. J Controlled Release. 2001, 75: 55-67. 10.1016/S0168-3659(01)00365-0.

33. Kost J, Langer R: Responsive Polymeric Delivery Systems. Advanced Drug Delivery Reviews. 2001, 46: 125-148. 10.1016/S0169-409X(00)00136-8.

34. Peppas N: Hydrogels and drug delivery. Current Opinion In Colloid & Interface Science. 1997, 2: 531-537.

35. Ista LK, Lopez GP: Lower critical solubility temperature materials as biofouling release agents. J Indust Microbiol Biotechnol. 1998, 20: 121-125. 10.1038/sj.jim.2900490.

36. Cunliffe D, Alarcon CD, Peters V, Smith JR, Alexander C: Thermoresponsive Surface-Grafted Poly(N-Isopropylacrylamide) Copolymers: Effect of Phase Transitions on Protein and Bacterial Attachment. Langmuir. 2003, 19: 2888-2899. 10.1021/la026358l.

37. Kwon OH, Kikuchi A, Yamato M, Okano T: Accelerated Cell Sheet Recovery by Co-Grafting of Peg With Pipaam Onto Porous Cell Culture Membranes. Biomaterials. 2003, 24: 1223-1232. 10.1016/S0142-9612(02)00469-6.

38. Yamato M, Konno C, Utsumi M, Kikuchi A, Okano T: Thermally responsive polymer-grafted surfaces facilitate patterned cell seeding and co-culture. Biomaterials. 2002, 23: 561-567. 10.1016/S0142-9612(01)00138-7.

39. Raucher D, Chilkoti A: Enhanced Uptake of a Thermally Responsive Polypeptide by Tumor Cells in Response to Its Hyperthermia-Mediated Phase Transition. Cancer Res. 2001, 61: 7163-7170.

40. Hoffmann J, Plotner M, Kuckling D, Fischer WJ: Photopatterning of Thermally Sensitive Hydrogels Useful for Microactuators. Sensors and Actuators a-Physical. 1999, 77: 139-144. 10.1016/S0924-4247(99)00080-1.

41. Szczubialka K, Jankowska M, Nowakowska M: «Smart» Polymeric Nanospheres as New Materials for Possible Biomedical Applications. Journal of Materials Science-Materials in Medicine. 2003, 14: 699-703. 10.1023/A:1024911716152.

42. Schild HG: Poly(N-Isopropylacrylamide): experiment, theory and application. Prog Polym Sci. 1992, 17: 163-249. 10.1016/0079-6700(92)90023-R.

43. Caliceti P, Veronese FM: Pharmacokinetic and biodistribution properties of poly(ethylene glycol)-protein conjugates. Advanced Drug Delivery Reviews. 2003, 55: 1261-1277. 10.1016/S0169-409X(03)00108-X.

44. Youngster S, Wang YS, Grace M, Bausch J, Bordens R, Wyss DF: Structure, Biology, and Therapeutic Implications of Pegylated Interferon Alpha-2b. Current Pharmaceutical Design. 2002, 8 : 2139-2157.

45. Maeda H: Smancs and Polymer-Conjugated Macromolecular Drugs: Advantages in Cancer Chemotherapy. Advanced Drug Delivery Reviews. 2001, 46: 169-185. 10.1016/S0169-409X(00)00134-4.

46. Ulbrich K, Strohalm J, Subr V, Plocova D, Duncan R, Rihova B: Polymeric Conjugates of Drugs and Antibodies for Site-Specific Drug Delivery. Macromolecular Symposia. 1996, 103: 177-192.

47. Omelyanenko V, Kopeckova P, Gentry C, Shiah JG, Kopecek J: Hpma Copolymer-Anticancer Drug-Ov-Tl16 Antibody Conjugates .1. Influence of the Method of Synthesis on the Binding Affinity to Ovcar-3 Ovarian Carcinoma Cells in Vitro. Journal of Drug Targeting. 1996, 3 : 357-373.

48. Duncan R, Spreafico F: Polymer Conjugates – Pharmacokinetic Considerations for Design and Development. Clinical Pharmacokinetics. 1994, 27: 290-306.

49. Chilkoti A, Chen GH, Stayton PS, Hoffman AS: Site-Specific Conjugation of a Temperature-Sensitive Polymer to a Genetically-Engineered Protein. Bioconjugate Chemistry. 1994, 5: 504-507.

50. Hoffman AS: Bioconjugates of Intelligent Polymers and Recognition Proteins for Use in Diagnostics and Affinity Separations. Clinical Chemistry. 2000, 46: 1478-1486.

51. Shimoboji T, Larenas E, Fowler T, Hoffman AS, Stayton PS: Temperature-Induced Switching of Enzyme Activity With Smart Polymer-Enzyme Conjugates. Bioconjugate Chemistry. 2003, 14: 517-525. 10.1021/bc025615v.

52. Shimoboji T, Larenas E, Fowler T, Kulkarni S, Hoffman AS, Stayton PS: Photoresponsive Polymer-Enzyme Switches. Proceedings of the National Academy of Sciences USA. 2002, 99: 16592-16596. 10.1073/pnas.262427799.

53. Pennadam SS, Lavigne M, Firman K, Mernagh D, Alexander C, Górecki DC: A molecular motor controlled by a thermoresponsive polymer switch. 6th International Symposium on Polymer Therapeutics. 2004, A52-

54. Pennadam SS, Lavigne M, Dutta CF, Firman K, Mernagh D, Górecki DC, Alexander C: Control of a multi-subunit DNA motor by a thermo responsive polymer switch. JACS. 2004,

CHAPTER 13

CONDUCTIVE POLYMERS: TOWARDS A SMART BIOMATERIAL FOR TISSUE ENGINEERING

Richard Balint[a], Nigel J. Cassidy[b], Sarah H. Cartmell[a]

[a]School of Materials, Faculty of Engineering and Physical Sciences, University of Manchester, Manchester M1 7HS, UK

[b]School of Physical and Geographical Sciences, Keele University, Stoke-on-Trent ST5 5BG, UK

ABSTRACT

Developing stimulus-responsive biomaterials with easy-to-tailor properties is a highly desired goal of the tissue engineering community. A novel type of electroactive biomaterial, the conductive polymer, promises to become one such material. Conductive polymers are already used in fuel cells, computer displays and microsurgical tools, and are now finding applications in the field of biomaterials. These versatile polymers can be synthesised alone, as hydrogels, combined into composites or electrospun into microfibres. They can be created to be biocompatible and biodegradable. Their physical properties can easily be optimized for a specific application through binding biologically important molecules into the polymer using one of the many available methods for their functionalization. Their conductive nature allows cells or tissue cultured upon them to be stimulated, the polymers' own physical properties to be influenced post-synthesis and the drugs bound in them released, through the application of an electrical signal. It is thus little wonder that these polymers are becoming very important materials for biosensors, neural implants, drug delivery devices and tissue engineering scaffolds. Focusing mainly on polypyrrole, polyaniline and poly(3,4-ethylenedioxythiophene), we review conductive polymers from the perspective of tissue engineering. The basic properties of conductive polymers, their chemical and electrochemical synthesis, the phenomena underlying their conductivity and the ways to tailor their properties (functionalization, composites, etc.) are discussed.

INTRODUCTION

Electroactive biomaterials are a part of a new generation of "smart" biomaterials that allow the direct delivery of electrical, electrochemical and electromechanical stimulation to cells [1], [2], [3] and [4]. The family of electroactive biomaterials includes conductive polymers, electrets, piezoelectric and photovoltaic materials [4]. Electrets and piezoelectric materials allow the delivery of an electrical stimulus without the need for an external power source, but the control over the stimulus is limited [4] and [5]. Conductive polymers, on the other hand, allow excellent control of the electrical stimulus, possess very good electrical and optical properties, have a high conductivity/weight ratio and can be made biocompatible, biodegradable and porous [1], [5], [6] and [7]. Furthermore, a great advantage of conductive polymers is that their chemical, electrical and physical properties can be tailored to the specific needs of their

Citation: Richard Balint, Nigel J. Cassidy, Sarah H. Cartmell, Conductive polymers: Towards a smart biomaterial for tissue engineering, Acta Biomaterialia, Volume 10, Issue 6, June 2014, Pages 2341-2353, ISSN 1742-7061, http://dx.doi.org/10.1016/j.actbio.2014.02.015.

application by incorporating antibodies, enzymes and other biological moieties [1], [4], [7] and [8]. Additionally, these properties can be altered and controlled through stimulation (e.g. electricity, light, pH) even after synthesis [3], [9], [10] and [11].

Considering the vast amount of new possibilities conductive polymers offer, we believe they will revolutionize the world of tissue engineering. Thus, we chose to gather together in this review all of the available information on the most commonly used conductive polymers, their biocompatibility, conductivity, synthesis, biomolecule doping and drug release applications.

The History of Conductive Polymers

First produced several decades ago [12], today there are over 25 conductive polymer systems [13]. (For a list of conductive polymers see Table 1.) They merge the positive properties of metals and conventional polymers – the ability to conduct charge, great electrical and optical properties – with flexibility in processing and ease of synthesis [13]. The early work on conductive polymers was triggered by the observation that the conductivity of polyacetylene, a polymer that is normally only semiconducting at best, increases by 10 million-fold when polyacetylene is oxidized using iodine vapour[12] and [14]. The underlying phenomenon was named "doping" and is essential for the conductivity of polymers, as only through this process do they gain their high conductivity (discussed in more detail in Section 4) [13]. As polyacetylene was difficult to synthesize and is unstable in air, the search for a better conductive polymer began [14]. Polyheterocycles since then have emerged as a family of conductive polymers with both good stability and high conductance [15]. This family contains all the currently generally researched conductive polymers: polypyrrole, polyaniline, and polythiophenes [2], [4],[6], [14], [16], [17], [18], [19] and [20].

Table 1: A list of conductive polymers and their abbreviations [172], [173], [174] and [175]

- Polypyrrole (Ppy)
- Polyaniline (Pani)
- Poly(3,4-Ethylenedioxythiophene) (Pedt, Pedot)
- Polythiophene (Pth)
- Polythiophene-Vinylene (Pth-V)
- Poly(2,5-Thienylenevinylene) (Ptv)
- Poly(3-Alkylthiophene) (Pat)
- Poly(P-Phenylene) (Ppp)
- Poly-P-Phenylene-Sulphide (Pps)
- Poly(P-Phenylenevinylene) (Ppv)
- Poly(P-Phenylene-Terephthalamide) (Ppta)
- Polyacetylene (Pac)
- Poly(Isothianaphthene) (Pitn)
- Poly(A-Naphthylamine) (Pna)
- Polyazulene (Paz)
- Polyfuran (Pfu)
- Polyisoprene (Pip)
- Polybutadiene (Pbd)
- Poly(3-Octylthiophnene-3-Methylthiophene) (Potmt)
- Poly(P-Phenylene-Terephthalamide) (Ppta)

Polypyrrole

Arguably the most studied conductive polymer, reflected by the amount of publication surrounding its properties and applications, is the conjugated polymer polypyrrole (PPy;Fig. 1) [9], [21], [22], [23], [24], [25] and [26]. PPy possesses many excellent qualities and stimulus-responsive properties, which make it a very promising "smart" biomaterial[9] and [21]. Most importantly, it has good in vitro and in vivo biocompatibility [8], [26],[27] and [28], good chemical stability in, for example, air and water [25] and [29], and reasonably high conductivity under physiological conditions [8], [26], [30], [31] and [32]. PPy can be easily and flexibly synthesized in large quantities at room temperature in a wide range of solvents, including water [8], [22], [26], [33] and [34]. It can be fabricated with a large surface area, with different porosities, and can easily be modified to make it more suitable for biomedical applications through the incorporation of bioactive molecules [8], [22], [26], [30], [35] and [36]. PPy is also stimulus responsive, allowing the dynamic control of its properties by the application of an electrical potential [21] and [31]. Unfortunately, PPy is very difficult to further process once synthesized [2], [25],[34] and [37], as its molecular structure makes it non-thermoplastic [25] and [28], mechanically rigid, brittle [2], [28] and [38] and insoluble after synthesis [39]. Today, PPy is used in numerous applications, including fuel cells [33] and [39], corrosion protection[40], computer displays [39], microsurgical tools [39], biosensors [2] and [10] and drug delivery systems [2] and [31], and as a biomaterial in neural tissue engineering [41], neural probes [42], nerve guidance channels [42] and [43] and blood conduits [44].

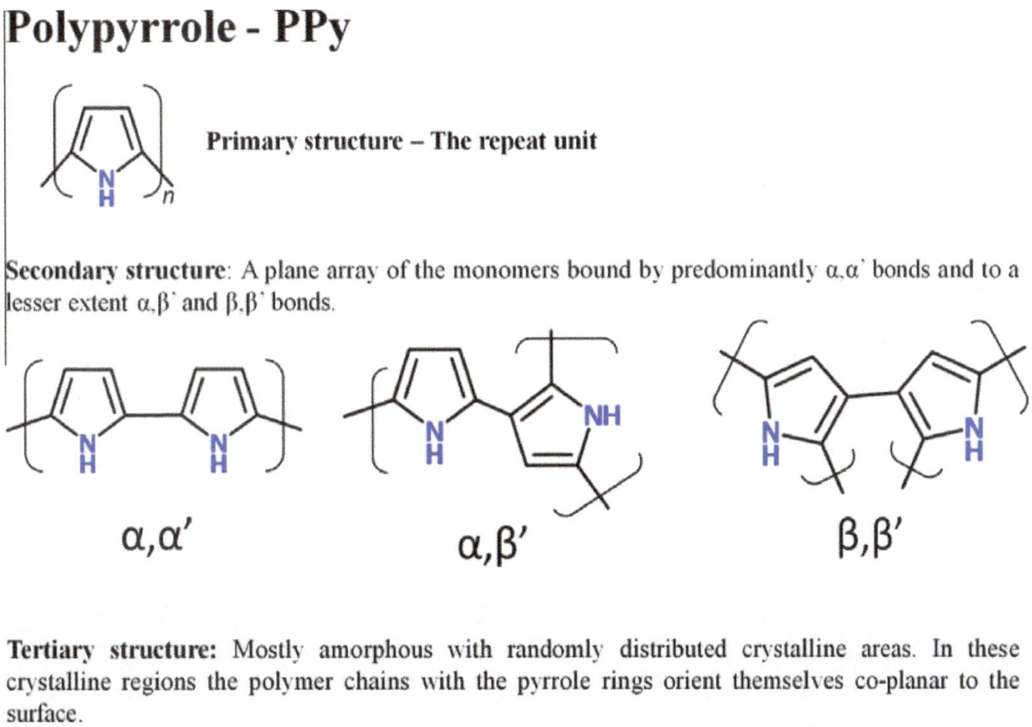

Figure 1: The structure of polypyrrole [172], [173] and [176]. Little information is known about the structures of most conductive polymers. This is a result of the difficulty in finding a solvent that produces single crystals of the polymer and the degradation of the polymer in X-ray diffraction studies [172], [173], [176],[177] and [182].

Polyaniline

The second most investigated conductive polymer after PPy is polyaniline (PANI), also known as aniline black [2], [14] and [45]. It exists in various forms based on its oxidation level: the fully oxidized pernigraniline base, half-oxidized emeraldine base and fully reduced leucoemeraldine base [2] and [14](Fig. 2). Of these, PANI emaraldine, is the most stable and conductive [2] and [14]. PANI has many advantages, such as ease of synthesis, low cost, good environmental stability, and the ability to be electrically switched between its conductive and resistive states [46], [47], [48], [49] and [50]. Unfortunately, its use in biological applications is limited by its low processibility, lack of flexibility and non-biodegradability, and has been noted to cause chronic inflammation once implanted [3], [47] and [51]. PANI has been investigated for biosensors, neural probes, controlled drug delivery and tissue engineering applications [20] and [52].

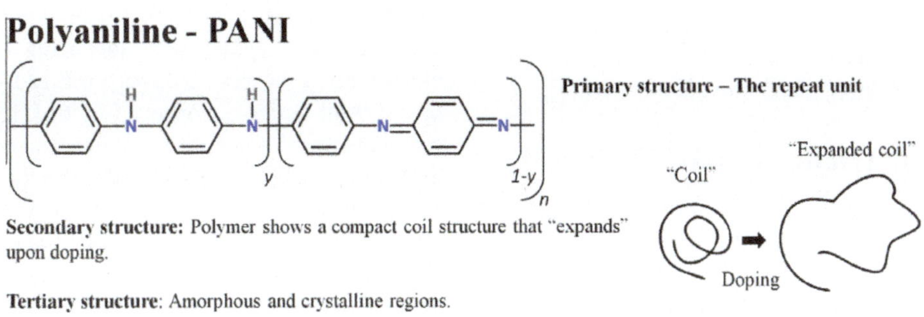

Figure 2: The structure of polyaniline [174], [178], [179], [180] and [181].

Polythiophene Derivatives

A third very interesting conjugated polymer is poly(3,4-ethylenedioxythiophene) (PEDOT), a polythiophene (PTh) derivative [2] and [53] (Fig. 3). PEDOT is formed by the polymerization of the bicyclic monomer 3,4-ethylenedioxythiophene [14]. Compared to PTh, PEDOT has a dioxyalkylene bridging group across the 3- and 4-positions of its heterocyclic ring (Fig. 4), greatly improving its properties by lowering its band gap, reduction and oxidation potential [14] and [54]. This is also what grants PEDOT its good electrical, chemical and environmental stability [53], and better conductivity and thermal stability than PPy [14] and [53].

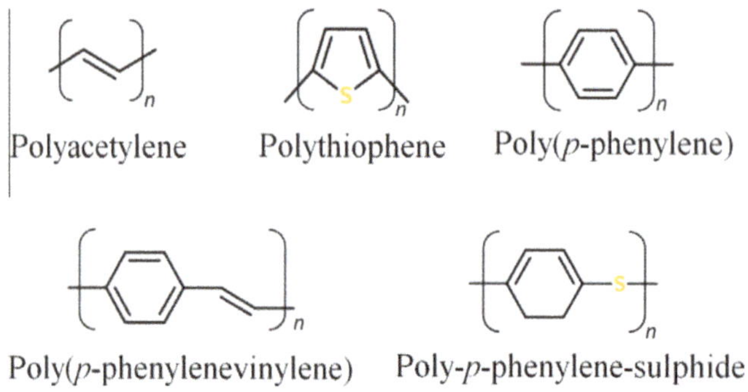

Figure 3: The repeat units of polythiophene and some further conductive polymers [63], [172], [173],[174] and [182].

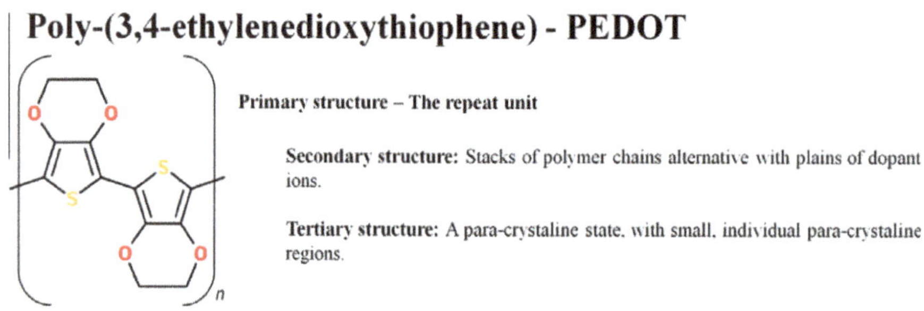

Figure 4: The structure of PEDOT [183].

Today, PEDOT is used in biosensing and bioengineering applications [2], e.g. in neural electrodes [14], [53] and [55], nerve grafts and heart muscle patches [53]. In one interesting example, a neural electrode was interfaced with the surrounding brain tissue through the in situ polymerization of PEDOT [56]. This formed PEDOT filaments extending far enough away from the electrode to breach the glial scar around it and forming sensitive contacts with the plasma membrane of the neurons [56]. PEDOT has also been polymerized within acellular muscle tissue, where it formed a network of elongated tubular structures throughout the tissue [53], in essence converting it into an extensive conductive three-dimensional substrate.

Another PTh derivate of interest is poly(3-hexylthiophene) (PHT) [57]. PHT possesses good solubility in organic solvents, excellent environmental stability and electrical conductivity [57], and has also been successfully electrospun with poly(lactide-co-glycolide) (PLGA) into nanofibres [57].

THE SOURCE OF THEIR CONDUCTIVITY

Simply put, conductive polymers can conduct charge thanks to the ease with which electrons jump within and between the chains of the polymer [2]. The conductivity in reality arises from a combination of a number of factors. The polymers possess a conjugated backbone (Fig. 5), meaning that it is formed by a series of alternating single and double bonds [2] and [12]. Single and double bonds both contain a chemically strong, localized σ-bond, while double bonds also contain a less strongly localized π-bond [58]. The *p*-orbitals in the series of π-bonds overlap each other, allowing the electrons to be more easily delocalized (i.e. they do not belong to a single atom, but to a group of atoms) and move freely between the atoms [2], [8], [12] and [59]. The final key to the conductivity of these polymers is the dopant [2], [20] and [58]. The polymer is synthesized in its oxidized, conducting form, and only in the presence of the dopant molecule (a negative charge/anion in most cases) is the backbone stabilized and the charge neutralized [60]. Parallel to this, the dopant introduces a charge carrier into this system by removing or adding electrons from/to the polymer chain and relocalizing them as polarons or bipolarons (a loosely held, but localized, electron surrounded by a crystal lattice distortion; Fig. 6) [2], [20], [58] and [61]. As an electrical potential is applied, the dopants start to move in or out of the polymer (depending on the polarity), disrupting the stable backbone and allowing charge to be passed through a polymer in the form of the above-mentioned polarons and bipolarons [2], [20] and [58]. For an elegant explanation of the phenomena behind the conductivity of these polymers, see the paper by Bredas and Streer [61].

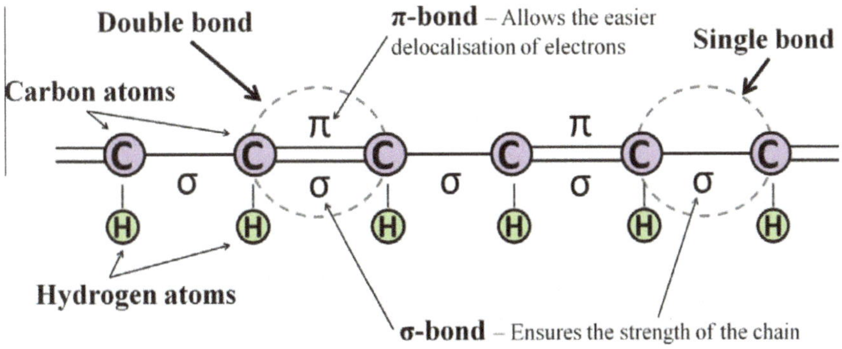

Figure 5: A simplified schematic of a conjugated backbone: a chain containing alternating single and double bonds.

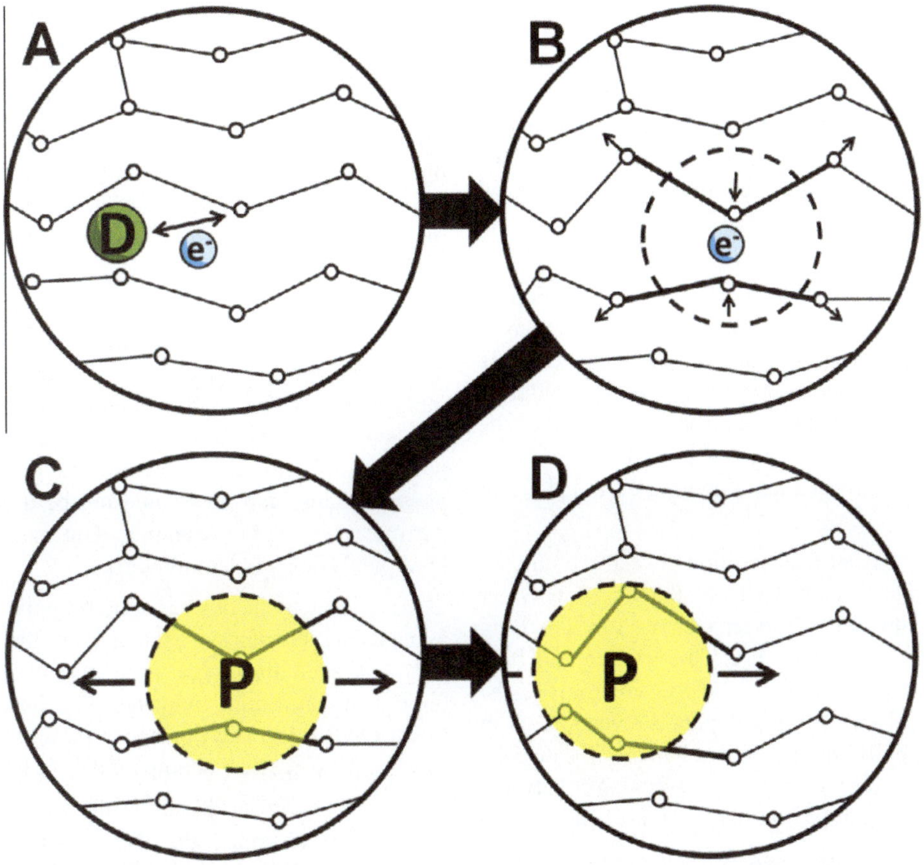

Figure 6: A simplified explanation of the electrical conductivity of conducting polymers. (A) The dopant removes or adds an electron from/to the polymer chain, creating a delocalized charge. (B) It is energetically favourable to localize this charge and surround it with a local distortion of the crystal lattice. (C) A charge surrounded by a distortion is known as a polaron (a radical ion associated with a lattice distortion [63]). (D) The polaron can travel along the polymer chain, allowing it to conduct electricity.

Conductivity in PPy specifically is due to *p*-type (bipolaron) conduction, the inter-chain hopping of electrons and the motion of anions or cations within the material [13],[39] and [62]. PPy can possess a conductivity of up to 7.5×10^3 S cm^{-1} (Table 2) [63]. The exact value depends on the charge transfer to adjacent molecules, the polaron, the chain and the conjugation length, and can be controlled by using different types and quantities of dopants [39] and [64]. The primary factor that limits the conductivity of PPy is the "disorder" (i.e. the defect sites) in the PPy backbone [54] and [62]. More of these defects can form as a result of redox switching or exposure to oxygen or water, resulting in the slow deterioration of conductivity [54] and [65].

Table 2: The conductivity values of some conductive polymers (adapted from Ref. [63])

Polymer	Conductivity (S cm^{-1})
Polypyrrole	10^2–7.5×10^3
Polyaniline	30–200
Polythiophene	10–10^3
Polyacetylene	10^3–1.7×10^5
Poly(*p*-phenylene)	10^2–10^3
Poly(*p*-phenylenevinylene)	3–5×10^3

Emaraldine base PANI has low conductivity in the range of 10^{-10} S cm^{-1}, while its salt, created by modifying the base's oxidative state, is conductive with 30 S cm^{-1}[41], [45],[46] and [63]. The exact conductivity depends on the method of preparation, and can be controlled by submerging the emaraldine base in aqueous solutions of picric, camphorsulfonic or phosphoric acids [45]. The increase in conductivity between the base and salt forms is explained by the polymer's molecular structure [66]: in its base state the polymer chains are coiled, while in its salt form the additional positive charges in the polymer repel each other extending the chains (Fig. 2) [66]. In its extended coil form electrons are easier to delocalize, thus resulting in an increased conductivity [66]. PANI was reported to maintain its conductivity, albeit reduced by three magnitudes, for 100 h under physiological conditions [67].

THE KEY: DOPING

As mentioned above, it is the doping process that introduces the charge carriers (polarons and bipolarons) into the polymer and renders it conductive [2] and [15]. Similarly to what happens in semiconductor technology, this can happen two ways: p-doping, where the polymer is oxidized and will have a positive charge, and n-doping, where the polymer is reduced and will possess a negative charge [2], [14], [58] and [66].

The doping process occurs during synthesis and can be carried out chemically, electrochemically or via photodoping [15], [16] and [68]. In most cases when the dopant is a biological molecule, as many of them are not capable of the redox chemistry that is necessary for chemical synthesis, the electrochemical method has to be used. In this case, the biological molecule has to be charged and placed with the monomer when the electrochemical synthesis occurs [15] and [68].

There is a proportional relationship between the amount of dopant used and the conductivity of the doped polymer [64]. The conductivity can be further increased by choosing a different dopant, but this will affect the surface and bulk structural properties (e.g. colour, porosity, volume) of the polymer [2], [15], [16] and [59]. The doping is reversible, as an electrical potential applied through the polymer will cause the dopant to leave or re-enter the polymer, switching it between its conductive and insulating redox states [3], [16] and [59].

Dopants can be separated into two categories based on their molecular size: small dopants (e.g. Cl^-) and largedopants (e.g. sodium polystyrenesulfonate, PSS) will behave and affect the polymer itself differently [2], [15], [20] and [65]. Both will affect the conductivity and structural properties of the polymer, but large dopants will affect the material properties more dramatically and can, for example, increase the density. Large dopants are more integrated into the polymer and will not be leached out with time or with the application of an electrical stimulus, granting the polymer greater electrochemical stability [20], [60] and [65]. With the application of a reducing electrical potential, due to their immobility rather than the large dopants expulsing, a cation (positive charge) is forced to enter the polymer [20] and [60]. Small dopants, on the other hand, can leave and re-enter the polymer with electrical stimulation [15] and [62], forming the basis of the drug release applications of conductive polymers. This also allows the physical properties of the polymer to be controlled through cycling between doping (oxidation) and dedoping (reduction) [15] and [62].

Surface roughness, morphology, wettability and stiffness are known to affect the adhesion and proliferation of multiple cell types [69], [70], [71] and [72]. It is therefore important to note the effect dopants can have on the bulk and surface material properties of conductive polymers [62], [73] and [74]. Indeed, the use of different dopants (and the amount used) has been observed to alter these material properties differently [75], [76],[77], [78] and [79]: for example, hyaluronic acid (HA) doped PPy is rougher and more brittle than PSS-doped PPy [73]. The roughness of chondroitin sulphate (CS) incorporated PPy increases as a function CS concentration [77]. When the effects of five biologically relevant dopants – dextran sulphate (DS), poly(2-methoxyaniline-5-sulfonic acid) (PMAS), para-toluenesulfonic (pTS) acid, HA and CS – were compared, it was found that the PMAS- and CS-doped films possessed a much lower surface roughness and Young's modulus than the films prepared with the other three dopants [76]. In the same study, it was also noted that the PMAS- and CS-doped films supported the adhesion and differentiation of skeletal myoblasts much better than their counterparts[76]. These observations show the connection between dopant, material property and cellular behaviour, and emphasize the importance of keeping this connection in mind when choosing the doping molecule [76].

Small and polymeric anions, buffer salts and biologically important proteins, enzymes and antibodies have all been used as dopants for PPy [14], [37], [73] and [80]. Some dopant molecules, such as poly(glutamic acid) (PGlu), can act as tethers, covalently binding additional molecules into the polymer [36]. This allows the binding of molecules that do not possess charge and thus could not be used as a dopant in a normal case [36]. For example, PGlu was successfully used to bind polylysine and laminin [36].

Similarly to the other conductive polymers, PEDOT needs a balancing counterion as a dopant for its polymerization [14]. One such dopant that is commonly used is PSS [14], but biologically relevant molecules like heparin, nerve growth factor (NGF), HA and fibrinogen [8] and [55] have also been used, improving the biocompatibility of PEDOT[55].

Doping with a biomolecule can have negative side effects. For example, the use of collagen was shown to result in bad film integrity [24], while HA, when applied to promote angiogenesis, was observed to reduce the electrical conductivity of the polymer [73].

SYNTHESIS AND PROCESSING

There are currently two main methods for synthesizing conductive polymers: chemical and electrochemical [7], [14], [39] and [58].

During chemical synthesis the monomer solution is mixed with an oxidizing agent (e.g. ferric chloride, ammonium persulfate) [81] and [82]. This process creates a powder or a thick film of the polymer, and allows its bulk production, which makes it the method of choice for commercial applications [82] and [83]. An additional advantage of chemical polymerization is that it can be used to create all types of conductive polymers, including some novel conducting polymers that cannot be synthesized with the electrochemical method [15]. Unfortunately, the conductivity of the polymers when synthesized using the chemical method has always been lower than their electrochemically synthesized counterparts

[83]. Additionally, the conductivity of the created polymer is known to be highly sensitive to the choice and purity of the solvent, the oxidant, the relative concentration of the reagents, reaction time, temperature, stirring rate, etc., making reliable and repeatable chemical synthesis a difficult thing to do [83], [84], [85] and [86].

The polymerization mechanism of the chemical synthesis of PPy was shown in a recent study [81]: as the first step, pyrrole is oxidized to give a radical cation. The radical cation reacts with a neutral monomer, followed by oxidation and deprotonation, giving a dimer (an oligomer of two monomers). The dimer is also oxidized, yielding a dimeric radical cation. This combines with a new neutral monomer, providing a trimer (an oligomer of three monomers). This reaction continues and the chain grows monomer by monomer[81]. The polymerization of PANI and PTh has been suggested to be similar [81].

Electrochemical polymerization (Fig. 7) occurs by applying an electrical current through electrodes placed into a solution containing the monomer of the polymer, the solvent and the doping agent [87], [88] and [89]. This method allows the deposition of a thin film of the polymer with a well-controlled thickness (down to 20 nm) and morphology [7],[15] and [88]. The electrical current causes the monomer to deposit and oxidize on the positively charged working electrode, forming insoluble polymer chains [15]. The properties of the synthesized film will be specified by the deposition charge and time, the temperature, the solvent, the doping agent and the electrode system [28], [68], [78], [90],[91], [92] and [93]. Electrochemical polymerization only allows the synthesis of the polymer, if its monomer can undergo oxidation in the presence of an electrical potential[15]. All of the main conductive polymers currently in use (e.g. PPy, PANI, PEDOT) fulfil this criterion [15].

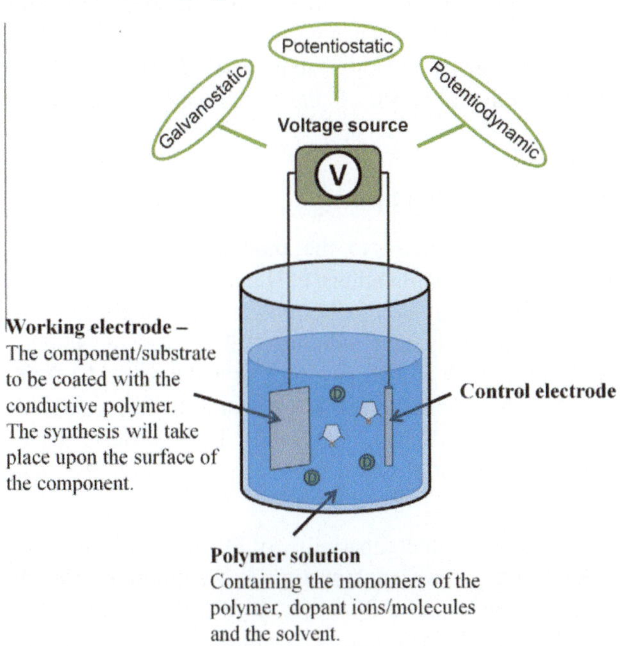

Figure 7: A schematic of the electrochemical synthesis set-up [7], [8], [14] and [15].

Electrochemical synthesis can be carried out using three techniques: the galvanostatic, the potentiostatic and potentiodynamic methods [7], [8] and [14]. In potentiostatic polymerization, the potential of the electrodes is controlled, while the current varies [7]. This protects the integrity of the component to be coated, making this method ideal for the manufacture of biosensors. The electrical current can vary depending on a number of factors (e.g. the electrode material, the plating conditions), thus a coulometer is necessary to control the amount of polymer that is deposited [7] and [14]. In galvanostatic polymerization, the electrical current is controlled instead of the potential. This means that the rate at which the polymer is deposited is steady and can be controlled accurately[7] and [14]. During potentiodynamic deposition, the polymerizing potential is swept between a low and high potential limit in cycles [94] and [95]. This causes the polymer to be deposited in layers, with each layer becoming electrically active before the next one is synthesized [96]. Potentiodynamic electrosynthesis has been observed to produce a different surface morphology to the potentiostatic and galvanostatic methods [94]. For example, in a study comparing PEDOT synthesized using the three methods, it was found that the potentiostatic and galvanostatic routes produced a porous, globular surface, while the potentiodynamic method generated a rod-like, fibrous morphology[94].

Electrochemical synthesis enables the rapid deposition of conductive polymers in situ, but the amount of bioactive molecules that can be doped into the polymer this way is limited [28] and [40]. Additionally, as this method requires an electrode for the polymer to be deposited upon, the range of shapes and quantities that can be synthesized will be restricted by the geometry and surface area of the electrode [65] and [97]. This requirement for an electrode also makes the creation of composites with this method difficult [65]. In contrast to this, chemical synthesis allows the easy creation of composites and was used to combine PPy with polymers such as poly(d,l-lactide) (PDLLA), poly(methyl methacrylate), polyvinylchloride, polystyrene and polyurethane[28] and [65].

Composites

One way to compensate for the shortcomings of a conductive polymer is to use it together with another polymer, combining the positive qualities of both materials [38].

For example, PPy is brittle after synthesis [2], [65] and [98]. To give it flexibility, PPy has been deposited onto polyester [97] and polyethylene terephthalate (PET) fabrics [98]. PPy-coated polyester fabrics were shown to be cytocompatible and support the growth of cells after an initial period of reduced adhesion [97]. Unfortunately, normal deposition will not covalently bind PPy and thus the coating formed will eventually release the surface[99]. In an attempt to overcome this, polyester fabrics were covalently combined with N-modified PPy, which is much less prone to delamination [98] and [99]. PPy was also combined with PDLLA – either deposited onto its surface as a film or into its matrix as nanoparticles [37], [65], [100] and [101] – yielding a flexible, biocompatible and biodegradable composite with improved conductivity compared to the PPy-coated polyester fabrics [35], [100] and [101]. PPy/PDLLA was able to maintain its electroactivity for up to 1000 h, and supported the growth of fibroblasts [65]. In vivo, when implanted in rats, it caused only minor inflammation [101]. Another biodegradable composite was created through the polymerization of PPy in the presence of the natural polymer dextran[102]. The nanocomposite produced was demonstrated to possess good conductivity and an interesting anti-microbial capability [102]. Paper-like nanocellulose–PPy composites have also been created [31]. These have excellent mechanical strength, flexibility and durability, while possessing high conductivity, high capacity and a fast ion exchange speed [31]. When PPy was combined with carbon nanotubes, it produced improved conductivity and gave excellent biocompatibility [62]. In another example, PPy was combined with poly(2-methoxy-5-aniline sulfonic acid) (PMAS), a water-soluble self-doped PANI [20]. PPy–PMAS is a hydrophobic composite with low electrical impedance, and was shown to support the adhesion and proliferation of PC-12 nerve cells [20]. PPy was even combined with animal tissue but, as the monomer was not able to permeate the tissue, it was only deposited on the surface [103] and [104].

Composites of PANI and polypropylene (PP) were created for neurobiological applications [46]. The PANI–PP fibres produced supported the adhesion of primary dorsal root ganglion neurons well, especially when coated with a ligand [46]. A similar composite was created from PANI and polycaprolactone for cardiac tissue regeneration[47]. PEDOT-based composites were synthesized by combining the polymer with carbon nanotubes to produce very promising high-performance neural electrodes[105] and [106].

Electrospinning

Electrospinning is a versatile process that allows the production of nano- and micrometre-scale fibres from a wide range of polymers [29] and [33]. During electrospinning, a high-voltage electrostatic field is used to draw a jet from a polymer solution [33]. As this jet travels toward a collector electrode, the solvent evaporates and a polymer fibre is formed [33].

Attempts have been made to process conductive polymers into nano- and microfibres through electrospinning [29] and [57]. Conductive polymers can be electrospun alone[29], [33] and [107], but this requires organic solvent-soluble PPy [33] or, in the case of PANI, chemical conditions (e.g. doping and dissolution in hot sulphuric acid) that might make the created fibres unsuitable for biological applications [50]. Thus conductive polymers are usually combined with spinnable polymers (e.g. polyethylene oxide, polystyrene) [2], [33] and [57].

One way to do this is to coat electrospun fibres with the conductive polymer [107]. PPy-coated fibroin fibres were demonstrated to support the adherence and proliferation of MSCs and fibroblasts [107]. PPy was also grown on PLGA fibres, and was shown to allow the growth and differentiation of PC-12 cells and hippocampal neurons [6].

Another method is to blend with a carrier material such as poly(ethylene oxide) or poly(vinly cinnamate) before the electrospinning process [33] and [50]. Using this method, PANI has been electrospun with polycaprolactone (PCL) to create a substrate for cardiac [47] and skeletal muscle tissue engineering [108]. Nanofibres of PANI–poly(L-lactide-co-e-caprolactone) (PLCL) have also been electrospun and reported to support the adhesion of human dermal fibroblasts, NIH-3T3 fibroblasts and C2C12 myoblasts[109]. PANI–PCL nanofibres were also successfully used to promote the fusion and maturation of C2C12 myoblasts [110]. Fibres of PANI–gelatine were observed to support the attachment and proliferation of H9c2 rat cardiac myoblasts to an extent similar to tissue culture plastic [19].

The blending affects the material properties of the resultant fibres: electrical conductivity will be compromised [2] and the diameter of the resultant fibres will be decreased[2] and [57]. Nonetheless, conductive, flexible and biocompatible nanofibrous scaffolds produced through electrospinning are very attractive substrates for neural tissue and other engineering applications [2], [33] and [48].

Hydrogels

Conductive polymers have also been successfully polymerized inside hydrogel networks[11], [55] and [111]. This allows the creations of electroactive hydrogels, which combine the redox switching capabilities of conductive polymers with the fast ion mobility and biocompatibility of hydrogels [11] and [112]. These electroactive hydrogels can be produced with a wide range of dimensions, binding bioactive molecules and nanotemplated to mimic the extracellular matrix [111] and [112]. These properties make them ideal for implantable biosensors, drug release devices and deep-brain stimulators[11] and [112]. Examples of electroactive hydrogels are the PANI–PVP, PANI–polyacrylamide and PPy/PANI–polyacrylamide hydrogels [11] and [111].

PPy has also been electrochemically grown within support hydrogels made out of poly(2-hydroxyethyl methacrylate) p(HEMA) [38] and [112] or made into a hydrogel using a PMAS dopant [68]. The PPy synthesized this way has an extremely large surface area, thus giving the PPy hydrogel an impedance that is much lower than that of a PPy film[17]. A similar hydrogel composite was created using PPy and complex mucopolysaccharides, and was shown to be able to release the proteins bound in it during synthesis upon the application of an electrical potential [80]. The hydrogel composites were shown to be non-cytotoxic, making them an ideal material for drug release, and neural and muscular tissue engineering [38] and [112].

BIOCOMPATIBILITY

Good cellular response to the biomaterial is essential for many biomedical applications[113] and [114]. Therefore it is important that many types of conductive polymers (e.g. PPy, PANI, PTh and polyethyleneimine (PEI)) have been shown to support the growth of a large variety of cell types [1], [2], [67] and [115]. Additionally, the biocompatibility of conductive polymers, if insufficient, can easily be improved through bonding biocompatible molecules, segments and side chains onto the polymer [116].

Although the biocompatibility of PPy has been questioned in some cases, it was demonstrated to support the in vitro adhesion, growth and differentiation of a wide range of cell types [1] and [2], including bone [1] and [100], neural [37], [43], [44], [73] and [100], glial [117], rat pheochromacytoma [8] and endothelial [73], [97] and [100] cells, fibroblasts [24] and [118], keratinocytes [24] and mesenchymal stem cells [40]. Schwann cells treated with a PPy powder solution showed no signs of acute toxicity, mutagenesis, pyretogen, haemolysis or allergic responses [40]. Composite meshes of PPy–PLGA were shown to be biocompatible with embryonic hippocampal neurons and PC-12 cells[6]. The biocompatibility of polyester fabrics was shown not to be affected by the presence of a PPy coating [98].

PPy's good biocompatibility was also shown in animal models [1], [21] and [39]: results indicate that PPy has no significant long-term effect in vivo [97], or induces only a minimal tissue response [114] and [116]. For example, when PPy films were implanted into the cerebral cortexes of rats, they were tolerated well and allowed the formation of complex neural networks [20]. In a study looking at chemically synthesized PPy, no cytotoxic or allergic response was found in mice [74]. Neither did the spleen, liver and kidney indexes of the mice change as a result of implanting PPy [74]. It was also demonstrated in mice that PPy does not cause haemolysis or changes in blood coagulation [97].

There are also a few reports of reduced biocompatibility. Human mesenchymal stem cells were shown to grow less on silk fibroin mats when those were coated with PPy[107]. Endothelial cells adhered less with increased PPy coating thickness [44]. In a similar investigation, endothelial cells were unable to synthesize DNA on neutral PPy, but no such problems were witnessed if the polymer was in the oxidized redox state [119]. In another study, bovine aortic endothelial cells only poorly adhered to PPy substrates that were not precoated with fibronectin [13]. PPy nanoparticles have also been observed to have an adverse affect on human lung fibroblast and mouse alveolar macrophage viability [120].

This variation in biocompatibility is theorized to be due to the different preparation protocols used in the experiments, as it was shown that rinsing, extraction and ageing all have a significant effect on biocompatibility [31]. If PPy is appropriately prepared with repeated steps of rinsing and extraction, the polymer should be completely cytocompatible [31] and [118]. These preparatory steps are necessary to remove the impurities, reactants, monomers and shorter oligomers (remnants of the synthesis) that are believed to be the cause of any reduced biocompatibility [13] and [31]. As mentioned above, the dopants and the synthesis conditions might also have affected the behaviour of cells in a negative way, as these change the surface topography of the polymer, which can result in altered cell behaviour [13].

The reports regarding PANI's biocompatibility are mixed: it was stated to support neural cell growth [2] and [66], provide acceptable proliferation and adhesion [52], maintain sufficient biocompatibility [2] and not cause significant inflammation [51]. Both the emeraldine base and emeraldine salt forms of PANI were reported to be cytocompatible regarding H9c2 cardiac myoblasts and not to result in inflammation in rodent models [2],[48] and [67]. In a similar investigation, none of the emeraldine, nigraniline and leucoemeraldine forms of PANI were found to invoke an inflammatory response in rats during a 90 day period [121]. It was also stated to be overtly noncytotoxic, but to require surface treatments to enhance its biocompatibility [46]. For example, PC-12 cells were reported to poorly adhere to untreated PANI, but this was greatly improved by coating the polymer with adhesive peptides (e.g. YIGSR) [3], [48] and [67].

Other investigations have reported poor cell adhesion and growth [15], tissue incompatibility [2] and fibrous tissue formation in rats [46]. Considerable cytotoxicity was observed regarding immortalized keratinocyte and hepatocellular carcinoma cell lines[52]. Similarly to PPy, the insufficient biocompatibility is theorized to be the result of small amounts of residual acid dopants and low-molecular-weight by-products still being present and leaking out of the polymer [52] and [67]. These by-products can be removed by additional curing and purification steps [52] and [67].

PEDOT has shown good biocompatibility, amongst others, with epithelial [2], neural [8]and neuroblastoma cells [122], L929 [123] and NIH3T3 fibroblasts [2]. For example, both PSS and tosylate anion-doped PEDOT films were demonstrated to support the adhesion and proliferation of fibroblasts [123]. Similarly, SH-SY5Y neuroblastoma cells adhered well and displayed a healthy morphology on PEDOT-coated PET fibres [122]. There are some reports that PEDOT shows light cytotoxicity, but the inflammatory response towards it in vivo is good [2] and [124]. In an attempt to enhance its compatibility with neural tissue, PEDOT was doped with NGF and indeed supported the growth of PC-12 cells [8]. The doping of PEDOT with multiwall carbon nanotubes yielded an excellent electrode material that, when implanted into rats for 6 weeks, produced a tissue response lower than that of platinum [125]. Similarly, PEDOT nanotube-coated electrodes implanted into the barrel cortex of rats were found to be accompanied by a better tissue response than their uncoated counterparts [126].

FUNCTIONALIZATION FOR A SPECIFIC APPLICATION

Optimizing the material properties (roughness, porosity, hydrophobicity, conductivity, degradibility) of conductive polymers and the binding of biological molecules (that makes conductive polymers so promising for biomedical applications) can be done through four major chemical ways (Fig. 8) [15]:

(A)Through adsorption. In this method, a solution of the functionalizing agent is placed in contact with the polymer after it has already been synthesized. The biomolecule is physically absorbed due the static interactions between the polymer matrix and the charge of the molecule [127]. Although this method is the simplest to carry out, the outcome is sensitive to pH, and the biomolecule is prone to leaching out and can compromise the conductivity of the polymer [15], [127] and [128].

(B)By entrapping the molecule inside the polymer [129]. This is achieved by mixing the functionalizing molecule with the monomer of the polymer, the dopant and the solvent prior to synthesis [129]. Upon electrochemical polymerization, the molecules of the functionalizing agent in the proximity of the electrode are incorporated into the growing polymer [129]. This technique is primarily applied to bind large molecules (e.g. enzymes, DNA), as these will be unable to leave the polymer once entrapped due to their size [130] and [131].

Both adsorption and entrapping are simple techniques that allow the incorporation of biomolecules without a chemical reaction that could affect their activity – which is something that could happen with, for example, covalent binding [129]. Therefore, these methods lend themselves to biosensor applications. For example, physical absorption has been used to successfully bind calf thymus DNA onto PPy in order to create a biosensor for toxicants [128]. Entrapping has been used to bind enzymes such as glucose oxidase to create glucose sensors [130], [132] and [133], and to bind DNA to detect aromatic amines, cDNA and Hep C virus [131]. Even live bacterial cells have been incorporated in an attempt to create a urea biosensor [134]. Absorption has also been used to improve the biocompatibility of neural electrodes: polylysine absorbed into PEDOT:PSS-coated electrodes was shown to promote the long-term survival of neural cells [135].

(C)By covalently bonding the molecule to the monomer of the polymer. With this method, the biological molecules will be strongly bound and will not be released, thereby enhancing the long-term stability of the polymer [23] and [136]. However, the conductivity of the polymer (as with absorption) might be compromised [15]. A good example of the covalent method is the binding of cysteines to the beta-positions on PPy with strong sulfide bonds. These cysteines can then serve as sites to covalently anchor further bioactive molecules [136] and [137]. Similar binding sites were created through the use of N-hydroxyl succinimidyl ester pyrrole [23] and [34], an intermediate photocrosslinker consisting of polyallylamine conjugated to an arylazido functional group [35], and poly(ethylene glycol) methacrylate graft copolymerization [26]. These binding sites were successfully used to bind NGF and heparin [23] and [26]. Such binding sites have been demonstrated to not always be necessary: in a novel study, Bax et al. [138] used plasma immersion ion implantation to modify the surface of PPy allowing the covalent binding of tropoelastin and collagen I without a chemical linking molecule.

(D)By exploiting the very doping process that renders conductive polymers conductive. This allows the bonding of a wide range of biomolecules as long as they are charged[8], [23], [139], [140], [141] and [142]. For example, growth factors, collagen, heparin, chitosan and ATP have already been successfully bound in conductive polymers via doping [2], [16] and [68]. Unfortunately, introducing bioactive molecules through doping allows only a relatively small amount of the molecules to be bound, while also having a greater negative effect on the polymer's conductivity than covalent bonding[2], [16] and [23].

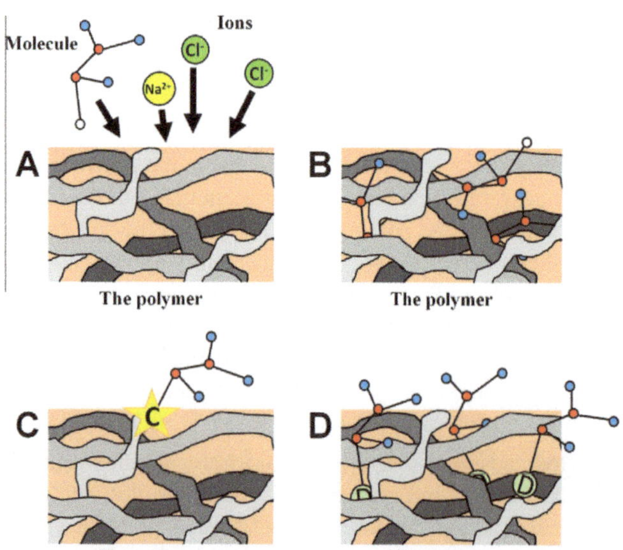

Figure 8: The four methods of functionalizing conductive polymers: (A) physical absorption, (B) entrapping, (C) covalent bonding and (D) exploiting the doping mechanism.

A wide range of biologically important molecules have been used to improve the bioactivity of PPy: for example, dermatan sulphate was used to increase keratinocyte viability [24], heparin was incorporated to promote endothelial cell proliferation[65] and [143], doping with laminin-derived peptides aided in neuron and astrocyte adhesion [143], NGF and poly-l-glutamic acid enhanced neuronal growth [144], while HA and CS were used for skeletal myoblast growth and differentiation [68]. In one study, PPy was doped using laminin-derived peptides p20 and p31 [144]. Both promoted neuroectoderm formation from human embryonic stem cells, but p20 enhanced neural differentiation more, while p31 better aided adhesion and spreading [144]. In other studies, PPy has been functionalized using HA to enhance vascularization[73] and [139], NGF to improve compatibility with neural cells [8], and heparin [28] and fibronectin [59] to promote cellular adhesion.

Covalently binding cell-adhesive agents, such as Arg–Gly–Asp (RGD), fibronectin, heparin and HA, onto the surface of PPy has also been used to enhance the polymer's cytocompatibility [17], [25], [34], [40] and [137].

Covalent modification has been used to improve the biocompatibility of PANI as well [48]. For example, the adhesive molecule RGD was covalently bound onto the surface of the polymer [48]. A substrate prepared this way was able to support PC-12 cells and induce neuritogenesis in the absence of neurotrophins [48].

Physical guidance cues, such as topographical features, can also be used to make a conductive polymer better suited for a certain application [42] – for example, through increasing the roughness of the surface or micropatterning [2] and [58]. Microchannel-patterned PPy was shown to increase the speed at which embryonic hippocampal neurons polarize [42].

BIODEGRADIBILITY

Although many conductive polymers (e.g. PPy and PTh) are not inherently biodegradable [2], [4] and [41], they can be made to be so [47], [58] and [145]. One of the three main routes is to synthesize the conductive polymer as a composite together with a biodegradable polymer [15] and [146]. For example, erodible PPy nanoparticle–polylactide (PLA) [18] and [65] and PPy–PLLA [28] and [147] composites have been successfully prepared. Although this combines the positive properties of both polymers[101], it does not solve the problem of removing the conductive polymer from the body after the degradable polymer has disappeared [18] and [146]. An advantage of this method is that the conductivity and rate of degradation can be controlled by choosing the right ratio of the two polymers [15]. The second route is to modify the conductive polymer itself. The addition of ionizable (butyric acid) or hydrolysable (butyric ester) side groups to the backbone of PPy has been reported to render it degradable [1], [11] and [148]. Again, choosing the amount of these groups allows researchers to set the rate of degradation[11]. As a third solution to the problem, small chains of PPy that can undergo gradual erosion and renal clearance simply due to their small size were electrochemically synthesized [148]. (For a more detailed description of the biodegradability of conductive polymers, see the excellent review by Guo et al. [145]).

DRUG DELIVERY

Many disciplines of science, including the medical, pharmaceutical and agricultural fields, require the controlled delivery of chemical compounds [11]. This has been a great challenge, but now the use of conductive polymers as a substrate material for controllable drug delivery devices promises to overcome this [11], [111] and [141]. Why are conductive polymers so promising? The molecules bound in such polymers through doping can be controllably expelled through the application of a reducing (negative) electrical potential [11], [111], [141] and [142]. The fact that they can be made porous and have delocalized charge carriers aids in the diffusion of the bound molecules, adding a further reason why conductive polymers are very suitable for drug release applications[11]. In experiments, many therapeutic drugs, such as 2-ethylhexyl phosphate [149], dopamine [150], naproxen [151], heparin [142], NGF [35] and dexamethasone [141], have already been bound and successfully released from these polymers [141]. Using PPy, neurotrophin-3 [116] and [140], brain-derived neurotrophic factor (BDNF) [140] and NGF were delivered successfully and initiated neural growth and differentiation [8],[21] and [23]. Heparin was released from a poly(vinyl alcohol) hydrogel formed on a PPy film [142], while in another experiment heparin was not released into the medium, but was exposed on the surface when an electrical potential between −0.4 and −0.7 V was applied for 90 s [9]. The anti-inflammatory drug dexamethasone was successfully released from a 50 nm thin film using cyclic voltammetry [141]. The release was proportional to the number of stimulation cycles [141]. The expulsion of the molecules generally seems to happen quite rapidly, in just a few minutes, which can be considered an advantage or a disadvantage of conductive polymers, depending on the application[27], [35] and [116].

There are, however, a few factors that limit the application of conductive polymers for drug release: for example, the molecules loaded into the polymer tend to leach out through diffusion, to be replaced by other molecules from the polymer's environment[27], [35] and [116]. This passive loss of load is further worsened by the fact that only a relatively small amount of drug can be bound in the polymer in the first place [27],[35] and [116]. Additionally, both charge and molecular weight restrict which molecules can be bound and released [27]. This can be easily overcome through the use of biotin–streptavidin coupling [27]. Biotin acts as the dopant, while the bioactive molecule is covalently bound to the biotin. The molecule is then released with electrical stimulation[27]. A further bonus is that biotin provides more uniform release kinetics [27].

A further hindering problem is the fatigue of conductive polymers with repeated cycles of electrical stimulation [11]: repeated cycles can cause irreversible oxidation in the polymer, which coincides with dedoping and reduced conductivity, which ultimately limits its useful lifetime [11] and [14]. For example, a PPy/PPS composite was shown to retain only 5% of its original conductivity after the application of 0.4 V for 16 h [14]. Repeated stimulation cycles not only cause functional problems, but also structural ones. The continuous movement of doping agents in and out of the polymer, and the subsequent swelling and deswelling, causes cracks, delamination and the overall degradation of the polymer [14].

ELECTRICAL STIMULATION DELIVERED THROUGH A CONDUCTIVE SCAFFOLD

Electrical stimulation (ES) has many noted beneficial effects, such as enhanced nerve regeneration in vivo [5]. Conductive polymers lend themselves as excellent novel scaffolds for a more efficient delivery of this stimulus type.

The most commonly reported result of ES delivered with this technique is greater neurite outgrowth in nerve cells [20], [23], [37], [116], [140] and [152]. For example, PC-12 cells have been reported to have 50% more neurites on NGF-doped PPy films when ES is applied [35]. Similarly, PC-12 stimulated with 10 mV cm^{-1} on PPy–PLGA scaffolds formed an increased number of neuritis, and the overall length of these neurites was also greater than without stimulation [6]. Very similar results were found on PPy–PDLLA [153]and PLLA–PANI scaffolds [49]. In a recent study by Weng et al. [154], ES was delivered through inkjet-printed collagen-coated PPy tracks to PC-12 cells [154]. The stimulation was shown to increase and direct neurite outgrowth parallel to the PPy tracks [154]. The reason behind this increase in neurite outgrowth has been postulated to be enhanced fibronectin adsorption onto the conductive polymer scaffolds [5], coupled with the effect that the electrical field has on the proteins and ion channels within the cell membrane[154]. ES has been demonstrated to have other desirable effects on nerve cells: 250 Hz biphasic current delivered via PPy/PMAS composite films was observed to increase neural differentiation in the presence of NGF [20]. Similarly, enhanced proliferation and neurite outgrowth was noted when neural cells were stimulated on nanofibrous PANI–PG scaffolds [66]. When ES was delivered through fibrillar collagen-coated PPy scaffolds, rat pheochromocytoma nerve cells displayed signs of increased neural differentiation [152]. Schwann cells cultured on chitosan–PPy composites showed greater viability and increased their NGF and BDNF mRNA expression when a stimulus of 100 mV mm^{-1} was applied [155].

Other cell and tissue types could also benefit from ES delivered through a conductive scaffold. The growth of NIH-3T3 fibroblasts was increased by ES delivered through a PANI-based electrospun scaffold [109]. This effect on fibroblast proliferation was also observed with DC current stimulation delivered through PPy–PDLLA scaffolds [65]. In another study, human cutaneous fibroblasts stimulated on PPy–PLLA films showed greatly increased viability, mitochondrial activity, and IL-6 and IL-8 secretion[146] and [147]. Aortic endothelial cells cultured on fibronectin-coated PPy spread out when an oxidizing potential was applied, but were rounded and synthesized DNA to a lesser extent when the polymer was reduced to its neutral state [59]. Cardiomyocytes cultured on PANI–PLGA composite nanofibres have been shown to synchronize their beating as a result of ES [156].

These beneficial effects have been theorized to be the result of negative ion release from and positive ion (e.g. Na^{+}) uptake by the polymer [5] and [15], electrophoretic redistribution of cell surface receptors [5] and the increased adsorption of ECM molecules like fibronectin onto the polymer [26] and [142]. It is also worth noting that applying an electrical current through the conductive polymer will gradually increase its resistivity, thereby limiting its useful life time [37], and that long-term exposure of cells to high electrical currents (in the range of 1 mA and above) can have a cytotoxic effect [118].

THE ELECTROMECHANICAL EFFECT

PPy, PANI and PEDOT can undergo considerable volume change when being switched between redox states, suggesting another two potential applications [157], [158],[159] and [160]: to serve as bioactuators and artificial muscles [14], [16] and [161]. Similarly to natural muscle, these polymers are capable of exerting a controlled mechanical force based on an electrochemical reaction and can work in physiological fluids [16]. Conductive polymers are already used in, for example, blood vessel sealers and microgrippers [16]. Conductive polymer-based actuators require only a small amount of electrical potential (in the range of 1 V) to function, can exert quite strong mechanical forces (1 MPa) with volume changes of up to 35% [162], are lightweight and can function at physiological temperatures [15] and [16]. Their greatest limitation is their slow reaction speed, thus methods to increase the ion mobility in such polymers (i.e. allowing faster doping/dedoping) are currently being sought [15] and [16]. There are two phenomena behind the electrically induced volume change: (i) changes in the structure of the polymer backbone; and (ii) the osmotic movement of solvents in and out of the polymer [161]. The osmotic movement is caused by the change in ionic content in the polymer upon doping/dedoping during the redox switch [161]. The extent of the volume change and whether the polymer expands or contracts during reduction or oxidation depends on the choice of dopant [76] and [157]. For example, it has been noted that CS-, DS-, HA- and PMAS-doped PPy films expand, while PPy–pTS films contract during reduction [76]. This is because there are two different ways in which the polymer can maintain its charge balance when an electrical potential is applied: it can happen by either expulsing the dopant or, if the dopant is not mobile, by incorporating cations/anions from the electrolyte surrounding the polymer [157] and [163]. Expulsion leads to contraction, while incorporation results in expansion [157]. Therefore the mobility of the dopant defines whether the polymer will undergo contraction or expulsion when reduced/oxidized [157] and [163].

Mechanical stimulation has well-known benefits for tissue engineering and is widely used to influence the behaviour of cells [164] and [165]. With this in mind, Svennersten et al.[166] developed a PPy-based microactuator system on a chip. This system was successfully used to deliver mechanical stimulation to individual renal epithelial cells. The stimulation resulted in increased internal calcium levels, a known cellular response to mechanical stimulation [166], [167] and [168]. To the best of the authors' knowledge, no study, other than the one conducted by Svennersten et al. [166], has been performed exploiting the electromechanical capability of PPy, PANI and PEDOT for tissue engineering purposes. This is very surprising, considering that the properties of these polymers (e.g. the ability to function in physiological fluids [16] or at a low working voltage [162]) lend themselves to such an application.

THE NEXT GENERATION OF CONDUCTIVE POLYMERS

There is an ongoing effort to develop a more biocompatible and inherently biodegradable conductive polymer [18]. Conductive quaterthiophene and biodegradable ester have been combined to create QAPE, a novel polymer that was shown to support Schwann cells [18]. Another novel conductive polymer, ATQD, was synthesized from the emeraldine form of amino-capped aniline trimers [48]. When combined with RGD, ATQD was observed to support PC-12 adhesion and proliferation, and to induce spontaneous neuritogenesis even in the absence of neurotrophic growth factors [48]. Polypyrrole-thiophene (PPy–PTh) oligomers were successfully combined with ester linkages to create another biodegradable conductive polymer [4]. The degradation of PPy–PTh produced oligomers that could be consumed by macrophages, lowering the chance of any long-term adverse effect in vivo [4]. PPy was modified to create poly(3,4-alkylenedioxypyrrole) (PXDOP) [54]. PXDOP possesses increased electrochemical stability, retaining a greater proportion of its conductivity even after 3000 reduction–oxidation cycles, increasing the useful lifetime of the polymer [54].

Polylactide (PLA) and aniline pentamer (AP) were combined to create PLA-b-AP-b-PLA (PAP) [116]. PAP was found to be biodegradable and biocompatible, to have excellent processibility and to possess a conductivity similar to PANI's [116]. However, the mechanical properties of PAP are not good enough for practical application; hence the polymer was modified to give an alternative combination of PLA and AP: PLAAP [3]. PLAAP has all the advantages of PAP but possesses improved mechanical properties[3].

AP has also been synthesized with chitosan, and its presence it the composite was shown to be essential for inducing PC-12 differentiation and neurite formation [169].

PAP was combined with glycine ethyl ester to increase cell attachment, producing polyphosphazene [51]. Polyphosphazene is conductive, soluble in common organic solvents, degrades in water into harmless products, and was shown to support RSC96 Schwann cell adhesion and proliferations in vitro [51] and [115]. A great advantage of polyphosphazene is that it can be made with side groups of amino acid ester, imidazole, alcohol and polyether, making grafting of further biomolecules a relatively simple task[51] and [115]. PANI has been modified to produce poly(aniline-co-ethyl 3-aminobenzoate) [170] and poly(aniline-co-3-aminobenzoic acid) [171]. These copolymers, when electrospun together with PLA, have been noted not only to support the proliferation of COS-1 fibroblasts, but also to possess a very high antibacterial activity, making them excellent materials for skin tissue engineering and wound dressing substrates [170] and [171].

Conductive polymers can now be synthesized in many versatile ways. Their biocompatibility has been well characterized. Many methods are available for rendering them biodegradable. What can still be improved?

The biocompatibility of these polymers has mostly been tested with cancer-derived cell lines. These cell lines are very resilient; therefore, in the authors' opinion, it would be desirable to more extensively test conductive polymers using primary and stem cells. These cells are much more sensitive to the culture conditions and are the ones that will be used in a clinical setting. Many promising composite, modified and co-polymer forms of PPy, PANI and PEDOT exist, but the synthesis of these is not always straightforward. An inexpensive, off-the-shelf, easy-to-use version of them is desirable – one that does not require cytotoxic or hazardous reagents for synthesis. The fatigue of conductive polymers with repeated ES [11] is a serious hindrance. Novel polymers, like PXDOP [54], show great promise, but more work is still necessary in this regard. Reliable electrical performance, together with the ability to bind large quantities of molecules over the long term and release them in a well controllable manner, would be very beneficial for drug release applications and tissue engineering.

CONCLUSION

The greatest advantage of conductive polymers is their vast versatility (Fig. 9). The key to this is the dopant. The choice of dopant defines the properties of the polymer and allows its functionalization for a specific application [8] and [15]. Using ES, the dopant can be expelled and incorporated again into the polymer, allowing the control of these preset physical properties [3] and [16]. If the chosen biomolecule cannot be used as the dopant, it can still be incorporated using an intermediating doping molecule [36]. If doping is not the right way to make the conductive polymer better suited for an application, physical adsorption, entrapment and covalent bonding offer alternative routes [15]. Synthesizing chemically or electrochemically, perhaps as composites, electrospun fibres or hydrogels, can also be used to improve the usefulness of the end product [39], [55] and [57]. The conductive polymer substrate will be biocompatible [1] and [67], and can be made biodegradable through various methods [47] and [58]. Applying an electrical signal through a conductive polymer substrate allows the behaviour of the cells or tissue cultured upon it to be influenced [20], [23] and [109] using an electrical or electromechanical stimulus [157]. These multiple levels, where conductive polymers can be easily modified, grant a degree of control over the properties of the material, before and after synthesis, that no other material can provide. This is why we believe that we will soon be seeing a lot more examples of the application of this smart material in tissue engineering, and we hope that we were able to convince the reader of the same.

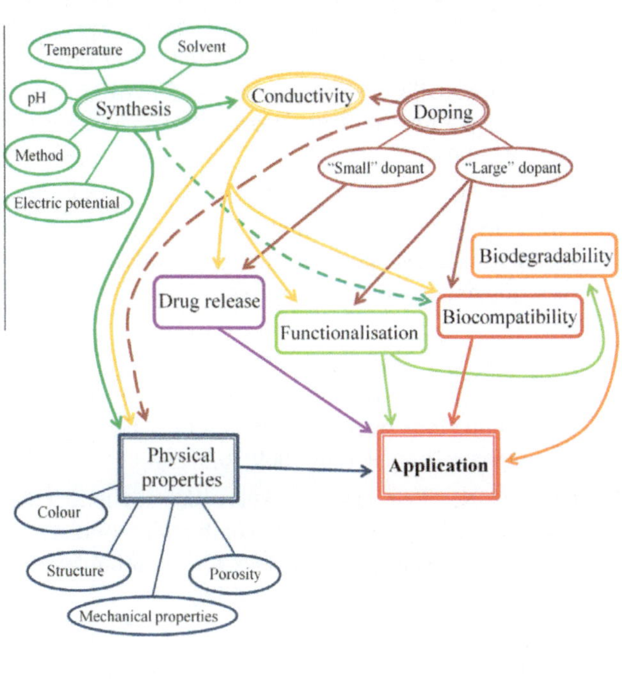

Figure 9: Everything is connected in the world of conductive polymers.

ACKNOWLEDGEMENTS

The authors thank Orthopaedic Research UK and the EPSRC DTA award for funding this work. We also thank Dr. Jimbo Dugan and Judit Balintne Farkas for their help in preparing the manuscript.

REFERENCES

1. Lakard B, Ploux L, Anselme K, Lallemand F, Lakard S, Nardin M, et al. Effect of ultrasounds on the electrochemical synthesis of polypyrrole, application to the adhesion and growth of biological cells. Bioelectrochemistry 2009;75:148–57.
2. Ghasemi-Mobarakeh L et al. Application of conductive polymers, scaffolds and electrical stimulation for nerve tissue engineering. J Tissue Eng Regen Med 2011;5:e17–35.
3. Huang L et al. Synthesis of biodegradable and electroactive multiblock polylactide and aniline pentamer copolymer for tissue engineering applications. Biomacromolecules 2008;9:850–8.
4. Rivers TJ, Hudson TW, Schmidt CE. Synthesis of a novel, biodegradable electrically conducting polymer for biomedical applications. Adv Funct Mater 2002;12:33–7.
5. Kotwal A, Schmidt CE. Electrical stimulation alters protein adsorption and nerve cell interactions with electrically conducting biomaterials. Biomaterials 2001;22:1055–64.
6. Lee JY, Bashur CA, Goldstein AS, Schmidt CE. Polypyrrole-coated electrospun PLGA nanofibers for neural tissue applications. Biomaterials 2009;30:4325–35.
7. Wallace GG, Smyth M, Zhao H. Conducting electroactive polymer-based biosensors. Trends Analyt Chem 1999;18:245–51.
8. Kim DH, Richardson-Burns SM, Hendricks JL, Sequera C, Martin DC. Effect of immobilized nerve growth factor on conductive polymers: electrical properties and cellular response. Adv Funct Mater 2007;17:79–86.
9. Garner B, Georgevich A, Hodgson AJ, Liu L, Wallace GG. Polypyrrole–heparin composites as stimulus-responsive substrates for endothelial cell growth. Inc. J Biomed Mater Res 1999;44:121–9.
10. Aoki T, Tanino M, Sanui K, Ogata N, Kumakura K. Secretory function of adrenal chromaffin cells cultured on polypyrrole films. Biomaterials 1996;17:1971–4.
11. Guiseppi-Elie A. Electroconductive hydrogels: synthesis, characterization and biomedical applications. Biomaterials 2010;31:2701–16.
12. Shirakawa H, Louis EJ, MacDiarmid AG, Chiang CK, Heeger AJ. Synthesis of electrically conducting organic polymers: halogen derivatives of polyacetylene, (CH)x. J Chem Soc Chem Commun 1977:578–80.
13. Ateh DD, Navsaria HA, Vadgama P. Polypyrrole-based conducting polymers and interactions with biological tissues. J R Soc Interface 2006;3:741–52.
14. Zhou DD, Cui XT, Hines A, Greenberg RJ. Conducting

polymers in neural stimulation applications. In: Zhou DD, Greenbaum E, editors. Implantable neural prostheses, vol. 2. Berlin: Springer; 2010. p. 217–52.

15. Guimard NK, Gomez N, Schmidt CE. Conducting polymers in biomedical engineering. Prog Polym Sci 2007;32:876–921.
16. Cortés MT, Moreno JC. Artificial muscles based on conducting polymers. ePolymers 2003;4:1–42.
17. Kim DH, Abidian M, Martin DC. Conducting polymers grown in hydrogel scaffolds coated on neural prosthetic devices. J Biomed Mater Res 2004;71A:577–85.
18. Guimard NKE, Sessler JL, Schmidt CE. Toward a biocompatible and biodegradable copolymer incorporating electroactive oligothiophene units. Macromolecules 2009;42:502–11.
19. Li M, Guo Y, Wei Y, MacDiarmid AG, Lelkes PI. Electrospinning polyanilinecontained gelatin nanofibers for tissue engineering applications. Biomaterials 2006;27:2705–15.
20. Liu X, Gilmore KJ, Moulton SE, Wallace GG. Electrical stimulation promotes nerve cell differentiation on polypyrrole/poly(2-methoxy-5 aniline sulfonic acid) composites. J Neural Eng 2009;6:1–10.
21. Garner B, Hodgson AJ, Wallace GG, Underwood PA. Human endothelial cell attachment to and growth on polypyrrole–heparin is fibronectin dependent. J Mater Sci Mater Med 1999;10:19–27.
22. Cui X, Hetke JE, Wiler JA, Anderson DJ, Martin DC. Electrochemical deposition and characterization of conducting polymer polypyrole/PSS on multichannel neural probes. Sens Actuator A-Phys 2001;93:8–18.
23. Lee JY, Lee JW, Schmidt CE. Neuroactive conducting scaffolds: nerve growth factor conjugation on active ester-functionalized polypyrrole. J R Soc Interface 2009;6:801–10.
24. Ateh DD, Vadgama P, Navsaria HA. Culture of human keratinocytes on polypyrrole-based conducting polymers. Tissue Eng 2006;12:645–55.
25. Bousalem S, Mangeney C, Chehimi MM, Basinska T, Miksa B, Slomkowski S. Synthesis, characterization and potential biomedical applications of Nsuccinimidyl ester functionalized, polypyrrole-coated polystyrene latex particles. Colloid Polym Sci 2004;282:1301–7.
26. Li Y, Neoh KG, Kang ET. Plasma protein adsorption and thrombus formation on surface functionalized polypyrrole with and without electrical stimulation. J Colloid Interf Sci 2004;275:488–95.
27. George PM, LaVan DA, Burdick JA, Chen CY, Liang E, Langer R. Electrically controlled drug delivery from biotin-doped conductive polypyrrole. Adv Mater 2006;18:577–81.
28. Meng S, Rouabhia M, Shi G, Zhang Z. Heparin dopant increases the electrical stability, cell adhesion, and growth of conducting polypyrrole/poly(L,Llactide) composites. J Biomed Mater Res 2008;87A:332–44.
29. Cetiner S, Kalaoglu F, Karakas H, Sarac AS. Electrospun nanofibers of polypyrrole–poly(acrylonitrile-co-vinyl acetate). Text Res J 2010;80:1784–92.
30. Akkouch A, Shi G, Zhang Z, Rouabhia M. Bioactivating electrically conducting polypyrrole with fibronectin and bovine serum albumin. J Biomed Mater Res 2010;92A:221–31.
31. Ferraz N, Strømme M, Fellstrom B, Pradhan S, Nyholm L, Mihranyan A. In vitro and in vivo toxicity of rinsed and aged nanocellulose–polypyrrole composites. J Biomed Mater Res A 2012;100a:2128–38.
32. Zhang X, Manohar SK. Bulk synthesis of polypyrrole nanofibers by a seeding approach. J Am Chem Soc 2004;126:12714–5.
33. Chronakis IS, Grapenson S, Jakob A. Conductive polypyrrole nanofibers via electrospinning: electrical and morphological properties. Polymer 2006;47:1597–603.
34. Bousalem S, Yassar A, Basinska T, Miksa B, Slomkowski S, Azioune A, et al. Synthesis, characterization and biomedical applications of functionalized polypyrrole-coated polystyrene latex particles. Polym Adv Technol 2003;14:820–5.
35. Gomez N, Schmidt CE. Nerve growth factor-immobilized polypyrrole: bioactive electrically conducting polymer for enhanced neurite extension. J Biomed Mater Res A 2007;81(1):135–49.
36. Song HK, Toste B, Ahmann K, Hoffman-Kim D, Palmore GTR. Micropatterns of positive guidance cues anchored to polypyrrole doped with polyglutamic acid: a new platform for characterizie neurite extension in complex environments. Biomaterials 2006;27:473–84.
37. Zhang Z et al. Electrically conductive biodegradable polymer composite for nerve regeneration: electricity-stimulated neurite outgrowth and axon regeneration. Artif Organs 2007;31(1):13–22.
38. Brahim S, Guiseppi-Elie A. Electroconductive hydrogels: electrical and electrochemical properties of polypyrrole–poly(HEMA) composites. Electroanalysis 2005;17:556–70.
39. Lee JW, Serna F, Nickels J, Schmidt CE. Carboxylic acid-functionalized conductive polypyrrole as a bioactive platform for cell adhesion. Biomacromolecules 2006;7:1692–5.
40. Castano H, O'Rear EA, McFetridge PS, Sikavitsas VI. Polypyrrole thin films formed by admicellar polymerization support the osteogenic differentiation of mesenchymal stem cells. Macromol Biosci 2004;4:785–94.
41. Bettinger CJ, Bruggeman JP, Misra A, Borenstein JT, Langer R. Biocompatibility of biodegradable semiconducting melanin films for nerve tissue engineering. Biomaterials 2009;30:3050–7.
42. Gomez N, Lee JY, Nickels JD, Schmidt CE. Micropatterned polypyrrole: a combination of electrical and topographical characteristics for the dtimulation of cells. Adv Funct Mater 2007;17:1645–53.
43. George PM et al. Fabrication and biocompatibility of polypyrrole implants suitable for neural prosthetics. Biomaterials 2005;26:3511–9.
44. Alikacem N, Marois Y, Zhang Z, Jakubiec B, Roy R,

King MW, et al. Tissue reactions to polypyrrole-coated polyesters: a magnetic resonance relaxometry study. Artif Organs 1999;23(10):910–9.

45. Blinova NV, Stejskal J, Trchova M, Prokes J. Control of polyaniline conductivity and contact angles by partial protonation. Polym Int 2008;57:66–9.

46. Cullen DK, Patel AR, Doorish JF, Smith DH, Pfister BJ. Developing a tissueengineered neural-electrical relay using encapsulated neuronal constructs on conducting polymer fibers. J Neural Eng 2008;5:374–84.

47. Borriello A, Guarino V, Schiavo L, Alvarez-Perez MA, Ambrosio L. Optimizing PANi doped electroactive substrates as patches for the regeneration of cardiac muscle. J Mater Sci: Mater Med 2011;22:1053–62.

48. Guo Y et al. Electroactive oligoaniline-containing self-assembled monolayers for tissue engineering applications. Biomacromolecules 2007;8:3025–34.

49. Prabhakaran MP, Ghasemi-Mobarakeh L, Jin G, Ramakrishna S. Electrospun conducting polymer nanofibers and electrical stimulation of nerve stem cells. J Biosci Bioeng 2011;112:501–7.

50. Yu QZ, Shi MM, Deng M, Wang M, Chen HZ. Morphology and conductivity of polyaniline sub-micron fibers prepared by electrospinning. Mater Sci Eng BSolid 2008;150:70–6.

51. Zhang QS, Yan YH, Li SP, Feng T. Synthesis of a novel biodegradable and electroactive polyphosphazene for biomedical application. Biomed Mater 2009;4:1–9.

52. Humpolicek P, Kasparkova V, Saha P, Stejskal J. Biocompatibility of polyaniline. Synth Met 2012;162:722–7.

53. Peramo A, Urbanchek MG, Spanninga SA, Povlich LK, Cederna P, Martin DC. In situ polymerization of a conductive polymer in acellular muscle tissue constructs. Tissue Eng Part A 2008;14:423–32.

54. Thomas CA, Zong K, Schottland P, Reynolds JR. Poly(3,4- alkylenedioxypyrrole)s as highly stable aqueous-compatible conducting polymers with biomedical implications. Adv Mater 2000;12:222–5.

55. Asplund M, Thaning E, Lundberg J, Sandberg-Nordqvist AC, Kostyszyn B, Inganäs O, et al. Toxicity evaluation of PEDOT/biomolecular composites intended for neural communication electrodes. Biomed Mater 2009;4: 1–12.

56. Richardson-Burns SM, Hendricks JL, Martin DC. Electrochemical polymerization of conducting polymers in living neural tissue. J Neural Eng 2007;4:L6–L13.

57. Subramanian A, Krishnan UM, Sethuraman S. Axially aligned electrically conducting biodegradable nanofibers for neural regeneration. J Mater Sci: Mater Med 2012;23:1797–809.

58. Ravichandran R, Sundarrajan S, Venugopal JR, Mukherjee S, Ramakrishna S. Applications of conducting polymers and their issues in biomedical engineering. J R Soc Interface 2010;7:S559–79.

59. Wong JY, Langer R, Ingber DE. Electrically conducting polymers can noninvasively control the shape and growth of mammalian cells. Proc Natl Acad Sci USA 1994;91:3201–4.

60. Wallace G, Spinks G. Conducting polymers – bridging the bionic interface. Soft Matter 2007;3:665–71.

61. Bredas JL, Streer GB. Polarons, bipolarons and solitons in conductive polymers. Acc Chem Res 1985;18:309–15.

62. Pelto J et al. Electroactivity and biocompatibility of polypyrrolehyaluronic acid multi-walled carbon nanotube composite. J Biomed Mater Res 2010;93A:1056–67.

63. Dai L. Conducting polymers. In: Dai L, editor. Intelligent macromolecules for smart devices: from materials synthesis to device applications. London: Springer; 2004. p. 41–80.

64. Kaynak A, Rintoul L, George GA. Change of mechanical and electrical properties of polypyrrole films with dopant concentration and oxidative aging. Mater Res Bull 2000;35:813–24.

65. Shi G, Rouabhia M, Wang Z, Dao LH, Zhang Z. A novel electrically conductive and biodegradable composite made of polypyrrole nanoparticles and polylactide. Biomaterials 2004;25:2477–88.

66. Ghasemi-Mobarakeh L, Prabhakaran MP, Morshed M, Nasr-Esfahani MH, Ramakrishna S. Electrical stimulation of nerve cells using conductive nanofibrous scaffolds for nerve tissue engineering. Tissue Eng Part A 2009;15:3605–19.

67. Bidez III PR, Li S, Macdiarmid AG, Venancio EC, Wei Y, Lelkes PI. Polyaniline, an electroactive polymer, supports adhesion and proliferation of cardiac myoblasts. J Biomater Sci Polym Ed 2006;17:199–212.

68. Gilmore KJ et al. Skeletal muscle cell proliferation and differentiation on polypyrrole substrates doped with extracellular matrix component. Biomaterials 2009;30:5292–304.

69. Ranella A, Barberoglou M, Bakogianni S, Fotakis C, Stratakis E. Tuning cell adhesion by controlling the roughness and wettability of 3D micro/nano silicon structures. Acta Biomater 2010;6:2711–20.

70. Huang HH, Ho CT, Lee TH, Lee TL, Liao KK, Chen FL. Effect of surface roughness of ground titanium on initial cell adhesion. Biomol Eng 2004;21:93–7.

71. Deligianni DD, Katsala ND, Koutsoukos PG, Missirlis YF. Effect of surface roughness of hydroxyapatite on human bone marrow cell adhesion, proliferation, differentiation and detachment strength. Biomaterials 2001;22:87–96.

72. Rosales-Leal JI et al. Effect of roughness, wettability and morphology of engineered titanium surfaces on osteoblast-like cell adhesion. Colloid Surf A: Physicochem Eng Aspects 2010;365:222–9.

73. Collier JH, Camp JP, Hudson TW, Schmidt CE. Synthesis and characterization of polypyrrole–hyaluronic acid composite biomaterials for tissue engineering applications. J Biomed Mater Res 2000;50:574–84.

74. Ramanaviciene A, Kausaite A, Tautkus S, Ramanavicius A. Biocompatibility of polypyrrole particles: an in vivo study in mice. J Pharm Pharmacol 2007;59:311–5.

75. Fonner JM, Schmidt CE, Ren P. A combined molecular dynamics and experimental study of doped polypyrrole. Polymer 2010;51:4985–93.

76. Gelmi A, Higgins MJ, Wallace GG. Physical surface and electromechanical properties of doped polypyrrole biomaterials. Biomaterials 2010;31:1974–83.

77. Moreno JS, Panero S, Artico M, Filippini P. Synthesis and characterization of new electroactive polypyrrole–chondroitin sulphate A substrates. Bioelectrochemistry 2008;72:3–9.

78. Han DH, Lee HJ, Park SM. Electrochemistry of conductive polymers. XXXV. Electrical and morphological characteristics of polypyrrole films prepared in aqueous media studied by current sensing atomic force microscopy. Electrochim Acta 2005;50:3085–92.

79. Silk T, Hong Q, Tamm J, Compton RG. AFM studies of polypyrrole film surface morphology. I. The influence of film thickness and dopant nature. Synth Met 1998;93:59–64.

80. Hodgson AJ, Gilmore K, Small C, Wallace GG, Mackenzie IL, Aoki T, et al. Reactive supramolecular assemblies of mucopolysaccharide, polypyrrole and protein as controllable biocomposites for a new generation of "intelligent biomaterials". Supramol Sci 1994;1:77–83.

81. Tan Y, Ghandi K. Kinetics and mechanism of pyrrole chemical polymerization. Synth Met 2013;175:183–91.

82. Armes SP. Optimum reaction conditions for the polymerization of pyrrole by iron (III) chloride in aqueous solution. Synth Met 1987;20:365–71.

83. Calvo PA, Rodriguez J, Grande H, Mecerreyes D, Pomposo JA. Chemical oxidative polymerization of pyrrole in the presence of m-hydroxybenzoic acid- and m-hydroxycinnamic acid-related compounds. Synth Met 2002;126:111–6.

84. Cao Y, Andreatta A, Heeger AJ, Smith P. Influence of chemical polymerization conditions on the properties of polyaniline. Polymer 1989;30:2305–11.

85. Kudoh Y, Akami K, Matsuya Y. Chemical polymerization of 3,4- ethylenedioxythiophene using an aqueous medium containing an anionic surfactant. Synth Met 1998;98:65–70.

86. Pron A, Genoud F, Menardo C, Nechtschein M. The effect of the oxidation conditions on the chemical polymerization of polyaniline. Synth Met 1988;24:193–201.

87. Martins NCT, Moura e Silva T, Montemora MF, Fernandes JCS, Ferreira MGS. Electrodeposition and characterization of polypyrrole films on aluminium alloy 6061-T6. Electrochim Acta 2008;53:4754–63.

88. Herrasti P, Díaz L, Ocón P, Ibáñez A, Fatas E. Electrochemical and mechanical properties of polypyrrole coatings on steel. Electrochim Acta 2004;49:3693–9.

89. Choi S, Park S. Electrochemistry of conductive polymers. XXVI. Effects of electrolytes and growth methods on polyaniline morphology. J Electrochem Soc 2002;149:E26–34.

90. Li CM, Sun CQ, Chen W, Pan L. Electrochemical thin film deposition of polypyrrole on different substrates. Surf Coat Technol 2005;198:474–7.

91. Kaplin DA, Qutubuddin S. Electrochemically synthesized polypyrrole films: effects of polymerization potential and electrolyte type. Polymer 1995;36:1275–86.

92. Sutton SJ, Vaughan AS. On the morphology and growth of electrochemically polymerized polypyrrole. Polymer 1995;36:1849–57.

93. Kaynak A. Effect of synthesis parameters on the surface morphology of conducting polypyrrole films. Mater Res Bull 1997;32:271–85.

94. Patra S, Barai K, Munichandraiah N. Scanning electron microscopy studies of PEDOT prepared by various electrochemical routes. Synth Met 2008;158:430–5.

95. Mondal SK, Prasad KR, Munichandraiah N. Analysis of electrochemical impedance of polyaniline films prepared by galvanostatic, potentiostatic and potentiodynamic methods. Synth Met 2005;148:275–86.

96. Girija TC, Sangaranarayanan MV. Investigation of polyaniline-coated stainless steel electrodes for electrochemical supercapacitors. Synth Met 2006;156:244–50.

97. Zhang Z, Roy R, Dugre FJ, Tessier D, Dao LH. In vitro biocompatibility study of electrically conductive polypyrrole-coated polyester fabrics. J Biomed Mater Res 2001;57:63–71.

98. Jiang X, Marois Y, Traoré A, Tessier D, Dao LH, Guidoin R, et al. Tissue reaction to polypyrrole-coated polyester fabrics: an in vivo study in rats. Tissue Eng 2002;8:635–47.

99. Tessier D, Dao LH, Zhang Z, King MW, Guidoin R. Polymerization and surface analysis of electrically-conductive polypyrrole on surface-activated polyester fabrics for biomedical applications. J Biomater Sci Polym Ed 2000;11:87–99.

100. Wang Z, Roberge C, Wan Y, Dao LH, Guidoin R, Zhang Z. A biodegradable electrical bioconductor made of polypyrrole nanoparticle/poly(D,L-lactide) composite: a preliminary in vitro biostability study. J Biomed Mater Res 2003;66A:738–46.

101. Wang Z et al. In vivo evaluation of a novel electrically conductive polypyrrole/poly(D,L-lactide) composite and polypyrrolecoated poly(D,Llactide-co-glycolide) membranes. Biomed Mater Res 2004;70A:28–38.

102. Zare EN, Lakouraj MM, Mohseni M. Biodegradable polypyrrole/dextrin conductive nanocomposite: synthesis, characterization, antioxidant and antibacterial activity. Synth Met 2014;187:9–16.

103. Khor E, Li HC, Wee A. In situ polymerization of pyrrole in animal tissue in the formation of hybrid biomaterials. Biomaterials 1995;16:657–61.

104. Khor E, Li HC, Wee A. Animal tissue–polypyrrole hybrid biomaterials: shrinkage temperature evaluation. Biomaterials 1996;17:1877–9.

105. Chen S et al. PEDOT/MWCNT composite film coated microelectrode arrays for neural interface improvement. Sens Actuator A 2013;193:141–8.

106. Luo X, Weaver CL, Zhou DD, Greenberg R, Cui XT. Highly stable carbon nanotube doped poly(3,4-

ethylenedioxythiophene) for chronic neural stimulation. Biomaterials 2011;32:5551–7.

107. Aznar-Cervantes S, Roca MI, Martinez JG, Meseguer-Olmo L, Cenis JL, Moraleda JM, et al. Fabrication of conductive electrospun silk fibroin scaffolds by coating with polypyrrole for biomedical applications. Bioelectrochemistry 2012;85:36–43.

108. Ku SH, Lee SH, Park CB. Synergic effects of nanofiber alignment and electroactivity on myoblast differentiation. Biomaterials 2012;33:6098–104.

109. Jeong SI, Jun ID, Choi MJ, Nho YC, Lee YM, Shin H. Development of electroactive and elastic nanofibers that contain polyaniline and poly(Llactide-co-e-caprolactone) for the control of cell adhesion. Macromol Biosci 2008;8:627–37.

110. Chen MC, Sun YC, Chen YH. Electrically conductive nanofibers with highly oriented structures and their potential application in skeletal muscle tissue engineering. Acta Biomater 2013;9:5562–72.

111. Small CJ, Too CO, Wallace GG. Responsive conducting polymer–hydrogel composites. Polym Gels Netw 1997;5:251–65.

112. Justin G, Guiseppi-Elie A. Electroconductive blends of poly(HEMA-coPEGMA-co-HMMAco- SPMA) and poly(Py-co-PyBA): in vitro biocompatibility. J Bioact Compat Pol 2010;25:121–40.

113. Zhang Q, Yan Y, Li S, Feng T. The synthesis and characterization of a novel biodegradable and electroactive polyphosphazene for nerve regeneration. Mater Sci Eng C 2010;30:160–6.

114. Huang L et al. Synthesis and characterization of electroactive and biodegradable ABA block copolymer of polylactide and aniline pentamer. Biomaterials 2007;28:1741–51.

115. Cui X, Lee VA, Raphael Y, Wiler JA, Hetke JF, Anderson DJ, et al. Surface modification of neural recording electrodes with conducting polymer/ biomolecule blends. Biomed Mater Res 2001;56:261–72.

116. Williams RL, Doherty PJ. A preliminary assessment of poly(pyrrole) in nerve guide studies. J Mater Sci Mater Med 1994;5:429–33.

117. Wang X et al. Evaluation of biocompatibility of polypyrrole in vitro and in vivo. J Biomed Mater Res 2004;68A:411–22.

118. Richardson RT et al. The effect of polypyrrole with incorporated neurotrophin-3 on the promotion of neurite outgrowth from auditory neurons. Biomaterials 2007;28:513–23.

119. Jakubiec B et al. In vitro cellular response to polypyrrole-coated woven polyester fabrics: potential benefits of electrical conductivity. J Biomed Mater Res 1998;41:519–26.

120. Kim S, Oh WK, Jeong YS, Hong JY, Cho BR, Hahn JS, et al. Cytotoxicity of, and innate immune response to, size-controlled polypyrrole nanoparticles in mammalian cells. Biomaterials 2011;32:2342–50.

121. Kamalesh S, Tan P, Wang J, Lee T, Kang ET, Wang CH. Biocompatibility of electroactive polymers in tissues. J Biomed Mater Res 2000;52:467–78.

122. Bolin MH, Svennersten K, Wang X, Chronakis IS, Richter-Dahlfors A, Jager EWH, et al. Nano-fiber scaffold electrodes based on PEDOT for cell stimulation. Sens Actuator B 2009;142:451–6.

123. Karagkiozaki V, Karagiannidis PG, Gioti M, Kavatzikidou P, Georgiou D, Georgaraki E, et al. Bioelectronics meets nanomedicine for cardiovascular implants: PEDOT-based nanocoatings for tissue regeneration. Biochim Biophys Acta 2013;1830:4294–304.

124. Luo SC, Ali EM, Tansil NC, Yu HH, Gao S, Kantchev EAB, et al. Poly(3,4- ethylenedioxythiophene) (PEDOT) nanobiointerfaces: thin, ultrasmooth, and functionalized PEDOT films with in vitro and in vivo biocompatibility. Langmuir 2008;24:8071–7.

125. Zhou H, Cheng X, Rao L, Li T, Duan YY. Poly(3,4-ethylenedioxythiophene)/ multiwall carbon nanotube composite coatings for improving the stability of microelectrodes in neural prostheses applications. Acta Biomater 2013;9:6439–49.

126. Abidian MR, Ludwig KA, Marzullo TC, Martin DC, Kipke DR. Interfacing conducting polymer nanotubes with the central nervous system: chronic neural recording using poly(3,4-ethylenedioxythiophene) nanotubes. Adv Mater 2009;21:3764–70.

127. Ahuja T, Mir IA, Kumar D, Rajesh. Biomolecular immobilization on conducting polymers for biosensing applications. Biomaterials 2007;28:791–805.

128. Arora K et al. Application of electrochemically prepared polypyrrole– polyvinyl sulphonate films to DNA biosensor. Biosens Bioelectron 2006;21:1777–83.

129. Cosnier S. Biomolecule immobilization on electrode surfaces by entrapment or attachment to electrochemically polymerized films. A review. Biosens Bioelectron 1999;14:443–56.

130. Genies EM, Marchesiello M. Conducting polymers for biosensors, application to new glucose sensors GOD entrapped into polypyrrole, GOD adsorbed on poly(3-methylthiophene). Synth Met 1993;55–57:3677–82.

131. Prabhakar N, Arora K, Singh SP, Singh H, Malhotra BD. DNA entrapped polypyrrole–polyvinyl sulfonate film for application to electrochemical biosensor. Anal Biochem 2007;366:71–9.

132. Nien PC, Tung TS, Hoa KC. Amperometric glucose biosensor based on entrapment of glucose oxidase in a poly(3,4-ethylenedioxythiophene) film. Electroanalysis 2006;18:1408–15.

133. Wang J, Myung NV, Yun M, Monbouquette HG. Glucose oxidase entrapped in polypyrrole on high-surface-area Pt electrodes: a model platform for sensitive electroenzymatic biosensors. J Electroanal Chem 2005;575:139–46.

134. Jha SK, Kanungo M, Nath A, D'Souza SF. Entrapment of live microbial cells in electropolymerized polyaniline and their use as urea biosensor. Biosens Bioelectron 2009;24:2637–42.

135. Collazos-Castro JE, Polo JL, Hernández-Labrado GR, Padial-Cañete V, GarcíaRama C. Bioelectrochemical

control of neural cell development on conducting polymers. Biomaterials 2010;31:9244–55.

136. De Giglio E, Sabbatini L, Zambonin PG. Development and analytical characterization of cysteine-grafted polypyrrole films electrosynthesized on Pt and Ti-substrates as precursors of bioactive interfaces. Biomater Sci Polym Ed 1999;10:845–58.

137. De Giglio E, Sabbatini L, Colucci S, Zambonin G. Synthesis, analytical characterization, and osteoblast adhesion properties on RGD-grafted polypyrrole coatings on titanium substrates. J Biomater Sci Polym Ed 2000;11:1073–83.

138. Bax DV et al. Cell patterning via linker-free protein functionalization of an organic conducting polymer (polypyrrole) electrode. Acta Biomater 2012;8:2538–48.

139. Cen L, Neoh KG, Li Y, Kang ET. Assessment of in vitro bioactivity of hyaluronic acid and sulfated hyaluronic acid functionalized electroactive polymer. Biomacromolecules 2004;5:2238–46.

140. Thompson BC, Richardson RT, Moulton SE, Evans AJ, O'Leary S, Clark GM, et al. Conducting polymers, dual neurotrophins and pulsed electrical stimulation – dramatic effects on neurite outgrowth. J Control Release 2010;141:161–7.

141. Wadhwa R, Lagenaur CF, Cui XT. Electrochemically controlled release of dexamethasone from conducting polymer polypyrrole coated electrode. J Control Release 2006;110:531–41.

142. Li Y, Neoh KG, Kang ET. Controlled release of heparin from polypyrrole– poly(vinyl alcohol) assembly by electrical stimulation. J Biomed Mater Res 2005;73A:171–81.

143. Stauffer WR, Cui XT. Polypyrrole doped with 2 peptide sequences from laminin. Biomaterials 2006;27:2405–13.

144. Zhang L, Stauffer WR, Jane EP, Sammak PJ, Cui XT. Enhanced differentiation of embryonic and neural stem cells to neuronal fates on laminin peptides doped polypyrrole. Macromol Biosci 2010;10:1456–64.

145. Guo B, Glavas L, Albertsson AC. Biodegradable and electrically conducting polymers for biomedical applications. Prog Polym Sci 2013;38:1263–86.

146. Shi G, Rouabhia M, Meng S, Zhang Z. Electrical stimulation enhances viability of human cutaneous fibroblasts on conductive biodegradable substrates. J Biomed Mater Res 2008;84A:1026–37.

147. Shi G, Zhang Z, Rouabhia M. The regulation of cell functions electrically using biodegradable polypyrrole–polylactide conductors. Biomaterials 2008;29:3792–8.

148. Zelikin AN, Lynn DM, Farhadi J, Martin I, Shastri V, Langer R. Erodible conducting polymers for potential biomedical applications. Angew Chem Int Ed 2002;41:141–4.

149. Massoumi B, Entezami AA. Electrochemically stimulated 2-ethylhexyl phosphate (EHP) release through redox switching of conducting polypyrrole film and polypyrrole/poly(N-methylpyrrole) or self-doped polyaniline bilayers. Polym Int 2002;51:555–60.

150. Miller LL, Zhou XU. Poly(N-methylpyrrolylium) poly(styrenesu1fonate). A conductive, electrically switchable cation exchanger that cathodically binds and anodically releases dopamine. Macromolecules 1987;20:1594–7.

151. Kontturi K, Pentti P, Sundholm G. Polypyrrole as a model membrane for drug delivery. J Electroanal Chem 1998;453:231–8.

152. Liu X, Yue Z, Higgins MJ, Wallace GG. Conducting polymers with immobilised fibrillar collagen for enhanced neural interfacing. Biomaterials 2011;32:7309–17.

153. Xu H et al. Conductive PPY/PDLLA conduit for peripheral nerve regeneration. Biomaterials 2014;35:225–35.

154. Weng B, Liu X, Shepherd R, Wallace GG. Inkjet printed polypyrrole/collagen scaffold: a combination of spatial control and electrical stimulation of PC12 cells. Synth Met 2012;162:1375–80.

155. Huang J, Hu X, Lu L, Ye Z, Zhang Q, Luo Z. Electrical regulation of Schwann cells using conductive polypyrrole/chitosan polymers. J Biomed Mater Res 2010;93A:164–74.

156. Hsiao CW et al. Electrical coupling of isolated cardiomyocyte clusters grown on aligned conductive nanofibrous meshes for their synchronized beating. Biomaterials 2013;34:1063–72.

157. Gandhi MR, Murray P, Spinks GM, Wallace GG. Mechanism of electromechanical actuation in polypyrrole. Synth Met 1995;73:247–56.

158. Kaneto K, Kaneko M, Min Y, MacDiarmid AG. "Artificial muscle": electromechanical actuators using polyanilinefilms. SynthMet 1995;71:2211–2.

159. Madden JD, Cush RA, Kanigan TS, Hunter IW. Fast contracting polypyrrole actuators. Synth Met 2000;113:185–92.

160. Okuzaki H, Suzuki H, Ito T. Electrically driven PEDOT/PSS actuators. Synth Met 2009;159:2233–6.

161. Bay L, Jacobsen T, Skaarup S. Mechanism of actuation in conducting polymers: osmotic expansion. J Phys Chem B 2001;105:8492–7.

162. Smela E, Gadegaard N. Surprising volume change in PPy(DBS): an atomic force microscopy study. Adv Mater 1999;11:953–7.

163. Jager EWH. Conducting polymer actuators for medical devices and cell mechanotransduction. In: IEEE/ASME international conference on advanced intelligent, mechatronics; 2013. p. 1661–6.

164. Chiu LLY, Radisic M. Cardiac tissue engineering. Curr Opin Chem Eng 2013;2:41–52.

165. Subramony SD, Dargis BR, Castillo M, Azeloglu EU, Tracey MS, Su A, et al. The guidance of stem cell differentiation by substrate alignment and mechanical stimulation. Biomaterials 2013;34:1942–53.

166. Svennersten K, Berggren M, Richter-Dahlfors A, Jager EWH. Mechanical stimulation of epithelial cells using polypyrrole microactuators. Lab Chip 2011;11:3287.

167. Demer LL, Wortham CM, Dirksen ER, Sanderson MJ. Mechanical stimulation induces intercellular calcium signalling in bovine aortic endothelial cells. Am J

Physiol-Heart C 1993;264:H2094–102.

168. Jager EWH, Bolin MH, Svennersten K, Wang X, Richter-Dahlfors A, Berggren M. Electroactive surfaces based on conducting polymers for controlling cell adhesion, signalling, and proliferation. In: International solid-state sensors, actuators and microsystems conference; 2009. p. 1778–81.

169. Hu J et al. Electroactive aniline pentamer cross-linking chitosan for stimulation growth of electrically sensitive cells. Biomacromolecules 2008;9:2637–44.

170. Gizdavic-Nikolaidis M, Ray S, Bennett J, Swift S, Bowmaker G, Easteal A. Electrospun poly(aniline-co-ethyl 3-aminobenzoate)/poly(lactic acid) nanofibers and their potential in biomedical applications. J Polym Sci Part A: Polym Chem 2011;49:4902–10.

171. Gizdavic-Nikolaidis M, Ray S, Bennett JR, Easteal AJ, Cooney RP. Electrospun functionalized polyaniline copolymer-based nanofibers with potential application in tissue engineering. Macromol Biosci 2010;10:1424–31.

172. Kumar D, Sharma RC. Advances in conductive polymers. Eur Polym J 1998;34:1053–60.

173. Wan M. Introduction of conducting polymers. In: Wan M, editor. Conducting polymers with micro or nanometer structure. Berlin: Springer; 2008. p. 1–15.

174. Epstein AJ. Electrical conductivity in conjugated polymers. In: Rupprecht L, editor. Conductive polymers and plastics in industrial applications. Norwich, NY: William Andrew Publishing/Plastics Design Library; 1999. p. 1–9.

175. Kirchmeyer S, Reuter K. Scientific importance, properties and growing applications of poly(3,4-ethylenedioxythiophene). J Mater Chem 2005;15:2077–88.

176. Ribo JM, Acero C, Anglada MC, Dicko A, Tura JM, Ferrer-Anglada N, et al. On the structure and transport properties of polypyrroles. Butll Soc Cat Cien 1992;13:335–51.

177. Street GB, Lindsey SE, Nazzal AI, Wynne KJ. The structure and mechanical properties of polypyrrole. Mol Cryst Liq Cryst 1985;118:137–48.

178. MacDiarmid AG, Epstein AJ. Secondary doping in polyaniline. Synth Met 1995;69:85–92.

179. MacDiarmid AG, Chiang JC, Richter AF. Polyaniline: a new concept in conducting polymers. Synth Met 1987;18:285–90.

180. Pouget JP, Jdzefowiczt ME, Epstein AJ, Tang X, MacDiarmid AG. X-ray structure of polyaniline. Macromolecules 1991;24:779–89.

181. Tourillon G, Garnier F. Morphology and crystallographic, structure of polythiophene and derivatives. Mol Cryst Liq Cryst 1985;118:221–6.

182. Batz P, Schmeisser D, Gopel W. Electronic structure of polypyrrole films. Phys Rev B 1991;43:9178–89.

183. Aasmundtveit KE, Samuelsent EJ, Pettersson LAA, Inganas O, Johansson T, Feidenhans R. Structure of thin films of poly(3,4-ethylenedioxythiophene. Synth Met 1999;101:561–4.

CHAPTER 14

POLYESTER-BASED (BIO) DEGRADABLE POLYMERS AS ENVIRONMENTALLY FRIENDLY MATERIALS FOR SUSTAINABLE DEVELOPMENT

Joanna Rydz[1,2], Wanda Sikorska[2], Mariya Kyulavska[1] and Darinka Christova[1]

[1]Bulgarian Academy of Sciences, Institute of Polymers, Acad. Georgi Bonchev St., Bl. 103A, Sofia 1113, Bulgaria

[2]Polish Academy of Sciences, Centre of Polymer and Carbon Materials, 34 M. Curie-Sklodowska St., Zabrze 41-800, Poland

ABSTRACT

This review focuses on the polyesters such as polylactide and polyhydroxyalkonoates, as well as polyamides produced from renewable resources, which are currently among the most promising (bio)degradable polymers. Synthetic pathways, favourable properties and utilisation (most important applications) of these attractive polymer families are outlined. Environmental impact and in particular (bio)degradation of aliphatic polyesters, polyamides and related copolymer structures are described in view of the potential applications in various fields.

INTRODUCTION

Polymeric materials currently play an important role in everyday life due to their unique range of properties. Compared to other traditional materials, plastics offer many advantages, such as performance and versatility, durability, lightness, resilience to corrosion, ease of processing, high productivity, low cost, environmental aspect, *etc.*, which determine their importance to society and enables even greater extent of industrial development in the coming years. Innovation in plastics will thus make a valuable contribution to increasing economic growth and quality of life and help to solve environmental problems. Now, there is a remarkable increase in the potential use of biodegradable polymers in various areas such as medicine, pharmacy, agrochemistry or the packaging industry. This fast growing scientific area relates directly to the most crucial health and social problems nowadays [1]. Introduction of new environmentally friendly and sustainable plastics in the packaging and end-user industries is the solution to major problems in waste management. This can be achieved through information dissemination and by identifying and removing the barriers to faster and more widespread use of sustainable plastics, particularly biodegradable and those based on renewable resources. Ecological concerns have resulted in an increase of interest in renewable resources-based products. In the perspective of sustainable development, biodegradable polymers can be considered as safe for the environment, and are an interesting alternative to conventional polymers [2,3]. A number of biodegradable polymers have been developed in recent decades. With regard to the origin of the raw material they are divided into two groups: (i) biodegradable polymers from renewable resources (e.g., polymers of microbiological origin as well as synthetic polymers from renewable monomers); and (ii) biodegradable polymers of non-renewable/fossil resources. [4]. However, it is important to underline that aliphatic polyesters synthesised either by polycondensation of hydroxy acids or diacids and

Citation: Joanna Rydz, Wanda Sikorska, Mariya Kyulavska and Darinka Christova, Polyester-Based (Bio)degradable Polymers as Environmentally Friendly Materials for Sustainable Development, doi: 10.3390/ijms16010564.

diols or by polymerisation of lactone-type heterocyles are not biopolymers [5]. Aliphatic polyesters play a predominant role as (bio)degradable polymers due to the potentially hydrolysable ester bonds and relatively short aliphatic chains present in the macromolecules and are the most representative examples of environmentally relevant polymeric materials [5,6]. On the other hand, they often lack good mechanical and physical properties, which can be compensated by development of copolymer structures such as poly(ester amide)s. In this review, aliphatic polyesters (in particular, polyhydroxyalkanoates and polylactides) as well as polyamides are summarised as the main representatives and most promising (bio)degradable polymers and their favourable properties, (bio)degradation and applications in various fields are described. The purpose of this review was not only to highlight the importance of polymers from renewable resources, such as polyhydroxyalkanoates, but also to emphasise their synthetic analogues such as isotactic, atactic and syndiotactic poly(3-hydroxybutyrate). Moreover, the possible preparation pathways and valuable properties of new copolymers, blends and composites based on polyhydroxyalkanoates from renewable resources containing synthetic poly(3-hydroxybutyrate) have not been covered broadly in the published surveys.

MAIN ROUTES OF POLYESTERS DEGRADATION

Generally, polymer degradation takes place mostly through scission of the main chains or side-chains of macromolecules, induced by thermal activation, oxidation, photolysis, radiolysis, or hydrolysis [7]. Polymer hydrolytic degradation may be defined as scission of chemical bonds in the polymer backbone, by water uptake, to form oligomers and finally monomers. In the first step, water molecules attack the water-labile bonds by either direct access to the polymer surface or by imbibition into the polymer matrix followed by bond hydrolysis [8]. In addition to nucleophilic attack by H_2O (neutral hydrolysis), the hydrolysis can be also catalysed with an acid, base or enzyme.

Alkali-Catalysed Polyester Hydrolysis

The first step of degradation under alkaline conditions is the attack of the hydroxide anion to the carbonyl carbon of the ester group, generating a tetrahedral intermediate. This step is reversible and the hydroxyl attached to the tetrahedral intermediate can leave, resulting in the regeneration of the ester. However, the ether connected to the tetrahedral intermediate (RO^-) can also leave, resulting in hydrolysis, *i.e.*, generation of an alcohol and carboxylic acid. The preference of the tetrahedral intermediate toward hydrolysis *versus* ester regeneration is determined by the ability of the leaving alcohol (ROH) to stabilise a negative charge; therefore, esters formed from acidic alcohols hydrolyse faster than ester formed from aliphatic alcohols (Figure 1) [9]. As a result, one hydroxyl and one carboxyl end group is generated [10].

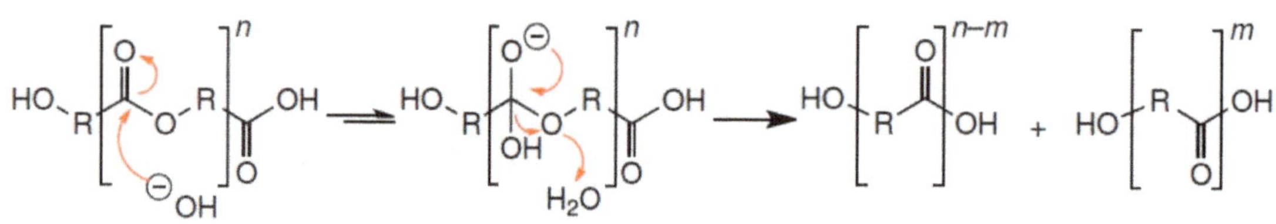

Figure 1: Alkaline-catalysed hydrolysis of polyesters, originally published in [9].

Higher degradation rates in alkaline medium were observed compared with those at acidic conditions [10,11,12].

Acid-Catalysed Polyester Hydrolysis

Under acidic conductions degradation of polyesters begins with protonation of the carbonyl oxygen of the ester group by a hydronium ion, which makes the carbonyl carbon more electrophilic due to the positive charge. This is followed by attack of water molecules on the carbonyl carbon, which generates a tetrahedral intermediate similar to the one generated during base-catalysed hydrolysis. The tetrahedral intermediate can then decompose into a carboxylic acid and an alcohol, or regenerate the original ester (Figure 2) [9]. In the acid hydrolysis, protonation of the chain oxygen atom of the ester group, followed by reaction with water produces one hydroxyl and one carboxyl end group [10].

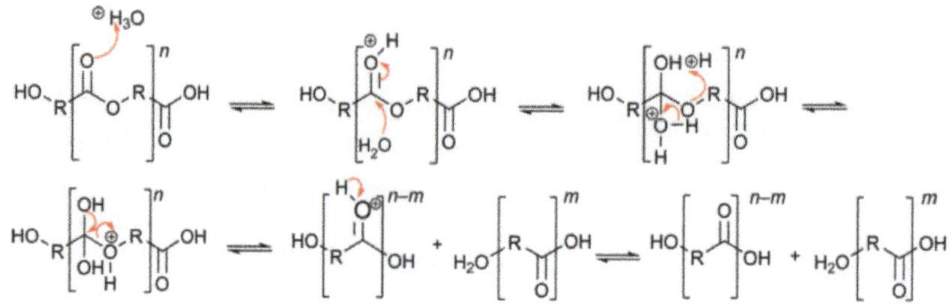

Figure 2: Acid-catalysed hydrolysis of polyesters, originally published in [9].

Enzymatic Degradation of Polyesters

Some polymers undergo degradation in biological environment when microorganisms are present around the polymer materials. Such environment includes soil, sea, river and lake as well as body of human being and animal [7]. Biodegradation of polymers is the attack of microorganisms on predominantly water insoluble polymer-based materials [13]. Due to water insolubility and high-molar mass of the polymer molecules, microorganisms are not able to incorporate the polymers, by outer cell membranes, directly into the cells where most of the biochemical processes take place. Therefore, excretion of extracellular enzymes occurs first which depolymerise the polymers outside the cells (Figure 3) [14]. When the molar mass of the polymer is sufficiently reduced to water-soluble intermediates, they can be introduced into the microorganisms' metabolic pathways. As a final result of these processes end products such as water, carbon dioxide, methane (in the case of anaerobic degradation), and new biomass are produced [14,15].

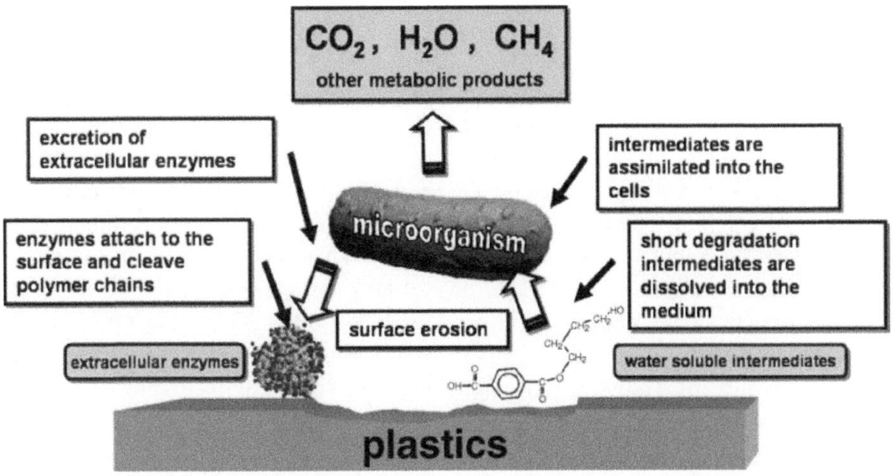

Figure 3: Scheme of general mechanism of enzymatic catalysed hydrolytic polymer degradation, originally published in [14].

POLYHYDROXYALKANOATES—POLYESTERS OF MICROBIOLOGICAL ORIGIN AND THEIR SYNTHETIC ANALOGUES

Natural polyhydroxyalkanoates (PHA)s are polymers from renewable resources which mean that the final polymer is directly converted from such non-fossil resource [16]. They are widely distributed in biological systems and attract increasing attention in many fields such as medicine, surgery, pharmacology, agriculture, packaging industry, biotechnology, polymer waste management, *etc.* [6,17]. Synthetic PHAs analogues offer greater advantages over natural polymers since they can be tailored to give a wider range of properties and to obtain tailor-made biodegradable material for specific applications in different areas [6,18,19].

Poly(3-hydroxyalkanoate)s from Renewable Resources—Synthesis and Properties

Poly(3-hydroxyalkanoate)s belong to a family of fully biodegradable, thermoplastic polyesters formed by several kinds of bacteria as carbon and energy reserves. They are produced not only by bacterial fermentation [20,21] (Figure 4), but also by transgenic microorganisms [22] and plants [23].

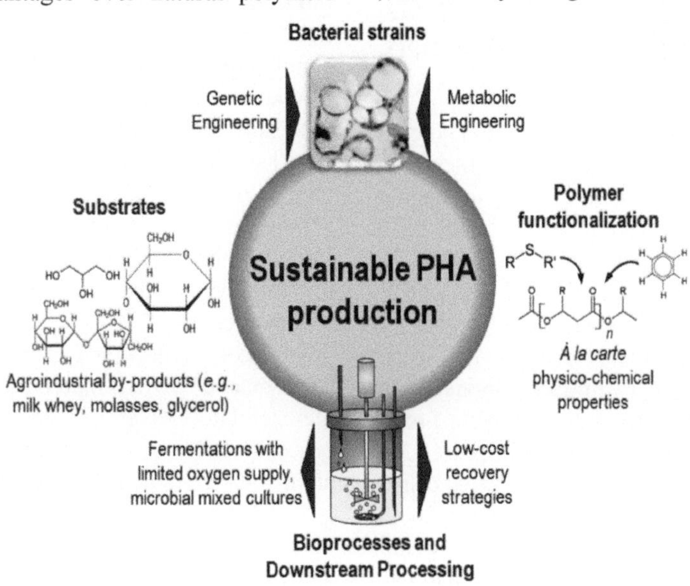

Figure 4: Strategies used to enhance the sustainability of polyhydroxyalkanoates (PHA) production processes, originally published in [21] under CC BY 3.0 license.

Plant cells can produce only a small quantity of PHA. In bacteria, PHAs are accumulated at 90% of the dry cell mass. Bacterial fermentation of PHAs uses sugar and fatty acids as carbon and energy sources [24]. This agricultural feeds are processed directly by the enzymatic activity to the polymer under balanced growth conditions when the cells become limited for an essential nutrient but are exposed to an excess of carbon [16,25].

Poly[(*R*)-3-hydroxybutyrate] was the first representative of PHAs described by Lemoigne in 1920s. Three types of natural poly[(*R*)-3-hydroxybutyrate] (PHB) with different numbers of 3-hydroxybutyric acid units (HB) and different functions have been identified in living organisms: (i) high-molar-mass storage PHB in cytosolic inclusion bodies of various microorganisms, consisting of 10,000 to >1,000,000 HB residues; (ii) low-molar-mass PHB in cell membranes (complexed with polyphosphates) consisting of 100–300 residues; and (iii) short-chain oligo[(*R,S*)-3-hydroxybutyrate] (OHB) (≤30 HB units) covalently conjugated to proteins, found even in human tissues [25,26,27,28].

The physical properties of PHAs vary from crystalline-brittle to soft-sticky materials depending on the length of the side aliphatic chain on β-carbon [29,30]. However, the high crystallinity of PHB limits their wide applications, especially as packaging. The melting point (T_m) ranges from 173–180 °C and the glass transition temperature (T_g) is around 5 °C [6]. The polymer is too brittle for most practical applications and materials obtained therefrom possess poor mechanical properties. To reduce the crystallinity and improve the properties, PHB can be copolymerised with the structural segments derived from selected natural PHAs. Copolymerisation of PHB with 5%–20% 3-hydroxyvaleric acid units (HV) allows the production of PHBV copolymer with improved mechanical properties and such material can be used as biodegradable packaging [31]. Functional PHAs can be prepared via biosynthetic procedures by incorporating various functional monomers or other metabolites in the polymer sequence. For example, fermentation of *Alcaligenes eutrophus*in a culture containing polyethylene glycol or polysaccharide has given a hydroxy-terminated block copolymer consisting of PHA [32,33].

Synthetic Poly(3-hydroxybutyrate)s—Polymers from Fossil Resources

Synthetic analogues of the natural PHB, with potential industrial importance are obtainable by direct copolymerisation of epoxides with carbon monoxide and via ring-opening polymerisation (ROP) of β-butyrolactone to isotactic, atactic and syndiotactic poly(3-hydroxybutyrate) [34,35,36]. The common anionic initiators activated by the addition of macrocyclic ligands such as crown ethers or cryptandes or by using bulky counter-ions or suitable highly polar aprotic solvents as e.g., DMSO are able to initiate polymerisation of β-butyrolactone [37,38,39,40]. The polymer chain growth proceeds regioselectively and stereoselectively entirely *via* carboxylate anions. Propagation on carboxylate active centres (much less sensitive to impurities than any other anionic species) enables the scaling up the anionic ring-opening polymerisation process up to industrial level. Short-chain PHB can be obtained from high-molar-mass PHB by degradation according to E1cB elimination mechanism which leads to oligomers with crotonate end groups or by thermal degradation [41,42,43,44,45]. The results of the fundamental research on the synthesis of homo- and copolymers have been patented [46,47].

The purity of β-butyrolactone used in the polymerisation process affects the molar mass of the obtained polymer. Synthesis of high-molar-mass PHB by anionic polymerisation must be carried out in a solvent-free system [39], and requires strict control of the reaction temperature and extraordinary care in the purification of the monomer. Regioselective anionic ring-opening polymerisation of β-lactone is accompanied by side reactions leading to the formation of polymers with unsaturated end groups. Unsaturated (crotonic) end groups during polymerisation of β-butyrolactone may be formed both in the initiation and propagation process. During isothermal degradation of synthetic atactic PHB (aPHB) with carboxylate end groups, a significant decrease of the molar mass and formation of oligomers with crotonic end groups as degradation products has been observed, in contrast to aPHB terminated by carboxy end groups. Therefore, it is very important to control the temperature during the polymerisation in order to avoid side reactions leading to a reduction of the polymer molar mass in relation to the assumed value. The recognition of this phenomenon allows for the synthesis of aPHB with a molar mass of more than 100,000 g/mol and its use as a component in poly(lactide-*co*-glycolide) (PLGA) compositions for the preparation of bioresorbable (degradable with proved elimination from the human body) fibres. This convenient and solvent-free method can also be used for the controlled degradation of natural PHA under mild conditions towards the preparation of slower degrading oligomers of predetermined molar mass and defined end groups. [48]

Copolymers, Blends and Composites Based on Microbial and/or Synthetic Poly(3-hydroxyalkanoate)s

Tailoring of the polymer architecture of biodegradable and biocompatible polymers and their synthetic analogues is the way to improve the properties and develop novel biodegradable materials [49]. PHA of natural origin can be modified to produce tailored block, graft and random copolymers containing structural segments derived from synthetic analogues and other synthetic polymers such as polycaprolactone (PCL), poly(methyl methacrylate) (PMMA), poly(lactic-*co*-glycolic acid) or synthetic PHB resulting in various materials with special features. Moreover, PHB graft copolymers were prepared by various polymerisation methods, for example, anionic polymerisation of β-butyrolactone on poly(methyl methacrylate) multifunctional macroinitiator to synthesise P(MMA-*graft*-aPHB) containing synthetic poly[(*R,S*)-3-hydroxybutyrate] side chains [50]. Atactic or isotactic PHB macromonomers functionalised with methacrylate end groups were copolymerised in a one-step procedure with several methacrylates via grafting through by atom transfer radical polymerisation (ATRP) [51]. Recently, a water-soluble brush copolymers composed of synthetic PHB and poly(ethylene glycol) (PEG) brushes were prepared applying a three-step procedure, including ATRP processes [52]. Such copolymers exhibit a wide range of interesting properties, with potential application in many fields especially in medicine and environmental protection. More recently, the controlled synthesis of PHB-PEG-PHB triblock copolymer by β-butyrolactone polymerisation on PEG macroinitiators via a crown ether-free anionic ring-opening polymerisation was reported providing good control on molar mass and molar-mass distribution of the final triblock copolymer [53,54].

Block or graft copolymers may be also synthesised by means of macroinititors suitable for further specific chemical reactions [55]. Controlled depolymerisation of natural PHA catalysed by KOH/18-crown-6 complex in $CHCl_3/H_2O$ system, due to the partial saponification of ester linkages followed by elimination reaction, leads to the formation of macroinitiator possessing unimodal molar-mass distribution and containing carboxylate end groups with K^+/18-crown-6 counterion [56,57]. Such oligomers have been used for the characterisation of several PHAs at the molecular level, and their sequence distribution has been established by ESI-MS^n fragmentation experiments [57,58,59,60]. The low-molar mass macroinitiators of natural poly(3-hydroxyalkanoate) containing olefinic and active 18-crown-6 ether carboxylic end groups, are suitable to initiate ring-opening polymerisation of racemic β-butyrolactone. By this method, new block copolymers were obtained which combine selected natural PHA with poly[(*R*,*S*)-3-hydroxybutyrate)] (Figure 5) [61].

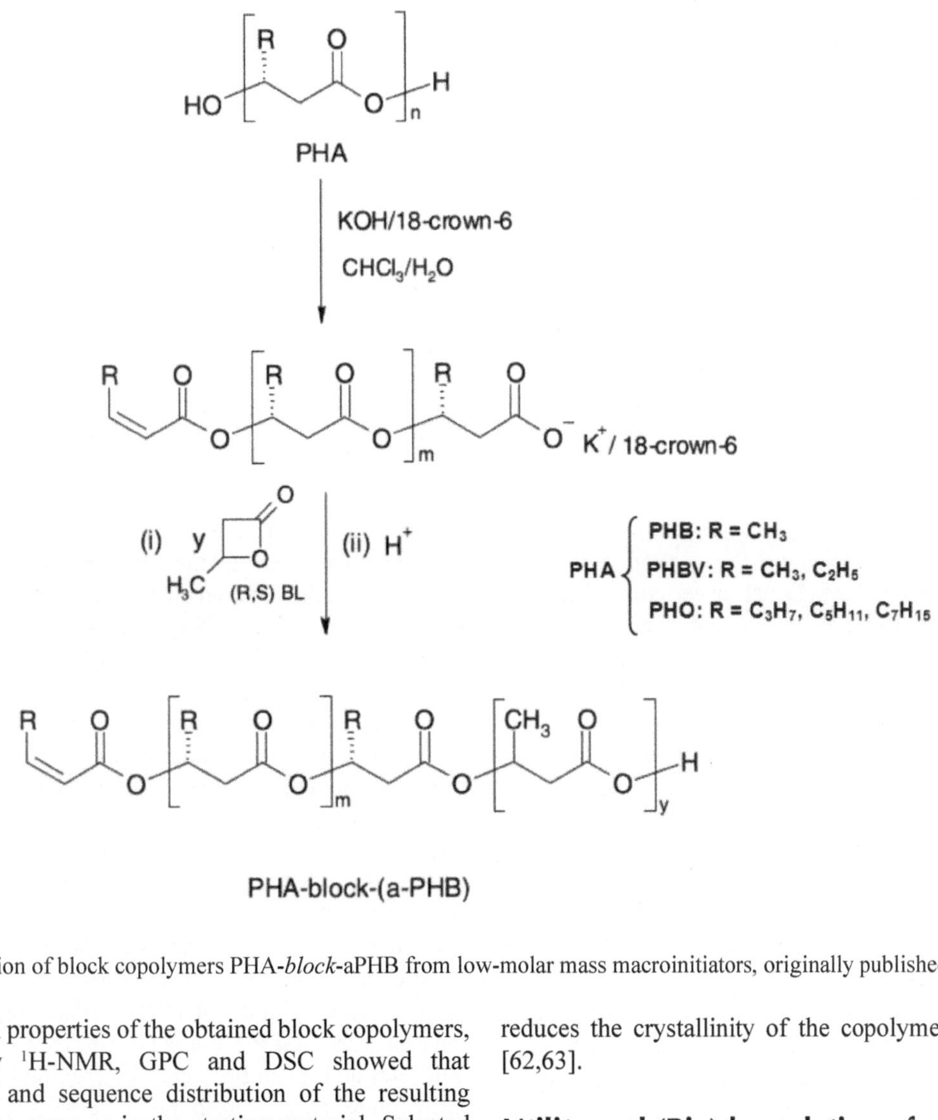

Figure 5: Preparation of block copolymers PHA-*block*-aPHB from low-molar mass macroinitiators, originally published in [61].

The physical properties of the obtained block copolymers, characterised by ^{1}H-NMR, GPC and DSC showed that the composition and sequence distribution of the resulting oligomers was the same as in the starting material. Selected copolymers were subjected to hydrolytic degradation studies. The possibility to apply the obtained PHO-*block*-aPHB oligomers (PHO—poly[(*R*)-3-hydroxyoctanoate]) as compatibiliser for PHO/aPHB blend was confirmed. Moreover, the suitability of obtained polymeric materials for cardiovascular engineering was positively verified. The studies of permeability (performed according Polish norm PN-79/P04884.03) showed that vascular prosthesis covered with PHB-*block*-aPHB copolymer was the more watertight as compared with pure prosthesis [61].

Generally, use of poly[(*R*,*S*)-3-hydroxybutyrate], which is atactic and amorphous synthetic analogues of natural PHB, reduces the crystallinity of the copolymer or blend systems [62,63].

Utility and (Bio)degradation of Poly(3-hydroxyalkanoate)s

Processing properties of natural poly[(*R*)-3-hydroxybutyrate] are similar to those of petroleum-based polypropylene (PP.) [64]. VINNEX®, poly(vinyl acetate) based resins, enhances the physical properties of the material containing PHB, which significantly simplifies the processing. This, combined with the high heat resistance of PHB, allows the possible use of this material in applications such as hot filling [65]. Many practical applications in medicine, agriculture and in the food industry take advantage of the biodegradability of the PHAs, especially PHB and its copolymers with (*R*)-3-hydroxyvalerate (PHBV, Biopol®, Zeneca BioProducts) or with (*R*)-3-hydroxyhexanate

(PHBH, Nodax™, P&G-Kaneka). Metabolix, Inc., an innovation-driven bioscience and engineering company, is developing and commercialising a family of high-performance PHA biopolymers targeted to the markets for performance additives, including film, toys, bags,*etc.* (Figure 6) [66].

Figure 6: Zoe B Beach Toys, Compostable Shopping Bag, based on Metabolix, Inc. [67].

In addition to the PHA producers from the U.S. (Metabolix Inc., Cambridge, MA, USA; Meredian Inc., Bainbridge, GA, USA) in Europe Biomer (Krailling, Germany) produces PHA under trade name of the Biomer® [68]. PHAs are also produced in Brazil by PHB Industrial SA (Sao Paulo, Brazil), in the amount of about 50 tons of PHB and PHBV per year from sugar cane. In recent times, new Chinese companies appeared on the PHA market such as Tianan Biologic Material Co., Ltd. (Ningbo, China) producing PHBV and Tianjin Green Bio-Science Co., Ltd. (Tianjin, China), which produces in collaboration with DSM Venturing 10,000 tons of poly(3-hydroxybutyrate-*co*-4-hydroxybutyrate) per year [69,70,71]. Selective use of biodegradable materials in certain applications may provide a solution to the above-mentioned environmental problems. Goods from PHA polyesters applied as green biodegradable packaging involve foils, bottles and containers [72]. The presented results on biodegradation PHBV showed that water is a prerequisite for enhanced microbial growth and the enzyme-substrate contact. The PHA was also found to biodegrade in the tropical climate and marine environment [73,74,75,76,77]. The degradation of aliphatic polyesters to common metabolites in a biological environment, including anaerobic and aerobic conditions, results from enzymatic attack or simple hydrolysis and depends on different abiotic parameters of degradation environment [78,79].

Synthetic poly[(*R,S*)-3-hydroxybutyrate] has been paid much attention as a suitable modifier in blends with such (bio)degradable polymers as brittle polylactide or highly crystalline natural poly[(*R*)-3-hydroxybutyrate]. Earlier studies demonstrated that selected crystalline polymers (natural PHAs, polypivalolactone, polycaprolactone, poly(l-lactide)) after blending with aPHB successfully underwent enzymatic hydrolysis [62,63,80,81,82,83]. Moreover, the hydrolytic degradation of aPHB, natural PHBV and thereof blends preferentially occurs in the amorphous state by random scission of polyester chains. The increase in the content of crystalline phase derived from PHBV significantly reduces the rate of hydrolytic degradation of the blends tested. Thus, the hydrolytic degradation rate also was dependent on degrees of crystallinity of the polymers. Synthetic amorphous PHB shows the fastest rate of hydrolytic degradation. Bacterial PHB, which has the highest crystallinity degree, degrades most slowly, while PHBV having a lower crystallinity degree than bacterial PHB has an intermediate rate of degradation [81]. In order to improve the mechanical or chemical properties and to reduce the price of the polymeric materials, while maintaining their biodegradability the composites of natural polymers and fillers are developed. The biodegradable composites very often contain natural fibres (e.g., hemp) or sawdust [84,85,86,87]. Previous results indicate that the natural PHBV and its binary blends containing aPHB degrade enzymatically in compost with activated sludge and in the dynamic Baltic Sea water. The degradation degree of samples incubated in compost is higher than those of the samples incubated in seawater. Moreover, the degradation degree of the blends was dependent on the aPHB content [88]. Degradation of aPHB binary blends with natural PHB and poly(l-lactide), respectively, has been also investigated in soil. The increase in the number of microorganisms observed particularly for the soil where binary blends were incubated, indicating the microbial degradation of aPHB. The terrestrial plant growth test (cress and barley) demonstrated no toxicity of the materials studied on the environment [89]. Recently, the PCL blends as food packaging with excellent durability and mechanical properties have been studied [90]. The synthesis and toxicity studies of well-defined oligo[(*R,S*)-3-hydroxybutyrate]s and their potential application as drug delivery carrier systems were also presented [91]. The chemical structure of OHBs is similar to that of high-molar-mass natural PHB, it is pure and does not contain toxic impurities and it can be useful for drug modification [40,92]. Furthermore, an anti-proliferative activity of ibuprofen-OHB conjugates against colorectal cancer cells *in vitro* was also demonstrated [93].

When biodegradability of a given material is combined with low toxicity, a biotechnological product of great commercial interest could be achieved. Biomedical applications of PHAs, such as biodegradable controlled antibiotic release systems, implants, sutures and other medical devices offer the highest growth during the last years. In medicine PHAs, in pure form or as composites with other materials, applications have been found as implants for targeted drug delivery and scaffolds as well as in tissue engineering. Oligohydroxyalkanoates can also be used as a calcium influx stimulant for drug applications [24,94].

In the case of low purity PHA, obtained from activated sludge or nutrient-rich wastewater, it could find application in the energy industry. Production of PHA by mixed-culture composition in the open system can reduce the production costs of PHA for biofuel applications [95]. Acid, alkali or enzyme-catalysed hydrolysis of PHAs as well as biotransformation or fermentation of appropriate strains leads to the corresponding methyl esters of hydroxyalkanoates. These compounds are good fuels, alone or as additives to other fuels. PHA biofuels possess high oxygen content and while not containing nitrogen and sulphur. Besides, PHA biofuels could be a good alternative for the traditional biofuels such as bioethanol or biomethanol which was produced from biomass [29,96].

According to Markets and Markets, a global market research and consulting company based in the United States, PHA market consumption will grow from an estimated 10,000 million tonnes in 2013 to 34,000 million tonnes by 2018. PHAs are currently used in food service, biomedical and agricultural applications, electronic, packaging, automotive and chemical industries, as well as in photographic applications and the printing industry. The largest area of PHA use is packaging followed by food services. The increasing demand for renewable and bio-based materials and shift in consumer preference for eco-friendly products is driving the global market of polyhydroxyalkanoate. The PHA market has only few companies but it has been making rapid technological advancement along with increased investments in R&D. Therefore, PHA will still be commercialised and used in a wide range such as packaging, food service, biomedical and agriculture [71].

POLYLACTIDE—SYNTHETIC POLYMER FROM RENEWABLE AND SYNTHETIC MONOMERS

Polylactide (PLA), an aliphatic degradable and biocompatible thermoplastic polyester is polymer with a high potential particularly for medical and packaging applications [97,98]. One of its most promising applications is its use for the production of (bio)degradable and biocompatible materials as an environmentally friendly alternative to non-biodegradable plastics derived from petrochemicals [99,100].

Synthesis of Polylactide and Polylactide-Based Copolymers, Blends and Composites

PLA can be obtained from natural or synthetic monomers. To produce natural monomer from renewable resources, biomass, in particular starch, cellulose and lignin, is converted into lactic acid (mostly l-isomer) in a fermentation process of carbohydrates (maltose, sucrose, lactose, *etc.*) and then chemically transformed to polymer [16,101]. Polylactide can be obtained by several methods. The conventional polycondensation of lactic acid does not increase sufficiently the molar mass of the resulting PLA due to water formation that negatively influences esterification equilibrium. Therefore, the most common method to obtain high-molar-mass polylactide is through ring-opening polymerisation of the cyclic lactide dimer [102,103,104,105,106].

Direct polycondensation is performed either in solution or in melt. It is possible to obtain PLA with a molar mass of 100,000 g/mol in one step polycondensation of lactic acid in solution using high-boiling-point solvents and molecular sieves as drying agents for the effective removal of water. It was also found that the tin oxide and chloride can effectively increase the PLA molar mass during melt condensation [107].

The ring-opening polymerisation requires the use of heavy-metal-based catalysts, such as zinc and stannous oxides, zinc and tin chlorides or stannous octoate, which often contaminate the resulting polymer. Therefore, polymerisation of lactide using zirconium(IV) acetylacetonate [$Zr(acac)_4$] as an initiator was investigated for medical applications [108]. The enzymatic synthesis is also considered an environmentally friendly method, when carried out under mild conditions and can provide adequate control of the polymerisation process [109]. Table 1 shows some of the catalysts commonly used for the synthesis of polylactide.

Table 1: The catalysts used for the preparation of PLA based on [102]

Polylactide *	Catalyst	Solvent	Molar Mass
PDLLA/PLLA	Aluminium isopropoxide	Toluene	$M_n = 90,000$
PDLLA	Stannous octoate	Alcohols	$M_W < 350,000$
PLLA	Stannous octoate	Alcohols, carboxylic acids	$M_n = 250,000$
PLLA	Stannous octoate, titanium or zirconium compounds	Toluene	$M_n = 40,000–100,000$
PDLA/PLLA/PDLLA	Stannous trifluoromethane sulfonate, scandium(III) trifluoromethane sulfonate	Ethanol	–
PLLA	alkoxides of Mg, Al, Zn, Ti	Methylene chloride	–
PLLA	Yttrium tris(2,6-di-tert butyl phenolate) in toluene	2-Propanol, butanol, ethanol	$M_n < 25,000$
PDLLA	Zn lactate	Bulk	$M_n = 212,000$
PDLLA/PLLA	Butylmagnesium halides (Grignard reagents)	Ethers	$M_n < 300,000$
PLLA	Potassium naphthalenide	THF, toluene	$M_n < 16,000$
PLLA	Complexes of iron with acetic, butyric, isobutyric or dichloroacetic acid	Bulk	$M_W = 150,000$

* PLLA—poly(L-lactide); PDLA—poly(D-lactide); PDLLA—poly(D,L-lactide).

The copolymerisation of lactide with other lactone-type monomers can improve properties such as high crystallinity, high melting point, and poor solubility [107]. PLA copolymers with PCL and/or polyglycolide (PGA) were synthesised by ring-opening polymerisation of corresponding cyclic lactones, *i.e.*, lactide, ε-caprolactone, or glycolide via anionic, cationic, coordination-insertion or enzymatic mechanism [6,110,111]. The synthesis of PGA/PCL copolymers were conducted using different conditions which lead to various chain microstructures and resulted in different degradation rate. The segmental microstructure of the copolymers chain which consists of the sequences containing longer lactidyl microblocks (GGGGG, LLLLL, GLGGG, GGGLG, LGGLG or GLGLG) and short LGL-segment, which originate from intermolecular transesterification were obtained during copolymerisation initiated with $Zr(acac)_4$ conducted in bulk at 110 °C [112,113,114]. Furthermore, PLGA can be synthesised by direct melt polymerisation of the hydroxy acids, lactic and glycolic acid. The copolymer poly(lactide-*co*-glycolide) is one of the most interesting polymers for medical applications. This bioresorbable, biocompatible and non-toxic polymer has also good degradability which can be modified by different copolymerisation ratio of the monomers [115]. Copolymers of l-lactide with 25%–70% glycolide are amorphous. There is no linear relationship between the copolymer composition and the mechanical and degradation properties of the copolymers. The rate of degradation depends not only on the LA/GA ratio but also on the molar mass and the shape and structure of the polymer matrix [18,19]. The poly(lactide-*co*-glycolide) are in glassy state during degradation process because their glass-transition temperature is higher than 37 °C, which accounts for the low water absorption and slow drug release [116]. To modify the biological and mechanical properties of PLGA composites with carbon fibres and hydroxyapatite was prepared [117,118,119,120].

Properties, Utility and (Bio)degradation of Polylactide

Polylactide has attracted much attention because of its excellent material properties which allow a wide range of potential applications in many fields as medical devices, textile, packaging applications and in sustainable development, recycling as well as environmental protection [98]. Polylactide can be a semicrystalline or fully amorphous material, depending on the stereochemical structure and thermal history, and may exhibit crystal polymorphism [6,121,122]. The T_g of polylactide ranges from 50–80 °C, and the T_m ranges from 130–180 °C [6,123]. The polymer architecture, crystallinity and molar mass are the most important parameters that affect the mechanical, chemical, and thermal properties of the polylactide materials [6,124]. PLA is suitable for large-scale production of materials with good mechanical and high barrier properties, resistance to fat action and UV rays, which are close to those of conventional oil-based polymers [6,125,126]. Although PLA is sensitive to processing conditions, it can be processed by injection molding, hot press molding, spinning, blow molding, foam molding or electrospinning [6,127]. Under extrusion, rapid decrease in viscosity and molar mass is observed due to macromolecular chain cleavage induced by increased shear level, temperature, and/or residence time [6,128]. Poly(l-lactide) is a highly crystalline (of about 37%), isotactic polymer with good tensile strength, low extension, high modulus (approximately 4.8 GPa), higher T_g and slower degradation time compare to the amorphous form. It is preferred in applications where higher mechanical strength and toughness are required [18]. On the other hand, poly(d,l-lactide) (PDLLA) is an amorphous polymer having a random distribution of both isomeric forms of lactic acid [6]. The degradation product of PLA, lactic acid, a normal human metabolic by-product is converted into water and carbon dioxide via the citric acid cycle [18].

Polylactide is widely considered to be a biodegradable polymer, however, poly(α-hydroxy acid)-type polyesters are now being proven to degrade *via* simple chemical hydrolysis or enzymatically, despite the fact that some exotic enzymes such as proteinase K can cleave their main chain. Only oligomers (oligomeric degradation by-products at the last stages of the hydrolytic degradation processes) undergo enzymatic degradation but the role of hydrolysis *versus* enzymatic degradation is still under discussion [5,98,129]. Many studies have shown that PLA degrades completely in different environments [130,131,132,133,134,135,136,137,138,139]. Even the degradation of PLA films occurred in paraffin due to presence of the residual water content [140,141]. In the human body, the temperature of degradation is generally around 37 °C and PLA may be degraded *in vivo* in 2–3 years, while in the environment PLA can be degraded completely in a short time under composting conditions where the temperature can go up to 70 °C, as well as remains almost stable in water at a temperature of 20 °C [5]. The degradation of polylactide in aqueous medium occurs through hydrolysis (the random scission) of ester bonds autocatalysed by carboxylic acid end groups, and the hydrolysis rate increases with degradation time. Water absorption is a critical factor. For a time the partially degraded macromolecules remain insoluble in the surrounding aqueous medium and the degradation proceeds homogeneously according to autocatalysis rules. During the first step of hydrolytic degradation of PLA, the molar-mass dispersity increased and the molar mass decreased rapidly with a slight mass loss. In the next step, when the molar mass loss slowed down, the molar-mass dispersity decreased since low-molar-mass compounds were removed from the polymer matrix, and the catalytic effect of the carboxylic acid groups was reduced [5]. Subsequently, the molar mass loss slowed down with increasing mass loss. [142]. Faster internal degradation of PLA polymers is regarded as a general phenomenon. Degradation process, that combines diffusion, chemical reaction, and dissolution phenomena, results in a differentiation between the rates of degradation at the surface and interior of the matrix [5,143,144]. It is also known that submillimeter films degrade homogeneously and more slowly (erosion is restricted at the surface) than large-sized devices [145]. Moreover, porous systems generally degrade at a slower rate than plain ones because the release of soluble degradation by-products, that are responsible for the autocatalysis, can escape more easily from the polymer matrix as a result of the higher surface. It has been shown that structural differences affect the degradation characteristic in the case of stereocopolymers of PLA synthesised in the presence of different initiating systems [5].

Despite the complexity of PLA degradation mechanism, bioresorbable and biorecyclable lactic acid-based materials are presently at the commercial stages as matrixes to make surgical devices (wound closure in form of sutures, surgical clips, surgical staples and adhesives, tissue repair and regeneration scaffold as bone plates, screws and filling material for bone reconstruction or plastic surgery), controlled drug delivery particulate systems in pharmacology, and packaging [5,146].

For many applications, PLA could be blended with other polymeric or non-polymeric components in order to achieve the desired properties. Due to the low moisture absorption and high wicking, low flammability and smoke generation, UV-light resistance, relatively low specific gravity and low refraction index, and provision of excellent colouring characteristics, PLA is suitable for many applications such as outdoor applications and textile-fibre applications (shirts, furniture and automobile fabrics, carpets, wipes, table clothes, curtains, underwear and sportswear). In addition, disposable garments, awnings, feminine hygiene products, diapers production can utilise PLA. Foamed PLA may be used as structural protective material as well as in loose-fill packaging and insulation. PLA is also usable for coated paper, high-value films, rigid thermoformed containers, bottles, dog poop bags and a variety of other packaging applications [147,148,149].

Polylactide is used in agricultural applications such as mulching films, plant pots, compostable yard bags, bindings, clips, and other devices. Plastic agricultural mulches provide many benefits to the cultivation of specialty crops such as vegetables, including weed prevention and increased soil temperature, which leads to an increase in crop yield. Poly(lactic acid)-based mulches prepared using nonwovens textile technology are the perfect plastic mulch which would be ploughed into the soil at the end of the growing season, and undergo full mineralisation within a few months [147,150].

PLA was commercialised on a large scale particularly as a degradable, short-term packaging and agriculture applications and as a fibre (Figure 7). Cargill has developed processes that use corn and other feedstock to produce different PLA grades (NatureWorks LLC) [97,151]. Polylactide water bottles, commercially available from Biota, were obtained from NatureWorks LLC PLA (Blair, NE, USA). In NatureWorks LLC Company, the production of PLA in 2012 was estimated to be 140,000 tons. Currently, this is world's largest production of biodegradable polyester with a price of less than 2 €/kg.

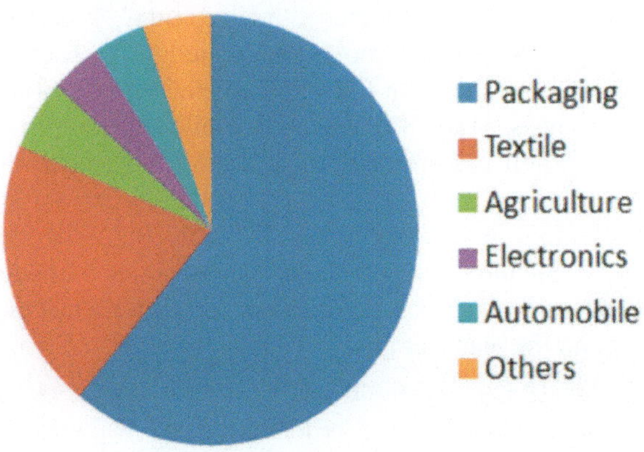

Figure 7: Polylactide market size in 2013 by application [154].

At present, the main European producers of PLA are Corbion Purac (Soesterberg, The Netherlands), Synbra Technology BV (Etten-Leur, The Netherlands), Pyramid (Guben, Germany) and Futerro (Escanaffles, Belgium, in 2007 joint venture between Galactic from Belgium and Total Petrochemicals from France) as well as US-based NatureWorks [97,101,152]. PLA consumption is around 200,000 tons per year. At present, only around 30% of lactic acid is used for PLA production [98]. However, the lactic acid market is projected to exceed $ 3577.5 million and poly(lactic acid) market to reach $ 4840.1 million by 2019, respectively [153].

ALIPHATIC POLYAMIDES

Synthetic Pathways and General Properties of Aliphatic Polyamides

Aliphatic polyamides, also known as nylons (e.g., Nylon-66; Nylon-612; Nylon-46; Nylon-6; Nylon-12;*etc.*) are among the most important commodity polymers [155]. Polyamides are heterochain polymers containing amide groups in the macromolecular backbone. This large polymer encompasses thermoplastics of extremely broad range of available properties which are used in the production of films and fibres, moulding compounds, *etc.* [156,157].

Polyamides are generally synthesised by three methods: (i) polycondensation between diamines and dibasic acids; (ii) polycondensation of amino carboxylic acids as bifunctional monomers; and (iii) ring-opening polymerisation of lactams—cyclic amide monomers containing 3–7 ring atoms [158,159]. Synthetic pathways to polyamide copolymers of various structures and compositions are well covered in the literature [158,160,161,162,163].

The majority of polyamides are semicrystalline and generally very tough materials with good thermal and chemical resistance. The different nylon types give a wide range of properties and are used in many applications due to the combination of outstanding mechanical and electrical properties, particularly toughness and wear resistance (Table 2). Nylons also have excellent chemical resistance and can be used in high temperature environments. Heat stabilised reinforced systems allow sustained performance at temperatures up to 185 °C [164].

Table 2: Most important commercial polyamides [165,166,167,168,169]

Name and Chemical Structure	Melting Point (°C)	Typical Application	Producer (Trade Name)
Nylon-6 (PA 6) –$[NH(CH_2)_5CO]_n$–	219–220	High-temperature applications; automotive industry; electrical and electronic industry (housings, plug and socket connectors, printed circuit boards); sports equipment.	BASF AG (Ultramid ® B); DSM (Akulon® and Novamid®); Honeywell Resins and Chemicals L.L.C. (Capron®); DuPont (Zytel®); Rhodia (Technyl®).
Nylon-66 (PA 66) –$[NH(CH_2)$-$NHCO$-$(CH_2)_4CO]_n$–	260–300	Industrial yarns and textile; automotive industry (radial tires; intake manifolds, engine covers, gears).	DuPont (Zytel®); BASF AG (Ultramid® A); DSM (Akulon® S); Rhodia (Technyl®); Ascent Performance Materials (Vydyne®).
Nylon-610 (PA 610) –$[NH(CH_2)_6$-$NHCO$-$(CH_2)_8CO]_n$–	211–236	Engineering and construction materials; industrial parts, tubings, rods and profiles; sheet applications.	BASF AG (Ultramid ® S); Rhodia (Special Technyl® grades); Toray Resin Company (Amilan®).
Nylon-612 (PA 612) –$[NH(CH_3)_6$-$NHCO$-$(CH_2)_{10}CO]_n$–	206–246	Pipes, bushings, electrical connectors, industrial parts, bristles, tubings, rods.	DuPont (Zytel® 150 series); Evonik (Vestamid® D series).

Nowadays, more attention is focused on the attractive class of bio-based polyamide thermoplastics, which are partly or wholly made from renewable resources [170,171]. In the production of such polyamides, bio-based monomers derived from castor oil or mass produced by fermentation are applied. Synthetic pathways to obtain bio-based polyamides are basically the same as to synthetic polyamides and there are a number of commercial products available on the market (Rilsan®11 of Arkema (Colombes, France); Ultramid Balance®of BASF (Ludwigshafen, Germany); Vestamid Terra® of Evonik (Essen, Germany); *etc.*) [172].

Poly(ester-amide)s are another fast developing family of thermoplastics that combine the valuable properties of both polyesters and polyamides, *i.e.*, polyester's biodegradability and polyamide's high thermal stability and high tensile strength [173]. Particular attention is focused on the poly(ester amide) s built of α-amino acids which became important materials in the biomedical field [174]. The presence of the α-amino acid contributes to better polymer-cell interactions and allows the introduction of functional groups thus enhancing the overall biodegradability of the material.

Utility and Biodegradation of Aliphatic Polyamides

Nylons are available for processing *via* injection moulding, rotational moulding, casting or extrusion into film or fibre. For industrial uses, polyamides persistently replace traditional materials in applications ranging from showpiece examples such as artificial organs and construction materials for moon-based space stations through to the more routine but equally important uses, including high load bearings, wear pads, support and guide wheels, buffer pads and gears, and many more [175].

Biodegradation of synthetic polyamides is generally known to be poor, although its chemical structure (presence of amide bonds in the main chain) resembles those of natural proteins and synthetic polypeptides [9,176,177]. The high resistance to degradation of synthetic polyamides is caused mainly by the high symmetry of their molecular structures and strong intermolecular cohesive force caused by hydrogen bonds between molecular chains, which results in highly crystalline morphology (Figure 8) [178,179,180]. However, several papers and reviews report on the degradation of oligomers and higher molar mass polyamides by different microorganisms and enzymes [178,181]. Nylon-66 and Nylon-6 are significantly degraded by enzymes from white-rot fungus *Bjerkandera adusta* [182,183,184]. Biodegradation of Nylon-6 oligomers is also observed with *Pseudomonas* and *Flavobacterium* [178]. Andreoni *et al.* [185] described three mixed cultures of aerobic bacteria able to grow on 6-aminocaproic acid polyamides of low molar mass (up to 6800 g/mol), but biodegradability of polyamides of molar mass higher than 11,000 g/mol was not reported. The extent of the bacterial growth was found to dependent on the amount of cyclic and linear oligomers present in the polymeric matrix.

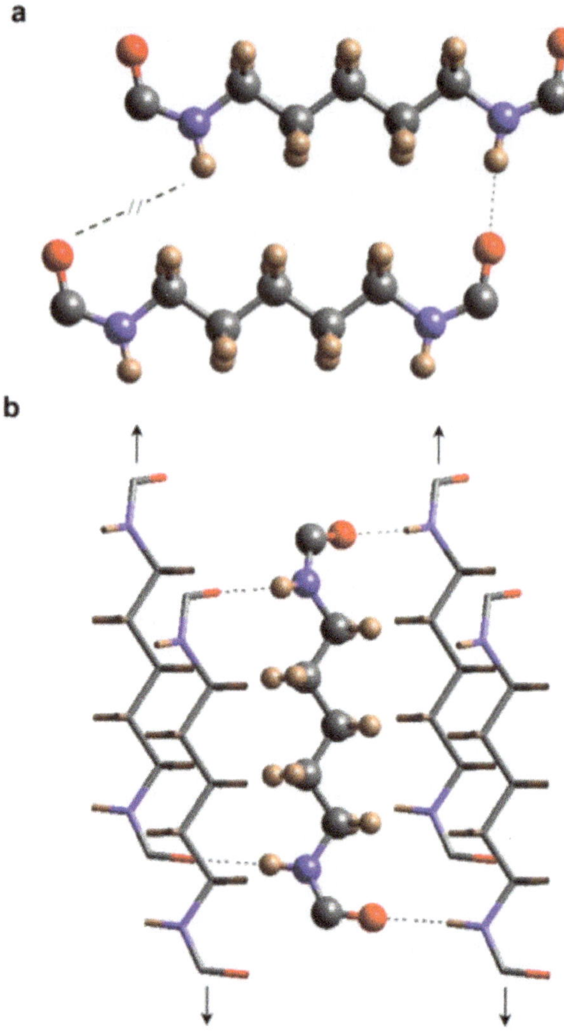

Figure 8: Establishment of hydrogen bonds between Nylon-56 molecular chains, (a) scheme of the unfavourable hydrogen-bond geometry between odd diamide units of Nylon-56 molecular chains with an all trans conformation and (b) scheme of the establishment of hydrogen ...

Although made of naturally occurring substances, bio-based polyamides are not necessarily readily biodegradable. Nevertheless, bio-based Nylon-4 and some end-group modified analogues are easily biodegraded by bacterium *Pseudomonas* sp. strain ND-11 isolated from activated sludge. It is shown that polyamide degradation in this artificial environment proceeds via hydrolysis [186,187]. Biodegradability of Nylon-4 in composted soil is also reported. [188,189,190]. In addition, Tachibana *et al.* [191] investigated the biodegradability of Nylon-4 films in seawater from the Tokyo Bay. The slower plastic degradation in marine environment is due to the lower temperature and plastics degradation into smaller pieces that persist into the seawater [192].

Nylon-12 and Nylon-66 could also biodegrade by fungi and bacteria. Deguchi *et al.* reported that the white rot fungi strain IZU-154, a kind of lignin-degrading microorganisms, degraded Nylon-66 films through oxidative processes [182,193]. *Geobacillus thermocatenulatus* could also provide Nylon-12 and Nylon-66 biodegradation [194]. Bacterial degradation of Nylon-12 is usually associated with the enzymatic hydrolysis of amine bonds, which is accompanied by the formation of 12-amino dodecanoic acid. Tomita *et al.*

[195] reported on a thermophilic strain isolated from 100 soil samples by enrichment culture technique at 60 °C which is capable of degrading Nylon-12. At this temperature, the strain grew on Nylon-12, accompanied by a marked decrease in molar mass of Nylon-12. The strain is also capable to degrade Nylon-6 as well as Nylon-12, but not Nylon-66.

Although Nylon-6 is generally regarded as a non-biodegradable polymer [196], the aforementioned fungi strain IZU-154 could also degrade Nylon-6 fibres to soluble oligomers [182]. Freidrich *et al.* [170] reported on the screening of a large number of fungi for their ability to degrade high molar mass Nylon-6. The study showed that the white rot *Basidiomycetes* is able to degrade Nylon-6 when grown on this polyamide as the only N-source. For the first time, this polymer was shown to be disrupted by *Bjerkandera adusta.* The remaining insoluble part of the nylon showed a decrease in number-average molar mass from 16,900–5600 g/mol during 60-day incubation. It was assumed that the insoluble polymer was partially solubilised and metabolised by the fungus.

Biodegradation of high molar mass polyamides could be increased by the introduction of various side groups as hydroxyl, methyl and benzyl or through copolymerisation [197]. In this context, poly(ester-amides)s are probably one of the most promising biodegradable structures among the various amide copolymers. General review on the synthesis, properties and degradation (although mostly enzymatic) of poly(ester-amide) s is given in many publications [173,174,198]. The increasing interest in poly(ester-amide)s as a biodegradable materials for environmental applications prompts a thorough study of the biodegradation process in nature, degradation products and its biocompatibility, *etc.* For example, Okada *et al.* [199] reported on biodegradation of eight different poly(ester amide)s obtained by solution polycondensation of the *p*-toluenesulfonic acid salts of *O,O'*-bis(α-aminoacyl)-1,4:3,6-dianhydro-d-glucitol and bis(*p*-nitrophenyl) esters of aliphatic dicarboxylic acids. The poly(ester-amide)s obtained ranged from amorphous to semicrystalline, and were all soluble in various polar solvents. Soil burial degradation tests of the films of investigated poly(ester amide)s, followed by SEM observation, indicated that all structures are biodegradable. Moreover, their biodegradability was dependent on the molecular structure. Comparison of the biodegradability of the poly(ester amide) s with that of the corresponding polyesters showed that the poly(ester amide)s were, in general, less readily degraded in composted soil than the corresponding polyesters having the same aliphatic dicarboxylic acid unit.

CONCLUSIONS

Currently, when the environmentally safe and sustainable development is crucial to the society research, production, and utilisation of (bio)degradable polymers becomes an increasingly important issue. In view of fact that (bio)degradable polymers derived from either renewable or fossil recourses attract increasing attention due to their immense impact in everyday life, (bio)degradation should be a key feature of these materials in order not to accumulate in the environment and reduce the ecological risk. The environmental impact of polyester-based (bio)degradable polymers is still under discussion. The most important advantages and limitations from our point of view are summarised in Table 3 [200,201,202].

Table 3: The environmental impact of polyester-based (bio)degradable polymers

Polymer	The Impact on Sustainable Environment	
	Advantages	**Limitations**
PHA	Polymer from renewable natural resources; produced by bacteria as storage material; commercial production by living organisms using biochemical processes; biodegradable polymers with lack of toxicity, converted to the same metabolites as in living organisms: water and carbon dioxide; reduction of fossil energy ("old carbon") usage.	Availability of waste collection systems and recycling methods; compostable plastic waste deposited on a landfill has a negative social environmental impact; approval of new bioplastics by society requires high level of customers' awareness which depends on capital and education expenditure; too energy-intensive extraction stage.
aPHB	Synthetic polymer biodegrade under appropriate conditions (in the presence of PhaZ7 depolymerase from *Paucimonas lemoignei*) to form of monomer, dimer and trimer.	End-of-life treatment problems; post-synthetic residues (e.g., organic solvents).
PLA	Synthesis from renewable monomer; easily hydrolytically degradable polymer; less greenhouse gases emission and less consumption of non-renewable energy than traditional polymers.	Availability of waste collection systems and recycling methods; compostable plastic waste deposited on a landfill has a negative social environmental impact; approval of new bio-based plastics by society requires high level of customers' awareness which depends on capital and education expenditure.
Polyamides	Possible synthesis from renewable monomers. Bio-based polyamide and poly(ester-amide)s thermoplastics of valuable properties.	Biodegradation possible only for polyamides of low molar mass; environmental impact still under evaluation.

In addition to the biodegradable polymers which are converted to carbon dioxide and water under appropriate conditions, there is an emerging class of oxo-degradable polymers, often incorrectly referred to as oxo-biodegradable polymers although they are only disintegrated, but not mineralised. Therefore, the various stages of biodegradation (biodeterioration, biofragmentation and assimilation) need to be precisely investigated and validated [6,203].

(Bio)degradation research on aliphatic polyesters, which are currently among the most promising groups of (bio) degradable polymers, is mainly focused on the degradation in soil or compost. However, aqueous environments are also very important, since they are the major routes of entry of polymeric packaging materials into the environment. For this reason, studies on the biodegradation pathways and conditions under which the biodegradation may occur when comparing different environmental compartments could considerably improve the development of (bio)degradable polymeric materials. It is also very important to ensure a real biodegradability and incorporation into geochemical life cycle rather than to provide materials that can disintegrated only to small particles [203].

ACKNOWLEDGMENTS

This work was supported by the European Commission under the Seventh Framework Programme, POLINNOVA project no. 316086 and by the European Regional Development Fund, MARGEN project No. POIG.01.03.01-00-018/08, "New generation packaging materials made from plastics subject to the organic recycling" in the framework of the Innovative Economy Operational Programme (IE OP). The outcomes of this research will be used to implement the PLASTiCE project (3CE368P1, "Innovation value chain development for sustainable plastics in central Europe", Central Europe Programme, co-financed by ERDF).

AUTHOR CONTRIBUTIONS

Joanna Rydz was responsible for chapter 1, 2, 4 and for organization of the text; Wanda Sikorska was responsible for chapter 3, Mariya Kyulavska and Darinka Christova were responsible for chapter 5. Darinka Christova was also responsible for corrections.

REFERENCES

1. Kowalczuk M. Europe for Sustainable Plastics. eBrochure of Launch conference PLASTiCE; Bologna, Italy: 2011. [(accessed on 25 September 2014)]. New Generation of the Polymeric Packaging Materials Susceptible to Organic Recycling; pp. 29–31. Available online: http://www.plastice.org/pl/publications.
2. Sikorska W., Dacko P., Sobota M., Rydz J., Musioł M., Kowalczuk M. Degradation study of polymers from renewable resources and their compositions in industrial composting pile. Macromol. Symp.2008;272:132–135. doi: 10.1002/masy.200851219.
3. Musioł M.T., Rydz J., Sikorska W.J., Rychter P.R., Kowalczuk M.M. A preliminary study of the degradation of selected commercial packaging materials in compost and aqueous environments. Pol. J. Chem. Technol. 2011;13:55–57. doi: 10.2478/v10026-011-0011-z.
4. Avérous L. Polylactic acid: Synthesis, properties and applications. In: Belgacem M.N., Gandini A., editors. Monomers, Polymers and Composites from Renewable Resources. Elsevier; Oxford, UK: 2008. pp. 433–450.
5. Vert M. Aliphatic polyesters: Great degradable polymers that cannot do everything. Biomacromolecules.2005;6:538–546. doi: 10.1021/bm0494702.
6. Rydz J., Zawidlak-Węgrzyńska B., Christova D. Degradable polymers. In: Mishra M.K., editor. Encyclopedia of Biomedical Polymers and Polymeric Biomaterials. CRC Press; Boca Ratón, FL, USA: 2015.
7. Ikada Y., Tsuji H. Biodegradable polyesters for medical and ecological applications. Macromol. Rapid Commun. 2000;21:117–132. doi: 10.1002/(SICI)1521-3927(20000201)21:3<117::AID-MARC117>3.0.CO;2-X.
8. Azevedo H.S., Reis R.L. Understanding the enzymatic degradation of biodegradable polymers and strategies to control their degradation rate. In: Reis R.L., Roman J.S., editors. Biodegradable Systems in Tissue Engineering and Regenerative Medicine. CRC Press; Boca Ratón, FL, USA: 2004. pp. 177–201.
9. Murthy N., Wilson S., Sy J.C. Biodegradation of polymers. In: Matyjaszewski K., Möller M., editors. Polymer Science: A Comprehensive Reference. Elsevier; Amsterdam, The Netherlands: 2012. pp. 547–560.
10. McIntire J.E. Polyester fibers. In: Lewin M., Pearce E.M., editors. Handbook of Fiber Chemistry. 2nd ed. CRC Press; New York, NY, USA: 1998. pp. 1–69.
11. Andersson S.R., Hakkarainen M., Inkinen S., Södergård A., Albertsson A.-C. Leads to higher hydrolytic stability but more acidic hydrolysis product pattern. Biomacromolecules. 2010;11:1067–1073. doi: 10.1021/bm100029t.
12. Cam D., Suong-Hyu H., Ikada Y. Degradation of high molecular weight poly(l-lactide) in alkaline medium. Biomaterials. 1995;16:833–843. doi: 10.1016/0142-9612(95)94144-A.
13. Swift G. Requirements for biodegradable water-soluble polymers. Polym. Degrad. Stable. 1998;59:19–24. doi: 10.1016/S0141-3910(97)00162-6.
14. Mueller R.-J. Biological degradation of synthetic polyesters—Enzymes as potential catalysts for polyester recycling. Process. Biochem. 2006;41:2124–2128. doi: 10.1016/j.procbio.2006.05.018.
15. Walter T., Augusta J., Muller R.-J., Widdecke H., Klein J. Enzymatic degradation of a model polyester by lipase from *Rhizopus. delemar*. Enzym. Microb. Tecnhnol. 1995;17:218–224. doi: 10.1016/0141-0229(94)00007-E.
16. Khemani K., Scholz C. Introduction and overview of degradable and renewable polymers and materials. In: Khemani K., Scholz C., editors. Degradable Polymers and Materials: Principles and Practice. 2nd ed. American Chemical Society; Washington, DC, USA: 2012. p. xi.
17. Philip S., Keshavarz T., Roy I. Polyhydroxyalkanoates: Biodegradable polymers with a range of applications. J. Chem. Technol. Biotechnol. 2007;82:233–247. doi: 10.1002/jctb.1667.
18. Nair L.S., Laurencin C.T. Biodegradable polymers as biomaterials. Prog. Polym. Sci. 2007;32:762–798. doi: 10.1016/j.progpolymsci.2007.05.017.
19. Middleton J.C., Tipton A.J. Synthetic biodegradable polymers as orthopedic devices. Biomaterials.2000;21:2335–2346. doi: 10.1016/S0142-9612(00)00101-0.

20. Anderson A.J., Dawes E.A. Occurrence, metabolism, metabolic role, and industrial uses of bacterial polyhydroxyalkanoates. Microbiol. Rev. 1990;54:50–72.

21. Gomez J.G.C., Méndez B.S., Nikel P.I., Pettinari M.J., Prieto M.A., Silva L.F. Making green polymers even greener: Towards sustainable production of polyhydroxyalkanoates from agroindustrial by-products. In: Petre M., editor. Advances in Applied Biotechnology. InTech Open Access Publisher; 2012. [(accessed on 16 September 2014)]. Available online: http://dx.doi.org/10.5772/31847.

22. Steinbüchel A., Valentin H.E., Schonebaum A.J. Application of recombinant gene technology for production of polyhydroxyalkanoic acids: Biosynthesis of poly(4-hydroxybutyric acid) homopolyester. J. Environ. Polym. Degrad. 1994;2:67–74. doi: 10.1007/BF02074775.

23. Poirier Y., Somerville C., Schechtman L.A., Satkowski M.M., Noda I. Synthesis of high-molecular-weight poly([R]-3-hydroxybutyrate) in transgenic *Arabidopsis thaliana* plant cells. Int. J. Biol. Macromol.1995;17:7–12. doi: 10.1016/0141-8130(95)93511-U.

24. Verlinden R.A.J., Hill D.J., Kenward M.A., Williams C.D., Radecka I. Bacterial synthesis of biodegradable polyhydroxyalkanoates. J. Appl. Microbiol. 2007;102:1437–1449. doi: 10.1111/j.1365-2672.2007.03335.x.

25. Doi Y. Microbial Polyesters. VCH Publishers; Weinheim, Germany: 1990.

26. Lenz R.W., Marchessault R.H. Bacterial polyesters: Biosynthesis, biodegradable plastics and biotechnology. Biomacromolecules. 2005;6:1–8. doi: 10.1021/bm049700c.

27. Reusch R.N., Bryant M.E., Henry D.N. Increased poly-(*R*)-3-hydroxybutyrate concentrations in streptozotocin (STZ) diabetic rats. Acta Diabetol. 2003;40:91–94. doi: 10.1007/s00592-003-0037-7.

28. Dedkova E.N., Blatter L.A. Role of β-hydroxybutyrate, its polymer poly-β-hydroxybutyrate and inorganic polyphosphate in mammalian health and disease. Front. Physiol. 2014;5:1–22. doi: 10.3389/fphys.2014.00260.

29. Gao X., Chen J.-C., Wu Q., Chen G.-Q. Polyhydroxyalkanoates as a source of chemicals, polymers, and biofuels. Curr. Opin. Biotechnol. 2011;22:768–774. doi: 10.1016/j.copbio.2011.06.005.

30. Kunasundari B., Sudesh K. Isolation and recovery of microbial polyhydroxyalkanoates. Express Polym. Lett. 2011;5:620–634.doi:10.3144/expresspolymlett.2011.60.

31. Modi S., Koelling K., Vodovotz Y. Assessment of PHB with varying hydroxyvalerate content for potential packaging applications. Eur. Polym. J. 2011;47:179–186. doi: 10.1016/j.eurpolymj.2010.11.010.

32. Shi F.Y., Gross R.A., Rutherford D.R. Microbial Polyester Synthesis: Effects of poly(ethylene glycol) on product composition, repeat unit sequence, and end group structure. Macromolecules. 1996;29:10–17. doi: 10.1021/ma950707j.

33. Shi F.Y., Ashby R., Gross R.A. Use of poly(ethylene glycol)s to regulate poly(3-hydroxybutyrate) molecular weight during *Alcaligenes. eutrophus* cultivations. Macromolecules. 1996;29:7753–7758. doi: 10.1021/ma960805k.

34. Jedliński Z., Kowalczuk M., Kurcok P., Adamus G., Matuszowicz A., Sikorska W., Gross R.A., Xu J., Lenz R.W. Stereochemical control in the anionic-polymerization of beta-butyrolactone initiated with alkali-metal alkoxides. Macromolecules. 1996;29:3773–3777. doi: 10.1021/ma951888s.

35. Jedliński Z., Kurcok P., Lenz R.W. First facile synthesis of biomimetic poly-(*R*)-3-hydroxybutyrate via regioselective anionic polymerization of (*S*)-β-butyrolactone. Macromolecules. 1998;31:6718–6720. doi: 10.1021/ma980663p.

36. Rydz J., Wolna-Stypka K., Adamus G., Janeczek H., Musioł M., Sobota M., Marcinkowski A., Krzan A., Kowalczuk M. Forensic engineering of advanced polymeric materials. Part 1—Degradation studies of polylactide blends with atactic poly[(*R*,*S*)-3-hydroxybutyrate] in paraffin. Chem. Biochem. Eng. Q. 2015accepted.

37. Jedliński Z., Kowalczuk M., Głowkowski W., Grobelny J., Szwarc M. Novel polymerization of β-butyrolactone initiated by potassium naphthalenide in the presence of a crown ether or a cryptand.Macromolecules. 1991;24:349–352. doi: 10.1021/ma00002a002.

38. KurcokP., KowalczukM., HennekK., JedlińskiZ. Anionic polymerization of beta-lactones initiated with alkali-metal alkoxides: Reinvestigation of the polymerization mechanism. Macromolecules.1992;25:2017–2020. doi: 10.1021/ma00033a027.

39. Kurcok P., Śmiga M., Jedliński Z. β-Butyrolactone polymerization initiated with tetrabutylammonium carboxylates: A novel approach to biomimetic polyester synthesis. J. Polym. Sci. Part A: Polym. Chem.2002;40:2184–2189. doi: 10.1002/pola.10285.

40. Juzwa M., Jedliński Z. Novel synthesis of poly(3-hydroxybutyrate) Macromolecules. 2006;39:4627–4630. doi: 10.1021/ma0602369.

41. Kawalec M., Sobota M., Scandola M., Kowalczuk M., Kurcok P. A convenient route to PHB macromonomers via anionically controlled moderate-temperature degradation of PHB. J. Polym. Sci. Part A.2010;48:5490–5497. doi: 10.1002/pola.24357.

42. Grassie N., Murray E.J., Holmes P.A. The thermal degradation of poly(-(D)-β-hydroxybutyric acid): Part 1-Identification and quantitative analysis of products. Polym. Degrad. Stable. 1984;6:47–61. doi: 10.1016/0141-3910(84)90016-8.

43. Kopinke F.D., Remmler M., Mackenzie K. Thermal decomposition of biodegradable polyesters I: Poly(β-hydroxybutyric acid) Polym. Degrad. Stable. 1996;52:25–38. doi: 10.1016/0141-3910(95)00221-9.

44. Nguyen S., Yu G., Marchessault R.H. Thermal degradation of poly(3-hydroxyalkanoates): Preparation of well-defined oligomers. Biomacromolecules. 2002;3:219–224. doi: 10.1021/bm0156274.

45. Kim J.K., Doi Y., Abe H. Effects of residual metal

compounds and chain-end structure on thermal degradation of poly(3-hydroxybutyric acid) Polym. Degrad. Stable. 2006;91:769–777. doi: 10.1016/j.polymdegradstab.2005.06.004.

46. Jedliński Z., Kurcok P., Kowalczuk M. Method of the Synthesis of Poly(3-Hydroxybutyrate) 199104.Polish Patent. 2008 Apr 16;

47. Kurcok P., Kowalczuk M., Kawalec M., Sobota M., Michalak M. Method of Puryfying β-Butyrolactone, in Particular for the Synthesis of poly(3-Hydroxybutyrate) and Its Copolymers. P-393751. Polish Patent Application. 2011 Jan 27;

48. Kawalec M., Adamus G., Kurcok P., Kowalczuk M., Foltran I., Focarete M.L., Scandola M. Carboxylate induced degradation of poly(3-hydroxybutyrate)s. Biomacromolecules. 2007;8:1053–1058. doi: 10.1021/bm061155n.

49. Arkin Z.G., Rydz J., Adamus G., Kowalczuk M. Watersoluble l-alanine and related oligopeptide conjugates with poly[(*R*,*S*)-3-hydroxybutanoic acid. Oligomers] Synthesis and structural studies by means of electrospray ionization multistage mass spectrometry. J. Biomater. Sci. Polym. Ed. 2001;12:297–305. doi: 10.1163/156856201750180825.

50. Kowalczuk M., Adamus G., Jedliński Z. Synthesis of new graft polymers via anionic grafting of β-butyrolactone on poly(methyl methacrylate) Macromolecules. 1994;27:572–575. doi: 10.1021/ma00080a035.

51. Neugebauer D., Rydz J., Goebel I., Dacko P., Kowalczuk M. Synthesis of graft copolymers containing biodegradable poly(3-hydroxybutyrate) chains. Macromolecules. 2007;40:1767–1773. doi: 10.1021/ma062251j.

52. Koseva N.S., Novakov Ch.P., Rydz J., Kurcok P., Kowalczuk M. Synthesis of aPHB-PEG brush co-polymers through ATRP in a macroinitiator–macromonomer feed system and their characterization. Des. Monomers Polym. 2010;13:579–595. doi: 10.1163/138577210X530675.

53. Kawalec M., Adamus G., Kurcok P., Kowalczuk M. Synthesis of Poly[(*R*,*S*)-3-hydroxybutyrate-block-ethylene glycol-block-(*R*,*S*)-3-hydroxybutyrate. Macromol. Symp. 2007;253:59–64. doi: 10.1002/masy.200750707.

54. Liu K.L., Goh S.H., Li J. Controlled synthesis and characterizations of amphiphilic poly[(*R*,*S*)-3-hydroxybutyrate]-poly(ethylene glycol)-poly[(*R*,*S*)-3-hydroxybutyrate] triblock copolymers. Polymer.2008;49:732–741. doi: 10.1016/j.polymer.2007.12.017.

55. Hazer B. Macromonomeric initiators. In: Salamone J.C., editor. Polymeric Materials Encyclopedia. CRC Press; Boca Ratón, FL, USA: 1996. pp. 3911–3918.

56. Jedliński Z., Kowalczuk M., Adamus G., Sikorska W., Rydz J. Novel synthesis of functionalized poly(3-hydroxybutanoic acid) and its copolymers. Int. J. Biol. Macromol. 1999;25:247–253. doi: 10.1016/S0141-8130(99)00039-2.

57. Adamus G., Sikorska W., Montaudo M., Scandola M., Kowalczuk M. Sequence distribution and fragmentation studies of bacterial copolyester macromolecules: Characterization of PHBV macroinitiator by electrospray ion-trap multistage mass spectrometry. Macromolecules. 2000;33:5797–5802. doi: 10.1021/ma000005g.

58. Adamus G., Sikorska W., Kowalczuk M., Noda I., Satkowski M.M. Electrospray ion-trap multistage mass spectrometry for characterisation of co-monomer compositional distribution of bacterial poly(3-hydroxybutyrate-*co*-3-hydroxyhexanoate) at the molecular level. Rapid Commun. Mass Spectrom.2003;17:2260–2266. doi: 10.1002/rcm.1190.

59. Adamus G. Aliphatic polyesters for advanced technologies structural characterization of biopolyesters with the aid of mass spectrometry. Macromol. Symp. 2006;239:77–83. doi: 10.1002/masy.200690113.

60. Zagar E., Krzan A., Adamus G., Kowalczuk M. Sequence Distribution in Microbial Poly(3-hydroxybutyrate-*co*-3-hydroxyvalerate) Co-polyesters determined by NMR and MS. Biomacromolecules.2006;7:2210–2216. doi: 10.1021/bm060201g.

61. Adamus G., Sikorska W., Janeczek H., Kwiecien M., Sobota M., Kowalczuk M. Novel block copolymers of atactic PHB with natural PHA for cardiovascular engineering: Synthesis and characterization. Eur. Polym. J. 2012;48:621–631. doi: 10.1016/j.eurpolymj.2011.12.017.

62. Kurcok P., Dubois Ph., Sikorska W., Jedliński Z., Jerome R. Macromolecular engineering of lactones and lactides. 24. Controlled synthesis of (*R*,*S*)-β-butyrolactone-b-ε-caprolactone block copolymers by anionic and coordination polymerization. Macromolecules. 1997;30:5591–5595. doi: 10.1021/ma970212p.

63. Scandola M., Focarte L., Gazzano M., Matuszowicz A., Sikorska W., Adamus G., Kurcok P., Kowalczuk M., Jedliński Z. Crystallinity-induced biodegradation of novel [(*R*,*S*)-β-butyrolactone]-b-pivalolactone copolymers. Macromolecules. 1997;30:7743–7748. doi: 10.1021/ma970989i.

64. Holmes P.A. Applications of PHB—A microbially produced biodegradable thermoplastic. Phys. Technol. 1985;14:32–36. doi: 10.1088/0305-4624/16/1/305.

65. Wacker Chemie A.G., editor. VINNEX®—Enabling the Next Generation of Bioplastics. Wacker Chemie AG; München, Germany: [(accessed on 25 September 2014)]. Available online:http://www.wacker.com/cms/media/publications/downloads/7002_EN.pdf.

66. Metabolix, Inc. Metabolix. to Highlight High Performance Biopolymers for Packaging Applications at Interpack 2014. [(accessed on 25 September 2014)]. Available online:http://ir.metabolix.com/releasedetail.cfm?ReleaseID=845609.

67. Metabolix, Inc. Reinvent Your Brand with Eco-Friendly Products. [(accessed on 25 September 2014)]. Available online: http://www.metabolix.com/brand-owner.

68. Website of Biomer: Krailling, Germany. [(accessed on 24 July 2014)]. Available online:http://www.biomer.de.

69. Chanprateep S. Current trends in biodegradable polyhydroxyalkanoates. J. Biosci.

Bioeng.2010;110:621–632. doi: 10.1016/j.jbiosc.2010.07.014.

70. Chen G-Q. A microbial polyhydroxyalkanoates (PHA) based bio- and materials industry. Chem. Soc. Rev. 2009;38:2434–2446.

71. Marketsandmarkets.com. Polyhydroxyalkanoate. (PHA) Market, by Application (Packaging, Food Services, Bio-Medical, Agriculture) & Raw Material—Global Trends & Forecasts to 2018. [(accessed on 14 March 2014)]. Available online: http://www.marketsandmarkets.com/Market-Reports/pha-market-395.html.

72. Matavulj M., Sad N., Molitoris H.P. Biodegradation of Polyhydroxyalkanoate-Based Plastic (BIOPOL) Under Different Environmental Conditions. Hoppea, Denkschr. Regensb. Bot. Ges. Bresinsky-Festschrift; Regensburg, Germany: 2000. pp. 735–749.

73. Tsuji H., Suzuyoshi K. Environmental degradation of biodegradable polyesters. 2. Poly(ε-caprolactone), poly[(*R*)-3-hydroxybutyrate] and poly(l-lactide) films in natural dynamic seawater. Polym. Degrad. Stable.2002;75:347–355. doi: 10.1016/S0141-3910(01)00240-3.

74. Volova T.G., Boyandin A.N., Vasiliev A.D., Karpov V.A. Biodegradation of polyhydroxyalkanoates (PHAs) in tropicalcoastalwatersandidentificationofPHA-degrading bacteria. Polym. Degrad. Stable.2010;95:2350–2359. doi: 10.1016/j.polymdegradstab.2010.08.023.

75. Akmal D., Azizan M.N., Majid M.I.A. Biodegradation of microbial polyesters P(3HB) and P(3HB-*co*-3HV) under the tropical climate environment. Polym. Degrad. Stable. 2003;80:513–518. doi: 10.1016/S0141-3910(03)00034-X.

76. O□Brine T., Thompson R.C. Degradation of plastic carrier bags in the marine environment. Mar. Pollut. Bull. 2010;60:2279–2283. doi: 10.1016/j.marpolbul.2010.08.005.

77. Bilck A., Grossmann M., Yamashita F. Biodegradable mulch films for strawberry production. Polym. Test. 2010;29:471–476. doi: 10.1016/j.polymertesting.2010.02.007.

78. Rutkowska M., Krasowska K., Heimowska A., Kowalczuk M. Degradation of the blends of natural and synthetic copolysters in different natural environment. Macromol. Symp. 2003;197:421–429. doi: 10.1002/masy.200350736.

79. Mohee R., Unmar G.D., Mudhoo A., Khadoo P. Biodegradability of biodegradable/degradable plastic materials under aerobic and anaerobic conditions. Waste Manag. 2008;28:1624–1629. doi: 10.1016/j.wasman.2007.07.003.

80. Freier T., Kunze C., Nischan C., Kramer S., Sternberg K., Sass M., Hopt U.T., Schmitz K.-P. *In vitro*and *in vivo* degradation studies for development of a biodegradable patch based on poly(3-hydroxybutyrate) Biomaterials. 2002;23:2649–2657. doi: 10.1016/S0142-9612(01)00405-7.

81. Scandola M., Focarete L., Adamus G., Sikorska W., Kowalczuk M., Jedliński Z., Baranowska I., Świerczek S., Gnatowski M. Polymer blends of natural poly(3-hydroxybutyrate-*co*-3-hydroxyvalerate) and synthetic atactic poly(3-hydroxybutyrate). Characterization and biodegradation studies. Macromolecules.1997;30:2568–2574. doi: 10.1021/ma961431y.

82. Abe H., Matsubara I., Doi Y. Physical properties and enzymic degradability of polymer blends of bacterial poly[(*R*)-3-hydroxybutyrate] and poly[(*R,S*)-3-hydroxybutyrate] stereoisomers. Macromolecules.1995;28:844–853. doi: 10.1021/ma00108a007.

83. Focarete M.L., Ceccorulli G., Scandola M., Kowalczuk M. Further Evidence of crystallinity-induced biodegradation of synthetic atactic poly(3-hydroxybutyrate) by PHB-depolymerase A from *Pseudomonas lemoignei*. Blends of atactic Poly(3-hydroxybutyrate) with crystalline polyesters. Macromolecules.1998;31:8485–8492. doi: 10.1021/ma981115e.

84. Yu L., Dean K., Li L. Polymer blends and composites from renewable resources. Prog. Polym. Sci.2006;31:576–602. doi: 10.1016/j.progpolymsci.2006.03.002.

85. Avella M., Martuscelli E., Raimo M. Review: Properties of blends and composites based on poly(3-hydroxy)butyrate (PHB) and poly(3-hydroxybutyrate-hydroxyvalerate) (PHBV) copolymers. J. Mater. Sci.2000;35:523–545. doi: 10.1023/A:1004740522751.

86. Zini E., Focarete M.L., Noda I., Scandola M. Bio-composite of bacterial poly(3-hydroxybutyrate-*co*-3-hydroxyhexanoate) reinforced with vegatable fibers. Compos. Sci. Technol. 2007;67:2085–2094. doi: 10.1016/j.compscitech.2006.11.015.

87. Avella M., La Rota G., Martuscelli E., Raimo M. Poly(3-hydroxybutyrate-*co*-3-hydroxyvalerate) and wheat straw fibre composites: Thermal, mechanical properties and biodegradation behaviour. J. Mater. Sci.2000;35:829–836. doi: 10.1023/A:1004773603516.

88. Rutkowska M., Krasowska K., Heimowska A., Adamus G., Sobota M., Musioł M., Janeczek H., Sikorska W., Krzan A., Žagar E., et al. Environmental degradation of blends of atactic poly[(*R,S*)-3-hydroxybutyrate] with natural PHBV in Baltic Sea water and compost with activated sludge. J. Polym. Environ. 2008;16:183–191. doi: 10.1007/s10924-008-0100-0.

89. Rychter P., Biczak R., Herman B., Smyłła A., Kurcok P., Adamus G., Kowalczuk M. Environmental degradation of polyester blends containing atactic poly(3-hydroxybutyrate). Biodegradation in soil and ecotoxicological impact. Biomacromolecules. 2006;7:3125–3131. doi: 10.1021/bm060708r.

90. Gumel A.M., Annuar M.S.M., Chisti Y. Recent advances in the production, recovery and applications of polyhydroxyalkanoates. J. Polym. Environ. 2013;21:580–605. doi: 10.1007/s10924-012-0527-1.

91. Piddubnyak V., Kurcok P., Matuszowicz A., Glowala M., Fiszer-Kierzkowska A., Jedliński Z., Juzwa M., Krawczyk Z. Oligo-3-hydroxybutyrates as potential carriers for drug delivery. Biomaterials.2004;25:5271–5279. doi: 10.1016/j.biomaterials.2003.12.029.

92. Juzwa M., Rusin A., Zawidlak-Węgrzyńska B., Krawczyk

Z., Obara I., Jedliński Z. Oligo(3-hydroxybutanoate) conjugates with acetylsalicylic acid and their antitumour activity. Eur. J. Med. Chem.2008;43:1785–1790. doi: 10.1016/j.ejmech.2007.11.004.

93. Zawidlak-Węgrzyńska B., Kawalec M., Bosek I., Łuczyk-Juzwa M., Adamus G., Rusin A., Filipczak P., Głowala-Kosińska M., Wolańska K., Krawczyk Z., et al. Synthesis and antiproliferative properties of ibuprofen-oligo(3-hydroxybutyrate) conjugates. Eur. J. Med. Chem. 2010;45:1833–1842. doi: 10.1016/j.ejmech.2010.01.020.

94. Chen G., Wang Y. Medical applications of biopolyesters polyhydroxyalkanoates. Chin. J. Polym. Sci.2013;31:719–736. doi: 10.1007/s10118-013-1280-1.

95. Johnson K., Kleerebezem R., van Loosdrecht M.C.M. Influence of ammonium on the accumulation of polyhydroxybutyrate (PHB) in aerobic open mixed cultures. J. Biotechnol. 2010;147:73–79. doi: 10.1016/j.jbiotec.2010.02.003.

96. Zhang X.J., Luo R.C., Wang Z., Deng Y., Chen G.Q. Application of (*R*)-3-hydroxyalkanoate methyl esters derived from microbial polyhydroxyalkanoates as novel biofuels. Biomacromolecules. 2009;10:707–711. doi: 10.1021/bm801424e.

97. Gruber P., O'Brien M. Polylactides "Natureworks™ PLA" In: Steinbüchel A., Doi Y., editors.Biopolymers: Polyesters III—Applications and Commercial Products. Volume 4. Wiley-VCH; Weinheim, Germany: 2002. pp. 235–239.

98. Avérous L., Pollet E. Biodegradable polymers. In: Avérous L., Pollet E., editors. Environmental Silicate Nano-Biocomposites. Springer-Verlag; London, UK: 2012.

99. Huang L.P., Jin B., Lant P., Zhou J. Biotechnological production of lactic acid integrated with potato wastewater treatment by *Rhizopus. arrhizus*. Chem. Technol. Biotechnol. 2003;78:899–906. doi: 10.1002/jctb.877.

100. Yao K., Tang C. Controlled polymerization of next-generation renewable monomers and beyond. Macromolecules. 2013;46:1689–1712. doi: 10.1021/ma3019574.

101. Vink E.T.H., Rábago K.R., Glassner D.A., Gruber P.R. Applications of life cycle assessment to NatureWorks polylactide (PLA) production. Polym. Degrad. Stable. 2003;80:403–419. doi: 10.1016/S0141-3910(02)00372-5.

102. Nampoothiri K.M., Nair N.R., John R.P. An overview of the recent developments in polylactide (PLA) research. Bioresour. Technol. 2010;101:8493–8501. doi: 10.1016/j.biortech.2010.05.092.

103. Okada M. Chemical syntheses of biodegradable polymers. Prog. Polym. Sci. 2002;27:87–133. doi: 10.1016/S0079-6700(01)00039-9.

104. Albertsson A-C., Varma I.K. Aliphatic polyesters: Synthesis, properties and applications. Adv. Polym. Sci. 2002;157:1–40.

105. Kowalczuk M., Adamus G., Sikorska W., Rydz J. Structural studies of biorelated polymers derived from natural PHA and their synthetic analogues with the aid of electrospray multistage mass spectrometry.Polym. Prepr. Am. Chem. Soc. Div. Polym. Chem. 2000;41:1626–1627.

106. Shuai X., Jedliński Z., Kowalczuk M., Rydz J., Tan H. Enzymatic synthesis of polyesters from hydroxyl acids. Eur. Polym. J. 1999;35:721–725. doi: 10.1016/S0014-3057(98)00164-5.

107. Cheng Y., Deng S., Chen P., Ruan R. Polylactic acid (PLA) synthesis and modifications: A review.Front. Chem. China. 2009;4:259–264. doi: 10.1007/s11458-009-0092-x.

108. Dobrzynski P. Initiation of L-lactide polymerization carried out with zirconium (IV) acetylacetonate. J. Polym. Sci. Part A. 2004;42:1886–1900. doi: 10.1002/pola.20026.

109. Lasprilla A.J.R., Martinez G.A.R., Lunelli B.H., Jardini A.L., Filho R.M. Poly-lactic acid synthesis for application in biomedical devices—A review. Biotechnol. Adv. 2012;30:321–328. doi: 10.1016/j.biotechadv.2011.06.019.

110. Dobrzynski P., Li S., Kasperczyk J., Bero M., Gasc F., Vert M. Structure-property relationships of copolymers obtained by ring-opening polymerization of glycolide and ε-caprolactone. Part 1. Synthesis and characterization. Biomacromolecules. 2005;6:483–488. doi: 10.1021/bm0494592.

111. Cabezas L.I., Mazarro R., Gracia I., de Lucas A., Rodríguez J.F. Optimizing the bulk copolymerization of d,l-lactide and glycolide by response surface methodology. Express Polym. Lett. 2013;7:886–894. doi: 10.3144/expresspolymlett.2013.86.

112. Dobrzynski P., Kasperczyk J., Janeczek H. Synthesis of biodegradable copolymers with the use of low toxic zirconium compounds. 1. Copolymerization of glycolide with l-Lactide initiated by $Zr(Acac)_4$.Macromolecules. 2001;34:5090–5099. doi: 10.1021/ma0018143.

113. Pamula E., Dobrzynski P., Bero M., Paluszkiewicz C. Hydrolytic degradation of porous scaffolds for tissue engineering from terpolymer of l-lactide, ε-caprolactone and glycolide. J. Mol. Struct. 2005;744–747:557–562.

114. Kasperczyk J., Li S., Jaworska J., Dobrzynski P., Vert M. Degradation of copolymers obtained by ring-opening polymerization of glycolide and ε-caprolactone: A high resolution NMR and ESI-MS study. Polym. Degrad. Stable. 2008;93:990–999. doi: 10.1016/j.polymdegradstab.2008.01.019.

115. D'Avila C., Erbetta C., Alves R.J., Resende J.M., de Souza Freitas R.F., de Sousa R.G. Synthesis and characterization of poly(d,l-lactide-*co*-glycolide) copolymer. J. Biomater. Nanobiotechnol. 2012;3:208–225. doi: 10.4236/jbnb.2012.32027.

116. Hu Y., He Y., Wei J., Fan Z., Dobrzynski P., Kasperczyk J., Bero M., Li S. Hydrolytic degradation of glycolide/l-lactide/ε-caprolactone terpolymers initiated by zirconium (IV) acetylacetonate. J. Appl. Pol. Sci.2007;103:2451–2456. doi: 10.1002/app.25529.

117. Chłopek J., Morawska-Chochol A., Paluszkiewicz C.,

Jaworska J., Kasperczyk J., Dobrzynski P. FTIR and NMR study of poly(lactide-*co*-glycolide) and hydroxyapatite implant degradation under *in vivo*conditions. Polym. Degrad. Stable. 2009;94:1479–1485. doi: 10.1016/j.polymdegradstab.2009.05.010.

118. Morawska-Chochol A., Jaworska J., Chlopek J., Kasperczyk J., Dobrzynski P., Paluszkiewicz C., Bajor G. Degradation of poly(lactide-*co*-glycolide) and its composites with carbon fibres and hydroxyapatite in rabbit femoral bone. Polym. Degrad. Stable. 2011;96:719–726. doi: 10.1016/j.polymdegradstab.2011.01.005.

119. Pamula E., Blazewicz M., Paluszkiewicz C., Dobrzynski P. FTIR study of degradation products of aliphatic polyesters-carbon fibres composites. J. Mol. Struct. 2001;596:69–75. doi: 10.1016/S0022-2860(01)00688-3.

120. Armentano I., Dottori M., Fortunati E., Mattioli S., Kenny J.M. Biodegradable polymer matrix nanocomposites for tissue engineering: A review. Polym. Degrad. Stable. 2010;95:2126–46. doi: 10.1016/j.polymdegradstab.2010.06.007.

121. Magon A., Pyda M. Study of crystalline and amorphous phases of biodegradable poly(lactic acid) by advanced thermal analysis. Polymer. 2009;50:3967–3973. doi: 10.1016/j.polymer.2009.06.052.

122. Calafel M.I., Remiro P.M., Cortázar M.M., Calahorra M.E. Cold crystallization and multiple melting behavior of poly(l-lactide) in homogeneous and in multiphasic epoxy blends. Colloid Polym. Sci.2010;288:283–296. doi: 10.1007/s00396-009-2156-3.

123. Kale G., Auras R., Singh S.P. Degradation of commercial biodegradable packages under real composting and ambient exposure conditions. J. Polym. Environ. 2006;14:317–334. doi: 10.1007/s10924-006-0015-6.

124. Bergeret A. Environmental-friendly biodegradable polymers and composites. In: Kumar S., editor.Integrated waste management. Volume I. InTech Open Access Publisher; 2011. [(accessed on 17 September 2013)]. Available online: http://www.intechopen.com/books/integrated-waste-management-volume-i/environmental-friendly-biodegradable-polymers-and-composites.

125. Zhang X., Espiritu M., Bilyk A., Kurniawan L. Morphological behaviour of poly(lactic acid) during hydrolytic degradation. Polym. Degrad. Stable. 2008;93:1964–1970. doi: 10.1016/j.polymdegradstab.2008.06.007.

126. Auras R.A., Harte B., Selke S., Hernandez R.J. Mechanical, physical, and barrier properties of poly(lactide) films. J. Plastic Film Sheet. 2003;19:123–135. doi: 10.1177/8756087903039702.

127. Ren J. Processing of PLA. In: Ren J., editor. Biodegradable Poly(Lactic Acid): Synthesis, Modification, Processing and Applications. Springer-Verlag; London, UK: 2011. pp. 142–207.

128. Ramkumar D.H.S., Bhattacharya M. Steady shear and dynamic properties of biodegradable polyesters.Polym. Eng. Sci. 1998;38:1426–1435. doi: 10.1002/pen.10313.

129. Bastioli C. Biodegradable materials-present situation and future perspectives. Macromol. Symp.1998;135:193–204. doi: 10.1002/masy.19981350122.

130. Höglund A., Odelius K., Albertsson A-C. Crucial differences in the hydrolytic degradation between industrial polylactide and laboratory-scale poly(l-lactide) Appl. Mater. Interfaces. 2012;23:2788–2793. doi: 10.1021/am300438k.

131. Cairncross R.A., Becker J.G., Ramaswamy S., O'Connor R. Moisture sorption, transport, and hydrolytic degradation in polylactide. Appl. Biochem. Biotechnol. 2006;131:774–785. doi: 10.1385/ABAB:131:1:774.

132. Krasowska K., Brzeska J., Rutkowska M., Dacko P., Sobota M., Kowalczuk M. The effect of poly(d,l-lactide) modification with poli[(*R*,*S*)-3-hydroxybutyrate] on the course of its degradation in natural environments. Polimery. 2008;53:730–736.

133. Torres A., Li S., Roussos S., Vert M. Poly(lactic acid) degradation in soil or under ncontrolled conditions. J. Appl. Polym. Sci. 1996;62:2295–2302. doi: 10.1002/(SICI)1097-4628(19961226)62:13<2295::AID-APP14>3.0.CO;2-2.

134. Gartiser S., Wallrabenstein M., Stiene G. Assessment of several test methods for the determination of the anaerobic biodegradability of polymers. J. Environ. Polymer Degrad. 1998;6:159–173. doi: 10.1023/A:1021869530253.

135. Kale G., Auras R., Singh S.P., Narayan R. Biodegradability of polylactide bottles in real and simulated composting conditions. Polym. Test. 2007;26:1049–1061. doi: 10.1016/j.polymertesting.2007.07.006.

136. Kolstad J.J., Vink E.T.H., de Wilde B., Debeer L. Assessment of anaerobic degradation of Ingeo polylactides under accelerated landfill conditions. Polym. Degrad. Stable. 2012;97:1131–1141. doi: 10.1016/j.polymdegradstab.2012.04.003.

137. Li S., Girard A., Garreau H., Vert M. Enzymatic degradation of polylactide stereocopolymers with predominant D-lactyl contents. Polym. Degrad. Stable. 2001;71:61–67. doi: 10.1016/S0141-3910(00)00152-X.

138. Hakkarainen M., Karlsson S., Albertsson A-C. Rapid (bio)degradation of polylactide by mixed culture of compost microorganisms-low molecular weight products and matrix changes. Polymer. 2000;41:2331–2338. doi: 10.1016/S0032-3861(99)00393-6.

139. Sikorska W., Richert J., Rydz J., Musioł M., Adamus G., Janeczek H., Kowalczuk M. Degradability studies of poly(l-lactide) after multi-reprocessing experiments in extruder. Polym. Degrad. Stable.2012;97:1891–1897. doi: 10.1016/j.polymdegradstab.2012.03.049.

140. Rydz J., Adamus G., Wolna-Stypka K., Marcinkowski A., Misiurska-Marczak M., Kowalczuk M.M. Degradation of polylactide in paraffin and selected protic media. Polym. Degrad. Stable. 2013;98:316–324. doi: 10.1016/j.polymdegradstab.2012.09.010.

141. Rydz J., Wolna-Stypka K., Musioł M., Szeluga U., Janeczek H., Kowalczuk M. Further evidence of polylactide degradation in paraffin and in selected protic media. A thermal analysis of eroded polylactide films. Polym. Degrad. Stable. 2013;98:1450–1457. doi: 10.1016/j.polymdegradstab.2013.05.005.

142. Hakkarainen M., Albertsson A.C., Karlsson S. Weight

losses and molecular weight changes correlated with the evolution of hydroxyacids in simulated *in vivo* degradation of homo- and copolymers of PLA and PGA. Polym. Degrad. Stable. 1996;52:283–291. doi: 10.1016/0141-3910(96)00009-2.

143. Siparsky G.L., Voorhees K.J., Miao F. Hydrolysis of polylactic acid (PLA) and polycaprolactone (PCL) in aqueous acetonitrile solutions: Autocatalysis. J. Polym. Environ. 1988;6:31–41. doi: 10.1023/A:1022826528673.

144. Lee M.W., Tan H.T., Chandrasekaran M., Ooi C.P. Synthesis and characterisation of PLLA by melt polycondensation using binary catalyst system. SIMTech Tech. Rep. 2005;6:40–45.

145. Grizzi I., Garreau H., Li S., Vert M. Hydrolytic degradation of devices based on poly(d,l-lactic acid) size-dependence. Biomaterials. 1995;16:305–311. doi: 10.1016/0142-9612(95)93258-F.

146. Jaworska J., Kasperczyk J., Dobrzynski P. Degradation process of bioresorbable copolyesters. Microstructure investigation by NMR and ESI-MS. Macromol. Symp. 2007;253:40–46. doi: 10.1002/masy.200750705.

147. Hakkarainen M., Finne-Wistrand A. Polylactide. In: Thomas S., Visakh P.M., editors. Handbook of Engineering and Specialty Thermoplastics: Polyethers and Polyesters. Volume 3. Wiley; New York, NY, USA: 2011. pp. 349–376.

148. Gruber P.R. Carbon Management: Implications for R & D in the Chemical Sciences and Technology. National Academies Press; Washington, DC, USA: 2001. Commodity polymers from renewable resources: Polylactic acid; pp. 166–184.

149. Ren J. Application in the field of commodity and industry product. In: Ren J., editor. Biodegradable Poly(Lactic Acid): Synthesis, Modification, Processing and Applications. Tsinghua University Press; Beijing and Springer-Verlag Berlin Heidelberg, China and Germany: 2010. pp. 208–239.

150. Hayes D., Dharmalingam S., Wadsworth L.C., Leonas K.K., Miles C., Inglis D.A. Biodegradable agricultural mulches derived from biopolymers. In: Khemani K., Scholz C., editors. Degradable Polymers and Materials: Principles and Practice. 2nd ed. Volume 1114. American Chemical Society; Washington, DC, USA: 2012. pp. 201–223. (ACS Symposium Series).

151. Rudnik E., Briassoulis D. Degradation behaviour of poly(lactic acid) films and fibres in soil under Mediterranean field conditions and laboratory simulations testing. Ind. Crop. Prod. 2011;33:648–658. doi: 10.1016/j.indcrop.2010.12.031.

152. Smock D. Bioplastics. Producers Ramp up Supply. 2012. [(accessed on 04 November 2014)]. Available online: http://www.mypurchasingcenter.com/commodities/commodities-articles/bioplastics-manufacturers-ramp-up-production.

153. Bioplastics Magazine New Market Research Projects Strong Growth for Lactic Acid. [(accessed on 25 September 2014)]. Available online: http://www.bioplasticsmagazine.com/en/news/meldungen/new-market-study-lactic-acid.php.

154. Lactic Acid Market Worth $3,577.5 Million and Polylactic Acid Market Worth $4,840.1 Million by 2019. [(accessed on 25 September 2014)]. Available online:http://www.marketsandmarkets.com/PressReleases/polylacticacid.asp.

155. Palmer R.J. Polyamides, Plastics. In: Mark H., editor. Encyclopedia of Polymer Science and Technology. 3rd ed. Volume 3. Wiley; New York, NY, USA: 2003. pp. 618–642.

156. Page I.B. Polyamides as Engineering Thermoplastic Materials. Smithers Rapra Technology; Shropshire, UK: 2000. *Rapra Review Reports* (Report 121)

157. Estes L.L., Schweizer M. Ullmann's Encyclopedia of Industrial Chemistry. Volume 14. Wiley-VCH; New York, NY, USA: 2011. Fibers, Polyamide fibers; p. 451.

158. Rulkens R., Koning C. Chemistry and Technology of Polyamides. In: Matyjaszewski K., Möller M., editors. Polymer Science: A Comprehensive Reference. Elsevier; Amsterdam, The Netherlands: 2012. pp. 431–467.

159. Hashimoto K. Ring-opening polymerization of lactams. Living anionic polymerization and its applications. Prog. Polym. Sci. 2000;25:1411–1462. doi: 10.1016/S0079-6700(00)00018-6.

160. Hashimoto K., Sudo E., Sugimura T., Inagaki Y. Synthesis of novel block copolymers containing polyamide4 segments and control of their biodegradability. J. Appl. Polym. Sci. 2004;92:3492–3498. doi: 10.1002/app.20348.

161. Kawasaki N., Nakayama A., Yamano N., Takeda S., Kawata Y., Yamamoto N., Aiba S. Synthesis, thermal and mechanical properties and biodegradation of branched polyamides 4. Polymer. 2005;46:987–993. doi: 10.1016/j.polymer.2005.06.092.

162. Nery L., Lefebvre H., Fradet A. Polyamide-polyester multiblock copolymers by chain-coupling reactions of carboxy-terminated polymers with phenylene and pyridylene bisoxazolines. J. Polym. Sci. Part A: Polym. Chem. 2005;43:1331–1341. doi: 10.1002/pola.20602.

163. Kyulavska M., Bryaskova R., Bozukova D., Mateva R. Synthesis, structure and behavior of new polycaprolactam copolymers based on poly(ethylene oxide)-poly(propylene oxide)-poly(ethylene oxide) macroactivators derived from Pluronic block copolymers. J. Polym. Res. 2014;21:471–485. doi: 10.1007/s10965-014-0471-y.

164. Prasad V.S., Hussain S.A., Pandurangadu V., PalaniKumar K. Modeling and Analysis of Spur Gear for Sugarcane Juice Machine under Static Load Condition by Using FEA. Int. J. Mod. Eng. Res. 2012;2:2862–2866.

165. Forman C. High.-Performance Films. CRC Press; Boca Ratón, FL, USA: 1996. p. 336.

166. McCoy M. Slowly changing how nylon is made. Chem. Eng. News. 2000;78:32–34. doi: 10.1021/cen-v078n040.p032.

167. Hoelderich W.F., Dahlhoff G. Developing Technolgy: The "greening" of nylon. The precursor to nylon 6, ε-caprolactam, is manufactured with toxic and corrosive chemicals and produces tons of byproducts.

Academic research suggests another approach based on heterogeneous catalysis. Chem. Inn. 2001;31:29–40.

168. Nylon 66 Market—Global Industry Analysis, Size, Share, Growth, Trends and Forecast, 2013–2019. [(accessed on 5 November 2014)]. Available online: http://www.transparencymarketresearch.com/nylon-66-market.html.

169. *A*dvantages of Welded Nylon for Powertrain Applications: Linear Vibration, Orbital Vibration and Hot Plate Welding Technologies. BASF Corporation. [(accessed on 5 November 2014)]. Available online:http://www2.basf.us//PLASTICSWEB/displayanyfile?id=0901a5e18000487b.

170. Brehmer B. Polyamides from biomass derived monomers. In: Kabasci S., editor. Bio-Based Plastics: Materials and Applications. John Wiley & Sons; Chichester, UK: 2013. pp. 275–293.

171. Kuciel S., Kuzniar P., Liber-Knec A. Polyamides from renewable sources as matricies of short fiber reinforced biocomposites. Polimery. 2012;57:627–634. doi: 10.14314/polimery.2012.627.

172. Thielen M. Basics of bio-polyamaides. Bioplastics Mag. 2010;5:50–53.

173. Fonseca A.C., Gil M.H., Simoes P.N. Biodegradable poly(ester amide)s—A remarkable opportunity for the biomedical area: Review on the synthesis, characterization and applications. Prog. Polym. Sci.2014;39:1291–1311. doi: 10.1016/j.progpolymsci.2013.11.007.

174. Rodriguez-Galan A., Franco L., Puiggali J. Degradable poly(esteramide)s for biomedical applications.Polymers. 2010;3:65–99. doi: 10.3390/polym3010065.

175. Herzog B., Kohan M.I., Mestemacher S.A., Pagilagan R.U., Redmond K. Ullmann's Encyclopedia of Industrial Chemistry. Volume 28. Wiley-VCH; Weinheim, Germany: 2013. Polyamides; p. 537.

176. Albertsson A.-C., Karlsson S. Biodegradable polymers. In: Allen G., Bevington J.C., editors.Comprehensive Polymer Science. Pergamon Press; Oxford, UK: 1992. pp. 285–297.

177. Williams D.F. Polymer degradation in biological environments. In: Allen G., Bevington J.C., editors. Comprehensive Polymer Science. Volume 6. Pergamon Press; Oxford, UK: 1989. pp. 607–719.

178. Negoro S. Biodegradation of nylon and other synthetic polyamides. In: Matsumura S., Steinbüchel A., editors. Biopolymers Online. Volume 9. Wiley-VCH; Weinheim, Germany: 2005. pp. 395–415.

179. Alexander M. Biodegradation and Bioremediation. Academic Press; San Diego, CA, USA: 1999. pp. 393–407.

180. Morales-Gámez L., Soto D., Franco L., Puiggalí J. Brill transition and melt crystallization of nylon 56: An odd-even polyamide with two hydrogen-bonding directions. Polymer. 2010;51:5788–5798. doi: 10.1016/j.polymer.2010.09.074.

181. Negoro S. Biodegradation of nylon oligomers. Appl. Microbiol. Biotechnol. 2000;54:461–466. doi: 10.1007/s002530000434.

182. Deguchi T., Kakezawa M., Nishida T. Nylon biodegradation by lignin degrading fungi. Appl. Environ. Microbiol. 1997;63:329–331.

183. Shimao M. Biodegradation of plastics. Curr. Opin. Biotechnol. 2001;12:242–247. doi: 10.1016/S0958-1669(00)00206-8.

184. Friedrich J., Zalar P., Mohorcic M., Klun U., Krzan A. Ability of fungi to degrade synthetic polymer nylon-6. Chemosphere. 2007;67:2089–2095. doi: 10.1016/j.chemosphere.2006.09.038.

185. Andreoni V., Baggi G., Manfrin P. Bacterial degradation of 6-aminocaproic acid polyamides (nylon 6) of low molecular weight. Int. Biodeter. Biodegr. 1993;31:41–53. doi: 10.1016/0964-8305(93)90013-R.

186. Yamano N., Nakayama A., Kawasaki N., Yamamoto N., Aiba S. Mechanism and characterization of polyamides 4 degradation by Pseudomonas sp. J. Polym. Environ. 2008;16:141–146. doi: 10.1007/s10924-008-0090-y.

187. Yamano N., Kawasaki N., Oshima M., Nakayama A. Polyamide 4 with long-chain fatty acid groups—Suppressing the biodegradability of biodegradable polymers. Polym. Degrad. Stable. 2014;108:116–122. doi: 10.1016/j.polymdegradstab.2014.06.011.

188. Hashimoto K., Hamano T., Okada M. Degradation of several polyamides in soils. J. Appl. Polym. Sci.1994;54:1579–1583. doi: 10.1002/app.1994.070541023.

189. Hashimoto K., Sudo E., Ohta K., Sugimura T., Yamada H., Aoki T. Biodegradation of nylon4 and its blends with nylon6. J. Appl. Polym. Sci. 2002;86:2307–2311. doi: 10.1002/app.11235.

190. Tachibana K., Hashimoto K., Yoshikawa M., Okawa H. Isolation and characterization of microorganisms degrading nylon 4 in the composted soil. Polym. Degrad. Stable. 2010;95:912–917. doi: 10.1016/j.polymdegradstab.2010.03.031.

191. Tachibana K., Urano Y., Numata K. Biodegradability of nylon 4 film in a marine environment. Polym. Degrad. Stable. 2013;98:1847–1851. doi: 10.1016/j.polymdegradstab.2013.05.007.

192. Eriksson C., Burton H. Origins and biological accumulation of small plastic particles in fur seals from Macquarie Island. Ambio. 2003;32:380–384.

193. Deguchi T., Kitaoka Y., Kakezawa M., Nishida T. Purification and characterization of a nylon-degrading enzyme. Appl. Environ. Microbiol. 1998;64:1366–1371.

194. Tomita K., Ikeda N., Uedo A. Isolation and characterization of a thermophilic bacterium, Geobacillus thermocatenulatus, degrading nylon 12 and nylon 66. Biotechnol. Lett. 2003;25:1743–1746. doi: 10.1023/A:1026091711130.

195. Tomita K., Hayashi N., Ikeda N., Kikuchi Y. Isolation of a thermophilic bacterium degrading some nylons. Polym. Degrad. Stable. 2003;81:511–514. doi: 10.1016/S0141-3910(03)00151-4.

196. Oppermann F.B., Pickartz S., Steinbüchel A. Biodegradation of polyamides. Polym. Degrad.

Stable. 1998;59:337–344. doi: 10.1016/S0141-3910(97)00175-4.

197. Mohanty A.K., Misra M., Hinrichsen G. Biofibres, biodegradable polymers and biocomposites: An overview. Macromol. Mater. Eng. 2000;276–277:1–24.

198. Lips P., Dijkstra P.J. Biodegradable Polyesteramides. In: Smith R., editor. Biodegradable Polymers for Industrial Applications. Woodhead Publishing Limited; Cambridge, UK: 2005. pp. 107–135.

199. Okada M., Yamada M., Yokoe M., Aoi K. Biodegradable polymers based on renewable resources. V. Synthesis and biodegradation behavior of poly(ester amide)s composed of 1,4:3,6-dianhydro-d-glucitol, α-amino acid, and aliphatic dicarboxylic acid units. J. Appl. Polym. Sci. 2001;81:2721–2734. doi: 10.1002/app.1718.

200. Slater S., Glassner D., Vink E., Gerngross T. Evaluating the environmental impact of biopolymers. In: Doi Y., Steinbüchel A., editors. Biopolymers Online. Volume 10. Wiley-VCH; Weinheim, Germany: 2005. pp. 473–492.

201. Papageorgiou A.C., Hermawan S., Singh C.B., Jendrossek D. Structural basis of poly(3-hydroxybutyrate) hydrolysis by PhaZ7 depolymerase from *Paucimonas. lemoignei*. J. Mol. Biol. 2008;382:1184–1194. doi: 10.1016/j.jmb.2008.07.078.

202. Fritz J., Link U., Braun R. Environmental impacts of biobased/biodegradable packaging. Starch. 2001;53:105–109. doi: 10.1002/1521-379X(200104)53:3/4<105::AID-STAR105>3.0.CO;2-I.

203. Eubeler J.P., Bernhard M., Zok S., Knepper T.P. Environmental biodegradation of synthetic polymers I. Test methodologies and procedures. Trend. Anal. Chem. 2009;28:1057–1072. doi: 10.1016/j.trac.2009.06.007.

CHAPTER 15

CHARACTERIZATION OF FLEXIBLE COPOLYMER OPTICAL FIBERS FOR FORCE SENSING APPLICATIONS

Marek Krehel[1,2], René M. Rossi[1] , Gian-Luca Bona[1,2] and Lukas J. Scherer[1]

[1]Empa, Swiss Federal Laboratories for Materials Science and Technology, Laboratory for Protection and Physiology, Lerchenfeldstrasse 5, St. Gallen 9014, Switzerland

[2]ETH Zurich, Swiss Federal Institute of Technology, Department of Information Technology and Electrical Engineering, Gloriastrasse 35, Zurich 8092, Switzerland

ABSTRACT

In this paper, different polymer optical fibres for applications in force sensing systems in textile fabrics are reported. The proposed method is based on the deflection of the light in fibre waveguides. Applying a force on the fibre changes the geometry and affects the wave guiding properties and hence induces light loss in the optical fibre. Fibres out of three different elastic and transparent copolymer materials were successfully produced and tested. Moreover, the influence of the diameter on the sensing properties was studied. The detectable force ranges from 0.05 N to 40 N (applied on 3 cm of fibre length), which can be regulated with the material and the diameter of the fibre. The detected signal loss varied from 0.6% to 78.3%. The fibres have attenuation parameters between 0.16–0.25 dB/cm at 652 nm. We show that the cross-sensitivies to temperature, strain and bends are low. Moreover, the high yield strength (0.0039–0.0054 GPa) and flexibility make these fibres very attractive candidates for integration into textiles to form wearable sensors, medical textiles or even computing systems.

INTRODUCTION

Optical fibres-based sensors have numerous advantages compared to electrical or chemical devices, e.g., insensitivity to electromagnetic fields, water and corrosion resistance, compact size and light weight [1,2]. Due to these benefits, optical fibre based sensors are used in numerous applications to detect changes in chemical and/or physical parameters [1,3–6]. Various methods to detect pressure using optical fibres are known. Wang *et al.* proposed a pressure sensor that employs the effect of photoelasticity [7]. Urban *et al.*designed a pressure sensor that relies on FBG lateral deformation [8]. Within the Seventh Framework Programme "Ofseth", optical fibres using optical time-domain reflectometry for distributed respiration measurement and optical fibres with long period gratings in microstructured POF were incorporated into elastic fabrics to measure the breathing rate and the breathing volume by measuring the changes in applied pressure [9–11]. The most common method to determine the pressure using optical fibres is by using fibre Bragg gratings [8,12–14]. To implement optical fibre based pressure sensors into textile fabrics, the fibres should be highly flexible, which is not the case for glass-based fibres most often used for fibre Bragg grating based sensors. Moreover, sensors based on fibre Bragg gratings have to be single mode and are dependent on temperature, strain and pressure simultaneously and require compensation of some of the parameters in order to sense one of them [15,16].

In this study, elastically deformable light pipes, so called multimode optical fibres which react to applied pressure by deflecting light in the fibre structure, are reported. Due to their

Citation: Marek Krehel, René M. Rossi, Gian-Luca Bona, and Lukas J. Scherer , Characterization of Flexible Copolymer Optical Fibers for Force Sensing Applications, doi:10.3390/s130911956.

high flexibility and their appropriate tensile strength (from 0.0033 to 0.0056 GPa), they meet the critical criteria to form medical textiles [17,18]. Long-term body monitoring solutions using textile fabrics have been intensively studied in recent years [19,20]. Plastic optical fibres have several advantages compared to textile electronics when wearing them close to the body, e.g., comfort, ease of movements, and reduced movement artefacts [17]. The electronics involved in the proposed solutions often influence negatively the haptic of medical textiles. Flexible and smart fibres allow the separation of the rigid electronics from the measured region and thus from the body. Possible body parameters which could be measured with the proposed solution are, e.g., muscles activity, motion detections or breath monitoring [21]. For technical applications, these pressure sensors can be used for counting occupied seats in airplanes, cars or for monitoring immobile patients in hospitals to prevent decubitus.

EXPERIMENTAL SECTION

Materials and Methods

All the polymers used were purchased from WackerChemie AG (München, Germany). The polymers were all from the Geniomer group and are copolymers containing a soft silicon part and a hard polyurethane part. Geniomer 100-HDS and Geniomer 100 consist of about 70% silicon and 30% polyurethane, while Geniomer 175 consists of only 10% polyurethane and 90% silicon and has a lower processing temperature than the two Geniomer 100 polymers. Geniomer 100-HDS is Geniomer 100 with the addition of highly dispersed silicate particles.

Fibres Extrusion with Melt Flow Index

The transparent light pipes were produced by a melt flow index apparatus (MFI), model 7085.15a, provided by ZWICK (Ulm, Germany) with an external drawing engine. In Table 1 the extrusion parameters at which the fibres were extruded are shown. All tested materials were firstly dried in a vacuum oven for 8 h at 80 °C. 7 g were used for each test. Fibres made of Geniomer 100 were extruded with three different diameters. Since this material showed the best melt flow characteristics of all three polymers, the diameter of the light pipes could be easily varied with this polymer. Three different fibre diameters made of this material were studied: 0.45 mm, 0.75 mm, and 0.85 mm. Consequently, these optical fibres can support thousands of optical modes although the fibre does not feature an outer cladding to protect the surface from scratches or other defects.

Table 1: Summary of parameters used for the fibre extrusions

Material type	Fibre Diameter [mm]	Temperature [°C]	Pre-worming time [s]	Load [kg]	Engine Speed [arbitrary units]
Geniomer 100	0.45	163	240	1.2	5–6
Geniomer 100	0.75	160	240	2.16	10
Geniomer 100	0.85	165	240	3.8	2–3
Geniomer 175	0.5	160	240	2.16	10
Geniomer 100-HDS	0.5	160	240	2.16	10

Light Loss

The optical loss measurements were performed with the cut-back method [22]. The medical laser used therefore had a wavelength of 652 nm with an intensity of 100 mW and was provided by AOL Medical Instruments. The 1 m long fibres were connected to the mode mixer (coupled to the light source) using F-SMA connectors from Thorlabs GmbH (Newton, MA, USA). The following equation was used to compute the light attenuation [23]. The optical loss measurements were repeated five times.

$$A = 10 \cdot log_{10}\left(\frac{I}{I_0}\right) \quad (1)$$

where: I - Input light intensity, I_0- Output light intensity.

The optical loss measurements were repeated five times. Intrinsic losses over a whole visible spectrum range from 300 nm to 750 nm of Geniomer 100 were performed using a Lambda 900 UV-VIS spectrophotometer provided by Perkin Elmer. The spectrometer measures continuously and independently (by means of beam splitting) the intensity of the reference beam and uses the value to compute reflection or transmission coefficients. To determine the intrinsic loss of the copolymers (Figure 1), a polymer cube with the dimension 50 × 10 × 5 mm^3 was formed. First the Fresnel reflection coefficient of the material was measured. This was done by measuring the light transmission through the thinnest part of the sample (5 mm), where the light absorption could be neglected. Afterwards, the light transmission through the 50 mm part was measured. The spectrophotometer was set to give the light transmission values from 300 nm to 750 nm in steps of 1 nm. For the force measurement setup, the extruded fibres were connected via a mode mixer to a Halogen-Deuterium Lamp model L10290 provided by Hamamatsu Photonics K.K. (Hamamatsu City, Japan), with SMA 905 connectors. In order to achieve flat fibre facets for good light coupling efficiency, fibres were cut with unused scalpel blades. Due to the elastic behaviour of the fibre material, polishing was not performed [17]. The ends of the fibres were connected by custom developed connectors to the measurement head RW3701-2 (Gigahertz Optik, Türkenfeld, Germany) and the outcoming light was measured using an Optometer P9710 detector (Gigahertz Optik). The force was applied with a Zwick INSTRON tensile testing machine model 4502 over the length of 3 cm. To study the influence of the temperature on the sensing system, the optical fibres were heated by a hot plate (Heidolph D-91126 type: MR Hei-Standard, Schwabach, Germany) and monitored by a temperature controller (Heidolph type: EHT Hei-Con). As a light source, an LED IFE97 and as a detector a photodiode IFD91 were used, both provided by Industrial Fiber Optics (Tempe, AZ, USA).

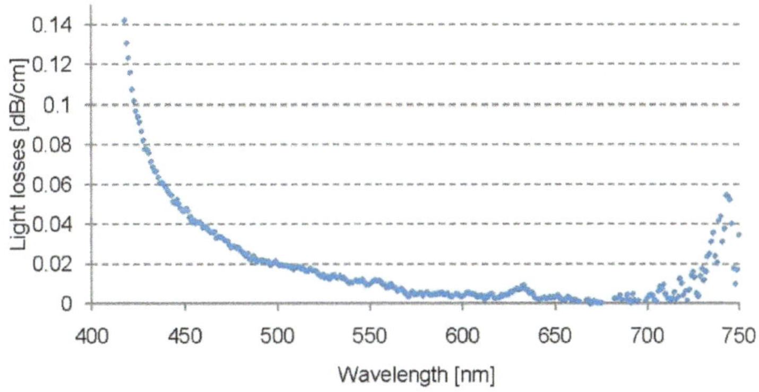

Figure 1: Intrinsic losses of the Geniomer 100-HDS polymer copolymer (measured from a polymer cube).

RESULTS AND DISCUSSION

Optical Properties of the Extruded Optical Fibres

The force sensor is based on the losses in light transmission along the light guides when the fibres are compressed. Thus it was essential to check whether the extruded fibres have reasonable light transmission parameters, *i.e.*, low optical losses. Since the final goal is to incorporate the fibres to medical textiles with a max. length of 1 m, the maximum light attenuation should not exceed 0.2–0.3 dB/cm at 652 nm. Light losses at this level allow the detection of the transmitted light of a 1 m long light guide.

Light Attenuation

In Table 2, a summary of light attenuation measurements at 652 nm is listed. No significant differences between the different materials and different diameters were observed. The optical signal loss varied between 0.16 and 0.25 dB/cm. That means that all fibres produced with the MFI apparatus were suitable for light transmissions over short distances of up to 1 m.

Table 2: Summary of light attenuation measurements of the fibres produced in this study

Type of a Fibre	Light Attenuation [dB/cm]	σ [1][dB/cm]
Geniomer 100-HDS d = 0.5 mm	0.25	0.09
Geniomer 175 d = 0.5 mm	0.19	0.07
Geniomer 100 d = 0.75	0.22	0.06
Geniomer 100 d = 0.85 mm	0.20	0.09
Geniomer 100 d = 0.45 mm	0.24	0.12

[1] Standard deviation.

Intrinsic Light Absorption Spectrum Measured at 652 nm

Figure 1 shows the intrinsic absorption spectrum of the light guide made of Geniomer 100-HDS as explained in Section 2.2. Since the overall attenuation of the light pipes was much higher (Table 2), the main light loss was due to extrinsic losses caused by the extrusion process (irregular fibre surface, bubbles or other inhomogeneities in the material). The absorption spectrum of Geniomer 100-HDS showed a local absorption band at around 635 nm, which corresponds to the 6th overtone of the C–H vibrations [24]. In the range of 660–700 nm, the absorption became lower than 0.01 dB/cm. Below 450 nm and above 700 nm the material starts to absorb strongly.

Force Sensing

The force sensing setup is presented below in Figure 2. The signal was measured from the middle of the three fibres. The outer fibres were used to support the weight.

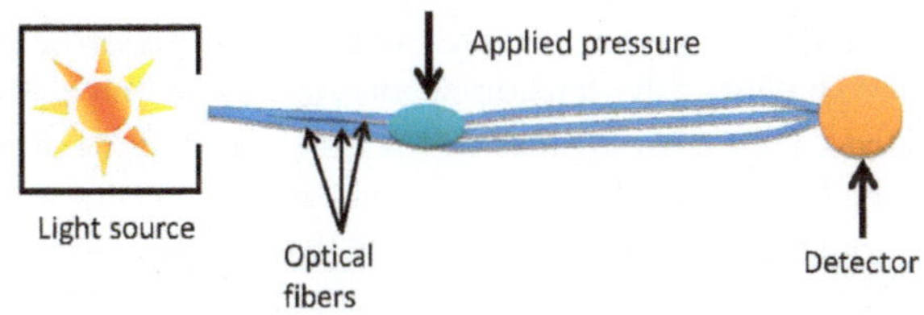

Figure 2: Force sensing setup.

In Figure 3, a scheme representing the pressure sensing principle is shown. The applied force resulted in an elliptical deformation of the fibre cross section. From the side projection a kind of cavity was observed, as schematically illustrated in Figure 3. This deflection increased the out-coupling of the light in the region under pressure due to the geometrical deformation of the fibre. Since the degree of deflection was directly related to the applied force, the force could be quantitatively determined.

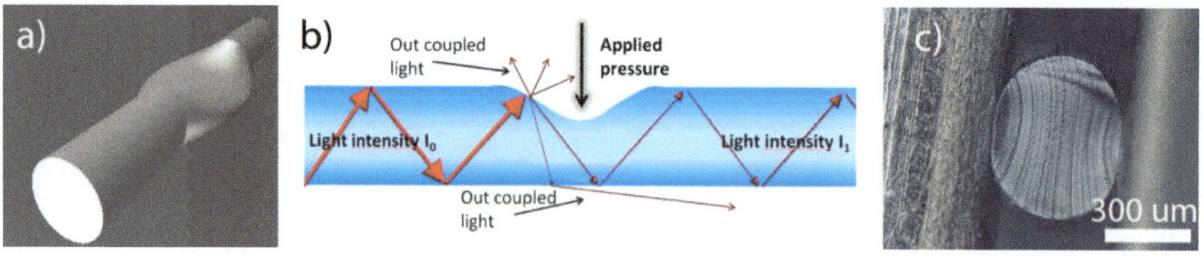

Figure 3: (a,b) Show schematically the influence of applied pressure on the elastic optical fibre. In (c) The cross-deformation of the fibre is demonstrated with a micrograph.

From the strain-stress curve (Figure 4) it can be concluded that the fibre made of Geniomer 175 had a smaller elastic modulus than the fibres made of the other two Geniomer materials and thus should have the highest response for applied pressure. The Young moduli of Geniomer 100 and Geniomer 100-HDS are similar and big differences in the light transmission could not be observed while applying pressure. However, the strain-stress curve was measured along the fibres and not perpendicularly (as the force was applied) which can cause a discrepancy between young moduli and the drop in light transmission.

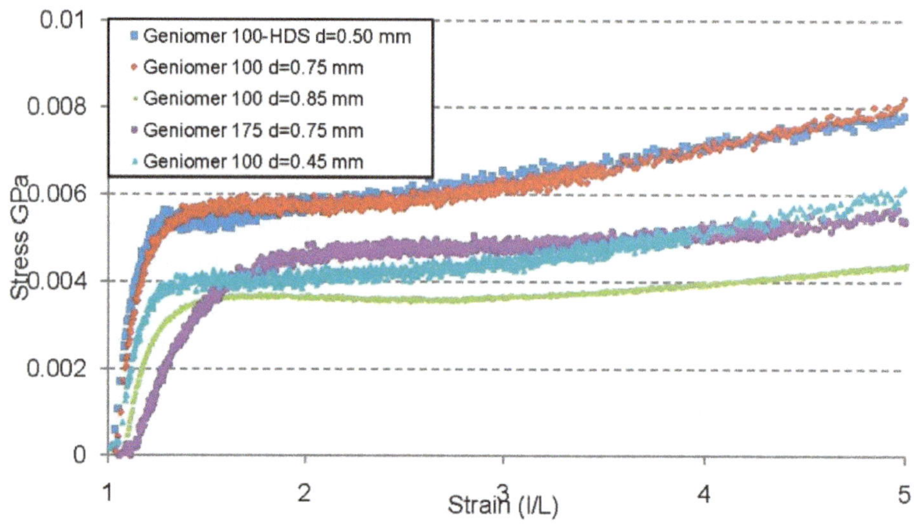

Figure 4: Stress strain curve of the different fibres produced in this study. The fibres made of Geniomer 175 show the lowest young's modulus.

The force sensing experiments were conducted with the three different types of fibre materials. In order to check the repeatability, the measurements for each force were repeated three times. Each time the force was applied, it was completely released afterwards to check if the geometry of the fibre relaxed completely to its initial state. Firstly, the force of 1 N was applied three times to check whether the fibre was sensitive to the force at this level. The forces were then increased with increments of 10 N until the normalized signal output after releasing was below 90% of the original signal. If the signal did not recover 100%, the fibre was deformed not only elastically but also plastically. If the original shape was not recovered it was a sign that plastic deformation occurred [25]. The stress which is needed to start the plastic deformation is defined as the yield strength which was different for the three materials (Table 3).

Table 3: Yield strength of the extruded fibres

Copolymer Type	Yield Strength [GPa]	σ [GPa]
Geniomer 100-HDS d = 0.5 mm	0.0054	0.0002
Geniomer 175 d = 0.5 mm	0.0045	0.0001
Geniomer 100 d = 0.45 mm	0.0039	0.0001

Depending on the copolymer material used for extrusion, the applied forces caused different light transmission responses (Table 4). This was expected due to the different yield strengths shown in Table 3. Geniomer 100-HDS showed full reversibility up to 20 N and was therefore the material which was best suited for large forces. At the level of 30 N the signal went back to 95.8% ± 1.1%. Although the fibre did not relax completely at 30 N and 40 N, the measured signal after applying pressure was completely reproducible (σ = 0.0%) at these forces. Geniomer 100 and Geniomer 175 showed plastic deformation already at

10 N. Geniomer 100 did only reach 91% ± 2.5% of the initial signal after releasing from 10 N and was therefore the material with the lowest applicable forces for the sensing system. The sensitivity of the applied force on the fibre deflection is related to the young's modulus of the material. As expected from the strain-stress curve, the fibre made of Geniomer 175 had the highest response to applied force, as shown in Table 4. The drop in the signal of 20.8% ± 0.5% for 1 N was twice as much than the fibres produced from Geniomer 100 and Geniomer 100-HDS as presented in Table 4 (Sections 2 and 4). Due to the highest sensitivity of the fibres extruded of Geniomer 175, these fibres were chosen to measure the smallest detectable force. Forces ranging from 0.05 N to 0.5 N were applied three times (over 3 cm fibre length). From Section 1 in Table 4 it can be concluded that all the measured forces were successfully detected and the measurements were fully reproducible (σ = 0.0%). The smallest detectable force was as low as 0.05 N.

Table 4: Summary of force influence on light transmission

	Polymer type	Diameter [mm]	Force [N]	Normalized signal	σ	Normalized signal after releasing force	σ
			0.05	99.4%	0.1%	100.0%	0.0%
			0.1	98.8%	0.1%	100.0%	0.0%
			0.2	97.8%	0.0%	100.0%	0.0%
1	Geniomer 175	0.5	0.5	94.3%	0.2%	100.0%	0.0%
			1	79.2%	0.5%	100.0%	0.0%
			10	42.0%	0.4%	100.0%	0.0%
			20	25.0%	0.7%	91.5%	1.7%
2	Geniomer 100	0.45	1	90.9%	1.0%	100.0%	0.0%
			10	61.6%	1.4%	91.0%	2.5%
			1	90.0%	0.3%	100.0%	0.0%
			10	55.1%	0.6%	100.0%	0.0%
3	Geniomer 100-HDS	0.5	20	40.6%	0.5%	100.0%	0.0%
			30	30.4%	0.0%	95.8%	1.1%
			40	21.7%	0.0%	90.6%	0.9%
			1	94.3%	0.2%	100.0%	0.0%
4	Geniomer 100	0.75	10	71.3%	0.6%	100.0%	0.0%
			20	53.0%	0.5%	90.7%	0.3%
			1	91.7%	1.4%	100.0%	0.0%
			10	71.4%	0.7%	100.0%	0.0%
5	Geniomer 100	0.85	20	54.5%	0.5%	100.0%	0.0%
			30	43.1%	0.8%	95.6%	1.0%
			40	31.9%	1.1%	91.7%	0.9%

Since the light losses caused by the applied forces are associated to the fibre deflection, not only the polymer material but also the diameter of the fibre influenced the measurement range. This was demonstrated with the polymer Geniomer 100. By decreasing the fibre diameter, the fibres became more sensitive to the applied force. The fibre with the diameter of 0.85 mm was not sensitive enough at low forces (1 N) to get accurate and repeatable values as shown in Table 4. However, the thicker fibre showed a larger measurement window; forces of up to 40 N could be measured reversibly, while the thinnest fibre (0.45 mm) could only be used for forces of up to 10 N.

Fibre Deflection against Signal Drop

In Figure 5 the relative fibre deflections in the force direction *vs.* the signal drop of all fibres (different polymers and different diameters) are presented. Only the measurements with strictly elastic deformations were included in the figure. The measurements of the fibre deflection were performed simultaneously with the force measurements using the tensile testing machine. Figure 5 clearly shows that the signal drop is correlated to the fibre deflection, independently of the polymer material and the fibre diameter. This shows that only the deflection in force direction influences the amount of the out-coupled light and that the elastic properties of the material has only a minor influence on the signal change. However, the sensitivity of the deflection towards the applied force is material dependent as shown above.

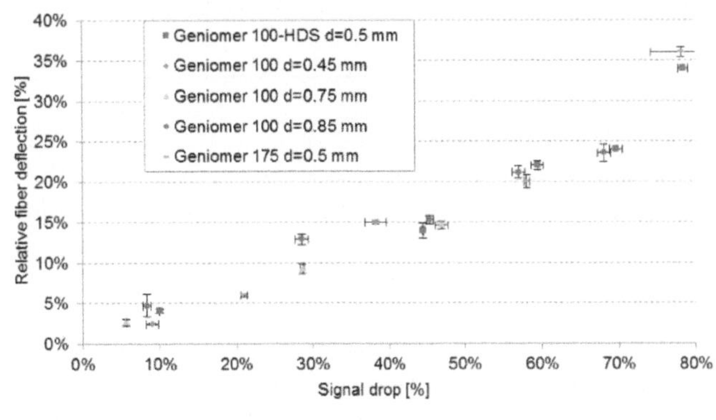

Figure 5: Fibre deflection *vs.* signal drop of fibres made of different materials and different diameters measured at 652 nm. Only the measurements with elastic deflections are shown.

Cross Sensitivity

Since this sensing technique relies on intensity measurements, it is essential to assess the influence of other parameters that can influence the light intensity. Below, we describe the influence of axial strain and bend losses on the fibre. The motion artifacts are not taken into account since they can be filtered out with signal processing.

- Axial Strain Influence on the Sensor Signal

In order to determinate whether the sensor is cross sensitive to axial strain, the fibre made of Geniomer 100 (d = 0.75 mm) was stitched into a woven textile (code name FRTT1022; provided by Forster Rohner, St. Gallen, Switzerland) as presented in Figure 6.

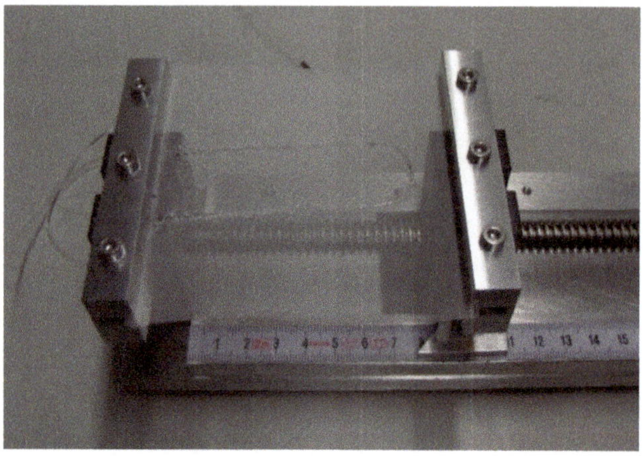

Figure 6: Measurement setup for axial strain influence.

According to the technical data sheet of this textile, the elastic deformation is 4%. The light intensity signal was measured simultaneously while the fabric was stretched with increments of 1.33%. In Table 5 the relative textile elongation *vs.* normalized signal is presented. It can be noticed that a loss in light transmission was detected when the fibre was stretched. At the maximum elongation of the fabric that was used, the signal dropped by 97% of the initial signal. This signal drop of 3% was lower than the signal drop of the smallest applied force in this study which was 1 N (signal drop of 5.7%). This leads us to the conclusion that axial strain would only influence the accuracy of the measurements at very low pressure. However, the influence of axial strain at low pressure can be bypassed by using a textile with a lower flexibility.

Table 5: Strain influence on the light transmission trough the sensor

Relative Textile Elongation	Normalized Signal
0.00%	100.0%
1.33%	99.6%
2.67%	99.0%
4.00%	97.0%

- Bend Losses

To determine the influence of bend losses, the textile presented in Figure 5 was placed on three cylinders with radii 3, 5 and 6.5 cm respectively. Those numbers were chosen as the representative of human body parts: forearm, arm and thigh. In Table 6, the light loss *vs.* bend radius is presented. The signal when bended with radii 5 and 6.5 cm decreased less than 1%, which is most likely caused by applied pressure and contact with the cylinder. When the fabric was stronger bended with a radius of 3 cm, the light loss increased by up to ~3.5%. The reason of the light loss was a combination of loss caused by increased imprinting of the matrix textile to the elastic optical fibre and of optical bend loss. This could be minimized by placing a cladding on the core of this waveguide, which in turn would protect the core from textile imprints. However, even at this level, the light loss was relatively small (detectable forces start at around 5.7% light loss).

Table 6: Influence of fibre bendings on the relative light loss

Bending Radius [cm]	Light Loss by Bending	σ	Light Loss when 1 N Force Applied	σ	Light Loss when 20 N Force Applied	σ
3	3.43%	0.24%	5%	0.4%	43.1%	0.2%
5	0.91%	0.26%	7.9%	0.4%	44.6%	0.3%
6.5	0.89%	0.93%	8.8%	0.4%	45.1%	0.6%

Table 6 shows the signal changes when a pressure was applied after the fibre was bent with a defined radius. As can be seen, the light loss was lower when the bending was narrower. The differences between the different radii became more significant when less force was applied.

- Temperature Dependency

All the three Geniomer materials used have a Tg of around −130 °C and therefore have no phase change at higher temperature due to the amorphic nature of these polymers. In order to check the temperature dependence of the sensing properties, the experiments were performed under different temperatures. A force of 3 N was always applied to the Geniomer 100-HDS fibre with a diameter of 0.5 mm to show the reversibility at a certain temperature. Since this sensing system is supposed to be integrated into medical textiles in close contact with human beings, the measurements of the light losses with applied forces were performed at temperatures ranging from 25 to 45 °C with intervals of 10 °C (Figure 7). As can be seen in Figure 7, the temperature had no major influence on the light loss of the fibre while a force was applied.

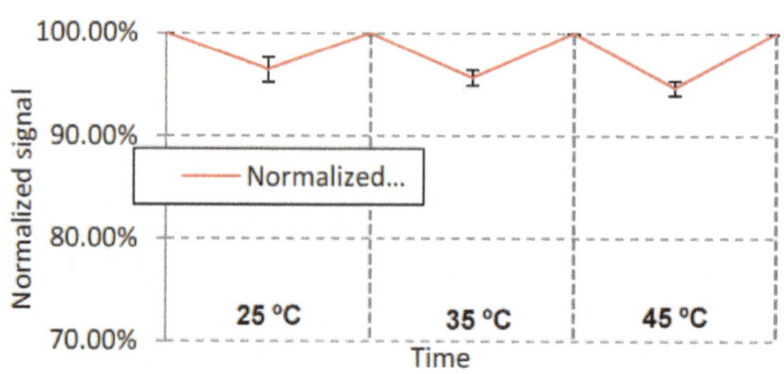

Figure 7: Signal dependency on the temperature. First, the attenuation was measured at a given temperature without applying any force. Then a force of 3 N was applied. The experiments were performed between 25 and 45 °C with 10 °C intervals and ...

- Inaccuracy

Applied strain, bends and temperature change can influence the data. However, the accumulated errors should not exceed 10% of the signal change under the tested conditions. To determine for instance the breathing rate or to monitor the seat occupation, 10% of signal change would be acceptable. When detecting very low forces, as it would be the case for heart rate monitoring, the cross sensitivity becomes more important. Influences on the signal change have to be controlled by using other sensors (e.g., temperature sensor) or by controlling the environment of the fibre (e.g., avoiding of strain on the textile by using a non-elastic textile matrix). Since the optical power of LEDs is temperature-dependent, a control measurement may be necessary. The power fluctuation of the LED can be monitored by installing another LED of the same type for reference measurements into the measurement setup.

CONCLUSIONS

We demonstrate a simple way to manufacture force sensors based on light pipes in the form of multimode optical fibres made of copolymers. The working principle of the sensor is the use of the deflection of the fibre structure when a force is applied. By using materials with different young's modulus, the sensitivity of the material can be tuned according to the desired application in the range of 0.05–40 N over a fibre length of 3 cm. The fibres produced from Geniomer 175 are best suited for low forces, while the Geniomer 100-HDS fibres are ideal sensors for larger forces. Moreover, due to the flexibility and the high mechanical strength of the material, the proposed sensor can be easily integrated into textiles to form textile-based force sensors with possible applications as seat occupation monitoring in automotive or aeroplanes or as force sensors for medical applications (prevention for decubitus or breathing monitoring). Since the flexible fibres and not the electronics have to be placed on the measuring place, this sensing principle has no negative influence on the flexibility and the haptic of the textile. Only one wavelength is used for the measurement, which simplifies the electronic system consisting of an LED and a photo detector like e.g., a photodiode. With a thin polymer cladding with a lower refractive index than the copolymer waveguide core, the robustness of the sensor could be improved further. Scratches and imperfections at the fibre surface would then not deteriorate or mask the sensing signal.

ACKNOWLEDGMENTS

This work was financially supported by NanoTera (TecInTex). Gratefully acknowledged is the support of Katrin Schöller, Benjamin Leuthold, Benno Wüst, Pierluigi Barbadoro, Alexander Haag, Youna Maussion, and Jörg Gschwend.

REFERENCES

1. Giallorenzi T.G., Bucaro J.A., Dandridge A., Sigel G.H., Cole J.H., Rashleigh S.C., Priest R.G. Optical fiber sensor technology. IEEE J. Quantum Elect. 1982;18:626–665.
2. Liehr S., Lenke P., Wendt M., Krebber K., Seeger M., Thiele E., Metschies H., Gebreselassie B., Munich J.C. Polymer optical fiber sensors for distributed strain measurement and application in structural health monitoring. IEEE Sens. J. 2009;9:1330–1338.
3. Ding J.Y., Shahriari M.R., Sigel G.H. Fiber optic ph sensors prepared by sol-gel immobilization technique. Electron. Lett. 1991;27:1560–1562.
4. Rothmaier M., Selm B., Spichtig S., Haensse D., Wolf M. Photonic textiles for pulse oximetry. Opt. Express. 2008;16:12973–12986.
5. Witt J., Narbonneau F., Schukar M., Krebber K., de Jonckheere J., Jeanne M., Kinet D., Paquet B., Depre A., D'Angelo L.T., et al. Medical textiles with embedded fiber optic sensors for monitoring of respiratory movement. IEEE Sens. J. 2012;12:246–254.
6. Rantala J., Hannikainen J., Vanhala J. Fiber optic sensors for wearable applications. Pers. Ubiquit. Comput. 2011;15:85–96.
7. Wang A.B., He S., Fang X.J., Jin X.D., Lin J.X. Optical fiber pressure sensor based on photoelasticity and its application. J. Lightwave Technol. 1992;10:1466–1472.
8. Urban F., Kadlec J., Vlach R., Kuchta R. Design of a pressure sensor based on optical fiber bragg grating lateral deformation. Sensors. 2010;10:11212–11225.
9. Grillet A., Kinet D., Witt J., Schukar M., Krebber K., Pirotte F., Depre A. Optical fiber sensors embedded into medical textiles for healthcare monitoring. IEEE Sens. J. 2008;8:1215–1222.
10. Witt J., Schukar M., Krebber K. Medicinal textiles with integrated polymer-optical fibers for respiration monitoring. Tm-Tech. Mess. 2008;75:670–677.
11. De Jonckheere J., Jeanne A., Grillet A., Weber S., Chaud P., Logier R., Weber J.L. Ofseth: Optical Fibre Embedded into Technical Textile for Healthcare, an Efficient Way to Monitor Patient under Magnetic Resonance Imaging. Proceedings of the 29th Annual International Conference of the IEEE Engineering in Medicine and Biology Society (EMBC'07); Lyon, France. 22–26 August 2007; pp. 3950–3953.
12. Zhang Z.C., Deng G., Dai Y.B., Liu Y.J., Leng J.S. A film pressure sensor based on optical fiber bragg grating. Smart Sens. Phenom. Technol. Netw. Syst. 2010;7648:764809:1–764809:7.
13. Tjin S.C., Hao J.Z., Lam Y.Z., Ho Y.C., Ng B.K. A pressure sensor using fiber bragg grating. Fiber Integr. Opt. 2001;20:59–69.
14. Kaczmarek T., Kaczmarek Z. Modified fiber bragg grating pulse pressure sensor. No. 66081e. Light. Their Appl. III. 2007;6608:66081E:1–66081E:4.
15. Tanaka N., Okabe Y., Takeda N. Temperature-compensated strain measurement using fiber bragg grating sensors embedded in composite laminates. Smart Mater. Struct. 2003;12:940–946.
16. Guan B.O., Tam H.Y., Chan H.L.W., Choy C.L., Demokan M.S. Discrimination between strain and temperature with a single fiber bragg grating. Microw. Opt. Technol. Lett. 2002;33:200–202.
17. Rothmaier M., Luong M.P., Clemens F. Textile pressure sensor made of flexible plastic optical fibers. Sensors. 2008;8:4318–4329.

18. Selm B., Gurel E.A., Rothmaier M., Rossi R.M., Scherer L.J. Polymeric optical fiber fabrics for illumination and sensorial applications in textiles. J. Intel. Mater. Syst. Struct. 2010;21:1061–1071.

19. Lymberis A., Dittmar A. Advanced wearable health systems and applications. IEEE Eng. Med. Biol. Mag. 2007;26:29–33.

20. Narbonneau F., Kinet D., Paquet B., Depré A., Jonckheere J.D., Logier R., Zinke J., Witt J., Krebber K. Smart textile embedding optical fibre sensors for healthcaremonitoring during MRI. Adv. Sci. Technol.2008;60:134–143.

21. Meyer J., Lukowicz P., Troster G. Textile pressure sensor for muscle activity and motion detection. Proceedings of 10th IEEE International Symposium on Wearable Computers; Montreux, Switzerland. 11–14 October 2006; pp. 69–72.

22. Ziemann O., Krauser J., Zamzow P.E., Daum W. Pof-handbook. Springer; Berlin, Germany: 2008.

23. Kuzyk M.G. Polymer Fiber Optics. CRC/Taylor & Francis; Boca Raton, FL, USA: 2007. pp. 97–98.

24. Koike Y., Koike K. Progress in low-loss and high-bandwidth plastic optical fibers. J. Polym. Sci. Pol. Phys. 2011;49:2–17.

25. Lubliner J. Plasticity Theory. Macmillan Publishing; New York, NY, USA: 2008.

INDEX

O

P

R

S

T

W